AF765137

Micro Fluidization

Micro Fluidization

Fundamentals and Applications

GUANGWEN XU
Shenyang University of Chemical Technology (SYUCT), Key Laboratory on Resources Chemicals and Materials of Ministry of Education, Institute of Process Engineering (IPE), Chinese Academy of Sciences (CAS), P.R. China

DINGRONG BAI
Shenyang University of Chemical Technology, P.R. China

MINGYAN LIU
Tianjin University, P.R. China

VLADIMIR ZIVKOVIC
School of Engineering, Newcastle University, United Kingdom

ELSEVIER CIP

Elsevier
Radarweg 29, PO Box 211, 1000 AE Amsterdam, Netherlands
The Boulevard, Langford Lane, Kidlington, Oxford OX5 1GB, United Kingdom
50 Hampshire Street, 5th Floor, Cambridge, MA 02139, United States

© 2023 Chemical Industry Press Co., Ltd. Published by Elsevier Inc., under an exclusive license with Chemical Industry Press Co., Ltd.

No part of this publication may be reproduced or transmitted in any form or by any means, electronic or mechanical, including photocopying, recording, or any information storage and retrieval system, without permission in writing from the publisher. Details on how to seek permission, further information about the Publisher's permissions policies and our arrangements with organizations such as the Copyright Clearance Center and the Copyright Licensing Agency, can be found at our website: www.elsevier.com/permissions.

This book and the individual contributions contained in it are protected under copyright by the Publisher (other than as may be noted herein).

Notices

Knowledge and best practice in this field are constantly changing. As new research and experience broaden our understanding, changes in research methods, professional practices, or medical treatment may become necessary.

Practitioners and researchers must always rely on their own experience and knowledge in evaluating and using any information, methods, compounds, or experiments described herein. In using such information or methods they should be mindful of their own safety and the safety of others, including parties for whom they have a professional responsibility.

To the fullest extent of the law, neither the Publisher nor the authors, contributors, or editors, assume any liability for any injury and/or damage to persons or property as a matter of products liability, negligence or otherwise, or from any use or operation of any methods, products, instructions, or ideas contained in the material herein.

ISBN: 978-0-443-18718-6

For Information on all Elsevier publications
visit our website at https://www.elsevier.com/books-and-journals

Publisher: Glyn Jones
Editorial Project Manager: Naomi Robertson
Production Project Manager: Bharatwaj Varatharajan
Cover Designer: Miles Hitchen

Typeset by MPS Limited, Chennai, India

Working together to grow libraries in developing countries
www.elsevier.com • www.bookaid.org

Contents

Preface

We are pleased to present this first monograph on micro-fluidization. Micro-fluidization, conceptually introduced in 2005, has emerged at the forefront of chemical engineering over the past decade. Micro-fluidized bed reactors combine the advantages of microreactors and particle bed reactors and thus potentially bring a revolutionary shift in the paradigm of particle reaction system development and scale-up. Compared with traditional large-scale fluidized beds, micro-fluidized beds have unique characteristics, including small-scale operation, high rates of heat and mass transfer, high selectivity, high conversion, high operation safety, and ease of operation. After more than 10 years of continuous fundamental research on hydrodynamics, gas back-mixing, solids mixing, modeling, and chemical reactions, researchers worldwide have published hundreds of peer-reviewed articles about micro-fluidized beds in various resources. We reviewed these publications and found that the available information was somewhat scattered, which motivated us to undertake some systematic and comprehensive analyses on almost all aspects of micro-fluidized beds, especially their fundamentals and applications. We believe that all these analyses, together with recent results reported by us and other research groups, would be helpful for researchers, engineers, and those concerned with micro-fluidized bed technology. For this reason, we wrote this book.

This monograph is composed of 12 chapters. Chapter 1 briefly introduces the research background and history of micro-fluidized beds and outlines their applications. Chapter 2 summarizes hydrodynamic characteristics of gas—solid micro-fluidized beds, focusing on the discussion of the wall effect. Chapter 3 examines gas and solids mixing in micro-fluidized beds. Chapter 4 reviews fluidization regimes and transitions in micro-fluidized beds with the proposed fluidization maps. Chapter 5 outlines investigations on micro-fluidized bed hydrodynamics based on the two-fluid and computational fluid dynamic discrete element models. Chapters 6—10 focus on applications of gas—solid micro-fluidized beds, especially characterizations of various physical and thermochemical reaction processes: Chapter 6 presents a brief description of different thermal analytic reactors commonly used in research of engineering thermochemistry field; Chapters 7 and 8 describe the design characteristics, analytical methodologies, and characteristics related to the micro-fluidized bed reaction analyzers (MFBRAs); Chapter 9 outlines a wide range of physical and chemical reaction processes investigated using MFBRAs; and Chapter 10 presents several advanced industrial processes that have been developed, designed, and commercialized with the assistance of

MFBRA. Chapters 11 and 12 are devoted to liquid–solid and gas–liquid–solid micro-fluidized beds.

A number of individuals have contributed to this monograph. Drs. Guangwen Xu and Dingrong Bai (Shenyang University of Chemical Technology, China) compiled Chapters 1–9; Drs. Ping An and Zhennan Han (Shenyang University of Chemical Technology, China) and Dr. Xi Zeng (Beijing Technology and Business University, China) prepared Chapter 10; Dr. Yi Zhang (Chinese Academy of Agricultural Sciences, China) and Dr. Vladimir Zivkovic (Newcastle University, United Kingdom) contributed to Chapter 11; and Dr. Mingyan Liu (Tianjin University, China) provided Chapter 12.

This monograph is intended for scientists and researchers who wish to broaden their knowledge in fluidization and powder technology, including micro-fluidization and its application processes, engineers and practitioners who wish to learn about the latest developments in micro-fluidization, and graduate students in chemical engineering who need to understand and follow the latest developments and publications of micro-fluidization. We hope that readers may find this book interesting and valuable.

By and large, the materials presented in this book represent a body of research contributed by our worldwide colleagues, including our former and present students, which have appeared in numerous publications. We acknowledge those whose dedicated research has made the writing of this book possible. We are especially grateful to those we have cooperated with in research projects on micro-fluidized bed research. We appreciate the members of our research groups for their assistance in figures, logistics, administrative details, revisions, and proofreading.

We thank the Chinese National Science Foundation, the Ministry of Science and Technology of China, and the Chinese Academy of Sciences for funding some of the expenses related to the preparation of this book. We appreciate their financial support of several studies that have helped us gain experience and expertise in micro-fluidization and related areas. Finally, we sincerely thank the Chemical Industry Press and Elsevier staff, especially Lan Yu and Yan Zhang, for their editorial contributions with constant encouragement and guidance.

Last but not least, we are incredibly grateful to our families for their love, support, patience, and sacrifice.

Guangwen Xu
Dingrong Bai
Mingyan Liu
Vladimir Zivkovic

Future prospects

In this monograph, we presented a comprehensive review and analysis of gas–solid, liquid–solid, and gas–liquid–solid micro-fluidized beds. In discussions of fundamentals and applications, we provided specific examples with comments to illustrate the characteristics and application potentials of micro-fluidized beds. We hope that this book will provide readers with a good understanding of this essential emerging microreactor technology.

In closing this book, we feel compelled to emphasize that compared with traditional fluidized beds in general and continuous-fluid microchannel reactors in particular, the research and application of micro-fluidized beds are still in their infancy. Some fundamental aspects, such as wall effects, mixing, and transport behavior, have not been well understood; the great application potential, especially in continuous and stable microchemical reaction processes, has not been fully explored. We attempt to summarize the fundamental and applied research needs and future opportunities for micro-fluidization in the following sections.

1 Fundamental research

1.1 Hydrodynamics

Knowledge of hydrodynamics lays the foundation for all multiphase reactors, and micro-fluidized beds are no exception. Although some studies have been carried out in micro-fluidization, the understanding of hydrodynamic properties and the underlying mechanisms is still inadequate. In addition to the research needs mentioned earlier in this book, the following aspects deserve special attention.

1.1.1 Wall effects

As we have repeatedly pointed out, wall effects are the decisive factor that makes micro-fluidized beds unique and advantageous over traditional macro-fluidized beds in many aspects. In Chapter 2, we identified three mechanisms of the wall effects: the wall friction force exerted directly on particles near the wall area, the bed voidage increase, and the radial voidage and gas velocity maldistributions. But we were unable to understand the relationship between them and quantify them separately. We noted that interparticle forces might affect the wall effects on the hydrodynamic characteristics in small-sized beds, complicating the understanding of wall effects. Therefore more extensive research is needed to better understand wall effects, interparticle forces, and their relationships.

1.1.2 Flow regimes

Based on some important hydrodynamic parameters, including bed pressure overshoot, minimum fluidization/bubbling/slugging/turbulent velocities, and gas backmixing, we have defined gas–solid micro-fluidized beds based on the bed-to-particle diameter ratio D_t/d_p, that is, $D_t/d_p = 25-150$ for Group A particles and $D_t/d_p = 10-150$ for Group B particles (Chapter 4). However, more extensive research is needed to explore the underlying mechanism and verify the validity of the criteria in a wide range of operating conditions.

1.1.3 Relative micro-fluidization

Nevertheless, we consider the bed-to-particle diameter ratio D_t/d_p a parameter that describes the relative importance of wall effects on flow behavior in fluidized beds. In this sense, when two fluidized beds of different diameters operate with particles of varying diameters (but both have the same D_t/d_p ratio), can we expect similar hydrodynamic behavior in these two fluidized beds? For example, consider two fluidized beds: a 500-mm-diameter bed fluidizing 5-mm-diameter particles and a 20-mm-diameter bed fluidizing 200 μm particles. Can we expect that they have the same close-to-plug gas flow pattern? If they have, a concept of "relative micro-fluidization" can be introduced to define micro-fluidization in terms of the effects of wall confinement rather than the bed size. Then, we not only expand the micro-fluidization from the traditional millimeter sizes to a wider range of sizes but also bring significant benefits to many large-scale fluidization industrialization processes, for example, effectively resolving the long-standing problem of intense gas backmixing that adversely affects the conversion and yield of chemical reactions. Therefore we encourage more future research to verify this concept of relative micro-fluidization through experiments and modeling simulations.

1.2 Heat and mass transfer

Micro-fluidized beds have been considered to have high heat and mass transfer rates. However, except for the limited research on mass transfer in liquid–solid and gas–liquid–solid micro-fluidized beds (see Chapters 11 and 12), there are no studies on heat and mass transfer in gas–solid micro-fluidized beds. In the future, we hope that more experiments and modeling simulations will be devoted to this field, and efforts will also be made to develop measurement techniques suitable for heat and mass transfer in micro-fluidized beds.

1.3 Solid mixing

Fluidized beds are characterized by exceptional mixing abilities between gas and solids and between particles of different sizes and densities when particles can move vigorously and freely in beds. In micro-fluidized beds the increased wall confinement may restrict the free motion of particles, especially when D_t/d_p is low. However, at present, we do not have an in-depth understanding of the effect of increased wall confinement on solid mixing in micro-fluidized beds because, except for a few studies based on numerical simulations (Chapters 3, 11, and 12), solid mixing in micro-fluidized beds has not been directly and comprehensively studied. Therefore research on solid mixing in micro-fluidized beds deserves more attention in the future.

1.4 CFD simulation

Computational fluid dynamics (CFD) modeling plays an essential role in understanding the characteristics of micro-fluidized beds, but the modeling results are sometimes difficult to be validated because experimental measurements are not available. For example, CFD modeling can help generate information on radial profiles of solid concentration and velocity, solid mixing, fluid residence time distribution, and effects of D_t/d_p on these critical parameters (Chapter 5). However, in some cases, there is no experimental data to validate CFD simulations, especially when the simulation results are sensitive to the selection of drag model, closure equations, and some empirical parameters. These issues need to be addressed in the future so that CFD modeling can continue to develop as an effective and reliable method for the research of micro-fluidized beds.

2 MFBRA

Mico-fluidized bed reaction analyzer (MFBRA) has served researchers well with the following unique capabilities: (1) obtaining close-to-intrinsic reaction kinetics of heterogeneous reactions, (2) discovering true reaction mechanisms, (3) accelerating screening of catalysts formulations, and (4) developing new reaction processes. However, some subjects or areas need further investigations, including but not limited to those described in the following sections.

2.1 Measurement of complete product profile

At present, MFBRA is only able to measure small-molecular chemical components (e.g., H_2, CH_4, CO, and CO_2), but not large-molecular products due to the lack of analyzers with fast response speeds and high resolutions. In the future, MFBRA designs need to be improved further to expand their capabilities, with the priority given to the development of rapid reaction analysis instruments to characterize not only the dynamics of individual products but the overall complex chemical reactions.

2.2 In situ real-time monitoring and analysis

MFBRA currently can sample solid particles during the reaction process at a specific time interval for offline sample characterizations. The information obtained in this method is helpful but not sufficient to provide a complete, real-time, and in situ dynamic reaction performance of solid particles in reaction processes. New generations of MFBRA need to solve this problem and contribute to future scientific and technological innovations.

2.3 Applications under extreme reaction conditions

Micro-fluidized beds can potentially be applied to many chemical reaction processes with extreme operating conditions, such as highly exothermic or endothermic, high reaction temperature and/or pressure, high degree of vacuum, and highly corrosive or toxic environments. Few thermal reaction analyzers today can operate under these extreme reaction conditions safely and cost-effectively, therefore future research is needed to expand the applications of micro-fluidized beds in these areas.

2.4 Multiphase reaction analysis

No MFBRA is currently available for characterizations of liquid—solid and gas—liquid—solid chemical reactions, although these multiphase micro-fluidized beds have been researched and applied in some reactions, such as wastewater processing, bioprocessing, and photocatalytic reactions. It is thus a great opportunity to make some breakthroughs in this exciting field.

3 MFB configurations

3.1 Continuous MFB

Micro-fluidized beds can operate continuously and stably to offer advantages of large processing capacity per unit reactor volume, high conversion rate, and increased product selectivity. However, up to now, micro-fluidized beds, especially with capabilities of continuous solid feedstock feeding and product discharging, have not been developed, which is highly anticipated in the future.

3.2 Micro-circulating fluidized beds

The research on micro-fluidized bed reactors has focused on low-velocity fluidized beds. To investigate some chemical reactions, such as coupled catalytic reaction and catalyst regeneration, chemical looping combustion or gasification, and coupled pyrolysis and gasification or combustion, it is necessary to develop micro-circulating fluidized beds. Compared to low-velocity micro-fluidized beds, micro-circulating fluidized beds have not received much attention, and thus more in-depth and systematic research is expected in the future.

3.3 External force-assisted MFBs

Processing submicron particles and nanoparticles in fluidized beds has attracted significant attention. These ultrafine particles have exceptionally large surface areas and are thus very active as reactants or catalysts, but they are very difficult to fluidize stably and uniformly because of strong interparticle forces. In the future, it is necessary to study the fluidization behavior of nanoparticles assisted by external forces, such as electrical, magnetic, and acoustic forces. The external forces-assisted micro-fluidized bed (MFB) MFB will be useful for industrial applications in cosmetics, food, plastics, catalysts, high performance, and other advanced materials synthesis.

4 Micro-fluidized beds for industrial applications

4.1 Numbering up

Scale-up of multiphase bed reactors has long been a considerable challenge due to the complex fluid dynamics and their nonlinear relationship with the reaction dynamics. The use of micro-bed reactors (MBRs) can significantly reduce the time and cost involved in developing and scaling up a reaction process using the numbering-up approach. In this approach the chemical reactions in each of the reactors are the same so that laboratory results can be achieved on an industrial scale. For now, however, the numbering-up approach is purely theoretical and has not been realized in the scaling-up of micro-fluidized beds. In the future, it is necessary to research all aspects related to the numbering-up approach, for example, technologies of uniformly distributing fluids and particles to each of multiple micro-fluidized bed reactors.

4.2 Continuous synthesis of specialty chemicals involving solid particles

Micro-fluidized beds have great potential to be used as continuously operating chemical reactors in various catalytic and noncatalytic reactions involving solid particles. The breakthrough in the continuous application of micro-fluidized beds may first be realized in the application of synthetic organic chemistry for the production of high value-added fine chemicals, pharmaceuticals, and natural products. In these applications, micro-fluidized beds are an attractive choice for the continuous processing of specialty chemicals involving solid particles, offering competitive advantages in yield, selectivity, cost, time, and operational flexibility for rapid batch-to-batch or product-to-product changes.

5 Micro-bed reactors

In Chapters 4, 11, and 12, we indicated that solid particles in a small-sized bed might be in a fixed, bubbling, slugging, turbulent, fast, or pneumatic transport fluidized bed state. We collectively refer to these small-sized beds of particles as "MBRs"—a new class of microchemical reactors (Figure B) that are clearly distinguishable from (Figure A) microchannel or microfluidic reactors. Conceptually, MBRs include all types of micro-scale fluid-solid (gas–solid, liquid–solid, and gas–liquid–solid) contacting reactors, in which solid particles (catalytic or reactant) can be in a fixed (i.e., micro-fixed beds) or a fluidized (i.e., micro-fluidized beds in various modes of fluidization) state depending on fluid velocities.

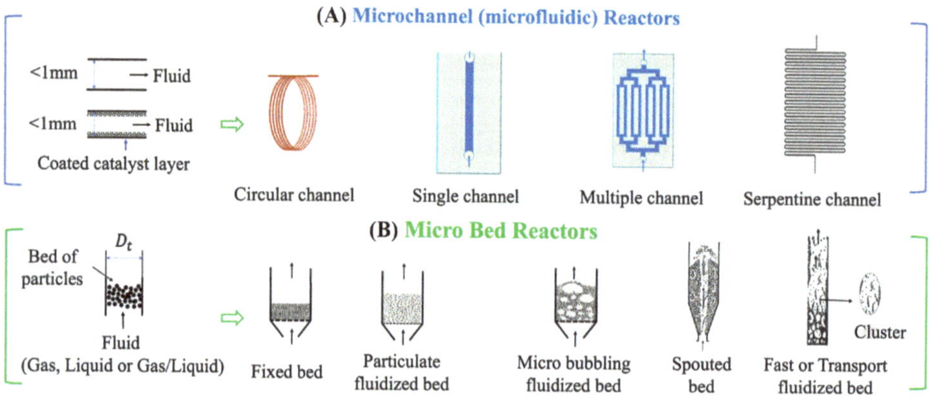

MBRs provide an effective way to deal with and manage the reaction process involving solid particles in the microreactor. We believe that MBRs, just like microchannel and microfluidic reactors, have significant potential to open a revolutionary and practical pathway in a wide range of process industries by extending the microreactor technology to the field of catalyst chemistry and engineering and diversified noncatalytic thermochemical conversions of solid particle reactants, including various fossil and naturally occurring resources.

Hopefully, our readers and we will have the opportunity to participate in and witness this exciting new era.

CHAPTER 1

Introduction

The miniaturization of fluidized beds opens the way for a revolutionary change to the century-old technology of fluidization, just as the miniaturization of electronics over the past few decades has transformed the way we live. For readers to comprehend the technical characteristics and advantages of micro-fluidized beds, this chapter first briefly reviews the development and status of traditional fluidized bed technology, then describes the concept of micro-fluidized beds and their development, and finally presents a comparison between micro- and traditional fluidized beds, as well as potential applications of micro-fluidized beds.

1.1 Fluidization and fluidized bed

Solid particles do not move freely in a packed state and are difficult to handle, transport, and process. However, they exhibit some fluid-like physical phenomena when suspended by a fluid (gas or liquid). The suspension state is called the fluidization of solid particles. While fluidization of solid particles has been known for hundreds of years and practiced in many processes, it was not, however, recognized as a part of the discipline of chemical engineering until fluidization was formally introduced in a textbook in 1950 [1]. Thereafter, extensive fundamental research has been carried out to enrich fluidization science and technology and extend fluidized beds to a wide range of industrial applications [2]. In particular, the successful large-scale operations of coal gasification in the 1920s and the catalytic cracking of petroleum in the 1940s are two significant milestones in the development of fluidization technology. Today, the fluidization of solid particles has become an indispensable area of study in many engineering disciplines and an important applied technology widely used in many industries [3].

Though common in nature, solid particles must be confined in a closed column or vessel to use fluidization phenomena purposely for different technological processes. In this case, the formed bulk entity of confined particles is called "bed" or "bed of particles." As a kind of reactor, the column or vessel is called a fixed, fluidized, or moving "bed reactor" when the bed of particles is in a fixed, fluidized, or moving state, respectively.

As schematically shown in Fig. 1.1A, a fluid (gas or liquid) flows into the bed of particles through a distributor. As the fluid moves upward between particles, a drag force occurs between the fluid and the particles. When the fluid velocity is relatively low, the resulting drag force is insufficient to counterbalance the gravity of particles,

Micro Fluidization
DOI: https://doi.org/10.1016/B978-0-443-18718-6.00001-2
© 2023 Chemical Industry Press Co., Ltd.
Published by Elsevier Inc., under an exclusive license with
Chemical Industry Press Co., Ltd.

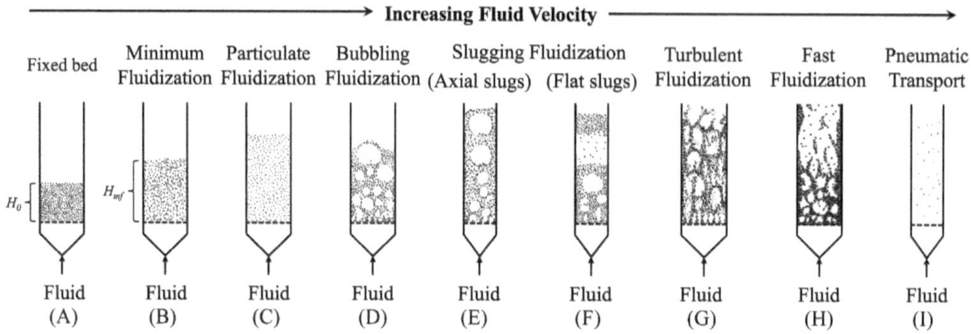

Figure 1.1 Schematic illustrations of the fluidization regimes.

and thus the particles remain still even though they may, to some extent, get loose. This state of particles is called "fixed bed" or "packed bed." As the fluid flow rate increases, the particles may move apart and vibrate a bit, leading to a phenomenon called "bed expansion." When the fluid velocity continuously rises and reaches a so-called minimum fluidization velocity U_{mf}, the resultant drag force balances the weight of particles. At this moment, all the particles become suspended and move freely in the upward-flowing fluid stream and behave like a fluid. Fig. 1.2 illustrates some of the fluid-like behavior of fluidized particles. The "minimum fluidization" is the onset of transition of the particles to the fluid-like behavior (Fig. 1.1B). After the minimum fluidization, several different flow patterns, also called fluidization regimes, can appear as the fluid velocity is increased further. In general, these include particulate fluidization, bubbling fluidization, slugging fluidization, turbulent fluidization, fast fluidization, and pneumatic transport, as schematically shown in Fig. 1.1C–I.

Transforming solid particles into a fluid-like state is helpful for a wide range of industrial physical and chemical processes [2]. The advantages offered by the fluid-like behavior of solid particles, for example, fast and easy transport, intimate contact between fluid and particles, and excellent heat transfer, are the main reasons for considering fluidization in many solids handling and reaction operations.

1.2 Typical $\Delta P_B \sim U_g$ relationship

The relationship between bed pressure drop and superficial gas velocity, that is, the $\Delta P_B \sim U_g$ diagram, is the most straightforward and particularly useful characteristic to reveal the gas-solid flow behavior for particulate systems. Here, the "superficial gas velocity" is calculated by dividing the total gas volumetric flow rate by the entire column area without considering the space occupied by particles. Now, consider a bed of uniformly sized particles, as shown in Fig. 1.3(A, left side). When a gas is introduced upward through the bed at a low velocity, the particles remain stationary,

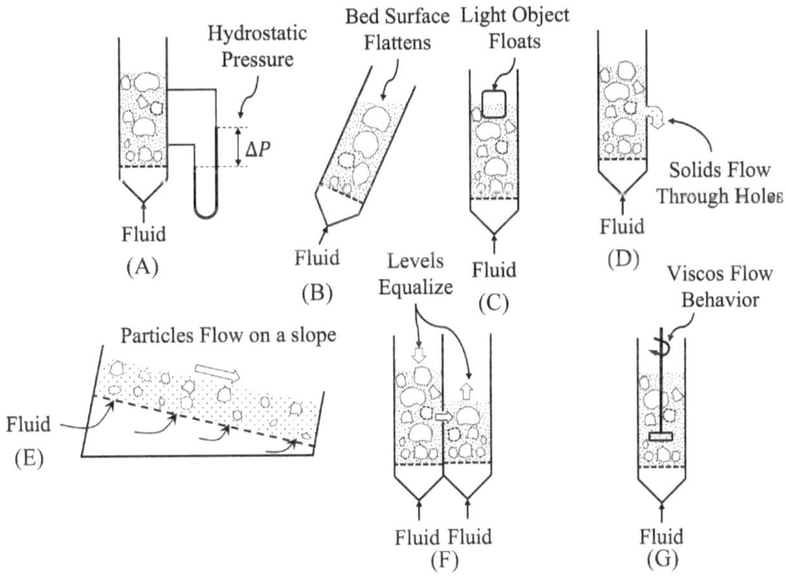

Figure 1.2 Schematic illustrations of the fluid-like behavior of particles under fluidization.

corresponding to the packed or fixed bed. As the gas velocity is further increased, the bed pressure drop increases correspondingly, as shown in Fig. 1.3(A, right side), until the gas flow reaches a velocity at which the drag force exerted by the gas on particles equals the net weight of the particles. At this point, the particles are entirely unlocked from each other and move freely like a fluid in the column. This point signals the minimum fluidization, and the corresponding gas velocity is called the minimum fluidization gas velocity U_{mf}. Beyond U_{mf}, the particles are called fluidized.

Once the bed is fluidized, the bed pressure drop should ideally remain constant; that is, the weight of the particles is fully supported by the gas; thus

$$\Delta P_B = mg/A \tag{1.1}$$

where m is the weight of particles, g is the acceleration of gravity, and A is the cross-sectional area of the bed or the column. Note that this equation is valid only for fluidized beds of large diameters where the wall effects are small enough to be negligible. As we will discuss later, it is not always correct for the so-called micro-fluidized beds because they are affected considerably by wall effects.

Now, the bed has been fully fluidized. Next, we reduce the gas velocity gradually to change the bed from fluidized to defluidized conditions. We find that even at the same gas velocity, the measured bed pressure drop is slightly lower when the gas velocity is descending than when the gas velocity is ascending. This phenomenon is explained in the next paragraph.

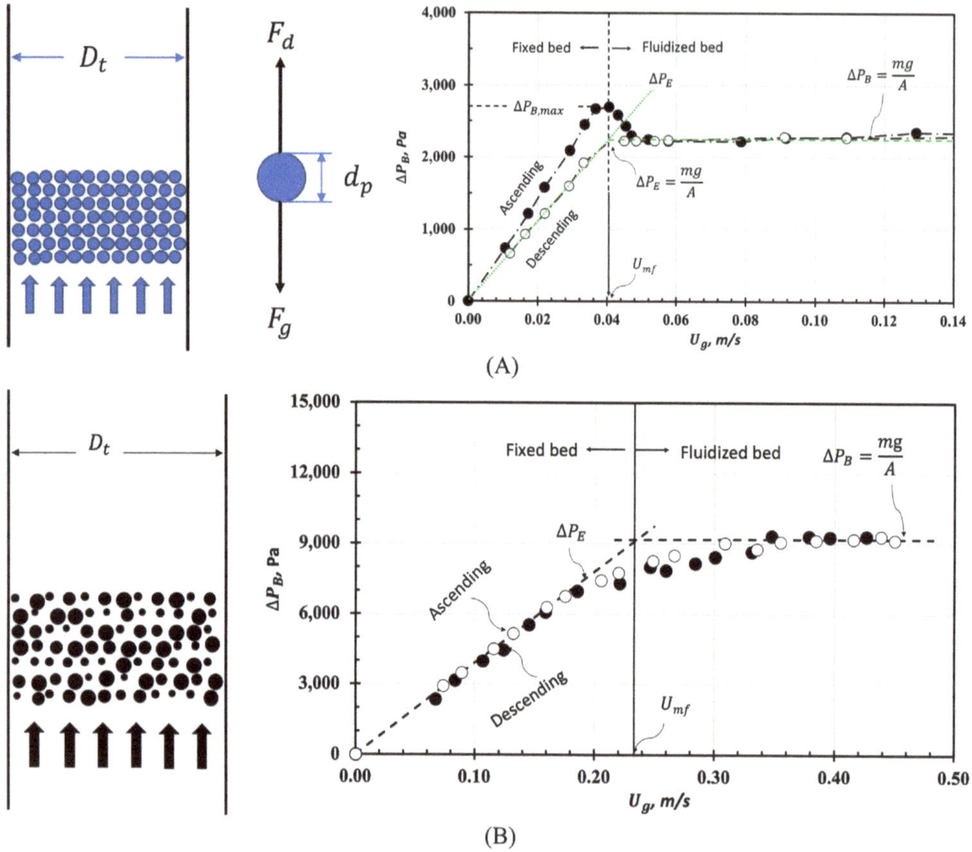

Figure 1.3 Variations of bed pressure drop measured by ascending and descending gas velocity methods for a typical fluidized bed with a diameter above tens of millimeters [2]. (A) Uniformly sized particles. (B) Wide size distribution of particles.

When the bed is in the path of ascending gas velocity, the gas needs to flow through the tightly locked particles; in other words, the bed must expand by increasing voidage. As a result, a greater peak pressure drop, called the "pressure drop overshoot," occurs when the bed bursts suddenly to become fluidized. In macro-fluidized beds, the extent of pressure drop overshoot is generally small. The gas velocity corresponds to the maximum ΔP_B or $\Delta P_{B,\max}$ and, in most cases, is identical or close to U_{mf}, as shown in Fig. 1.3A. Comparatively, in the path of descending the gas velocity, the bed is already aerated so the bed has a higher voidage and thus a slightly lower pressure drop, and the transition from the fluidized bed to the fixed bed is smooth rather than abrupt. Consequently, a hysteresis appears in the ΔP_B-versus-U_g curve. However, the hysteresis is insignificant when particles with a wide size distribution are fluidized in a macro-fluidized bed (Fig. 1.3B), where small-sized particles function as a lubricating medium to help fluidize large-sized particles. In this case, the small particles are first partly fluidized (suspended) to break the "compactness" of the packed particle bed.

For large diameter beds, for example, over 100 mm, numerous works have confirmed that the pressure drop measured by the "descending gas velocity" method in the fixed bed regime can be well presented by the Ergun equation [4]:

$$\frac{\Delta P_E}{H_s} = 150\frac{(1-\varepsilon_0)^2}{\varepsilon_0{}^3}\frac{\mu_g U_g}{\phi^2 d_p{}^2} + 1.75\frac{(1-\varepsilon_0)}{\varepsilon_0{}^3}\frac{\rho_g U_g{}^2}{\phi d_p} \quad (1.2)$$

where d_p: mean particle diameter, H_s: static bed height, ΔP_E: bed pressure drop estimated by Ergun equation, U_g: superficial gas velocity, ε_0: bed voidage at static condition, μ_g: gas viscosity, ρ_g: gas density, and ϕ: particle sphericity

The first term in the above equation represents the viscous force contribution, dominant in the laminar flow regime subject to low Reynolds numbers. The second term accounts for the kinetic contribution corresponding to higher Reynolds numbers. It is worth pointing out that the Ergun equation is valid only when the wall frictional force is negligible, and the gas distribution across the bed cross section is uniform.

1.3 Geldart powder classification

Gas-solid fluidization behavior depends largely on the particles fluidized. Therefore, it is helpful to understand the relationship between fluidization quality and the properties of particles. By carefully observing the fluidization of many sorts of solids [5–7], Geldart [8] classified particles into four recognizable groups depending on their fluidization characteristics, described by density difference ($\rho_p - \rho_g$) versus mean particle size d_p. Each group of particles exhibits distinguishable gas-solid fluidization behavior, as shown in Fig. 1.4. The four-particle groups are classified as follows:

- *Group A particles*: These are easily aeratable particles and can be fluidized smoothly. After minimum fluidization, the bed continues to expand homogeneously until gas bubbles are formed at the minimum bubbling velocity U_{mb}, that is, $U_{mb}/U_{mf} > 1$.
- *Group B particles*: These particles lead gas bubbles to form readily and can be fluidized with good quality. The gas bubbles form immediately at the minimum fluidization, that is, $U_{mb}/U_{mf} = 1$.
- *Group C particles*: These are cohesive powders that are extremely difficult to fluidize naturally.
- *Group D particles*: These are coarse particles that can be fluidized often in spouted beds.

Geldart's classification is clear and easy to use, but it is based on the experimental data obtained from large-scale (macro) fluidized beds. As we will show later, the same

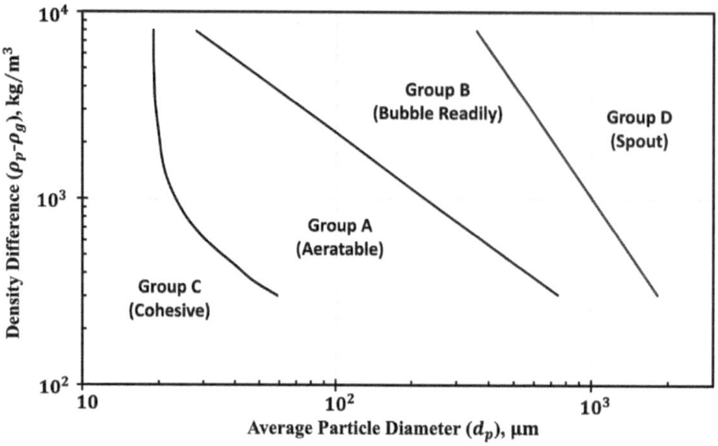

Figure 1.4 The Geldart classification of particles at ambient conditions.

particles will exhibit different fluidization behavior in micro-fluidized beds; for example, the minimum bubbling velocity U_{mb} can be smaller than U_{mf} for Group B particles, and there are cases where $U_{mb}/U_{mf} = 1$ for Group A particles.

1.4 Gas-solid fluidization regimes

As mentioned above, gas-solid flow behavior in a micro-fluidized bed is different from that generally observed in a macro-fluidized bed. We will briefly summarize fluidization regimes that are typically encountered in macro-fluidized beds and outline the influencing factors that determine the transitions of fluidization regimes. This knowledge is necessary for understanding the unique characteristics of micro-fluidized beds, as discussed later.

Fig. 1.5 shows the most common fluidization patterns encountered with increasing gas velocity depending on the properties of particles. When the superficial gas velocity U_g is increased from a fixed bed (Fig. 1.5A) to the minimum fluidization (Fig. 1.5B), the bed pressure drop can be expressed as

$$150\frac{(1-\varepsilon_{mf})^2}{\varepsilon_{mf}{}^3}\frac{\mu_g U_g}{\phi^2 d_p{}^2} + 1.75\frac{(1-\varepsilon_{mf})}{\varepsilon_{mf}{}^3}\frac{\rho_g U_g{}^2}{\phi d_p} = \left[\left(\rho_p - \rho_g\right)g(1-\varepsilon_{mf})\right] \quad (1.3)$$

where ε_{mf} is the bed voidage at the minimum fluidization condition, and ρ_p is the particle density. Wen and Yu [9] found that for most particles ϕ and ε_{mf} can be related by

$$\phi\varepsilon_{mf}^3 \approx \frac{1}{14}, \text{ and } \frac{(1-\varepsilon_{mf})}{\phi^2\varepsilon_{mf}^3} \approx 11 \quad (1.4)$$

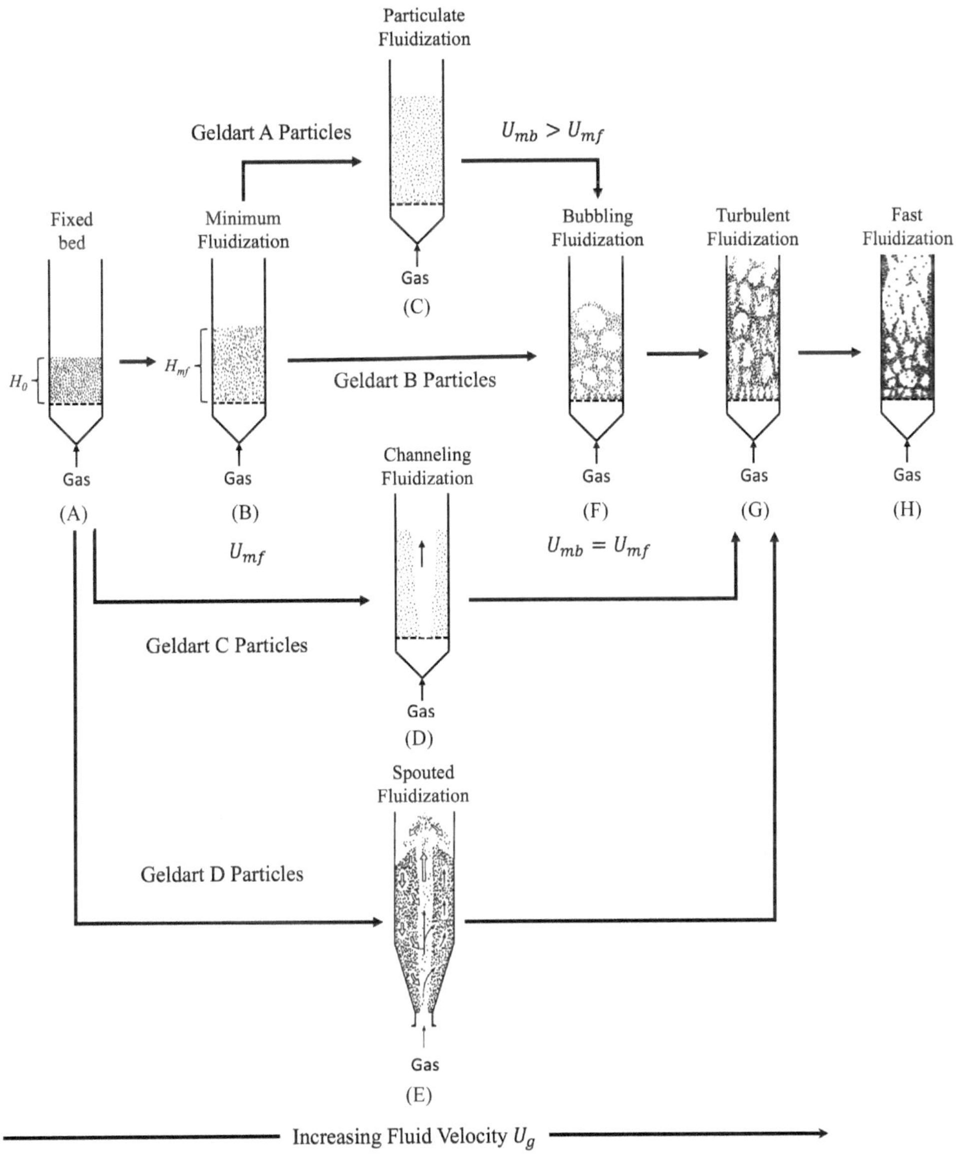

Figure 1.5 Gas-solid flow patterns for different Geldart group particles as the gas velocity increases.

Solving Eq. (1.4) yields $\phi = 0.67$ and $\varepsilon_{mf} = 0.47$. Thus, the following equation to estimate U_{mf} in macro-fluidized beds can be obtained:

$$U_{mf} = \frac{\mu_g}{\rho_g d_p}\left(\sqrt{33.7^2 + 0.0408Ar} - 33.7\right) \tag{1.5}$$

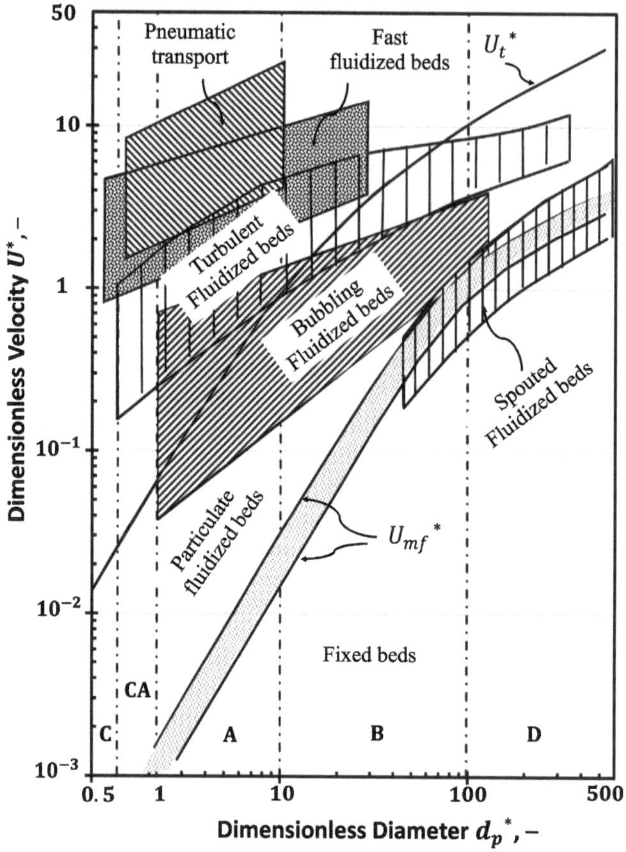

Figure 1.6 A general fluidization regime diagram for macro gas-solid fluidized beds.

Therefore, the minimum fluidization velocity in macro-fluidized beds is apparently unaffected by the bed diameter, but dependent on the properties of gas and solids.

As the velocity increases from the minimum fluidization, the bed of Group A particles can continue homogeneous expansion (i.e., the particulate fluidization, Fig. 1.5C) until gas bubbles are formed that make the bed enter the bubbling fluidization regime (Fig. 1.5F). For Group B particles, the bed enters the bubbling fluidization regime directly at the point of minimum fluidization. For Group C particles, the strong interparticle forces make the bed hard to fluidize smoothly. When the gas is forced to flow through, the bed may form rat-holes or channels (Fig. 1.5D), and with further increasing gas velocity, the bed will enter the turbulent fluidization regime (Fig. 1.5G). For Group D particles, the bed becomes spouted when it is fluidized (Fig. 1.5E). Increasing gas velocity further may bring the bed to the turbulent fluidization regime and, eventually, the fast fluidization regime (Fig. 1.5H).

Fig. 1.6 presents a general fluidization regime diagram for macro-fluidized beds. This flow regime diagram is created by combining several similar charts reported in the literature. In this figure, the dimensionless gas velocity $U^* = \mathrm{Re}/Ar^{\frac{1}{3}}$, the dimensionless particle diameter $d_p{}^* = Ar^{1/3}$, the Archimedes number $Ar = \left(\rho_p - \rho_g\right)\rho_g g(\phi d_p)^3/\mu_g{}^2$, and the Reynolds number $\mathrm{Re} = U_g d_p \rho_g/\mu_g$. The diagram shows that the conditions for different flow regimes may overlap because other factors, which are not included in the above two dimensionless parameters, also affect the flow regime transitions. Readers are advised to refer to the literature for more information about the onset velocities of these different fluidization regimes [10–12]. Note that except for the minimum slugging velocity, the transition velocities in macro-fluidized beds are essentially independent of bed sizes. As we will discuss below, this is not the case for micro-fluidized beds, in which all these transition velocities are strongly affected by bed diameters.

1.5 Fluidization applications

Fluidized beds have been commercially used to deal with a broad range of fluid-solid contacting operations in chemicals, petrochemical, metallurgy, material, mineral, energy, environmental, food, pharmaceutical, and many other industries. Particles to be fluidized can be reactive or unreactive, such as catalysts, fuels, and mineral powders. The most common fluidized bed applications include drying, adsorption, coating, granulation, synthesis, thermal/catalytic cracking, pyrolysis, gasification, combustion, oxidation, and reduction. Several commercial designs of fluidized beds are illustrated in Fig. 1.7. For these applications, fluidized beds offer one or more of the following advantages:

- *Ease of solids handling*: Fluidization enables solid particles to flow, transport, and process stably in continuous operations, in a way that fluid does.
- *High rates of heat and mass transfer between fluid and particles*: Intimate contacts between fluid and fluidized particles make heat and mass transfer between them very fast—highly beneficial to the control of reaction temperature, reduction of transfer limitation, and intensification of reaction processes.
- *High rate of heat transfer between bed and immersed objects/(bed walls)*: Vigorous movements of particles significantly improve the heat transfer between bed particles and surfaces of submerged objects and/or bed walls. This feature permits the accomplishments of strong exothermic or endothermic reactions in relatively small fluidized beds.
- *Rapid and extensive mixing of bed materials*: Intensified movements of solid particles result in quick and comprehensive mixing of bed materials, leading to a uniform temperature distribution throughout the reactor and thus remarkably simplifying process control and operation.
- *Large thermal flywheel*: The fluidized particles provide a large thermal storage capacity, which can substantially neutralize or cushion rapid and abrupt changes in operating conditions and increase the margin of safety by preventing bed temperature from runaways for highly exothermic or endothermic reactions.

Fluidized bed reactors have operated commercially since the 1920s when fluidized bed coal gasifiers were massively implemented in Germany. Fluid catalytic cracking of crude oils for producing high-octane gasoline and synthesis of phthalic anhydride in

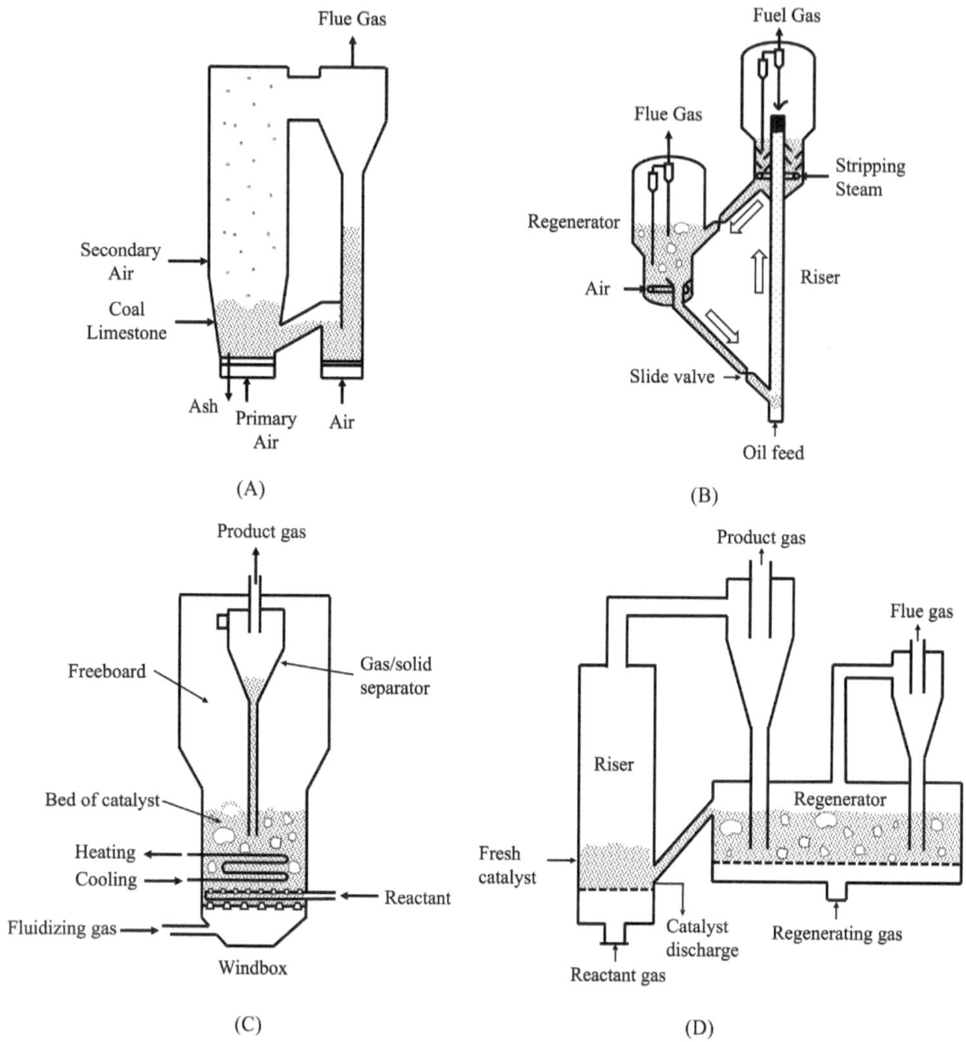

Figure 1.7 Several typical fluidized bed reactor designs. (A) A typical circulating fluidized bed boiler. (B) A typical FCC unit. (C) A typical bubbling fluidized bed reactor. (D) A typical circulating fluidized bed reactor.

fluidized beds have also been in operation since the 1940s (Fig. 1.7A). Today, major gasifiers and combustors of coal, biomass, and organic waste are fluidized beds (Fig. 1.7B). Fluidized beds have also been expanded from traditional bubbling reactors (Fig. 1.7C) to circulating fluidized bed reactors (Fig. 1.7D) and are widely used for mineral processing (e.g., iron reduction, calcination of aluminum hydroxide). Today, we can say that each fluidization regime shown in Fig. 1.1 can be found in numerous industrial applications of fluidized beds.

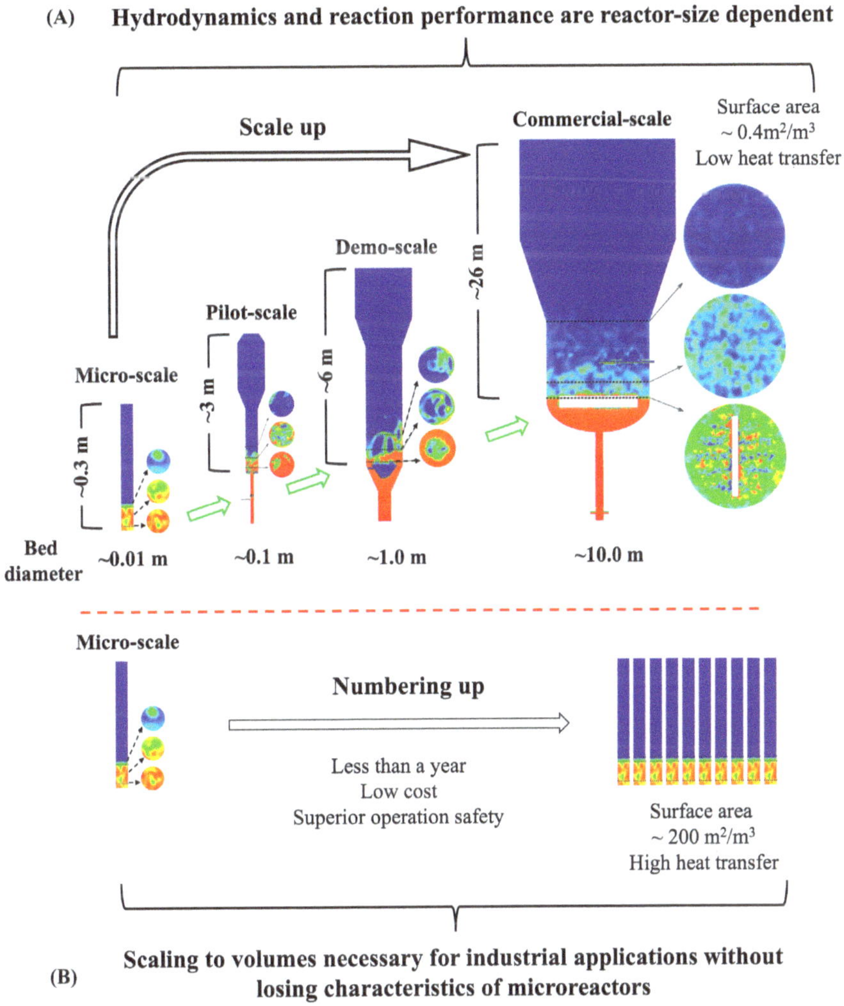

Figure 1.8 Conceptual illustrations of (A) traditional scale-up and (B) advanced numbering up of fluidized bed reactors.

From the history of fluidization research, we see that the research so far has been focusing on large-scale fluidized beds, since industries have sought economic advantage through economies of scale. It is generally agreed that laboratory-scale fluidized beds should have diameters of at least 50 mm so that the experimental results can be considered useful for guiding the design of large-scale industrial fluidized beds. Nevertheless, the scale-up of fluidized bed reactors is still a considerable challenge due to the complex fluid dynamics and its nonlinear relationship to the reaction kinetics [13,14]. As shown in Fig. 1.8A, developing a commercial fluidized bed reactor requires extensive experiments on the laboratory scale, pilot scale, demonstration scale, and commercial scale, which is extremely time-consuming and costly [15]. In many cases, large-scale fluidized beds are likely to compromise the plant performance in one or more of the following aspects: product selectivity and yield, production cost, energy consumption, operation controllability, and system safety. In contrast, micro-fluidized bed reactors can increase the product throughput simply by increasing the number of microreactors in parallel or a module (Fig. 1.8B), which significantly reduces the time and cost of the process scale-up [16]. The numbering-up approach allows scaling reaction processes to volumes necessary for industrial-scale applications without losing the characteristics of microreactors.

For the above reason, researchers in countries such as China, the Netherlands, Canada, Australia, and the UK have started investigations on micro-fluidization at the beginning of the new century. The gained experiences to date have demonstrated that the micro-fluidization technology opens new pathways for developing economic, innovative, and intensified processes, which enable mass production through the numbering-up method.

1.6 Miniaturization of fluidized beds

In recent years, the miniaturization of chemical reactors has attracted a great deal of attention from academia and industry [17–25]. The microreactor technology is potentially helpful for producing high-value-added fine chemicals. Using microreactors offers the following benefits:

- *Controllability of temperature and fluid flow*: This feature is critical for reaction processes to be carried out with increased safety, selectivity, and efficiency. It allows microreactors to be operated safely and efficiently under extreme conditions such as substantial exothermic or endothermic, high temperature and pressure, explosive, corrosive, or use of toxic agents.
- *Uniform temperature distribution*: Due to intensive mixing, high heat transfer rate, and high specific heat transfer area, temperature distribution uniformity in microreactors is greatly improved. This feature allows reactions to take place in a desirable homogenized temperature field. Therefore, the product selectivity and yield are increased significantly because the uniform temperature results in fewer by-products.
- *Close-to-plug flow of gas*: This feature, characterized by a narrow residence time distribution of gas, is unique for microreactors compared to large-scale reactors that sometimes suffer from serious gas backmixing. A plug flow of gas permits control of the reaction time precisely to minimize unwanted secondary reactions, and hence, the best reaction performance, that is, conversion and selectivity, can be attained.

- *High reaction rate*: Microreactors can have the process parameters precisely controlled to best suit the reaction requirements. Consequently, reactions can occur under the most favorable conditions to achieve a maximum reaction rate that approaches the thermodynamic equilibrium. Compared to large-scale reactors, microreactors complete the same reaction with high conversion, selectivity, and safety.
- *Flexible operation at low cost*: Microreactors are safe to operate and maintain. When necessary, they can be prepared to meet various needs, such as changing reaction parameters, using a different feedstock, and switching to a different reaction process in an uninterrupted operation. This feature is remarkably beneficial for rapid screening and optimization of process conditions, catalyst formulations, process additives, etc. It not only increases the effectiveness of process optimizations but saves a considerable amount of time, material, energy, and cost.
- *Simple scale-up*: Conventional large-scale reactors are scaled up through a long process from laboratory-scale experiments to pilot-scale tests, pre-commercial demonstrations, and finally commercial applications. It is because the reactor size and the involved process parameters (e.g., hydrodynamics, heat and mass transfer characteristics, and properties of chemical reactions) are not linearly correlated. With microreactors, the process magnification is achieved by simply increasing the number of microreactors. In this case, the large-scale operation is a sum of many microreactors, since each of the reactors operates and performs the same way as it is individually.

With the rapid development of microchemical reactors, micro-fluidized beds, as a type of microchemical reactors in general and micro-bed reactors in particular, have recently attracted much attention globally, even though the use of mini-scale gas-solid fluidized beds could have started at least as early as the 1970s [26]. Gross et al. [26] used a fluidized bed of 40.6 mm diameter and 355.6 mm tall to obtain kinetic data of catalytic cracking of three gas oil types. The kinetic data obtained with the same commercial FCC zeolite catalysts in the micro-sized fluidized and fixed beds separately were found to be different noticeably. In 1979, Tyler [27] investigated the devolatilization behavior of finely ground (<0.2 mm) brown coal under rapid heating conditions using a small-scale fluidized bed pyrolyzer (~50 mm i.d., ~350 mm high), as shown in Fig. 1.9A. In order to minimize secondary reactions, the reactor was designed to reduce the residence time of volatile products in the hot zone by introducing a purging and quenching stream of nitrogen at the surface of the bed. Later on, the researchers at the University of Waterloo [28–30] employed small-scale fluidized beds to investigate biomass flash pyrolysis and coal gasification. The reactors used in these investigations were updated from the design of Tyler [31], as shown in Fig. 1.9B. Funazukuri et al. [32,33] used a small fluidized bed reactor to investigate the pyrolysis of cellulose and explicitly called the reactor the "micro-fluidized bed" or "micro fluidized bed." Zhang et al. [34] also reported the use of a "micro-fluidized

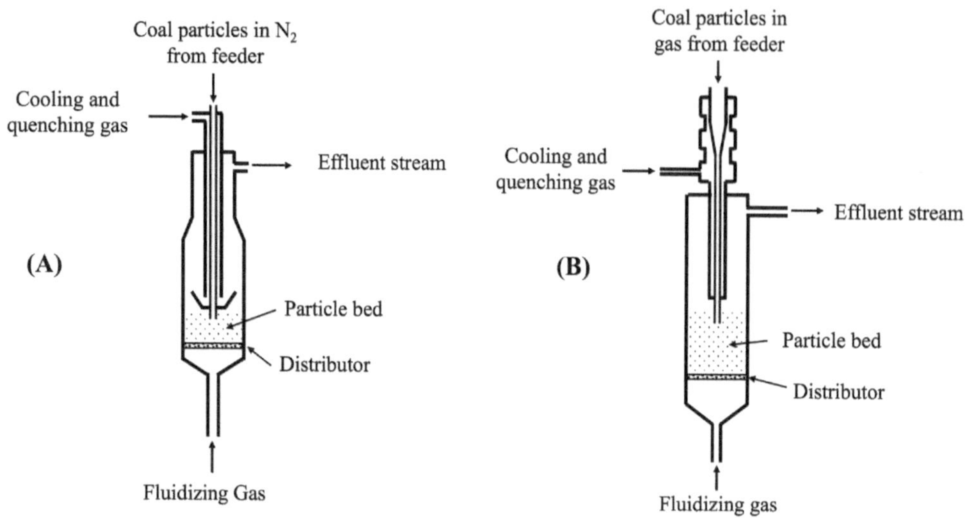

Figure 1.9 The early-stage micro-sized fluidized bed reactors used for pyrolysis studies. (A) Micro-sized fluidized bed pyrolysis reactor. (B) Modified micro-sized fluidized bed pyrolysis reactor.

bed" reactor to investigate hydrogen production from biogas through steam reforming. Focusing primarily on the reactions studied, these early researchers referred to these laboratory-bench reactors as mini-, small- or micro-sized fluidized beds without paying attention to the characteristics of these small-scale reactors. In particular, they did not recognize that the reactor size could significantly impact experimental results.

In a paper submitted to Fuel in 2005 and published in 2007 [35], Murakami et al. investigated biomass pyrolysis using a laboratory-scale fluidized bed reactor of 80 mm in i.d. The results showed that the completion of pyrolysis reactions needed approximately 45 s after injecting 1.0 g of fuel sample into the reactor at 800°C, as indicated in Fig. 1.10A. They noticed that the reaction completion time of 45 s was much longer than anticipated because biomass pyrolysis was known as a fast reaction [31,32]. They suspected that it was due to the use of a big-size reactor. Xu, the corresponding author of this paper and one of the authors of this book, realized that the size of the reactor could have played a significant role in determining the rate of the reaction, which was later verified by his group [31,32], as illustrated in Fig. 1.10B. This realization then motivated Xu to initiate a research proposal to develop a gas-solid reaction analyzer based on the concept of micro-fluidized beds. Two research projects were granted to him: one by the Chinese Academy of Science [37] in 2005 and another by the Ministry of Science and Technology of China [38] in 2011. The objective of the projects was to develop a standardized micro-fluidized bed reaction analyzer (MFBRA) to measure the close-to-intrinsic kinetics of various gas-solid reactions by taking the advantages of both microreactors and fluidized beds, which include the

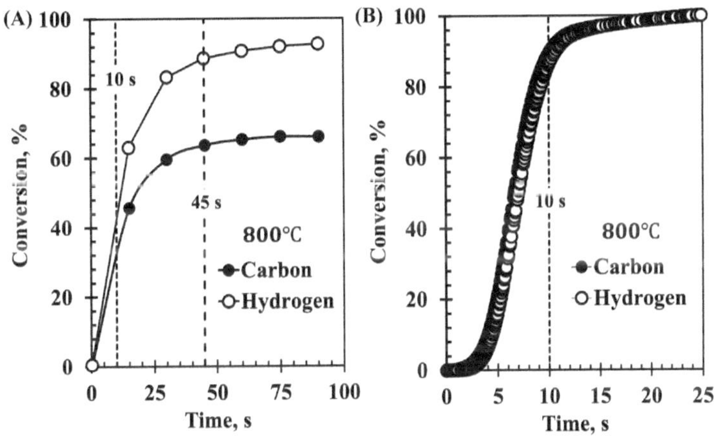

Figure 1.10 Comparison of biomass pyrolysis in two different-sized fluidized beds. (A) 80 mm i.d. fluidized bed; 1.0 g sample (coffee grounds), 800°C [35]. (B) 20 mm i.d. fluidized bed; 20 mg sample (beer lees), 800°C [36].

close-to-plug flow of gas, fast heating rate, high rates of mass and heat transfer, and intensive particle mixing in fluidized beds. After, Xu and coworkers conducted extensive investigations into the fundamentals and applications of gas-solid micro-fluidized beds—including wall effects and operability [39], gas backmixing [40], solid mixing [41], simulation [42–44], and measurements of various reactions [45,46].

In 2005, Potic et al. [47] also used a cylindrical quartz reactor with an internal diameter of only 1 mm to investigate the gasification of biomass and waste streams in compressed hot water at temperatures up to 500°C and pressures over 244 bar. They realized that conducting the research-needed tests under such severe conditions would be expensive and labor-intensive if the conventional laboratory- or pilot-scale experimentation approaches were adopted. To deal with this challenge, they applied the reactor miniaturization principle to fluidized bed research and introduced the concept of micro-fluidized beds (MFBs) as a new experimental method, which is fast, safe, and inherently cost-effective.

The independent studies of Xu [37] and Potic et al. [47] can be considered the starting point of micro-fluidization research. Since then, Xu and his colleagues have continued research on micro-fluidized beds [48]. In addition to fundamental research, they have broadened the application of gas-solid MFBs in areas such as gasification, pyrolysis, and combustion. Zivkovic et al. [49–52] and Liu et al. [53–59] then promoted the research and development of liquid–solid and gas-liquid-solid micro-fluidized beds. Hundreds of research papers on the micro-fluidized bed have been published, covering many aspects of fundamental research and various applications in the gas-solid, liquid-solid, and gas-liquid-solid systems. These studies have greatly enriched our scientific understanding of micro-fluidized beds (Table 1.1).

Table 1.1 A brief comparison between micro- and macro-fluidized beds.

	Macro-fluidized bed	Micro-fluidized bed
Typical design illustration	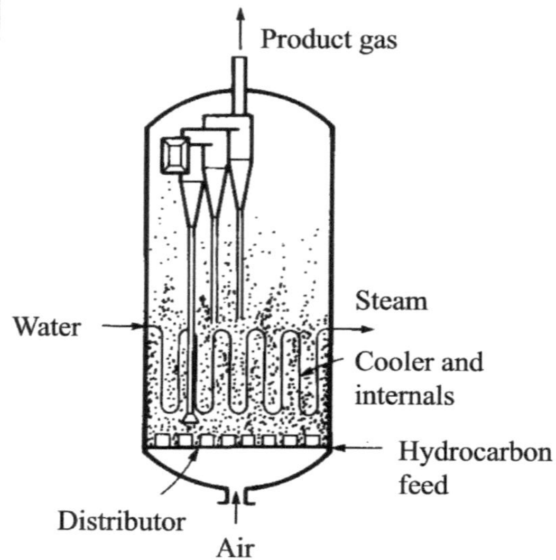	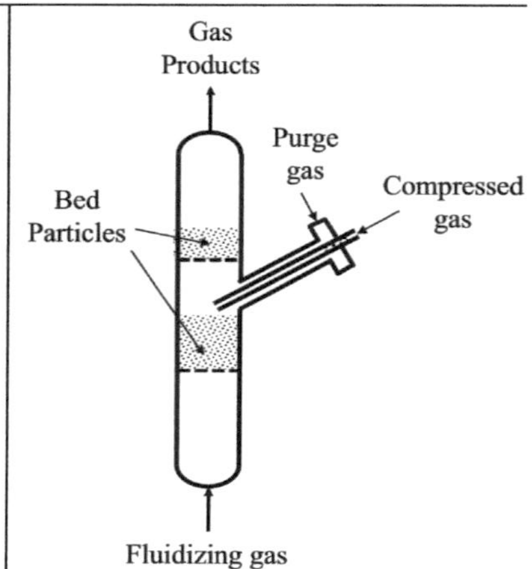
Bed diameter	≥ 50mm	< 50mm
Bed-to-particle diameter ratio	>150	<150
Feed rate	~kg/h	~g/h or 10–50mg/feed
Wall effect	Negligible	Significant
Gas flow	Serious backmixing	Close-to-plug flow
Bed-surface-to-volume ratio	Small	Large
Conversion and selectivity	Low	High
Scale-up complexity	Complex	Simple
Construction and operating costs	High	Low

1.7 Micro-fluidized bed applications

Numerous experimental and theoretical investigations have demonstrated that micro-fluidized beds, conceptually introduced in 2005 [37,47], have unique features compared to macro-fluidized beds due to the increased wall effects. The wall effects, which act as an external force on the movements of particles, cause significant delays in the onset of minimum, bubbling, and slugging fluidization regimes and advances in the transitions to turbulent and fast fluidization regimes [39,60–62]. The wall effects also promote the closeness of gas flow to ideal plug flow in micro-fluidized beds [40,63,64], fundamentally differentiating the micro- from macro-fluidized beds [65].

Based on extensive research on hydrodynamics, gas backmixing, method of sample feeding, data sampling, and analysis associated with MFBs, a series of standardized micro-fluidized bed reaction analyzers (MFBRAs) has been developed. Up to now, MFBRAs have been successfully applied to numerous reactions, including thermal or catalytic decomposition [66–68], thermal cracking [26,68–70], non-catalytic gas-solid reactions [71–74], catalytic reactions [26,75–79], pyrolysis [36,69,70,77,80–86], gasification [87–99], combustion [100–102], and reduction [75,103–113]. These applications involve various biomass materials, coal, oil shale, organic waste, and mineral materials.

For all these investigations, the MFBRA has been well proven to be an efficient and reliable analytical tool with low capital and operational costs, low energy consumption, high safety, and high efficiency in reaction analysis and characterization. Recent work has further demonstrated that the MFB provides a unique tool to obtain the correct kinetic data of reactions producing a gas product that forms an inhibitory atmosphere around the reacting particles, such as the decomposition of carbonate ores [114]. The fundamentals, reaction rate, and kinetic data obtained using MFBRA have been successfully applied to develop several new industrial technologies [115]. The reaction kinetics of magnesite calcination obtained using MFBRA provided the basic data for the engineering design of a 400 kt/a caustic calcination commercial plant [66]. The superior capability of tar in reducing NOx, recognized by research using MFBRA [111], successfully guided the design and operation of a low-NOx dual fluidized bed decoupling combustion plant treating 60,000 t/a distilled spirit lees [116].

1.8 Sources of information on micro-fluidization

As mentioned above, the study of micro-fluidized beds has attracted more attention in recent years, and more than 100 papers have been published in scientific journals so far. These papers are primarily published in the major journals of chemical engineering, energy and fuels, resources, and environment, such as *Chemical Engineering Journal*, *AIChE Journal*, *Chemical Engineering Science*, *Industrial and Engineering Chemistry Research*,

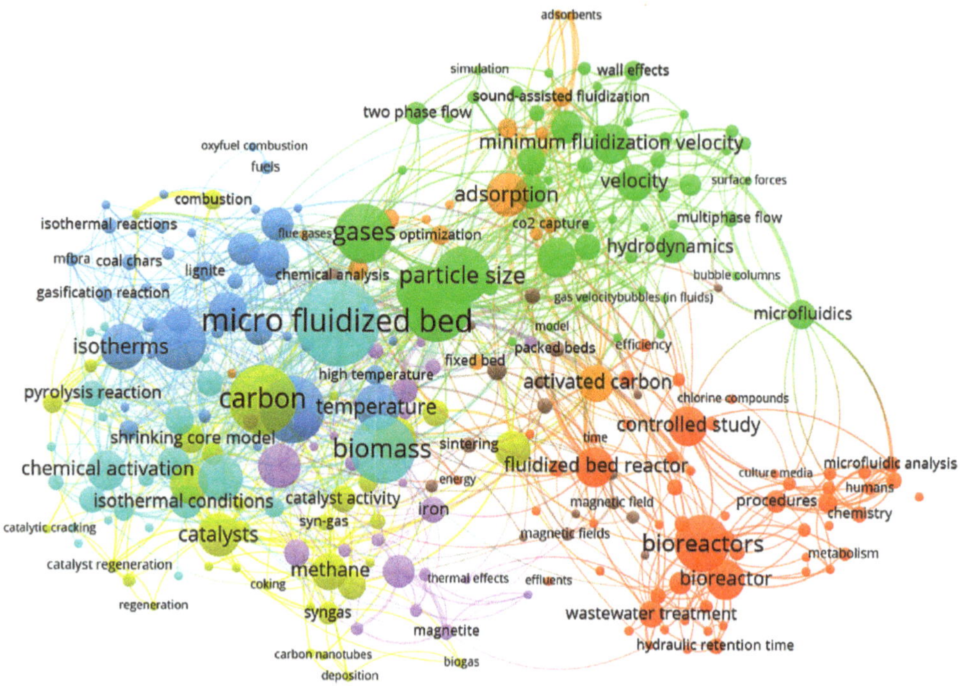

Figure 1.11 Bibliographic co-occurrences of MFB topics based on the Scopus database with bubble and font size proportional to the relative frequency of terms and connection size proportional to the relative frequency of co-occurrences of terms [48].

CIESC Journal, Applied Energy, Fuel, Energy and Fuels, Fuel Processing Technology, International Journal of Hydrogen Research, Environmental Science and Technology, Carbon Resources Conversion. In addition, a chapter on micro-fluidized bed reaction analyzers is presented in the book titled "*Fundamentals and Technologies of Decoupling Thermochemistry Conversion*" published by Science Press in 2018. A few reviews on micro-fluidization science and technologies are also available [48,115,117,118]; one can refer to these reviews on the state-of-the-art hydrodynamics and applications of MFBs.

The bibliographic network of MFB studies from the Scopus database is analyzed based on the fuzzy boundary, as shown in Fig. 1.11 [48]. According to this chart, it is clear that MFB techniques have been intensively studied in terms of the fundamental characteristics (illustrated by green and orange clusters) and applications, including chemical conversions (indicated by yellow and blue clusters) and bioprocessing (presented by red clusters).

Abbreviation

MFB micro-fluidized bed
MFBRA micro-fluidized bed reaction analyzer

Nomenclature

A	cross-sectional area of bed (m^2)
Ar	Archimedes number ($Ar = \frac{(\rho_p - \rho_g)\rho_g g(\phi d_p)^3}{\mu_g^2}$)
d_p	mean particle diameter (m)
d_p^*	dimensionless particle diameter
D_t	bed diameter (m)
g	acceleration of gravity (m/s^2)
H_S	static bed height (m)
m	weight of particles (kg)
Re	Reynolds number ($Re = U_g d_p \rho_g / \mu_g$)
U^*	dimensionless gas velocities
U_g	superficial gas velocity (m/s)
U_{mb}	minimum bubble velocity (m/s)
U_{mf}	minimum fluidization velocity (m/s)
U_{mf}^*	dimensionless minimum fluidization velocity
U_t^*	dimensionless terminal velocity
ΔP_B	bed pressure drop (Pa)
$\Delta P_{B,max}$	maximum bed pressure drop overshoot (Pa)
ΔP_E	bed pressure drop estimated by Ergun equation (Pa)

Greek letters

ε_0	bed voidage at static condition
ε_{mf}	bed voidage at minimum fluidization condition
μ_g	gas viscosity ($Pa \cdot s$)
ρ_g	gas density (kg/m^3)
ρ_p	particle density (kg/m^3)
ϕ	particle sphericity

References

[1] Kwauk M, Li H. Handbook of fluidization. Beijing: Chemical Industry Press; 2007.

[2] Kunii D, Levenspiel O. Fluidization engineering. 2nd ed. Boston: Butterworth—Heinemann; 1991.

[3] Jin Y, Zhu J-X, Wang Z, et al. In: Jin Y, Zhu J-X, Wang Z, et al., editors. Principles of fluidization engineering. 1st ed. Beijing: Tsinghua University Press; 2001.

[4] Ergun S. Fluid flow through packed columns. Chemical Engineering Progress 1952;48:1179−84.

[5] Cranfield RR, Geldart D. Large particle fluidization. Chemical Engineering Science 1974;29:935−47.

[6] Geldart D. The effect of particle size and size distribution on the behaviour of gas-fluidised beds. Powder Technology 1972;6:201−15.

[7] Geldart D, Cranfield RR. The gas fluidisation of large particles. Chemical Engineering Journal 1972;3:211−31.

[8] Geldart D. Types of gas fluidization. Powder Technology 1973;7(5):285−92.

[9] Wen CY, Yu YH. A generalized method for predicting the minimum fluidization velocity. AIChE Journal 1966;12(3):610−12.

[10] Wu C, Cheng Y. Downer reactors. Grace JR, Bi X, Ellis N. Essentials of fluidization technology. Wiley-VCH Verlag GmbH & Co. KGaA, Boschstr. 12, 69469 Weinheim, Germany: 499−530.

[11] Bai D, Jin Y, Yu Z. Flow regimes in circulating fluidized beds. Chemical Engineering & Technology 1993;16:307−13.

[12] Bi HT, Ellis N, Abba IAA, et al. A state-of-the-art review of gas-solid turbulent fluidization. Chemical Engineering Science 2000;55(21):4789–825.
[13] Knowlton TM. Fluidized bed reactor design and scale-up. Fluidized Bed Technologies for Near-Zero Emission Combustion and Gasification. Woodhead Publishing Limited; 1962. p. 481–523.
[14] Rüdisüli M, Schildhauer TJ, Biollaz SMA, et al. Scale-up of bubbling fluidized bed reactors - a review. Powder Technology 2012;217:21–38.
[15] Lu B, Zhang J, Luo H, et al. Numerical simulation of scale-up effects of methanol-to-olefins fluidized bed reactors. Chemical Engineering Science 2017;171:244–55.
[16] Zhang J, Wang K, Teixeira AR, et al. Design and scaling up of microchemical systems: a review. Annual Review of Chemical and Biomolecular Engineering 2017;8:285–305.
[17] Wegeng RS, Drost MK, Brenchley DL. Process intensification through miniaturization of chemical and thermal systems in the 2rt century. Ehrfeld W. IMRET3: proceedings of the third international conference on microreaction technology. Frankfurt: Springer: 2–13.
[18] Kobayashi J, Mori Y, Okamoto K, et al. A microfluidic device for conducting gas-liquid-solid hydrogenation reactions. Science 2004;304(5675):1305–8.
[19] Elvira KS, I Solvas XC, Wootton RCR, et al. The past, present and potential for microfluidic reactor technology in chemical synthesis. Nature Chemistry 2013;5(11):905–15.
[20] Aminian M, Bernardi F, Camassa R, et al. How boundaries shape chemical delivery in microfluidics. Science 2016;354(6317):1252–6.
[21] Kolb G, Hessel V. Micro-structured reactors for gas phase reactions. Chemical Engineering Journal 2004;98(1–2):1–38.
[22] Kiwi-Minsker L, Renken A. Microstructured reactors for catalytic reactions. Catalysis Today 2005;110(1–2):2–14.
[23] Watts P, Haswell SJ. The application of microreactors for small scale organic synthesis. Chemical Engineering & Technology 2005;28(3):290–301.
[24] Hessel V, Löwe H. Organic synthesis with microstructured reactors. Chemical Engineering & Technology 2005;28(3):267–84.
[25] Sovago J-N, Benke M. Microreactors: a new concept for chemical synthesis and technological feasibility (review). Materials Science and Engineering 2014;39(2):89–101.
[26] Gross B, Nace DM, Voltz SE. Application of a kinetic model for comparison of catalytic cracking in a fixed bed microreactor and a fluidized dense bed. Industrial and Engineering Chemistry Process Design and Development 1974;13(3):199–203.
[27] Tyler RJ. Flash pyrolysis of coals. Devolatilitation bituminous coals in a small fluidized-bed reactor. Fuel 1980;59:218–26.
[28] Scott DS, Piskorz J. The flash pyrolysis of aspen-poplar wood. The Canadian Journal of Chemical Engineering 1982;60(5):666–74.
[29] Scott DS, Piskorz J. Flash pyrolysis of biomass. Klass DL, Emert GH. Fuels from biomass and wastes. Michigan: Ann Arbor Science: 421–434.
[30] Zhang ZG, Scott DS, Silveston PL. Steady-state gasification of an Alberta subbituminous coal in a microfluidized bed. Energy and Fuels 1994;8(3):637–42.
[31] Tyler RJ. Flash pyrolysis of coals. 1. Devolatilization of Victorian brown coal in a small fluidized-bed reactor. Fuel 1979;58:680–6.
[32] Funazukuri T, Hudgins RR, Silveston PL. Product distribution in pyrolysis of cellulose in a microfluidized bed. Journal of Analytical and Applied Pyrolysis 1986;9(2):139–58.
[33] Funazukuri T, Hudgins RR, Silveston PL. Flash pyrolysis of cellulose in a micro fluidized-bed. Abstracts of Papers of the American Chemical Society 1984;188(Aug) 89-FUEL.
[34] Zhang ZG, Effendi A, Chen X, et al. Hydrogen production from biogas through steam reforming followed by water gas shift reaction and purification with Selective Oxidation reaction. Takahashi J, Young BA. 1st International Conference on Greenhouse Gases and Animal Agriculture. Obihiro, Japan: 275–279.
[35] Murakami T, Xu G, Suda T, et al. Some process fundamentals of biomass gasification in dual fluidized bed. Fuel 2007;86:244–55.

[36] Yu J, Yao C, Zeng X, et al. Biomass pyrolysis in a micro-fluidized bed reactor: characterization and kinetics. Chemical Engineering Journal 2011;168(2):839–47.
[37] Xu G. Research and development of a differential reaction analyzer based on micro fluidized bed, Project #2005014, Chinese Academy of Science, China. http://www.rcmlab.net/enclosure-mfbra.
[38] Xu G, Yu J, Yao M, et al. Progress in the research, development, and application demonstration of the isothermal micro fluidized bed (gas) solid reaction analyzer. Management and Research of Scientific and Technological Achievements 2016;10:63–6.
[39] Liu X, Xu G, Gao S. Micro fluidized beds: wall effect and operability. Chemical Engineering Journal 2008;137(2):302–7.
[40] Geng S, Han Z, Yue J, et al. Conditioning micro fluidized bed for maximal approach of gas plug flow. Chemical Engineering Journal 2018;351:110–18.
[41] Yang X, Liu Y, Yu J, et al. Numerical simulation of mixing characteristics of trace sample and bed material in micro fluidized bed reaction analyzer. CIESC Journal 2014;65(9):3323–30.
[42] Liu X, Su J, Qian Y, et al. Comparison of two-fluid and discrete particle modeling of gas-particle flows in micro fluidized beds. Powder Technology 2018;338:79–86.
[43] Xu Y, Li T, Musser J, et al. CFD-DEM modeling the effect of column size and bed height on minimum fluidization velocity in micro fluidized beds with Geldart B particles. Powder Technology 2017;318:321–8.
[44] Liu X, Zhu C, Geng S, et al. Two-fluid modeling of Geldart A particles in gas-solid micro-fluidized beds. Particuology 2015;21:118–27.
[45] Yu J, Zeng X, Yue J, et al. Micro fluidized bed reaction analysis and its applications. Kuipers JAM, Mudde RF, Van Ommen JR, et al. The 14th international conference on fluidization - from fundamentals to products. ECI Symposium Series.
[46] Wang F, Zeng X, Geng S, et al. Distinctive hydrodynamics of a micro fluidized bed and its application to gas-solid reaction analysis. Energy and Fuels 2018;32(4):4096–106.
[47] Potic B, Kersten SRAA, Ye M, et al. Fluidization with hot compressed water in micro-reactors. Chemical Engineering Science 2005;60(22):5982–90.
[48] Zhang Y, Goh K-L, Ng YL, et al. Process intensification in micro-fluidized bed systems: a review. Chemical Engineering and Processing - Process Intensification 2021;164:108397.
[49] do Nascimento OL, Reay DA, Zivkovic V. Influence of surface forces and wall effects on the minimum fluidization velocity of liquid-solid micro-fluidized beds. Powder Technology 2016;304:55–62.
[50] Zivkovic V, Kashani MN, Biggs MJ. Experimental and theoretical study of a micro-fluidized bed// AIP Conference Proceedings. 1542: 93–96.
[51] Zivkovic V, Ridge N, Biggs MJ. Experimental study of efficient mixing in a micro-fluidized bed. Applied Thermal Engineering 2017;127:1642–9.
[52] Zivkovic V, Biggs MJ. On importance of surface forces in a microfluidic fluidized bed. Chemical Engineering Science 2015;26:143–9.
[53] Tang C, Liu M, Li Y. Experimental investigation of hydrodynamics of liquid–solid mini-fluidized beds. Particuology 2016;27:102–9.
[54] Li Y, Liu M, Li X. Flow regimes in gas-liquid-solid mini-fluidized beds with single gas orifice. Powder Technology 2018;333:293–303.
[55] Li X, Liu M, Li Y. Bed expansion and multi-bubble behavior of gas-liquid-solid micro-fluidized beds in sub-millimeter capillary. Chemical Engineering Journal 2017;328:1122–38.
[56] Li X, Liu M, Li Y. Hydrodynamic behavior of liquid–solid micro-fluidized beds determined from bed expansion. Particuology 2018;38:103–12.
[57] Wang X, Liu M, Yang Z. Coupled model based on radiation transfer and reaction kinetics of gas–liquid–solid photocatalytic mini-fluidized bed. Chemical Engineering Research and Design 2018;134:172–85.
[58] Li X, Liu M, Ma Y, et al. Experiments and meso-scale modeling of phase holdups and bubble behavior in gas-liquid-solid mini-fluidized beds. Chemical Engineering Science 2018;192:725–38.
[59] Li Y, Liu M, Li X. Single bubble behavior in gas-liquid-solid mini-fluidized beds. Chemical Engineering Journal 2016;286:497–507.

[60] Wang F, Fan LS. Gas-solid fluidization in mini- and micro-channels. Industrial and Engineering Chemistry Research 2011;50(8):4741–51.
[61] Mcdonough JR, Law R, Reay DA, et al. Fluidization in small-scale gas-solid 3D-printed fluidized beds. Chemical Engineering Science 2019;200:294–309.
[62] Quan H, Fatah N, Hu C. Diagnosis of hydrodynamic regimes from large to micro-fluidized beds. Chemical Engineering Journal 2020;391.
[63] Dang TYNYN, Gallucci F, Van Sint Annaland M. Gas back-mixing study in a membrane-assisted micro-structured fluidized bed. Chemical Engineering Science 2014;108:194–202.
[64] Guo Y, Zhao Y, Meng S, et al. Development of a multistage in situ reaction analyzer based on a micro fluidized bed and its suitability for rapid gas – solid reactions. Energy and Fuels 2016;30:6021–33.
[65] Yu J, Geng S, Liu J, et al. A multi-channel micro fluidized bed and its applications, Chinese Patent No. CN105921082B. 2016.
[66] Jiang W, Hao W, Liu X, et al. Characteristic and kinetics of light calcination of magnesite in micro fluidized bed reaction analyzer. CIESC Journal 2019;70(8):2928–37.
[67] Geng S, Han Z, Hu Y, et al. Methane decomposition kinetics over Fe_2O_3 catalyst in micro fluidized bed reaction analyzer. Industrial and Engineering Chemistry Research 2018;57(25):8413–23.
[68] Gai C, Dong Y, Fan P, et al. Kinetic study on thermal decomposition of toluene in a micro fluidized bed reactor. Energy Conversion and Management 2015;106:721–7.
[69] Gai C, Dong Y, Lv Z, et al. Pyrolysis behavior and kinetic study of phenol as tar model compound in micro fluidized bed reactor. International Journal of Hydrogen Energy 2015;40(25):7956–64.
[70] Gao W, Farahani MR, Jamil MK, et al. Kinetic modeling of pyrolysis of three Iranian waste oils in a micro-fluidized bed. Petroleum Science and Technology 2017;35(2):183–9.
[71] Yu J, Zeng X, Zhang G, et al. Kinetics and mechanism of direct reaction between CO_2 and Ca (OH)2 in micro fluidized bed. Environmental Science and Technology 2013;47(13):7514–20.
[72] Prajapati A, Renganathan T, Krishnaiah K. Kinetic studies of CO_2 capture using K_2CO_3/activated carbon in fluidized bed reactor. Energy and Fuels 2016;30(12):10758–69.
[73] Li X, Wang L, Jia L, et al. Numerical and experimental study of a novel compact micro fluidized beds reactor for CO_2 capture in HVAC. Energy and Buildings 2017;135:128–36.
[74] Amiri M, Shahhosseini S. Optimization of CO_2 capture from simulated flue gas using K_2CO_3/Al_2O_3 in a micro fluidized bed reactor. Energy and Fuels 2018;32(7):7978–90.
[75] Liu Y, Guo F, Li X, et al. Catalytic effect of iron and nickel on gas formation from fast biomass pyrolysis in a micro fluidized bed reactor: a kinetic study. Energy and Fuels 2017;31(11):12278–87.
[76] Samih S, Chaouki J. Catalytic ash free coal gasification in a fluidized bed thermogravimetric analyzer. Powder Technology 2017;316:551–9.
[77] Liu Y, Wang Y, Guo F, et al. Characterization of the gas releasing behaviors of catalytic pyrolysis of rice husk using potassium over a micro-fluidized bed reactor. Energy Conversion and Management 2017;136:395–403.
[78] Chen K, Zhao Y, Zhang W, et al. The intrinsic kinetics of methane steam reforming over a nickel-based catalyst in a micro fluidized bed reaction system. International Journal of Hydrogen Energy 2020;45:1615–28.
[79] Wang Q, Zhang C, Zhu Z, et al. Comparison study for the oxidative dehydrogenation of isopentenes to isoprene in fixed and fluidized beds. Catalysis Today 2016;276:78–84.
[80] Guo F, Liu Y, Guo C, et al. Influence of AAEM on kinetic characteristics of rice husk pyrolysis in micro-fluidized bed reactor. CIESC Journal 2017;68(10):3795–804.
[81] Dai C, Ma S, Liu X, et al. Study on the pyrolysis kinetics of blended coal in the fluidized-bed reactor. Procedia Engineering 2015;102:1736–41.
[82] Guo F, Liu Y, Wang Y, et al. Characterization and kinetics for co-pyrolysis of Zhundong lignite and pine sawdust in a micro fluidized bed. Energy and Fuels 2017;31(8):8235–44.
[83] Guo F, Dong Y, Lv Z, et al. Pyrolysis kinetics of biomass (herb residue) under isothermal condition in a micro fluidized bed. Energy Conversion and Management 2015;93:367–76.
[84] Guo F, Peng K, Zhao X, et al. Influence of impregnated copper and zinc on the pyrolysis of rice husk in a micro-fluidized bed reactor: characterization and kinetics. International Journal of Hydrogen Energy 2018;3:21256–68.

[85] Mao Y, Dong L, Dong Y, et al. Fast co-pyrolysis of biomass and lignite in a micro fluidized bed reactor analyzer. Bioresource Technology 2015;181:155–62.
[86] Yu J, Zhu J, Guo F, et al. Reaction kinetics and mechanism of biomass pyrolysis in a micro fluidized bed reactor. Journal of Fuel Chemistry and Technology 2010;38(6):666–72.
[87] Wang F, Zeng X, Wang Y, et al. Non-isothermal coal char gasification with CO_2 in a micro fluidized bed reaction analyzer and a thermogravimetric analyzer. Fuel 2016;164:403–9.
[88] Wang F, Zeng X, Wang Y, et al. Comparison of non-isothermal coal char gasification in micro fluidized bed and thermogravimetric analyzer. CIESC Journal 2015;66(5):1716–22.
[89] Zhang Y, Yao M, Gao S, et al. Reactivity and kinetics for steam gasification of petroleum coke blended with black liquor in a micro fluidized bed. Applied Energy 2015;160:820–8.
[90] Zhang Y, Yao M, Sun G, et al. Characteristics and kinetics of coked catalyst regeneration via steam gasification in a micro fluidized bed. Industrial and Engineering Chemistry Research 2014;53(15):6316–24.
[91] Wang F, Zen X, Wang Y, et al. Investigation on in/ex-situ coal char gasification kinetics in a micro fluidized bed reactor. Journal of Fuel Chemistry and Technology 2015;43(4):393–401.
[92] Wang F, Zeng X, Han J, et al. Comparison of char gasification kinetics studied by micro fluidized bed and by thermogravimetric analyzer. Journal of Fuel Chemistry and Technology 2013;41(4):407–13.
[93] Gao W, Farahani MR, Rezaei M, et al. Kinetic modeling of biomass gasification in a micro fluidized bed. Energy Sources, Part A: Recovery, Utilization and Environmental Effects 2017;39(7):643–8.
[94] Guo Y, Zhao Y, Gao D, et al. Kinetics of steam gasification of in-situ chars in a micro fluidized bed. International Journal of Hydrogen Energy 2016;41(34):15187–98.
[95] Li Y, Wang H, Li W, et al. CO_2 gasification of a lignite char in micro fluidized bed thermogravimetric analysis for chemical looping combustion and chemical looping with oxygen uncoupling (CLC/CLOU). Energy and Fuels 2019;33:449–59.
[96] Wang F, Zeng X, Shao R, et al. Isothermal gasification of in situ/ex situ coal char with CO_2 in a micro fluidized bed reaction analyzer. Energy and Fuels 2015;29(8):4795–802.
[97] Wang F, Zeng X, Wang Y, et al. Characterization of coal char gasification with steam in a micro-fluidized bed reaction analyzer. Fuel Processing Technology 2016;141:2–8.
[98] Zeng X, Wang F, Wang Y, et al. Characterization of char gasification in a micro fluidized bed reaction analyzer. Energy and Fuels 2014;28(3):1838–45.
[99] Zhang Y, Sun G, Gao S, et al. Regeneration kinetics of spent FCC catalyst via coke gasification in a micro fluidized bed. Procedia Engineering, 102. Elsevier Ltd; 1758–65.
[100] Zhang W, Wang P, Sun S, et al. Effects of demineralization methods on structure and reactivity of Zhundong sub-bituminous coal. CIESC Journal 2017;68(8):3291–300.
[101] Fang Y, Luo G, Chen C, et al. Combustion kinetics of In-situ char and cold char in micro-fluidized bed. Journal of Combustion Science and Technology 2016;22(2):148–54.
[102] Zheng M, Shen L, Feng X. Study on chemical-looping combustion of coal with $CaSO_4$ oxygen carrier assisted by CaO addition. Journal of Fuel Chemistry and Technology 2014;42(4):399–407.
[103] Chen H, Zheng Z, Shi W. Investigation on the kinetics of iron ore fines reduction by CO in a micro-fluidized bed. Procedia Engineering 2015;102:1726–35.
[104] Lin Y, Guo Z, Tang H. Effect of atmosphere on degree of reduction in micro-fluidized bed. Journal of Iron and Steel Research 2014;26(4):18–23.
[105] Wang Q, Shao J, Lin Y, et al. An experimental study on the kinetics of iron ore fine reduced by CO in micro fluidized bed. Journal of Iron and Steel Research 2012;24(4):6–9.
[106] Chen H, Zheng Z, Chen Z, et al. Multistep reduction kinetics of fine iron ore with carbon monoxide in a micro fluidized bed reaction analyzer. Metallurgical and Materials Transactions B 2017;48B:841–52.
[107] Chen H, Zheng Z, Chen Z, et al. Reduction of hematite (Fe_2O_3) to metallic iron (Fe) by CO in a micro fluidized bed reaction analyzer: a multistep kinetics study. Powder Technology 2017;316:410–20.

[108] Li J, Liu X, Zhou L, et al. A two-stage reduction process for the production of high-purity ultra-fine Ni particles in a micro-fluidized bed reactor. Particuology 2015;19:27–34.
[109] Lin Y, Guo Z, Tang H. Reduction behavior with CO under micro-fluidized bed conditions. Journal of Iron and Steel Research International 2013;20(2):8–13.
[110] Lin Y, Guo Z, Tang H, et al. Kinetics of reduction reaction in micro-fluidized bed. Journal of Iron and Steel Research International 2012;19(6):6–8.
[111] Song Y, Wang Y, Yang W, et al. Reduction of NO over biomass tar in micro-fluidized bed. Fuel Processing Technology 2014;118:270–7.
[112] He K, Zheng Z, Chen Z. Multistep reduction kinetics of Fe_3O_4 to Fe with CO in a micro fluidized bed reaction analyzer. Powder Technology 2020;360:1227–36.
[113] Yu J, Han Y, Li Y, et al. Mechanism and kinetics of the reduction of hematite to magnetite with $CO-CO_2$ in a micro-fluidized bed. Minerals 2017;7(11):1–12.
[114] Liu X, Hao W, Wang K, et al. Acquiring real kinetics of reactions in an inhibitory atmosphere containing the product gases using a micro fluidized bed. AIChE Journal 2020; Article 17325.
[115] Han ZNZ, Yue JRJ, Zeng X, et al. Characteristics of gas-solid micro fluidized beds for thermo-chemical reaction analysis. Carbon Resources Conversion 2020;3:203–18.
[116] Han Z, Geng S, Zeng X, et al. Reaction decoupling in thermochemical fuel conversion and technical progress based on decoupling using a fluidized bed. Carbon Resources Conversion 2018;1 (2):109–25.
[117] Qie Z, Alhassawi H, Sun F, et al. Characteristics and applications of micro fluidized beds (MFBs). Chemical Engineering Journal 2022;428(June 2021):131330.
[118] Han Z, Yue J, Geng S, et al. State-of-the-art hydrodynamics of gas-solid micro fluidized beds. Chemical Engineering Science 2021;232:116345.

CHAPTER 2

Fundamentals of gas-solid micro-fluidization

Knowledge of the hydrodynamic characteristics of micro-fluidized beds (MFBs) is critical to the proper design and operation of MFB-based applications. This chapter introduces the hydrodynamics of MFBs, with an emphasis on the impact of bed size and operating parameters on bed pressure drop. These fluidization characteristics are compared with those usually observed in large-sized fluidized beds to illustrate the increased importance of wall effects in MFBs. Therefore, to understand the uniqueness of MFBs, readers are encouraged to review the hydrodynamic behavior of large-scale fluidized beds [1–3].

2.1 Bed pressure drops in micro-fluidized beds

Liu et al. [4] and McDonough et al. [5] experimentally measured the relationships between ΔP_B and U_g in MFBs of different dimensions using Group A and B particles. Fig. 2.1 shows the typical experimental results. It shows that the bed pressure drop increases with increasing gas velocity in fixed beds and reaches a plateau when the bed is fluidized ultimately. This $\Delta P_B \sim U_g$ relationship seems similar to that recorded in macro-fluidized beds. However, on careful inspection of the figure, we can identify at least three considerable differences between them, that is,

1. the bed pressure drop **overshoot** when the beds are about to fluidize,
2. the bed pressure drop **offset** in fully fluidized beds, and
3. the $\Delta P_B \sim U_g$ relationship in fixed beds **deviates** from the estimates of the Ergun equation.

Next, we will discuss these three issues in detail.

2.1.1 The bed pressure drop overshoot

Refer to the $\Delta P_B \sim U_g$ curve measured by the ascending gas velocity method in Fig. 2.1. When the superficial gas velocity increases to a critical value U_{cr}, a remarkable pressure drop overshoot ($\Delta P_{B,\max}$) occurs when the bed is still in the fixed bed regime but is about to fluidize. A significant pressure drop overshoot in the mini- or MFB has been observed by several researchers [4–7]. If we use $\Delta P_{B,\max}/(mg/A)$ to present the degree of pressure drop overshoot, it is about 40% higher than mg/A in Fig. 2.1A and 20% higher than mg/A in

Micro Fluidization
DOI: https://doi.org/10.1016/B978-0-443-18718-6.00002-4
© 2023 Chemical Industry Press Co., Ltd.
Published by Elsevier Inc., under an exclusive license with Chemical Industry Press Co., Ltd.

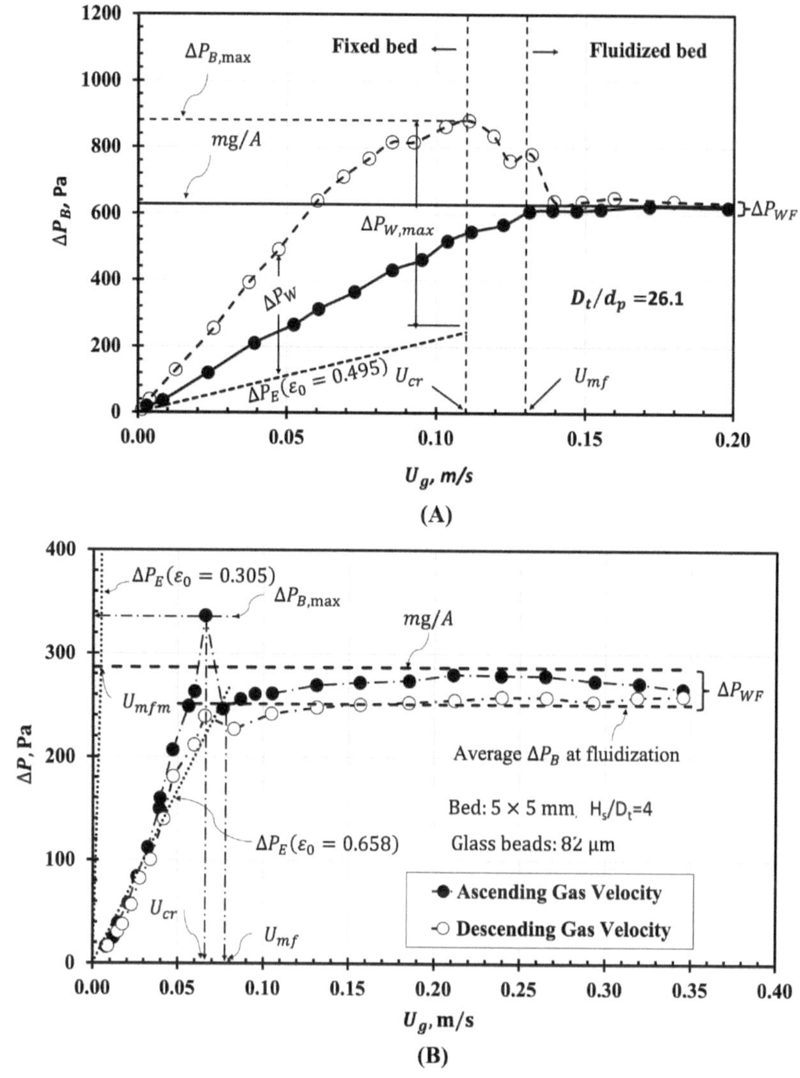

Figure 2.1 Variations of pressure drops of MFBs experimentally measured by the ascending and descending gas velocity methods. (A) $D_t = 12$ mm, $d_p = 460.6$ μm, and $H_s = 50$ mm [4]. (B) 5 × 5 mm, $d_p =$ 82 μm, and $H_s = 20$ mm [5].

Fig. 2.1B. Compared to Fig. 1.1, it is apparent that the pressure drop overshoot is comparatively more significant in micro-fluidized beds than in MFBs. To understand the underlying mechanism that increases pressure drop overshoot, we conceptually illustrate the force balances in the processes of ascending gas velocity (i.e., the bed expansion) and descending gas velocity (i.e., the bed consolidation) in Fig. 2.2. In the case of bed expansion (Fig. 2.2A), both the gravity F_g and the downward wall frictional force F_w exerted on the gas and

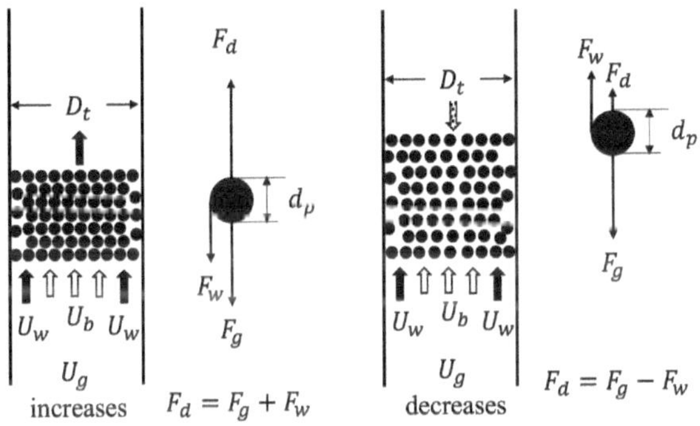

Figure 2.2 Conceptual illustrations of force balances in small-diameter beds during processes of (A) bed expansion and (B) consolidation.

particles prevent particles (particularly those in contact with or adjacent to the wall) from being moved by the upward gas drag force F_d. Because particles in a fixed bed are in close contact, the wall frictional force is progressively transmitted to the adjacent particles toward the bed center. Consequently, the wall friction, which acts as an external force, restricts the movements of particles. As such, mobilizing and fluidizing particles require a more significant drag force from the gas, thus a higher bed pressure drop. When the gas reaches the critical velocity U_{cr}, the bed suddenly expands and the pressure drop registers a remarkable overshoot consequently. Then, as the bed is fully fluidized, the pressure drop reduces and stabilizes. In the process of bed consolidation, the wall frictional force exerted on the particles directs upward, as shown in Fig. 2.2B. This upward friction offsets a portion of the gas drag force, reducing the bed pressure drop. Therefore, even at the same gas velocity, the bed pressure drop measured during the bed expansion is higher than that during the bed consolidation. In addition, since the particles are fully aerated and stay unlocked, the bed pressure drop does not show an overshoot in the consolidation process.

To quantitatively describe the wall effect on the maximum pressure drop overshoot $\Delta P_{B,\max}$, Loezos et al. [7] defined a normalized pressure drop, denoted by $\delta_{\Delta P} = \frac{\Delta P_{B,\max}}{mg/A} - 1$. According to this definition, $\delta_{\Delta P} = 0$ indicates no pressure drop overshoot; $\delta_{\Delta P} > 0$ indicates that a pressure drop overshoot occurs; the larger the value of $\delta_{\Delta P}$, the greater extent of the maximum pressure drop overshoot. To examine the effect of ratio D_t/d_p on the maximum pressure drop overshoot, the experimental data from the literature [4–8] are collected and calculated, and the calculated results are shown in Fig. 2.3. It shows that when the ratio D_t/d_p is higher than 200, $\delta_{\Delta P}$ is approximately 10% or less and changes little. When D_t/d_p is equal to or less than 200, $\delta_{\Delta P}$ increases as the ratio D_t/d_p decreases. Finally, when the D_t/d_p ratio is further reduced to 20 or lower, $\delta_{\Delta P}$ increases sharply, reaching 40%–70%.

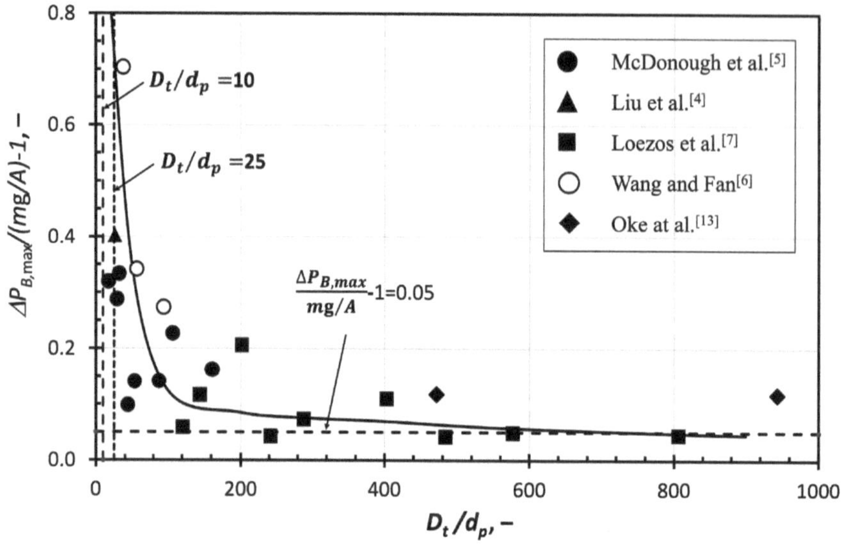

Figure 2.3 Variation in the extent of the maximum bed pressure drop overshoot with D_t/d_p.

2.1.2 The bed pressure drop offset

Under the full fluidization conditions, the measured bed pressure drop may be lower than the weight of particles, as shown in Fig. 2.1B. The difference between the measured bed pressure drop (ΔP_B) and the weight of particles (mg/A), that is, $\Delta P_{WF} = \Delta P_B - mg/A$, is called the "**pressure drop offset**." The extent of the pressure drop offset can be quantified by $\Delta P_{WF}/(mg/A)$, or equivalently by $\Delta P_B/(mg/A)-1$, or $\Delta P_B/(mg/A)$. The pressure drop offset is caused mainly by the wall effect of small beds, although interparticle forces may also play a role for fine particles [8]. In these fully fluidized beds, solid particles may tend to move downward near walls, which can lead to an upward force of friction between solid particles and walls, partially counteracting the weight of particles. Therefore, a smaller drag force is required to support the weight of particles (Fig. 2.2). As a result, the bed pressure drop (proportional to F_d) becomes lower than the weight of particles (proportional to F_g). To validate this analysis, we collected and analyzed the experimental pressure drop data of different-sized MFBs reported by Guo et al. [9], McDonough et al. [5], and Quan et al. [10]. The results are shown in Fig. 2.4, where the normalized bed pressure drop $\Delta P_B/(mg/A)$ is plotted against a normalized gas velocity U_g/U_{mfm} (here, U_{mfm} is the minimum fluidization velocity under the wall effect-free and homogenous flow conditions in fluidized beds). Unless specified, U_{mfm} is calculated by Wen and Yu equation [11]. Based on the definition, the bed pressure drop offset occurs if $\Delta P_B/(mg/A) < 1$. For typical Group A particles (i.e., FCC), Fig. 2.4A shows that the bed pressure drop approximates the weight of particles for beds of inner diameters >15.5 mm (i.e., D_t/d_p >186.7). The pressure drop offset becomes evident when the bed inner diameter is smaller than

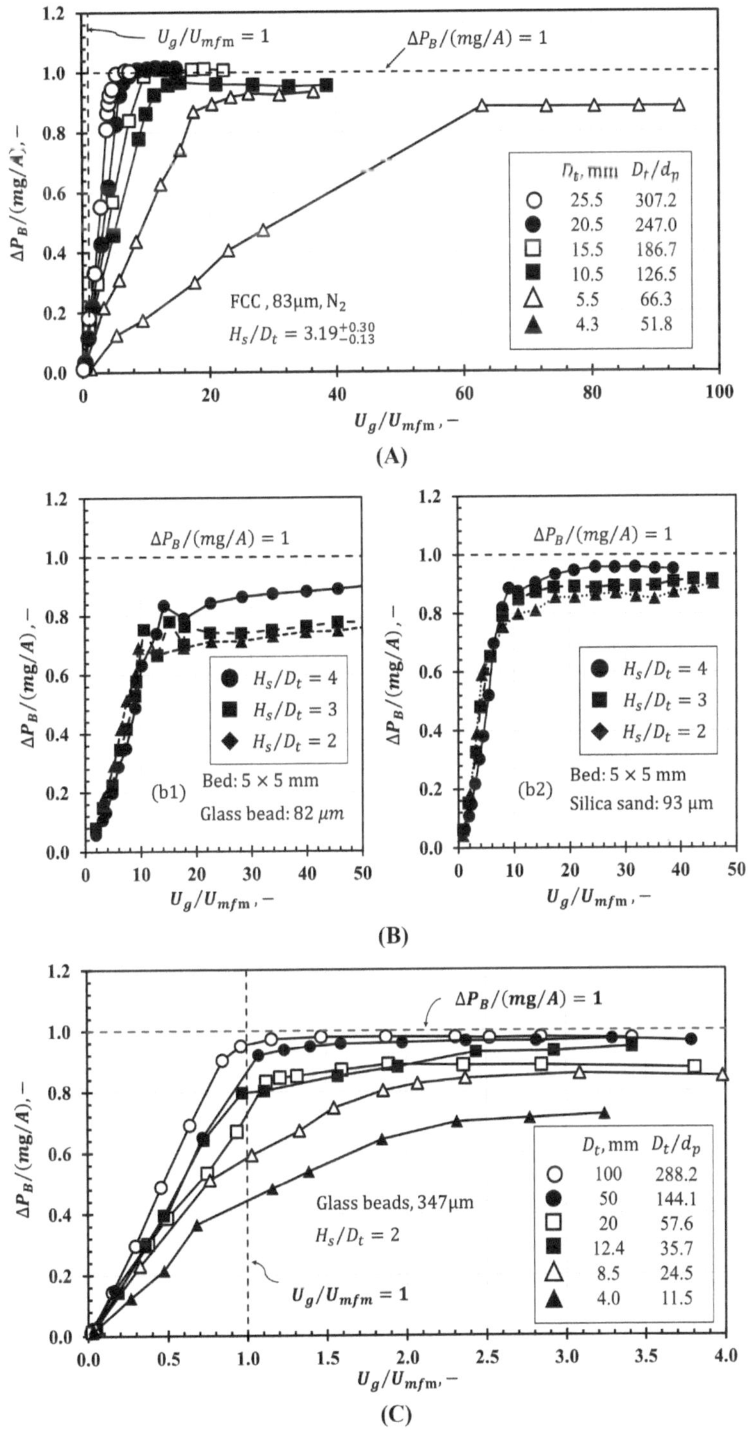

Figure 2.4 $\Delta P_B/(mg/A)$ versus U_g/U_{mfm} for typical Group A and B particles in various-sized fluidized beds. (A) FCC particles (Group A) [9]. (B) Glass beads and silica sand particles (Group A) [5]. (C) Glass bead particles (Group B) [10].

15.5 mm. The extent of the pressure drop offset increases as bed diameter reduces. For typical Group B particles, the bed pressure drop offset is also significant in small-diameter beds corresponding to $D_t/d_p \leq 144$, as shown in Fig. 2.4C.

Fig. 2.4 also displays another typical characteristic of MFBs, that is, delayed fluidization. When the ratio D_t/d_p reduces, the corresponding normalized minimum fluidization velocity U_{mf}/U_{mfm} increases. The value of U_{mf}/U_{mfm} ranges from 10 to 50 for Group A particles (Fig. 2.4A and B) and from 1 to 5 for Group B particles (Fig. 2.4C), respectively. It is evident that the minimum fluidization is delayed significantly in MFBs compared to large-scale fluidized beds. We will discuss this later in detail.

To demonstrate explicitly how the bed diameter affects the bed pressure drop offset, we plot the experimental results of $\Delta P_{wF}/(mg/A)$ calculated from the literature data [5,9,10] against D_t/d_p ratio in Fig. 2.5. It shows that the value of $\Delta P_{wF}/(mg/A)$ is below 5% when the ratio D_t/d_p is large (approx.>150). Even this offset is insignificant, however, it is unexpected in macro-fluidized beds. Although measurement errors may cause such deviations, we recommend careful analysis of the data to ensure that they are not caused by measurement errors but by other causes, such as shallow bed height. Indeed, we note from Fig. 2.5 that the static bed height or H_s/D_t ratio influences the extent of pressure drop. Although there are some exceptions, the general trend shows that the smaller the H_s/D_t ratio, the higher the pressure drop offset. Note that the bed pressure drop offset is still around 2%–4% for D_t/d_p above 200, which is unexpected in beds with such high D_t/d_p. If we analyze the data further, we can find that these data points correspond to H_s/D_t ratios from 0.59 to 2, indicating that the

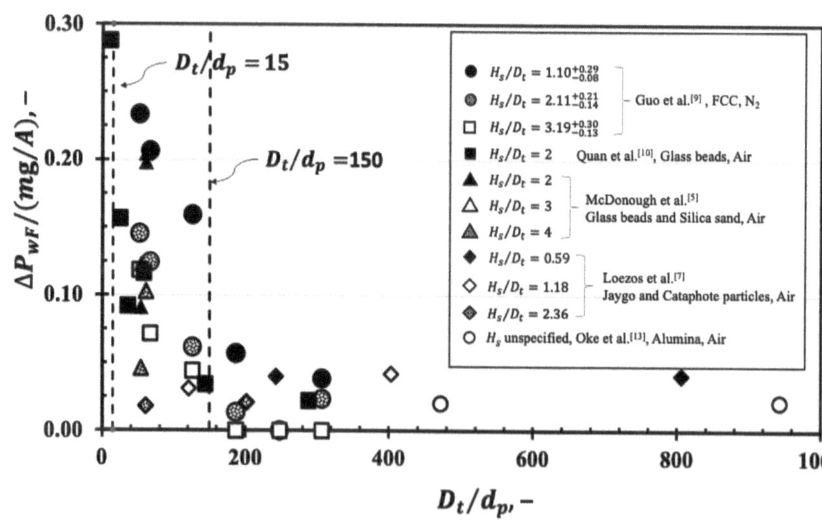

Figure 2.5 Bed pressure drop offset as a function of ratio D_t/d_p.

beds used to obtain these data are shallow. It is well known that shallow beds are prone to form nonuniform gas-solid flows (e.g., channeling, rathole flowing), which lead to a reduced bed pressure drop or a fake pressure drop offset. Indeed, several researchers [7,12–14] have observed the offset between the experimental and theoretical pressure drops in larger diameter (or larger D_t/d_p ratios) beds. These researchers found that the normalized pressure drop (i.e., $\Delta P_B/(mg/A)$) in fluidization conditions was slightly less than 1, and the offset from unity increased with bed diameter (or decreased H_s/D_t). They believed that such an offset was an indication that the bed of particles was not completely fluidized by the gas.

In fully fluidized micro-beds, the pressure drop offset is notably high. When the ratio D_t/d_p is below 150, $\Delta P_{wF}/(mg/A)$ increases with decreasing D_t/d_p. As the ratio D_t/d_p reduces further to approach 15, $\Delta P_{wF}/(mg/A)$ rises sharply. A large pressure drop offset indicates that a large portion of the weight of bed particles is supported by the wall frictional force in MFBs. Consider a fully fluidized bubbling bed, in which gas bubbles move up through the bed particles and burst at the top bed surface. The particles entrained by the bubbles fall back to the bed and move downward near the wall. This forms an internal circulation flow structure characterized by a relatively bubbles-rich upward-moving core surrounded by a downward-moving annular region near the wall [15]. The downward-moving particles in the wall region are partially supported by the upward wall frictional force, reducing the bed pressure drop.

2.1.3 Deviation from the Ergun equation

One of the fundamental characteristics of differentiating between micro- and macro-beds is whether the $\Delta P_B \sim U_g$ curve in a fixed bed follows the Ergun equation. It has been well known that the Ergun equation can describe the $\Delta P_B \sim U_g$ relationship reasonably well for fixed beds with uniform gas flow and negligible wall effects. If the bed-to-particle diameter ratio D_t/d_p is not large enough, the bed pressure drop calculated by the Ergun equation (ΔP_E) would considerably differ from the experimental measurements (ΔP_B) [16,17]. To explicitly present the deviation between ΔP_B and ΔP_E, Liu et al. [4] defined a specific wall effect—the difference between the experimentally measured and the Ergun equation estimated pressure drops per unit volume of bed particles:

$$\frac{\Delta P_w}{V_B} = \frac{\Delta P_B - \Delta P_E}{\frac{\pi}{4} D_t^2 H_B} = \frac{\Delta P_B/H_B - \Delta P_E/H_B}{\frac{\pi}{4} D_t^2} \tag{2.1}$$

where H_B is the bed height. Since the Ergun equation (Eq. 1.2) is only applicable to the fixed beds with the homogenous flow and negligible wall effect, the bed average voidage ε would be measured in the freely packed bed of particles.

As shown in Fig. 2.1A, with the ratio D_t/d_p as low as 26 (as discussed later, this is the approximate lower boundary of the MFB), the experimentally measured pressure drops are higher than the estimates by the Ergun equation. The difference $\Delta P_w = \Delta P_B - \Delta P_E$ increases with increasing superficial gas velocity, reaching a maximum (i.e., $\Delta P_{w,\max}$) at U_{cr}. As a typical indicator of the specific wall frictional force, variations of $\Delta P_{w,\max}/V_B$ with the bed diameter for three particles are presented in Fig. 2.6. It indicates that $\Delta P_{w,\max}/V_B$ decreases with increasing D_t and increases with increasing both d_p and H_s. The results suggest that smaller beds, larger particles, and higher static bed heights increase the specific wall frictional force.

To better understand the dependency of $\Delta P_{w,\max}/V_B$ on the bed-to-particle diameter ratio D_t/d_p, the experimental data reported by Liu et al. [4] and McDonough et al. [5] are plotted in Fig. 2.7. It shows that $\Delta P_{w,\max}/V_B$ increases with decreasing D_t/d_p and increasing H_s/D_t generally. Note that data at lower D_t/d_p ratios appear somewhat scattered because of varying bed heights and particle properties involved. For Group B particles, $\Delta P_{w,\max}/V_B$ approaches zero, showing a typical characteristic of macro-fluidized beds when the ratio D_t/d_p is above 150. This result suggests that $D_t/d_p \sim 150$ can be considered an approximate boundary to differentiate the transition between macro- and MFBs. This transition criterion seems to be in general agreement with the discussions to date. It is also supported by the experimental observation of Guo et al. [9]. They found that when the bed diameter was smaller than 15.5 mm, the measured bed pressure drop began to deviate from the results calculated by the Ergun equation. This indicates that the wall effect begins to play a more prominent role in contributing to the bed pressure drop. Based on the average particle size used

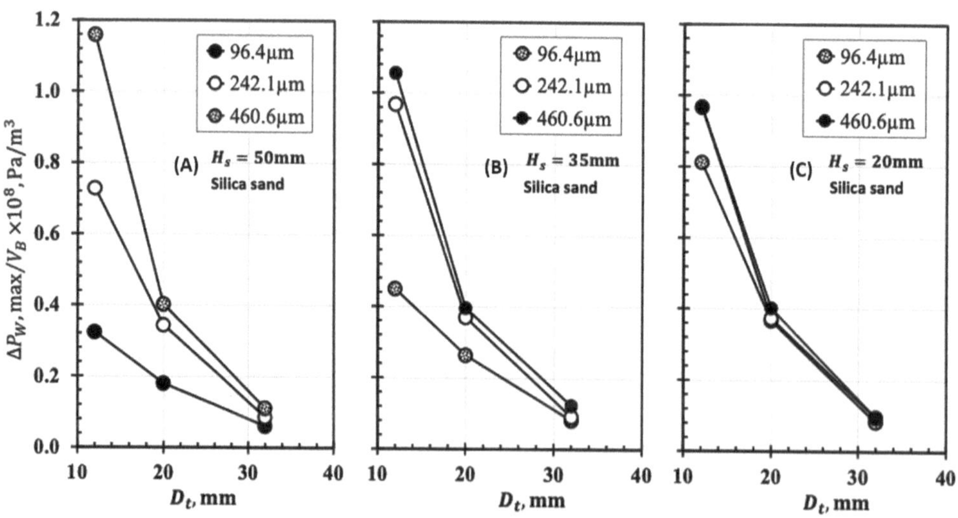

Figure 2.6 Variations of $\Delta P_{w,\max}/V_B$ with the bed diameter for the three particles.

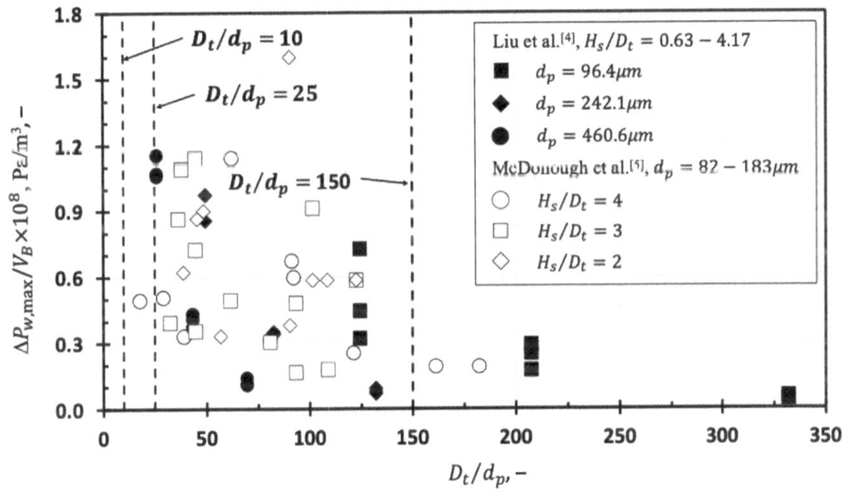

Figure 2.7 Variation of the maximum $\Delta P_{w,\max}/V_B$ with the ratio D_t/d_p.

in their experiments, the corresponding ratio D_t/d_p is 187, close to the 150 defined above.

Vanni et al. [18] and Ansart et al. [19] investigated the impact of bed diameter on fluidization hydrodynamics by using a tungsten powder of 70 μm in mean diameter and 19,300 kg/m^3 in density in columns of internal diameter 20–50 mm. They found that only the 20 mm diameter column had shown wall effects, as evidenced by an increase in the hysteretic behavior between the gas velocity ascending and descending resultant pressure drop curves, an increase in the minimum fluidization velocity, and a decrease in bed voidage. In absolute terms, this diameter of 20 mm is in general consistent with the diameters of MFBs used by Liu et al. [4], Guo et al. [9], and Quan et al. [10], but the corresponding D_t/d_p ratio of 286 is larger than 150 defined above. Therefore, the transition between macro- and MFBs may be defined not only by a sole parameter D_t/d_p but also by other influencing factors (e.g., ρ_p/ρ_g). Due to insufficient experimental data, a complete set of criteria has not been developed for a clear distinction between micro- and MFBs. So further studies are needed in the future.

Based on Eq. (2.1), Liu et al. [4] further calculated $\Delta P_w/V_B$ as a function of the superficial gas velocity in the fixed bed regime for three sizes of silica sand particles under different bed diameters and static bed heights. The results are replotted in Fig. 2.8. It can be seen that $\Delta P_w/V_B$ initially increases with increasing U_g/U_{mf}, reaching a maximum at $U_g/U_{mf} \sim 0.6$ and then declining with further increase in U_g/U_{mf}. $\Delta P_w/V_B$ seemingly increases with decreasing bed diameter, or increasing static bed height, and/or increasing particle diameter. When the bed diameter is greater than 20 mm, $\Delta P_w/V_B$ becomes relatively small and independent of d_p and H_s.

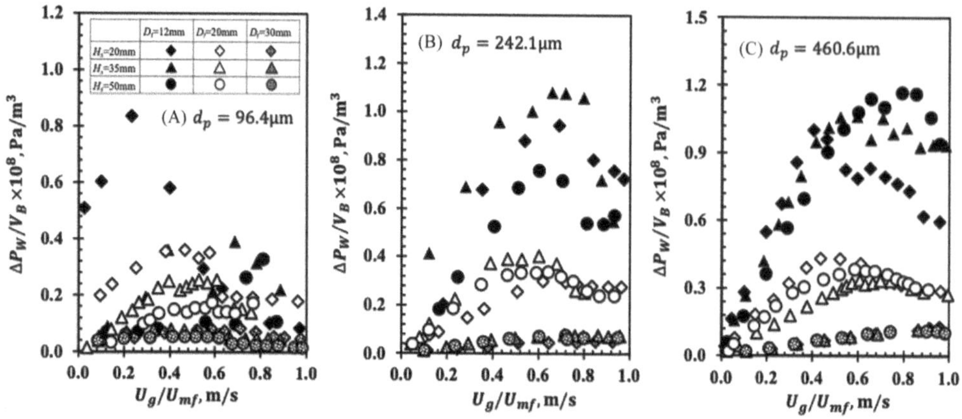

Figure 2.8 Variations of $\Delta P_w/V_B$ with U_g/U_{mf} for three silica sand particles under different static bed heights [4].

2.2 Mechanistic analysis of the wall effects

It is noteworthy that the above discussions are based mainly on the works conducted by Liu et al. [4]. They found that the experimentally measured pressure drops were higher than those estimated by the Ergun equation. Now, consider the experimental data reported by other researchers using different bed sizes and particles, such as Guo et al. [9] and McDonough et al. [5]. We can see that the experimentally measured pressure drops are lower than those estimated by the Ergun equation, as demonstrated previously in Fig. 2.1B, which is just opposite to the trend shown in Fig. 2.1A. Fig. 2.1 shows that the wall effects may increase or decrease the pressure drop in fixed beds depending on operating conditions. To provide deeper insights into the matter, more experimentally measured pressure drops are compared with the estimations of the Ergun equation in Fig. 2.9. For clarity, the weights of particles mg/A in each test condition are also indicated in the figures. The results are obtained with an FCC particle, which has a particle density ρ_p of 1807.5 kg/m^3 and a bulk density ρ_b of 846.5 kg/m^3 (at random packing conditions). Therefore, the voidage at the random packing condition $\varepsilon_0 = 1 - \rho_b/\rho_p = 0.532$ is used in the Ergun equation. In order to obtain the precise mass of particles loaded in the bed (m), the bed is first filled with the particles to the predetermined static height (H_s), and then, the particles are emptied from the bed and weighed. Then, we calculate the actual particle bulk densities (ρ_b) and the average bed voidages (ε) under the static condition based on the mass balance equations: $m = \rho_b A H_s = \rho_p(1 - \varepsilon)AH_s$. The results indicate that the actual bulk density and average bed voidage are in the ranges of 745.4~782.9 kg/m^3 and 0.567~0.587, respectively. It is clear now that in smaller beds, the wall frictional force prevents the particles from densely consolidating when particles are loaded (see Fig. 2.2B), leading

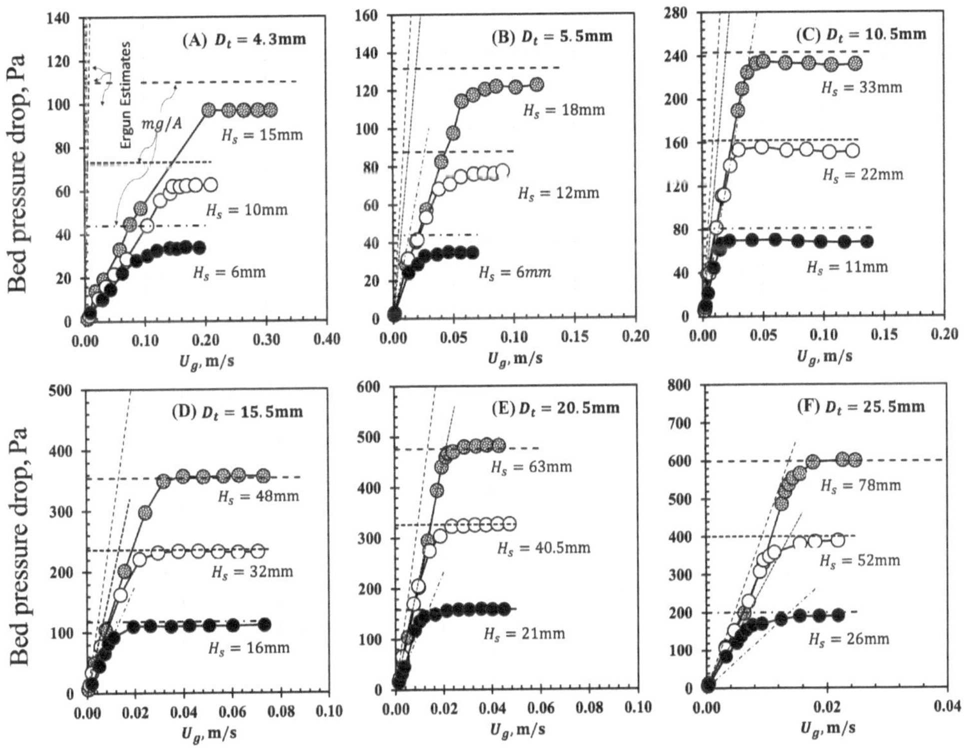

Figure 2.9 Bed pressure drops for six MFBs under different static bed heights for 83 μm FCC particles [9].

to a reduced bulk density or increased bed voidage. In other words, the wall friction supports a fraction of the weight of bed particles.

Furthermore, Fig. 2.9 shows that the measured pressure drops are significantly lower than those predicted by the Ergun equation in small-diameter beds or in low D_t/d_p ratios (Fig. 2.9A–C). When the bed diameter increases to 15.5 mm (Fig. 2.9C) and beyond, the predictions of the Ergun equation approach the experimentally measured results in shallow beds. It seems that for larger and deeper beds, the Ergun equation predictions become lower than the experimental measurements (Fig. 2.9D, $H_s = 21$ mm, Fig. 2.9F, $H_s = 26$ mm)—like what was observed in Fig. 2.1A. On the other hand, the differences between the pressure drop measured by the experiment and that calculated by the Ergun equation show a tendency to increase for smaller and shallower beds.

We can conclude that the above experimental observations provide us an insight into the complex mechanism of the wall effect; that is, the wall effect is related to the wall frictional force and other factors caused by the wall confinement. These factors would play a compromise-in-competitive role in determining the hydrodynamics of the micro-beds. Indeed, for small-diameter fixed beds, researchers have presented some compelling analyses that the effect of confining walls can have multiple interrelated impacts on the gas-solid flow performance [16,17,20–30], including:

- The frictional force that the confining wall directly exerts on adjacent particles,
- Inhomogeneities in the bed voidage and gas velocity, and
- Increased bed voidage.

Next, we will discuss these factors in detail.

2.2.1 The wall frictional force

The small-diameter beds lead to a significant increase in the wall confinement and restriction to the gas and particle flow. Apparently, it is related to the increased specific wall surface area per bed volume for small beds. To quantitatively estimate the wall frictional force, consider a gas-solid flow in a wall-confined column, as shown in Fig. 2.10. The force balance for the particles can be mathematically expressed as

$$\frac{d\sigma_v}{dz} \pm \frac{4}{D_t}\mu_{eff} j\sigma_v = \left[\left(\rho_p - \rho_g\right)\varepsilon_s g - C_d\frac{U_g}{1-\varepsilon_s}\right] \tag{2.2}$$

where z is the axial coordinate, σ_v is the vertical component of stress, μ_{eff} is the wall friction coefficient, j is the Janssen coefficient, and ε_s is the averaged solid volume fraction. The corresponding force balance on the gas is written by

$$\frac{dP}{dz} = C_d\frac{U_g}{1-\varepsilon_s} \tag{2.3}$$

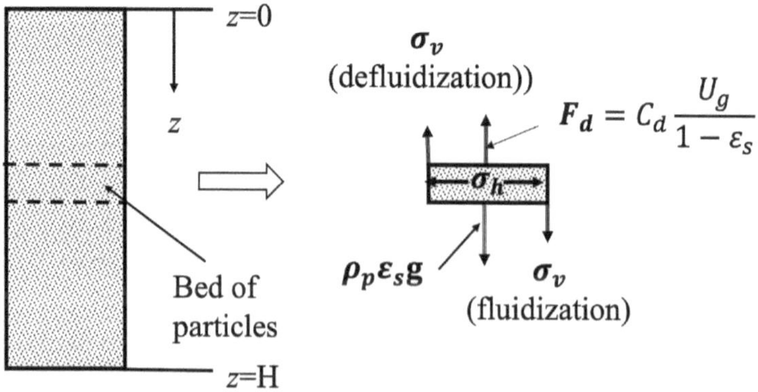

Figure 2.10 Schematic of the force balance for a gas-solid flow in a wall-confined column.

Assuming that the Richardson-Zaki equation can describe the bed expansion [31], the drag coefficient C_d is given by

$$C_d = \frac{\rho_p \varepsilon_s g}{U_t(1-\varepsilon_s)^{n-1}} \tag{2.4}$$

where g is the gravitational acceleration, n is the Richardson-Zaki exponent, and U_t is the single-particle terminal velocity. Accordingly, the wall frictional force is

$$F_w = \frac{4}{D_t}\mu_{eff} j \sigma_v \tag{2.5}$$

where the values of j and μ_{eff} are related to the angle of internal friction φ as follows:

$$j = \frac{1-\sin\varphi}{1+\sin\varphi}, \mu_{eff} = \tan\varphi \tag{2.6}$$

Thus, the wall frictional force F_w can be estimated if the angle of internal friction φ and the vertical component of stress σ_v are known. The angle of internal friction φ is usually measured by a rheometer. The following procedure is usually followed to determine the compressive yield stress of the particles [7,13,32]. First, a weighed mass (m) of particles is filled in a column of diameter D_t and a gas is introduced upward through the column at a rate that enables the particles to fluidize thoroughly and gently. Then, the gas flow rate is reduced slowly to zero. After the bed reaches defluidization completely, the settled bed height (H_s) is measured and recorded. Thus, the average solid volume fraction (ε_s) is calculated by

$$\varepsilon_s = \frac{m}{\rho_p H_s} \tag{2.7}$$

Next, a second batch of the weighed particles is added. Perform the above experiment again, and obtain the second $\varepsilon_s \sim H_s$ dataset. Repeating this measurement procedure several times produces a set of data points ε_s as a function of H_s. An illustrative example is shown in Fig. 2.11A. It shows that ε_s starts from a minimum $\varepsilon_{s,\min}$ at $H_s/D_t = 0$ and then increases rapidly with increasing H_s/D_t when $H_s/D_t < 2$ before reaching a maximum $\varepsilon_{s,\max}$ at the bed exit. Note that in the defluidized conditions, $\rho_p \int_0^{H_s} \varepsilon_s(z)dz = m$.

The compressive yield stress is correlated to the average solid volume fraction as follows [33]:

$$\sigma_v = \begin{cases} c\dfrac{(\varepsilon_s - \varepsilon_{s,\min})^a}{(\varepsilon_{s,\max} - \varepsilon_s)^b}, \varepsilon_{s,\min} < \varepsilon_s < \varepsilon_{s,\max} \\ 0, \varepsilon_s \le \varepsilon_{s,\min} \end{cases} \tag{2.8}$$

where $\varepsilon_{s,\min}$, $\varepsilon_{s,\max}$, a, b, and c are positive constants. Note that this empirical functional form hypothesizes that constant contact between particles occurs only when $\varepsilon_s > \varepsilon_{s,\min}$. It also assumes that the compressive yield stress diverges as $\varepsilon_s \rightarrow \varepsilon_{s,\max}$. Now, if we combine Eqs. (2.2) and (2.8) to fit the experimental data of ε_s versus H_s numerically, the parameters $\varepsilon_{s,\min}$, $\varepsilon_{s,\max}$, a, b, and c can be obtained. Table 2.1 lists these parameters obtained based on fitting this model to the data shown in Fig. 2.11A. We can calculate the compressive yield stress with these parameters, as illustrated in Fig. 2.11B.

We can now estimate the compressive yield stress based on the parameters listed in Table 2.1 and then calculate the wall frictional force for the fine Jaygo particles ($\varphi = 19.5°$). The results are plotted in Fig. 2.12. It shows that the wall frictional force

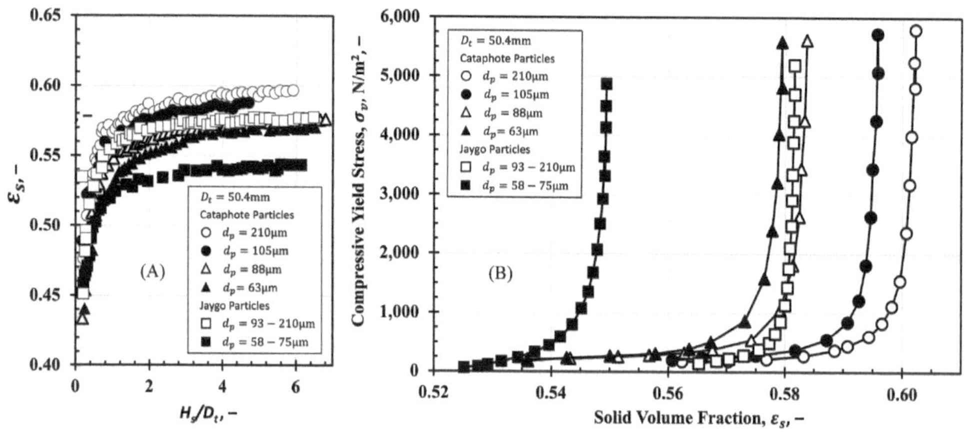

Figure 2.11 Variations of (A) average solid volume fraction with bed height and (B) the estimated compressive yield stress [7].

Table 2.1 The compressive yield stress model parameters.

Particle	a	b	c (N/m²)	$\varepsilon_{s,\min}$	$\varepsilon_{s,\max}$	n
Jaygo particles, $\rho_p = 2550$ kg/m³						
Fine (58–75 μm)	1.25	0.70	2500	0.520	0.550	4.94
Coarse (93–210 μm)	1.74	0.93	2500	0.555	0.582	
Cataphote particles, $\rho_p = 2460$ kg/m³						
63 μm	1.53	0.80	2500	0.540	0.580	4.61
88 μm	1.59	0.81	2500	0.540	0.584	4.30
105 μm	1.77	0.83	2500	0.540	0.596	4.05
150 μm	1.80	0.85	2500	0.540	0.598	
210 μm	1.95	0.91	2500	0.540	0.602	

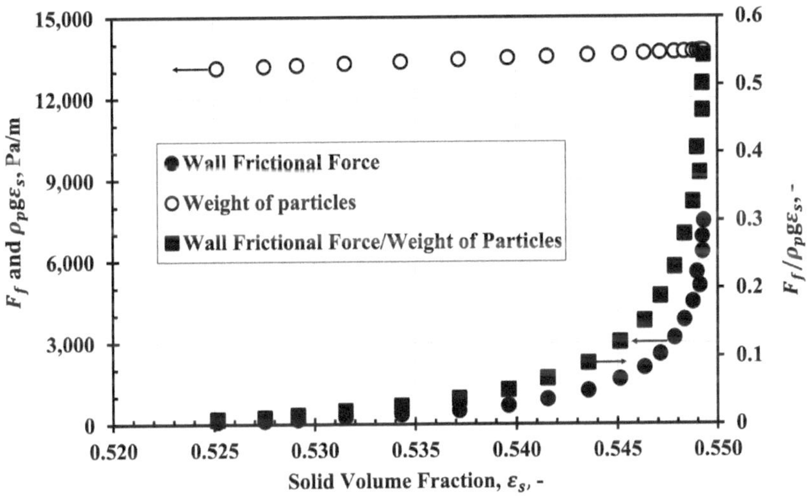

Figure 2.12 The estimated wall frictional force and its comparison with the weight of particle in a 50.8 mm diameter bed.

increases with increasing solid volume fraction. This is understandable because of more sustained contacts between particles and walls at higher solid volume concentrations. Compared with the weight of particles, the support force provided by the wall friction increases with increasing solid volume fraction. In this illustrative case, the wall frictional force can provide up to 55% of the force required to support the weight of particles when the solid volume fraction reaches 0.55. Comparatively, when the solid volume fraction is below 0.545, the ratio of the wall frictional force to the weight particles is lower than 10%.

Liu et al. [15] and Xu et al. [34] investigated gas-solid flows in MFBs using the two-fluid and discrete particle models. They found that the total particle-wall tangential force was positive during the defluidization process, indicating that the wall frictional force partially supported the particles. The model simulations revealed that the total particle-wall tangential force increased compared to the weight of particles as the superficial gas velocity decreased (correspondingly, the solid volume fraction increased), a trend consistent with the above observations. However, the simulations also found that the total particle-wall tangential force was slightly negative during fluidization, indicating an upward particle movement. The negative particle-wall friction force would increase the pressure drop in MFBs, contradictory to the observed pressure drop offsets due to the wall effects. This inconsistency could result from different gas-solid flow structures (i.e., solids internal circulation) or improper selection of model parameters, so further research is needed to better understand the underlying mechanism.

In the light of the above discussion, we conclude that the wall frictional force is affected by both the cohesiveness of particles (i.e., the angle of internal friction) and the compressive yield stress of particle beds. The compressive yield stress of particle beds is a function of the solid volume fraction. Compared to the weight of particles, the wall frictional force is negligible at lower solid volume fractions but could become large at higher solid volume fractions. Under the conditions of fluidization, the wall friction increases the bed pressure drop, but the effect of wall friction on the pressure drop is just the opposite in the defluidization process.

It is important to note that the analysis above assumed that particles moved downward at the wall, which is not always true for beds with different height-to-diameter ratios. As illustrated in Fig. 2.13A, Kunii and Levenspiel [1] explicitly explained that for shallow beds ($H_B/D_t \leq 1$), the emulsion solids circulate as a vortex ring with upflow at the wall and downflow at the bed axis at low gas velocities; but the flow pattern reverses at high gas velocities (Fig. 2.13B). As H_B/D_t approximates 2, emulsion solids begin to move down the wall near the bed surface (Fig. 2.13C). A second vortex ring forms in deeper beds above the original vortex ring. The movement of particles can be dominated by the upper or lower vortex ring depending on the gas velocity (Fig. 2.13D). Multiple circulation loops may be formed for shallow beds with uniform or tuyere distributors (Fig. 2.13E and F). Therefore, when the downward flow of particles at the wall dominates, the wall friction will increase the bed pressured drop, as predicted by Xu et al. [34].

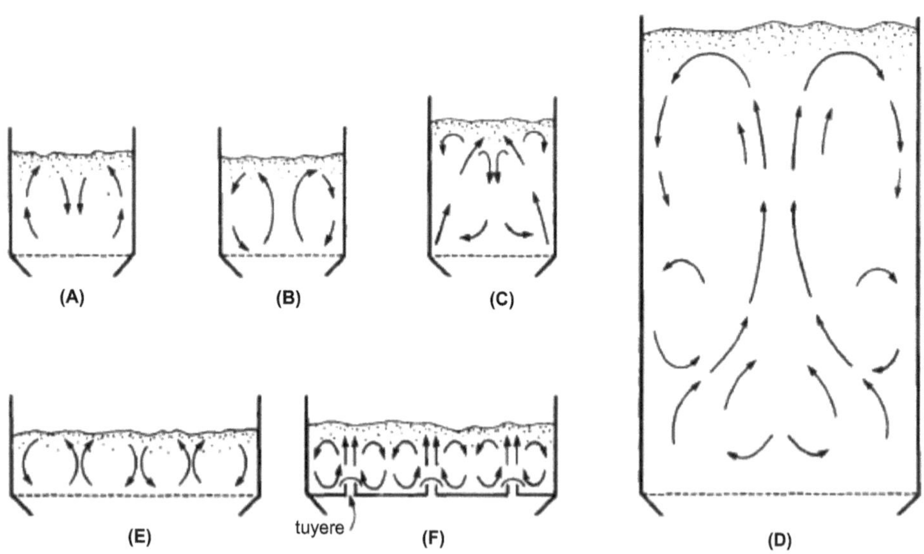

Figure 2.13 Movement of solids in bubbling fluidized beds [1]: (A) $H_B/D_t \leq 1$, low U_g; (B) $H_B/D_t \leq 1$, high U_g; (C) $H_B/D_t \sim 2$, high U_g; (D) general pattern in deep beds; (E) shallow bed, uniform distributor; (F) shallow bed, with tuyeres.

2.2.2 Increase in bed voidage

Bed voidage is generally measured by the defluidization method. During the defluidization process, the wall friction provides partial support for particles to prevent them from being fully compacted. Therefore, the average bed voidage increases with decreasing bed diameter [9,19,20]. In principle, higher voidage provides better permeability for gas flow and thus leads to a lower bed pressure drop. In addition, the bed pressure drop is very sensitive to voidage, especially in the range of $\varepsilon < 0.6$. Therefore, an increase in voidage due to the wall effect can significantly reduce the bed pressure drop.

2.2.2.1 *Micro-fixed beds*

Fig. 2.14 shows the solid volume fraction as a function of D_t/d_p for Group A [9] and B [10] particles in fixed beds. For the typical Group B particles (i.e., glass beads), the solid volume fraction reduces from its bulk value (i.e., without the wall effect) ε_{s0} with decreasing D_t/d_p ratios when D_t/d_p is smaller than 150 approximately. For Group A particles (i.e., FCC), Fig. 2.14 shows that the solid volume is lower than ε_{s0} even as D_t/d_p is up to 250. Adopting the same formula used by Zou and Yu [35], data in Fig. 2.14 can be correlated as follows:

$$\varepsilon_0 = \varepsilon_b + 80.109\left[\exp\left(\frac{0.012}{D_t/d_p}\right) - 1\right] \tag{2.9}$$

where $\varepsilon_b = 0.383$ for Group B particles and $\varepsilon_b = 0.572$ for Group A particles. Indeed, the validity of the above correlation for other particle systems needs further verification because the data used to develop this correlation are limited.

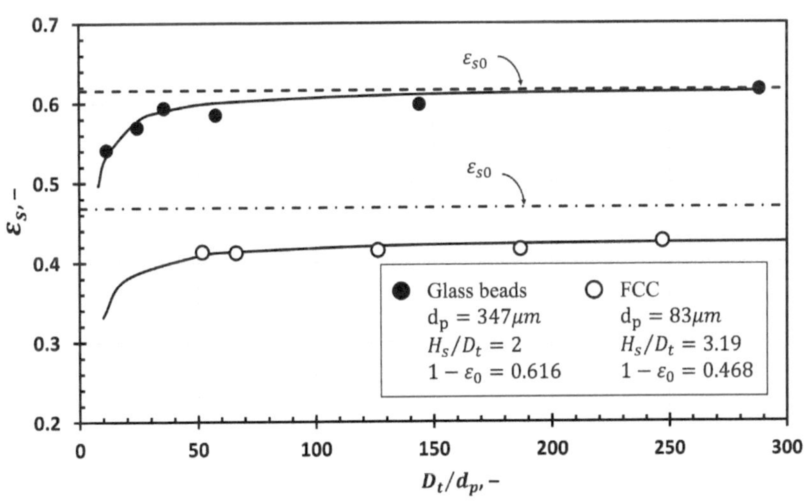

Figure 2.14 The average bed solid fraction at the fixed bed state [10].

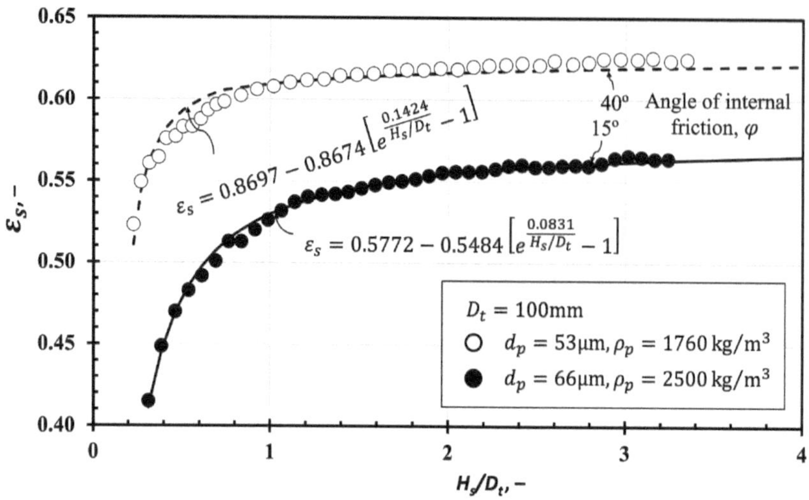

Figure 2.15 The average bed solid volume fraction as a function of ratio H_s/D_t in a fixed bed [13].

Note that the wall effect and the interparticle force contribute to keeping particles less dense in fixed beds for fine cohesive particles. As shown in Fig. 2.15, the more cohesive particles characterized by a higher angle of internal friction ($\varphi = 40$ degrees) produce higher solid volume fraction than the lesser cohesive particles with a lower angle of internal friction ($\varphi = 15$ degrees), even though the two particles are close in diameter [13].

2.2.2.2 Micro-fluidized beds

Fig. 2.16 shows the wall effect on solid volume fractions at minimum fluidization, $\varepsilon_{s,mf}$, for Group A and B particles in MFBs beds. It shows that $\varepsilon_{s,mf}$ is larger for Group B particles than for Group A particles. When D_t/d_p is high, $\varepsilon_{s,mf}$ approximates 0.60 and 0.46 for Group B and A particles, respectively. However, $\varepsilon_{s,mf}$ declines with decreasing D_t/d_p at low D_t/d_p ratios. Based on the published data [4,5,9,10,19,36], two empirical correlations are developed to estimate $\varepsilon_{s,mf}$ [37]:

$$\varepsilon_{s,mf} = 0.5851 - 10.906\left(\frac{H_S}{D_t}\right)^{0.275}\left[\exp\left(\frac{1}{D_t/d_p}\right) - 1\right] \text{for Group A particles} \quad (2.10)$$

$$\varepsilon_{s,mf} = 0.5963 - 6.057\left(\frac{H_S}{D_t}\right)^{0.577}\left[\exp\left(\frac{0.457}{D_t/d_p}\right) - 1\right] \text{for Group B particles} \quad (2.11)$$

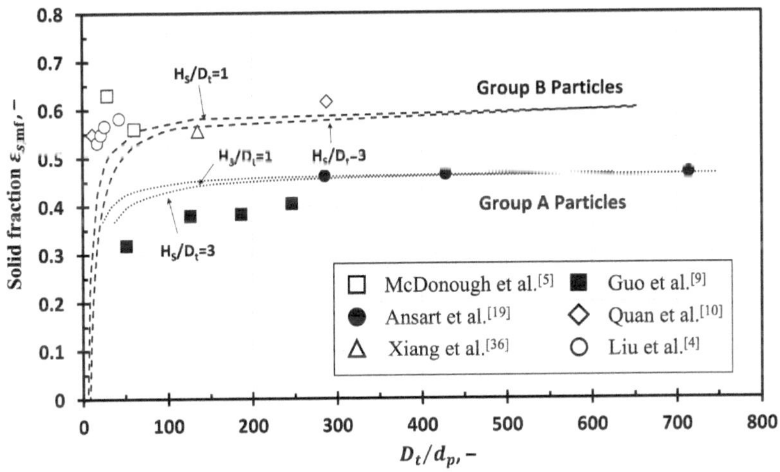

Figure 2.16 Variation of the solid volume fraction at the minimum fluidization with ratio D_t/d_p for Group A and B particles in a MFB.

Note that the above equations may only be valid if $D_t/d_p = 20 \sim 332$ and $H_s/D_t = 0.6 \sim 4.2$ for Group A particles and $D_t/d_p = 10 \sim 714$ and $H_s/D_t = 1.0 \sim 4.2$ for Group B particles. In Fig. 2.16, the predicted variations in $\varepsilon_{s,mf}$ for typical Group A and B particles are plotted. It shows that the predictions agree in general with the trends of experimental data. It is also interesting to note that the estimated $\varepsilon_{s,mf}$ reduces sharply when D_t/d_p for Group A and B particles is reduced to 25 and 10, respectively. The lowest limit of $\varepsilon_{s,mf}$ could be below $0.2 \sim 0.3$ mathematically, which is practically unrealistic. Therefore, we consider that fluidization will not be attainable under these conditions.

2.2.3 Inhomogeneous flow

The wall confinement redistricts the movements of particles and consequently results in an inhomogeneous flow inside small-sized columns, which can be characterized by significantly nonuniform radial distributions of voidage and gas velocity [27]. These radial inhomogeneities are favored hydrodynamically because they provide the optimal permeability for gas flow and hence the least pressure drop [20,30].

Fig. 2.17 shows typical radial profiles of the local voidage and axial gas velocity in a micro-fixed bed [30]. It indicates that the local voidage is close to or equal to 1 at the wall. In the region of about three-particle diameter (i.e., $3d_p$) from the wall, the voidage decreases in a damped oscillatory pattern. Except in the vicinity of the wall, the variations of gas velocity and voidage follow a similar pattern; that is, both of them move toward higher or lower at the same location. This result complicates the wall effect on the bed pressure drop, since the effects of gas velocity and voidage on the

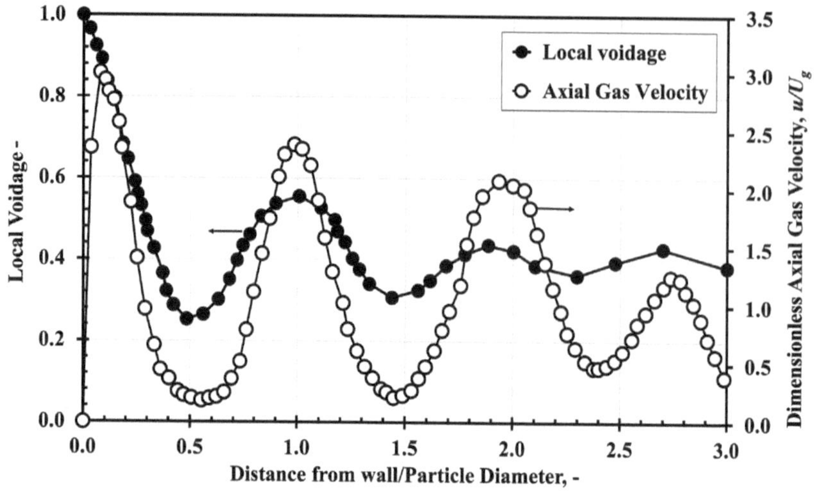

Figure 2.17 Variations of local voidage and gas velocity in a micro-fixed bed ($d_p = 26.5$ μm, $D_t/d_p = 15.1$) [30].

pressure drop are opposite. In principle, the bed pressure drop increases with increasing gas velocity or reducing voidage. Therefore, the wall effect can increase or decrease the bed pressure drop depending on which will be the dominant factor between the gas velocity and voidage.

To quantitatively investigate the specific effects of wall friction and gas maldistributions on the pressure drop in fixed beds, Winterberg and Tsotsas [23] presented a model, which reads

$$-\frac{\partial P}{\partial z} = 150\frac{[1-\varepsilon(r)]^2}{\varepsilon(r)^3}\frac{\mu_g u(r)}{d_p^{\,2}} + 1.75\frac{1-\varepsilon(r)}{\varepsilon(r)^3}\frac{\rho_g u(r)^2}{d_p} - \frac{\mu_{eff}}{r}\frac{\partial}{\partial r}\left(\frac{\partial u(r)}{\partial r}\right) \tag{2.12}$$

The boundary conditions are as follows:

$$r = 0, \frac{\partial u(r)}{\partial r} = 0, \text{and } r = R, u(r) = 0$$

In Eq. (2.12), $\varepsilon(r)$ and $u(r)$ are radial distributions of bed voidage and gas velocity, respectively. When the wall friction and flow maldistribution are neglected (i.e., $\varepsilon(r) = \varepsilon$, $u(r) = U_g$, and $\mu_{eff} = 0$), Eq. (2.12) is simplified to the Ergun equation. Assuming that $\varepsilon(r)$ is expressed as

$$\varepsilon(r) = \varepsilon_0\left[1 + 1.36\exp\left(-2.5\frac{D_t}{d_p}\left(1-\frac{r}{R}\right)\right)\right] \tag{2.13}$$

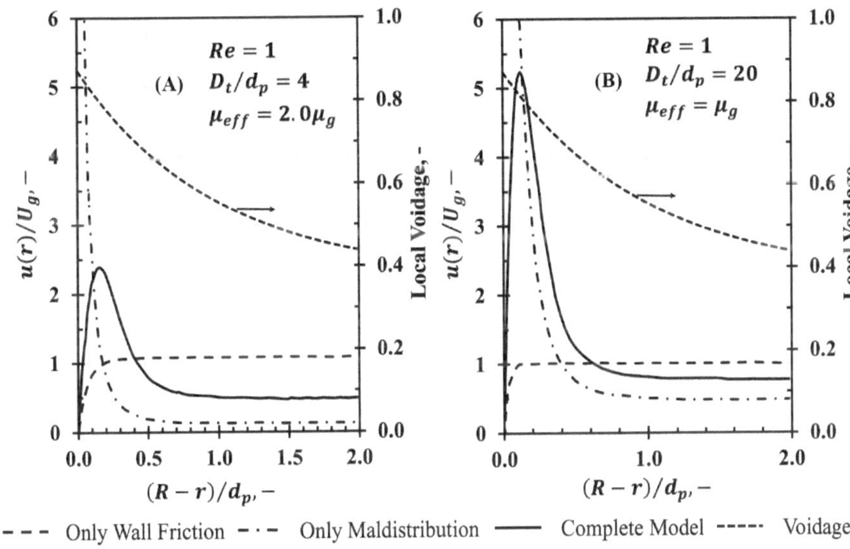

Figure 2.18 Reduced gas velocity profiles for (A) $Re=1$, $D_t/d_p=4$, $\mu_{eff}=2\mu_g$ and (B) $Re=1$, $D_t/d_p=20$, $\mu_{eff}=\mu_g$ [23].

Then, Eq. (2.12) can be solved to yield the following three kinds of pressure drop:

- ΔP_w: the pressure drop caused by the wall friction only,
- ΔP_m: the pressure drop caused by the maldistribution of voidage only,
- ΔP_c: the pressure drop caused by the wall friction and maldistribution of voidage.

Fig. 2.18 shows the profiles of bed voidage and gas velocity predicted by the model under two typical operating conditions with the same Reynolds number $Re=1$ ($Re=U_g d_p \rho_g/\mu_g$). It shows that the calculated bed voidage reduces continuously with the distance from the wall. It also reveals that the wall friction produces a strong drag force that slows down the gas flow near the wall, but this effect is short-ranged. The gas velocity reaches the maximum when the radial location is about $0.3d_p$ away from the wall. With only the flow maldistribution considered, the model predicts that the gas velocity is high near the wall and quickly declines toward the bed center. When the wall friction and flow maldistribution are considered, the model (i.e., complete model) predicts a velocity profile that provides an optimal permeability for gas flow—a high peak near the wall. An increase in the wall friction (i.e., μ_{eff}) reduces the peak height of gas velocity but increases the velocity in the bed center.

Fig. 2.19 shows the model predicted $\Delta P_B/\Delta P_E$ as a function of D_t/d_p, in which ΔP_E is predicted by the Ergun equation and used as the basis for comparison. It shows clearly that the wall friction and the flow maldistribution affect the bed pressure drop oppositely. ΔP_B is higher than ΔP_E (i.e., $\Delta P_B/\Delta P_E > 1$) due to the wall friction, but lower than ΔP_E (i.e., $\Delta P_B/\Delta P_E < 1$) due to the flow maldistribution. With

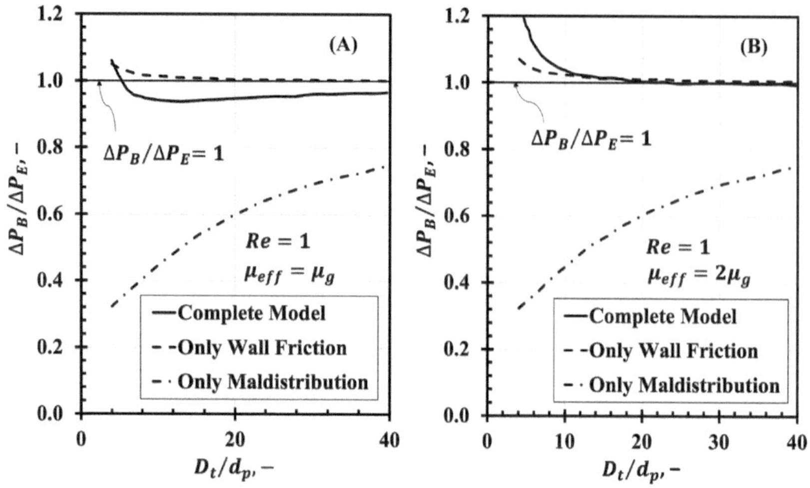

Figure 2.19 The model-predicted $\Delta P_B/\Delta P_E$ as a function of ratio D_t/d_p for (A) $Re = 1$, $\mu_{eff} = \mu_g$ and (B) $Re = 1$, $\mu_{eff} = 2\mu_g$ [23].

decreasing D_t/d_p, the effects of wall friction and the flow maldistribution on the pressure drop increase, which can be seen by the fact that the ratio of $\Delta P_B/\Delta P_E$ constantly deviates from $\Delta P_B/\Delta P_E = 1$. With the combined effects of the wall friction and gas maldistribution, ΔP_B may increase or decrease as a function of D_t/d_p depending on the balance of the two opposing competing factors. Under low wall friction (Fig. 2.19A, $\mu_{eff} = \mu_g$), ΔP_B is lower than ΔP_E (i.e., $\Delta P_B/\Delta P_E < 1$) as $D_t/d_p > 4$, suggesting that the gas maldistribution plays a dominant role in reducing the bed pressure drop under these operating conditions. When $D_t/d_p < 4$, the confining wall friction shows a comparative or even more pronounced effect, so ΔP_B becomes higher than ΔP_E (i.e., $\Delta P_B/\Delta P_E > 1$). As the wall friction increases, as shown in Fig. 2.19B with $\mu_{eff} = 2\mu_g$, larger values of $\Delta P_B/\Delta P_E$ are predicted by the wall friction and the complete models. Consequently, the intersection point where $\Delta P_B/\Delta P_E$ across the line $\Delta P_B/\Delta P_E = 1$ occurs at a relatively high value of D_t/d_p (i.e., $D_t/d_p = 20$ in Fig. 2.19B compared to $D_t/d_p = 4$ in Fig. 2.19A).

Di Felice and Gibilaro [17] presented a simple two-zone model for micro-fixed beds to further illustrate the effect of wall friction-induced nonuniform flow on the bed pressure drop. The model divided the gas flow through the bed cross-section into two zones: a center bulk zone and a wall friction zone, as schematically illustrated in Fig. 2.20A.

The model assumes that the Ergun equation applies to the bulk zone, which is considered unaffected by the wall friction. Under a steady state, $\Delta P_B = \Delta P_c = \Delta P_w$ is applied based on the principle of pressure balance. Therefore, the bed pressure drop ΔP_B can be estimated by ΔP_c based on the Ergun equation if the gas velocity and

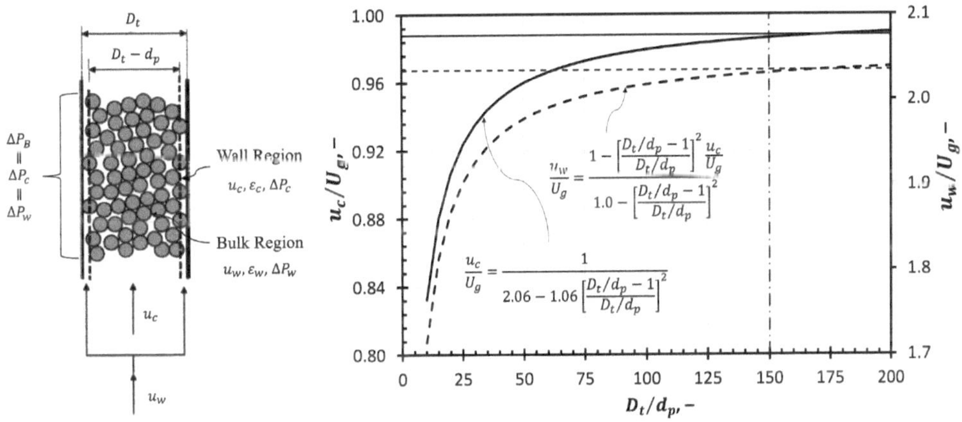

Figure 2.20 A simple two-zone model and its calculated gas velocities in the wall and central bulk zones as a function of ratio D_t/d_p [17]. (A) Two-zone model. (B) The gas velocities in the two zones.

voidage in the bulk zone are known. The model assumes that the bulk zone's bed voidage and gas velocity are constants. Based on the mass balance, the gas velocity u_c can be calculated by

$$\frac{u_c}{U_g} = \frac{1}{2.06 - 1.06\left[\frac{D_t/d_p - 1}{D_t/d_p}\right]^2} \tag{2.14}$$

and the bed voidage can be estimated based on the following empirical correlation [35]:

$$\varepsilon = \varepsilon_c + 0.01\left[\exp\left(\frac{10.686}{D_t/d_p}\right) - 1\right], \varepsilon_c = 0.4 \tag{2.15}$$

The calculation results based on Eq. (2.14) are plotted in Fig. 2.20B. It indicates that the gas velocity u_c in the bulk zone increases sharply with increasing D_t/d_p ratio initially, and then reaches a plateau when D_t/d_p approaches 150 approximately.

Substituting Eqs. (2.14) and (2.15) into the Ergun equation gives

$$\Delta P_B = \Delta P_c = 150\frac{(1-\varepsilon_c)^2}{\varepsilon_c^3}\frac{\mu_g u_c}{\phi^2 d_p^2} + 1.75\frac{(1-\varepsilon_c)}{\varepsilon_c^3}\frac{\rho_g u_c^2}{\phi d_p}$$

$$\Delta P_E = 150\frac{(1-\varepsilon)^2}{\varepsilon^3}\frac{\mu_g U_g}{\phi^2 d_p^2} + 1.75\frac{(1-\varepsilon)}{\varepsilon^3}\frac{\rho_g U_g^2}{\phi d_p} \tag{2.16}$$

Consequently, $\Delta P_B/\Delta P_E$ can be calculated as a function of D_t/d_p, as shown in Fig. 2.21. For comparison, Fig. 2.21 also provides the experimental data reported in

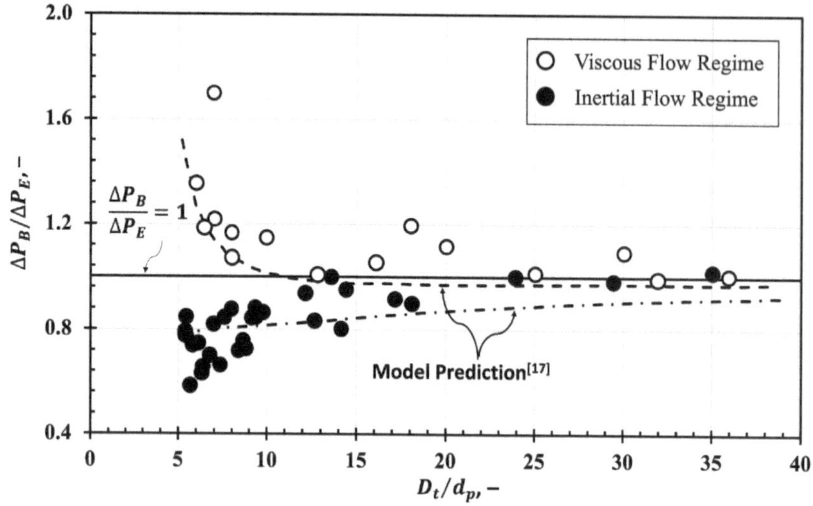

Figure 2.21 The ratio of bed pressure drop to the Ergun prediction in fixed beds.

the literature. The experimental data are grouped into two categories: the points obtained under the viscous flow regime and those under the inertial flow regime. It shows that the model predictions generally agree with the experimental data. As D_t/d_p decreases, which leads to an increase in the wall effect, $\Delta P_B/\Delta P_E$ increases for beds operating in the viscous flow regime but decreases for beds operating in the inertial flow regime. $\Delta P_B/\Delta P_E$ approaches unity as D_t/d_p increases regardless of the flow regime.

2.3 Discussions on the wall effects

The wall effect mechanisms discussed above are briefly summarized in Table 2.2. It is evident that due to the multiple effects of the wall confinement, the ultimate effect of the confining wall on the bed pressure drop for micro-sized gas-solid systems cannot be determined straightforwardly. Nevertheless, based on the understanding presented so far, we can provide some insights into the experimentally observed bed pressure drop variations, which have been confusing, questionable, or incomprehensible.

Fig. 2.22 shows that the Ergun equation predicts a higher bed pressure drop than the experimental data (Fig. 2.22A) and a lower bed pressure drop than the experimental data (Fig. 2.22B). The experimental data are measured by the descending gas velocity method in both of these experiments. The wall effects discussed previously—pressure drop offset, increased packing voidage, deviations from the Ergun equation due to wall friction, voidage increase, and nonuniform flow—are well demonstrated in these two figures.

Table 2.2 Summary of the wall effects in small-diameter beds.

	Bed expansion	Bed consolidation
Gas-solid flow patterns	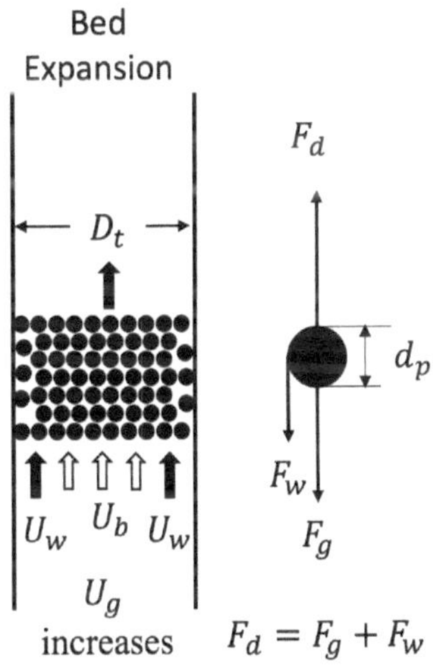	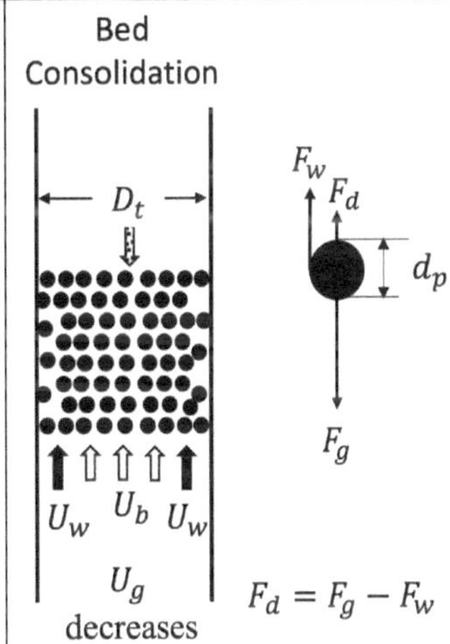
Wall frictional force	Particles flow upward at the wall dominated: $\Delta P_B \uparrow$ Particles flow downward at the wall dominated: $\Delta P_B \downarrow$	
Voidage increase	$\Delta P_B \downarrow$	
Inhomogeneous flow	Voidage increase: $\Delta P_B \downarrow$ Gas velocity increase: $\Delta P_B \uparrow$ Viscous flows: $\Delta P_B \uparrow$ Inertial flows: $\Delta P_B \downarrow$	

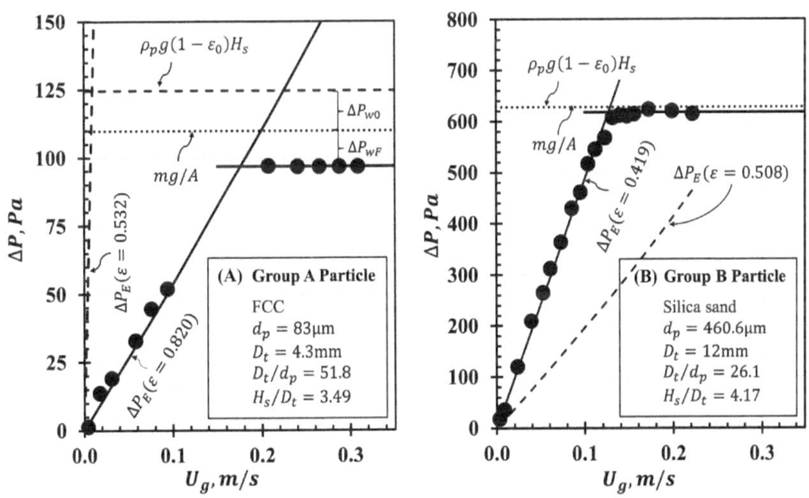

Figure 2.22 The wall effects observed in small-diameter beds for (A) a typical Group A particle and (B) a typical Group B particle.

For FCC particles used in Fig. 2.22A, the particle and bulk densities are 846.5 and 1807.5 kg/m^3, respectively. Therefore, the average voidage at a wall effect-free condition is $\varepsilon_0 = 846.5/1807.5 = 0.532$. Comparatively, the average voidage corresponding to the actual mass loaded to the bed is 0.587 ($=1 - mg/A\rho_p$), which is approximately 10% higher than ε_0. Substituting the voidage ε_0 into the Ergun equation gives the ΔP_B corresponding to the homogeneous and wall friction-free fixed bed condition as a function of superficial gas velocity. It can be seen that the results are significantly higher than the experimental measurements at the same gas velocity. Assuming that if the wall friction only causes a change in the bed voidage and the Ergun equation still applies, an average voidage of 0.820 is required to provide the best fitting between the Ergun equation and the experimental data, which is shown in Fig. 2.22A by the straight line $\Delta P_E(\varepsilon = 0.820)$. Obviously, this voidage is too high to be practical at fixed bed conditions. Therefore, as previously discussed, the significant reductions in fixed bed pressure drop should be caused by the nonuniform flow across the bed cross section. When the bed is fully fluidized, particles move vigorously in an internal circulation mode and the wall friction impacts particle flow. As a result, the pressure drop offset $\Delta P_{wF} = 13.0$ Pa (i.e., 11.8% of mg/A) is obtained.

We then draw the following conclusions for typical Group A particles: (1) the wall confinement and the interparticle cohesiveness prevent particles from being densely packed, leading to a high average voidage when particles are initially loaded to beds; (2) the wall effect induces voidage variation and flow maldistribution, which are the dominant factors responsible for the significantly reduced pressure drops in fixed bed compared with the wall-friction-free condition; and (3) the wall effect and the interparticle cohesiveness produce a notable pressure drop offset ΔP_{wF} in fluidization regime.

Fig. 2.22B shows the bed pressure drop in a 12 mm diameter bed with silica sand particles of 460.6 μm. For this typical Group B particle, the particle and bulk densities are 2600 and 1280 kg/m^3, respectively. Correspondingly, the voidage ε_0 is calculated to be 0.508. Since the interparticle cohesion is usually negligible, a relatively dense packing can be obtained when the H_s/D_t is large. In this case, the actual measured mass of particles mg/A is close to $\rho_p g(1 - \varepsilon_0)$, and there is no significant pressure drop offset in the fluidization operation. However, if the voidage $\varepsilon_0 = 0.508$ is used in the Ergun equation, the calculated pressure drops in fixed beds are lower than those experimentally measured. To make the Ergun equation fit the experimental pressure drops in fixed bed conditions, the average voidage must be adjusted to 0.419. This voidage is lower than the bulk value (i.e., 0.508) due to the strong axial compressive yield strength of large particles. Therefore, for the Group B particles, the wall friction only dominates in the fixed bed regime and is negligible in the fluidization regime.

2.4 More influencing factors

2.4.1 Influence of particle diameter

Fig. 2.23 displays variations of bed pressure drop with gas velocity for FCC particles of 30 and 83 μm in a 4.3-mm-diameter MFB [9]. Both particles belong to Geldart Group A classification, but the 30 μm particles would be more cohesive, since they are closer to the Group C classification. Fig. 2.23A indicates that when the superficial gas velocity increases further, the bed pressure drop appears to be higher for smaller particles. Comparatively, at first glance, the effect of particle size on the bed pressure drop is not apparent at low gas velocities. To see it clearly, we reproduce the fixed bed pressure drop data at low gas velocity segment and use $\Delta P_B(\phi d_p)^2/(150\,\mu_g H_s)$ to represent the pressure drop, as shown in Fig. 2.23B. Apparently, if the Ergun equation applies, the relationship between $\Delta P_B(\phi d_p)^2/(150\,\mu_g H_s)$ and U_g should be a straight line with the slope corresponding to $(1-\varepsilon)^2/\varepsilon^3$, based on which the average bed voidage in the fixed bed can be estimated. As shown in Fig. 2.23B, $\Delta P_B(\phi d_p)^2/(150\,\mu_g H_s)$ varies with U_g almost linearly under the experimental conditions. The calculated average voidages are shown in the figure accordingly. For the particles of 30 μm, the obtained voidages are between 0.84 and 0.88, which are significantly higher than the bulk voidage of 0.538. For the particles of 83 μm, the average voidages are from 0.750 to 0.785, also considerably higher than the bulk voidage of 0.547. As discussed above, it is irrational for a fixed bed to have such a high average voidage. The obtained bed voidage can only be considered to be an indicative parameter reflecting the total wall friction force and its induced variations in

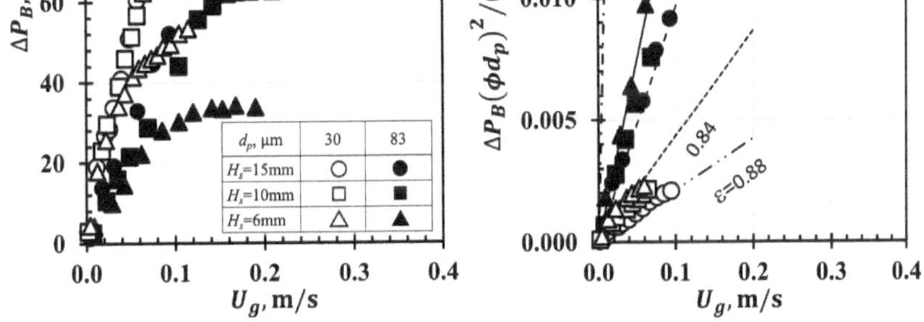

Figure 2.23 The bed pressure drops across two different FCC particles in a 4.3-mm-diameter MFB [9].

bed flow nonuniformity. In other words, the higher this indicative voidage, the more significant the total wall effect. The data shown in Fig. 2.23B, therefore, reveal that the total wall effect of the particles of 30 μm is more prominent than those of 83 μm. Gas flowing through the particles of 30 μm is less uniform radially owing to strong interparticle cohesiveness. Consequently, the pressure drop of the fine particle bed deviates more from the wall-friction-free Ergun equation at a given bed size.

Fig. 2.24 shows the experimental results of the FCC particles of 30 and 83 μm in a bed of diameter 10.5 mm, corresponding to D_t/d_p ratios of 186.7 and 350.0, respectively. It shows that under such large D_t/d_p ratios, the effect of particle diameter on the bed pressure drop is small in the fluidization regime as the mass of particles dominates the bed pressure drop. For the 83 μm particles, the pressure drop variation with gas velocity in fixed beds is very close to the prediction of the Ergun equation with the average voidage of 0.55–0.64, although lateral gas maldistribution likely occurs in the bed with height of 11 mm because it is too shallow. For the fine particles of 30 μm, the estimated average voidage is 0.75–0.80, suggesting that the existence of lateral inhomogeneity of bed voidage and velocity distributions contributes significantly to the reduction in the bed pressure drop.

For Group B particles, the influence of particle diameter on the pressure drop is shown in Fig. 2.25A. It shows that the pressure drop at a given gas velocity increases with particle size in fixed beds. As shown in Fig. 2.25B, the $\Delta P_B(\phi d_p)^2/(150\mu_g H_s) \sim U_g$ relationships for the three particles follow almost a single straight line corresponding to an average voidage of 0.576. This voidage is approximately the same as the bulk voidage, suggesting that the wall effect of Group B particles is insignificant due to varying particle diameter.

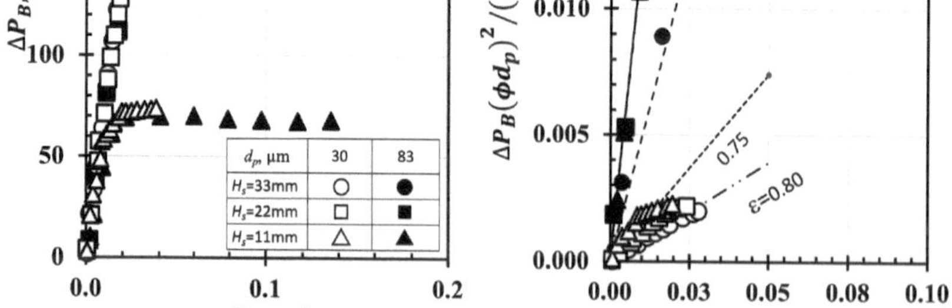

Figure 2.24 The bed pressure drops across two different diameters FCC particles in a 10.5-mm-diameter MFB [9].

2.4.2 Influence of gas properties

Fig. 2.26 shows the variation of bed pressure drop with the type of fluidizing gas. It seems that the influence of gas property on the bed pressure drop is not readily distinguishable. Generally, the bed pressure drop is positively proportional to the gas

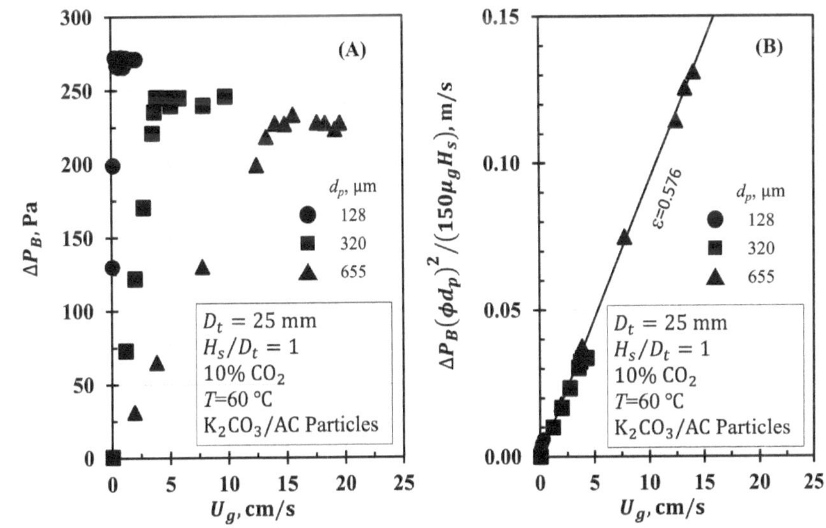

Figure 2.25 The bed pressure drops across three different diameter particles in a MFB [38].

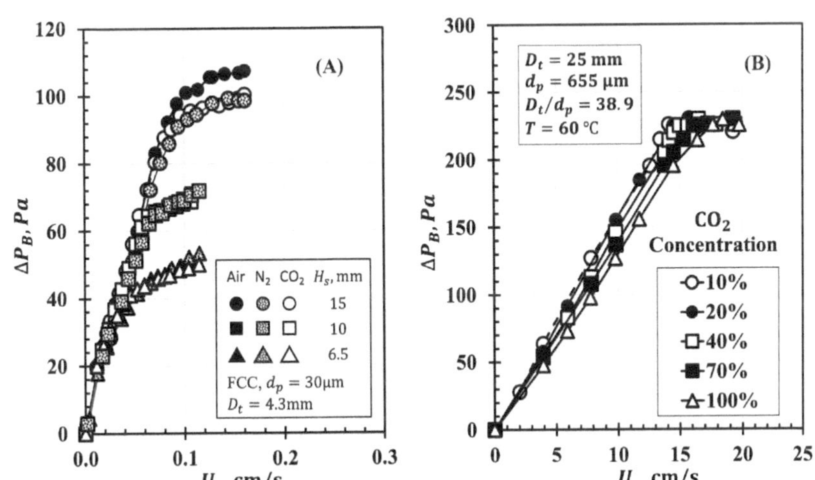

Figure 2.26 The bed pressure drops with (A) different gases [9] and (B) a gas mixture with varying CO_2 concentration [38].

viscosity in the laminar flow regime in fixed beds. As shown in Fig. 2.26A for the static bed height of 15 mm ($H_s/D_t = 3.49$), the average bed pressure drop is reduced by 20% and 10% when CO_2 and N_2 are used to replace the air as the fluidizing gas, respectively. The change in bed pressure drop is in proportion to the variation in the gas viscosity. However, this may not be true if the static bed height is too shallow, for example, 10 mm ($H_s/D_t = 2.33$) and 6.5 mm ($H_s/D_t = 1.51$) because of uneven distribution of gas flow in the shallow beds.

Fig. 2.26B shows the bed pressure drops in a 25-mm-diameter MFB using air with different CO_2 concentrations as the fluidizing gases. It shows that the bed pressure drop decreases with increasing CO_2 concentration at a given gas velocity in fixed beds. The CO_2 concentration in the gas mixture is varied from 10% to 100% on a dry basis. This result is consistent with the variation trend in gas viscosity, since the viscosity of the gas mixture decreases with the increase in the CO_2 concentration at a given temperature. In fluidization conditions, the bed pressure drops are much the same regardless of the CO_2 concentration in the gas mixture.

2.4.3 Influence of temperature

Fig. 2.27 shows the variation of bed pressure drop with the gas velocity at different bed temperatures. It indicates that at the same gas velocity, the pressure drop in fixed beds increases with increasing bed temperature due primarily to the increase in the viscosity of the fluidizing gas with temperature. Increasing bed temperature also decreases the gas density, but the gas density has no or negligible effect on the bed pressure drop because the gas flow in a small-sized bed is mainly in a laminar flow regime, in

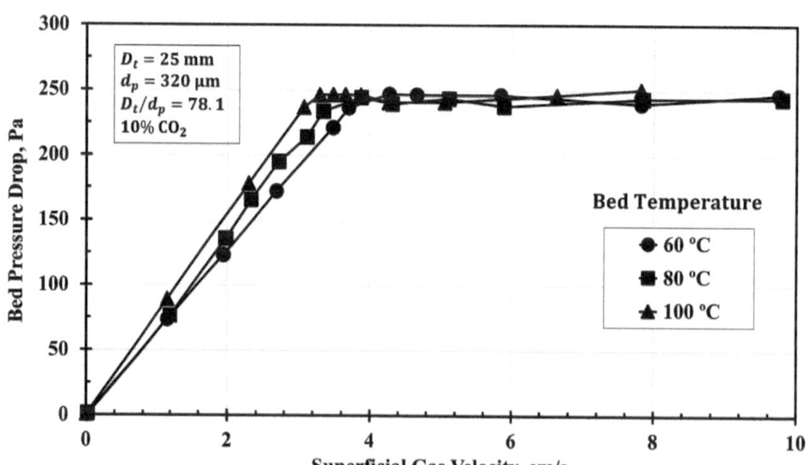

Figure 2.27 The bed pressure drops in a MFB operating at different temperatures. *Data from Prajapati [38].*

which the pressure drop is independent of the gas density according to the Ergun equation. At the fluidization state, the pressure drops remain constant at all bed temperatures, suggesting that the bed pressure drop depends only on the weight of the particles.

In summary, MFBs are different from macro-fluidized beds in many hydrodynamic aspects. In terms of the $\Delta P_B \sim U_g$ relationship, MFBs are distinguishably featured by pressure drop overshoot, pressure drop offset, and deviation from estimates of the Ergun equation. These characteristics are the collective results of wall effects. Three mechanisms of the wall effects have been identified: the wall friction force exerted directly on particles near the wall area, the bed voidage increase, and the radial voidage and gas velocity maldistributions. In addition to the wall effects caused by the particle movement confinement in MFBs, the interparticle cohesion forces also contribute to the unique flow behavior when fine particles are fluidized in small beds. The wall effects vary depending on the operating conditions (e.g., bed diameter, bed height, gas velocity, gas viscosity) and particle properties (e.g., density, diameter, shape).

Abbreviation

FCC fluid catalytic cracking
MFB micro-fluidized bed

Nomenclature

A	cross-sectional area of bed (m^2)
C_d	drag coefficient
d_p	mean particle diameter (m)
D_t	bed diameter (m)
F_d	gas drag force (N)
F_f	friction force (N)
F_g	gravitational force (N)
F_w	wall frictional force (N)
g	acceleration of gravity (m/s^2)
H_B	bed height (m)
H_S	static bed height (m)
j	the Janssen coefficient
m	weight of particles (kg)
n	Richardson-Zaki exponent
P	pressure (Pa)
R	bed radius (m)
Re	Reynolds number ($Re = U_g d_p \rho_g / \mu_g$)
U_b	gas velocity at the bed center bulk zone (m/s)
U_{cr}	gas velocity corresponding to the maximum bed pressure drop overshoot $\Delta P_{B,max}$ (m/s)
U_g	superficial gas velocity (m/s)
U_{mf}	minimum fluidization velocity (m/s)

U_{mfm}	minimum fluidization velocity under the wall effect-free and homogenous flow conditions (m/s)
U_t	single-particle terminal velocity (m/s)
u_w	gas velocity at the wall zone (m/s)
u_c	gas velocity in the center zone (m/s)
V_B	volume of bed particles (m^3)
z	axial coordinate (m)
ΔP_B	bed pressure drop (Pa)
$\Delta P_{B,max}$	maximum value of ΔP_B
ΔP_E	bed pressure drop estimated by Ergun equation (Pa)
ΔP_{WF}	pressure drop offset (Pa), $\Delta P_{WF} = \Delta P_B - mg/A$ (Pa)
ΔP_W	difference between ΔP_B and ΔP_E (Pa)
$\Delta P_{W,max}$	maximum value of ΔP_W (Pa)
ΔP_{W0}	ΔP_W corresponding to ε_0 (Pa)
ΔP_m	the pressure drop caused by maldistribution of voidage only (Pa)
ΔP_c	the pressure drop caused by both wall friction and maldistribution of voidage (Pa)

Greek letters

$\delta_{\triangle P}$	normalized pressure drop
σ_v	the vertical component of stress
σ_h	the horizontal component of stress
φ	the angle of internal friction (°)
ε	average bed voidage
ε_0	bed voidage at static condition
ε_b	voltage at the bed center bulk zone
ε_c	a constant voidage
ε_s	the averaged solid volume fraction
$\varepsilon_{s,min}$	the minimum value of ε_s
$\varepsilon_{s,max}$	the maximum value of ε_s
$\varepsilon_{s,mf}$	solid volume fraction at minimum fluidization
ρ_p	particle density (kg/m^3)
ρ_b	bulk density (kg/m^3)
μ_{eff}	coefficient of wall friction ($Pa \cdot s$)
μ_g	gas viscosity ($Pa \cdot s$)

References

[1] Kunii D, Levenspiel O. Fluidization engineering. 2nd ed. Boston: Butterworth—Heinemann; 1991.
[2] Grace JR. Hydrodynamics of gas fluidized beds. Basu P. Fluidized Bed Boilers: Design and Application. Pergamon Press Canada Ltd 1984;13—30.
[3] Grace JR, Bi X, Ellis N. Essentials of fluidization technology, 12. Weinheim, Germany: Essentials of Fluidization Technology, Wiley-VCH Verlag GmbH & Co. KGaA, Boschstr; 2020. p. 69469.
[4] Liu X, Xu G, Gao S. Micro fluidized beds: wall effect and operability. Chemical Engineering Journal 2008;137(2):302—7.
[5] Mcdonough JR, Law R, Reay DA, et al. Fluidization in small-scale gas-solid 3D-printed fluidized beds. Chemical Engineering Science 2019;200:294—309.
[6] Wang F, Fan LS. Gas-solid fluidization in mini- and micro-channels. Industrial and Engineering Chemistry Research 2011;50(8):4741—51.
[7] Loezos PN, Costamagna P, Sundaresan S. The role of contact stresses and wall friction on fluidization. Chemical Engineering Science 2002;57:5123—41.

[8] Soleimani I, Elahipanah N, Shabanian J, et al. In-situ quantification of the magnitude of interparticle forces and its temperature variation in a gas-solid fluidized bed. Chemical Engineering Science 2021;232:116349.
[9] Guo Q, Xu Y, Yue X. Fluidization characteristics in micro-fluidized beds of various inner diameters. Chemical Engineering and Technology 2009;32(12):1992−9.
[10] Quan H, Fatah N, Hu C. Diagnosis of hydrodynamic regimes from large to micro-fluidized beds. Chemical Engineering Journal 2020;391.
[11] Wen CY, Yu YH. A generalized method for predicting the minimum fluidization velocity. AIChE Journal 1966;12(3):610−12.
[12] Srivastava A, Sundaresan S. Role of wall friction in fluidization and standpipe flow. Powder Technology 2002;124(1−2):45−54.
[13] Oke O, Lettieri P, Mazzei L. An investigation on the mechanics of homogeneous expansion in gas-fluidized beds. Chemical Engineering Science 2015;127:95−105.
[14] Tsinontides SC, Jackson R. The mechanics of gas fluidized beds with an interval of stable fluidization. Journal of Fluid Mechanics 1993;255:237−74.
[15] Liu X, Su J, Qian Y, et al. Comparison of two-fluid and discrete particle modeling of gas-particle flows in micro fluidized beds. Powder Technology 2018;338:79−86.
[16] Cheng NS. Wall effect on pressure drop in packed beds. Powder Technology 2011;210:261−6.
[17] Di Felice R, Gibilaro LG. Wall effects for the pressure drop in fixed beds. Chemical Engineering Science 2004;59:3037−40.
[18] Vanni F, Caussat B, Ablitzer C, et al. Effects of reducing the reactor diameter on the fluidization of a very dense powder. Powder Technology 2015;277:268−74.
[19] Ansart R, Vanni F, Caussat B, et al. Effects of reducing the reactor diameter on the dense gas-solid fluidization of very heavy particles: 3D numerical simulations. Chemical Engineering Research and Design 2017;117:575−83.
[20] Tian FY, Huang LF, Fan LW, et al. Wall effects on the pressure drop in packed beds of irregularly shaped sintered ore particles. Powder Technology 2016;301:1284−93.
[21] Cairns EJ, Prausnitz JM. Velocity profiles in packed and fludized beds. Industrial and Engineering Chemistry 1959;51(12):1441−4.
[22] Allen KG, von Backström TW, Kröger DG. Packed bed pressure drop dependence on particle shape, size distribution, packing arrangement and roughness. Powder Technology 2013;246:590−600.
[23] Winterberg M, Tsotsas E. Impact of tube-to-particle-diameter ratio on pressure drop in packed beds. AIChE Journal 2000;46(5):1084−8.
[24] Mehta D, Hawley MC. Wall effect in packed columns. Industrial and Engineering Chemistry Process Design and Development 1969;8(2):280−2.
[25] Eisfeld B, Schnitzlein K. The infuence of confining walls on the pressure drop in packed beds. Chemical Engineering Science 2001;56(14):4321−9.
[26] Stephenson JL, Stewart WE. Optical measurements of porosity and fluid motion in packed beds. Chemical Engineering Science 1986;41(8):2161−70.
[27] Bey O, Eigenberger G. Fluid flow through catalyst filled tubes. Chemical Engineering Science 1997;52(8):1365−76.
[28] Chu CF, Ng KM. Flow in packed tubes with a small tube to particle diameter ratio. AIChE Journal 1989;35(1):148−58.
[29] Nield DA. Alternative model for wall effect in laminar flow of a fluid through a packed column. AIChE Journal 1983;29(4):688−9.
[30] Navvab Kashani M, Elekaei H, Zivkovic V, et al. Explicit numerical simulation-based study of the hydrodynamics of micro-packed beds. Chemical Engineering Science 2016;145:71−9.
[31] Richardson JF, Zaki WN. Sedimentation and fluidization: part I. Transactions of the Institution of Chemical Engineers 1954;32:35−53.
[32] Valverde JM, Ramos A, Castellanos A, et al. The tensile strength of cohesive powders and its relationship to consolidation, free volume and cohesivity. Powder Technology 1998;97:237−45.
[33] Johnson PC, Nott PR, Jackson R. Frictional−collisional equations of motion for particulate flows and their application to chutes. Journal of Fluid Mechanics 1990;210:501−36.

[34] Xu Y, Li T, Musser J, et al. CFD-DEM modeling the effect of column size and bed height on minimum fluidization velocity in micro fluidized beds with Geldart B particles. Powder Technology 2017;318:321–8.
[35] Zou RP, Yu AB. The packing of spheres in a cylindrical container: the thickness effect. Chemical Engineering Science 1995;50:1504–7.
[36] Xiang J, Zhang Y, Li Q. Effect of bed size on the gas-solid flow characterized by pressure fluctuations in bubbling fluidized beds. Particuology 2019;47:1–9.
[37] Han Z, Yue J, Geng S, et al. State-of-the-art hydrodynamics of gas-solid micro fluidized beds. Chemical Engineering Science 2021;232:116345.
[38] Prajapati A, Renganathan T, Krishnaiah K. Kinetic studies of CO_2 capture using K_2CO_3/activated carbon in fluidized bed reactor. Energy and Fuels 2016;30(12):10758–69.

CHAPTER 3

Gas and solid mixing

Fluidized beds offer distinct operational advantages over other gas–solid contact patterns, such as fixed and moving beds, because the distributions of temperature and flow in beds are relatively uniform due to vigorous mixing between gas and solids as well as between solids. The uniform temperature and flow distribution provide the necessary conditions for high-quality products with consistent properties. Good mixing also improves the contact efficiency between gas and solid particles. These characteristics are some of the reasons that fluidized beds are chosen in many practical solids handling applications. Gas and solids mixing characteristics affect reaction performances, such as conversion and selectivity [1–4]. Therefore, a good knowledge of gas and solid mixing is essential to understand, evaluate, predict, scale up, and optimize gas–solid fluidized bed processes. This chapter focuses on gas and solid mixing in micro-fluidized beds.

3.1 Experimental and analytic techniques

In principle, gas flow in a fluidized bed is somewhere between ideal plug flow and mixed flow, and gas mixing occurs in both axial and radial directions. In this chapter, we focus only on axial mixing characteristics in micro-fluidized beds, since it is reasonable to assume that radial mixing will not be as critical in micro-fluidized beds due to their small diameters as in macro-fluidized beds of large diameters. When the radial dispersion is ignored, the axial dispersion coefficient and the backmixing coefficient become equal [5]. Gas mixing in fluidized beds can be characterized experimentally by measurements of gas residence time distribution (RTD) and theoretically by the axial dispersion model analysis.

3.1.1 Gas residence time distribution

Residence time distribution is a powerful yet convenient diagnostic tool to evaluate the deviation from ideal flow in reactors. Gas backmixing in fluidized beds can be caused by gas flows in opposite directions, circulations of gas and solids, nonuniform flow, channeling, and other reasons. A direct cause of gas backmixing is that gas flow exhibits a certain residence time distribution. Therefore, gas residence time distribution is an important characteristic that provides essential insights into gas mixing behavior from the viewpoint of chemical reaction engineering.

Micro Fluidization
DOI: https://doi.org/10.1016/B978-0-443-18718-6.00003-6
© 2023 Chemical Industry Press Co., Ltd.
Published by Elsevier Inc., under an exclusive license with
Chemical Industry Press Co., Ltd.

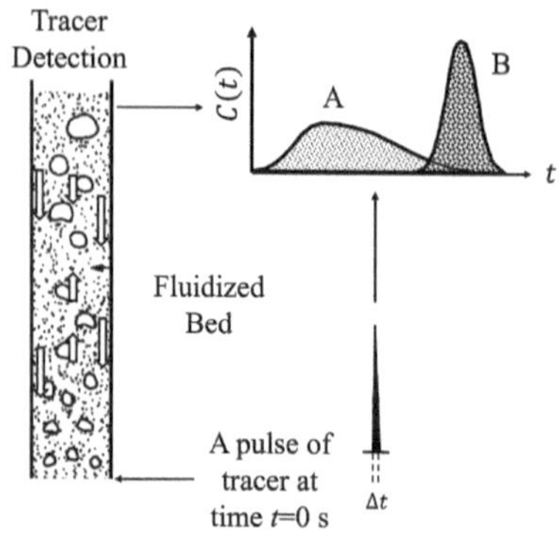

Figure 3.1 A typical pulse tracer experiment to determine the axial gas dispersion coefficient in a fluidized bed. (A) A typical broad and nonsymmetrical distribution, (B) a typical narrow and symmetrical distribution.

Consider a typical stimulus–response (or pulse injection) test to determine the gas mixing behavior in a gas–solid fluidized bed (other alternative test methods can be found in the reference [6]). As shown in Fig. 3.1, a close-to-ideal pulse of inert and nonabsorbable tracer gas is injected into the bed bottom. The evolution of the tracer at bed exit is continuously recorded as a function of time by thermal conductivity detectors, gas chromatography, or mass spectrometry. The recorded trace gas concentration $C(t)$ versus time t is used to calculate the following:

The gas residence time distribution function

$$E(t) = \frac{C(t)}{\int_0^\infty C(t)dt} = \frac{C(t_i)}{\sum C(t_i)\Delta t_i} \tag{3.1}$$

The mean residence time

$$\bar{t} = \frac{\int_0^\infty tE(t)dt}{\int_0^\infty E(t)dt} = \frac{\sum t_i E(t_i)\Delta t_i}{\sum E(t_i)\Delta t_i} \tag{3.2}$$

The variance

$$\sigma_t^2 = \frac{\int_0^\infty t^2 E(t)dt}{\int_0^\infty E(t)dt} - \bar{t}^2 \tag{3.3}$$

The variance σ_t^2 represents the spread of the gas residence time distribution. It helps evaluate the extent of gas mixing in fluidized beds from experimentally measured $E(t)$ curves. If the pulse response is broad and nonsymmetrical, as shown in case A, the value of σ_t^2 is large, indicating that the gas flows through the bed with a large extent of mixing. In contrast, if the $E(t)$ curve is narrow, symmetrical, and close to the normal distribution, as shown in case B, the value of σ_t^2 is small, indicating that the gas flow does not deviate very much from plug flow.

For the sake of simplicity and generality, the parameters describing a residence time distribution can also be presented in dimensionless forms as follows:

$$\begin{aligned} E(\theta) &= E(t)\bar{t} \\ \sigma_\theta^2 &= \frac{\sigma_t^2}{\bar{t}^2} \\ \bar{\theta} &= 1 \end{aligned} \tag{3.4}$$

Because the bed diameter and height are too small for micro-fluidized beds, it is difficult to inject the tracer gas into the bed during RTD measurements, so the tracer gas is usually injected into the gas input line. Fig. 3.2A shows a schematic of an experimental system for measuring gas residence time distributions in micro-fluidized beds. In this exemplary system, air and helium are employed as fluidizing gas and tracer gas, respectively. The flow rates of air and helium are controlled by mass flowmeters. The tracer gas is injected into the pipeline leading to the fluidized bed by a pulse through a six-way valve. The introduced volume of helium is significantly smaller than that of the fluidizing gas such that the disturbance to the gas flow caused by the helium injection is negligible. After injection, the tracer gas flows through the inlet pipeline, enters the bed and flows through it, and finally flows through the outlet pipeline to a gas analyzer.

A quick-response mass spectrometer (MS) is used to monitor the tracer gas evolution, as shown in Fig. 3.2B. The measured gas residence time distribution in this way

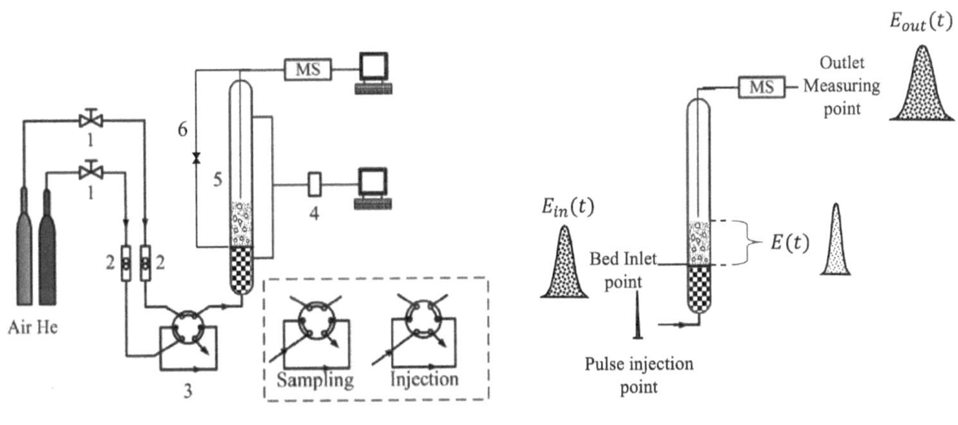

Figure 3.2 (A) Schematic of an experimental system for measuring gas time distribution in micro-fluidized beds. (1) Pressure regulator; (2) mass flowmeter; (3) six-way valve; (4) pressure transducer; (5) micro-fluidized bed; (6) bypass line for inlet RTD measurements. (B) An illustration of the relationship between $E_{in}(t)$, $E(t)$ and $E_{out}(t)$.

is $E_{out}(t)$, that is, the gas residence time distribution from the tracer gas injection point to the measuring point. Clearly, $E_{out}(t)$ is different from $E(t)$, the actual gas residence time distribution corresponding to the tracer gas flow through the bed of particles. It is necessary to measure the residence time distribution corresponding to the tracer gas flow from the injection point to the bed bottom, $E_{in}(t)$, so that $E(t)$ can be calculated by the following relationship:

$$E_{out}(t) = \int_0^t E(t')E_{in}(t - t')dt' \tag{3.5}$$

3.1.2 Axial dispersion model

Since the diameter of micro-fluidized beds is small, the radial gas diffusion is negligible, and the gas flow through the bed deviates little from plug flow. Thus, the one-dimensional diffusion model can be used to describe the gas flow through micro-fluidized beds

$$\frac{\partial C(t)}{\partial \theta} = \frac{D_{a,g}}{uH}\frac{\partial^2 C(t)}{\partial^2 z} - \frac{\partial C(t)}{\partial z} \tag{3.6}$$

where $D_{a,g}$ is the axial gas dispersion coefficient, $z = (ut + y)/H$, $\theta = t/\overline{\tau}$, $u = U_g/\varepsilon$. The dimensionless group $D_{a,g}/uH$, called the bed dispersion number, is the parameter that measures the extent of axial gas dispersion. The reverse of the bed dispersion number is called the Peclet number, that is, $Pe = uH/D_{a,g}$. If $D_{a,g}/uH \to 0$, a gas flow can be considered a plug flow with negligible gas dispersion. If $D_{a,g}/uH \to \infty$, a substantial gas dispersion occurs. The gas dispersion is relatively small in micro-fluidized beds with an ideal tracer pulse injection. When $D_{a,g}/uH < 0.01$, solving Eq. (3.6) gives

$$E(\theta) = \frac{1}{\sqrt{4\pi \frac{D_{a,g}}{uH}}} \exp\left[-\frac{(1-\theta)^2}{4\frac{D_{a,g}}{uH}}\right] \tag{3.7}$$

This equation represents a typical Gaussian curve or Normal distribution. Alternatively, in the actual parameter terms, we have

$$\begin{aligned} E(t) &= \frac{E(\theta)}{\overline{\tau}} = \sqrt{\frac{u^3}{4\pi D_{a,g}H}} \exp\left[-\frac{(H-ut)^2}{4\frac{D_{a,g}H}{u}}\right] \\ \overline{\tau} &= \frac{H}{u}, \overline{\theta} = 1 \\ \sigma_\theta^2 &= \frac{\sigma_t^2}{\overline{\tau}^2} = 2\frac{D_{a,g}}{uH}, \sigma_t^2 = 2\frac{D_{a,g}H}{u^3} \end{aligned} \tag{3.8}$$

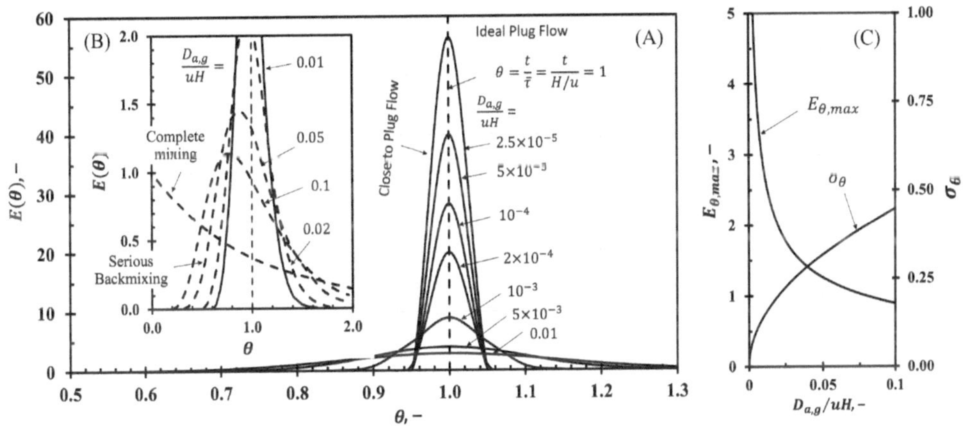

Figure 3.3 (A,B) Relationship between $E(\theta)$ and θ for different values of $D_{a,g}/uH$. (C) Variations of $E_{\theta,max}$ and $\sigma\theta$ with $D_{a,g}/uH$.

Here, we want to clarify that the average residence time $\overline{\tau}$ is the time needed for all gas flowing through the height H of bed particles at a constant interstitial gas velocity u. It is valid for a system with a constant fluid density and uniformly distributed velocity. For systems with varying fluid density or nonuniform velocity, the actual average residence time is different from $\overline{\tau} = H/u$. We will get back to this point later.

Fig. 3.3A shows the curves of Eq. (3.7) for different values of $D_{a,g}/uH$. It shows that with decreasing $D_{a,g}/uH$, the height of the distribution ($E_{\theta,\max}$) increases, but the width of the distribution (σ_θ) decreases (Fig. 3.3C). When $D_{a,g}/uH$ is lower than 0.001 and continues to decrease, the peak height of the distribution continues to increase, while the width of the distribution reduces slowly. As $D_{a,g}/uH$ is greater than 0.01, the extent of gas dispersion becomes so large that the $E(\theta) \sim \theta$ curves start to deviate from the normal distribution noticeably (Fig. 3.3B). Further analyzing Eq. (3.7) gives

$$E_{\theta,max} = \frac{1}{\sqrt{4\pi \frac{D_{a,g}}{uH}}} \qquad \sigma_\theta = \sqrt{2\frac{D_{a,g}}{uH}} \tag{3.9}$$

Fig. 3.4 shows the above relationships. Based on this chart, one can calculate the axial dispersion coefficient by using either $E_{\theta,\max}$ or σ_θ determined from experimentally measured RTD curves when $D_{a,g}/uH < 0.01$.

Further, multiplying $E_{\theta,\max}$ by σ_θ in Eq (3.9) yields

$$E_{\theta,\max}\sigma_\theta = E(t)_{\max}\sigma_t = \frac{1}{\sqrt{2\pi}} \sim 0.4 \tag{3.10}$$

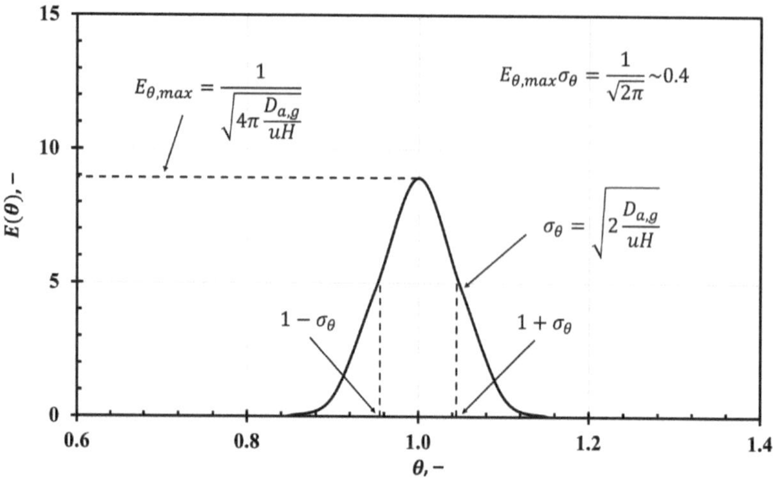

Figure 3.4 Relationship between $E_{\theta,\max}$ and σ_θ for the small extent of dispersion.

The above equation is useful practically when evaluating experimental results. It can serve as a diagnostic tool to determine whether or not it is appropriate to describe the experimental RTD curves by the one-dimensional dispersion model. It can be ensured that using a one-dimensional dispersion model is appropriate when gas back-mixing is small, namely, that the experimental data of $E_{\theta,\max}$ and σ_θ satisfy Eq. (3.10). If Eq. (3.10) is not satisfied, $E(\theta)$ curves should appear asymmetrical; thus, the axial dispersion cannot be estimated from the above equations. In this case, $D_{a,g}/uH > 0.01$, and the solutions to Eq. (3.6) depend on the boundary conditions. Under the closed boundary conditions, no analytical expression for the E curve is available, but the variance can be calculated numerically by

$$\frac{\sigma_t^2}{\tau^2} = 2\frac{D_{a,g}}{uH} - 2\left(\frac{D_{a,g}}{uH}\right)^2\left(1 - e^{-uH/D_{a,g}}\right) \tag{3.11}$$

Under the open boundary conditions, the analytical expressions for the E curve are

$$\begin{aligned} E(\theta) &= \frac{1}{\sqrt{4\pi\frac{D_{a,g}}{uH}}}\exp\left[-\frac{(1-\theta)^2}{4\theta\frac{D_{a,g}}{uH}}\right] \\ E(t) &= \frac{u}{\sqrt{4\pi D_{a,g}t}}\exp\left[-\frac{(H-ut)^2}{4D_{a,g}t}\right] \\ \sigma_\theta^2 &= \frac{\sigma_t^2}{\tau^2} = 2\frac{D_{a,g}}{uH} + 8\left(\frac{D_{a,g}}{uH}\right)^2 \end{aligned} \tag{3.12}$$

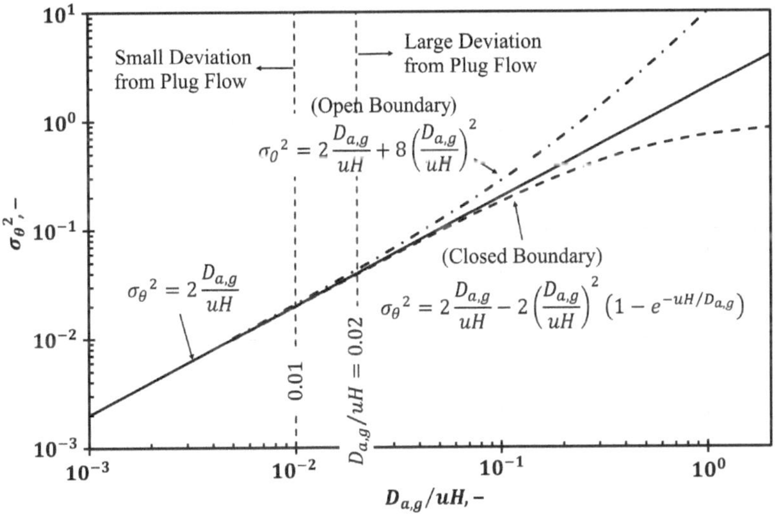

Figure 3.5 The relationships between Eqs. (3.9), (3.11), and (3.12).

Fig. 3.5 displays the variances calculated by Eqs. (3.9), (3.11), and (3.12) as a function of $D_{a,g}/uH$. It shows that for small values of $D_{a,g}/uH$ (i.e., <0.01), the curves under different boundary conditions approach the "small deviation" curve of Eq. (3.9). At large values of $D_{a,g}/uH$ (>0.01), the curves differ more and more as $D_{a,g}/uH$ increases. In micro-fluidized beds, the inlet and outlet pipelines are typically designed in a way that the gas flows through them essentially in plug flow. Therefore, micro-fluidized beds can be considered "closed" systems from the viewpoint of RTD analysis. So, if $E(t)$ curves appear asymmetric, Eq. (3.11) should be used to estimate $D_{a,g}/uH$.

3.2 Gas mixing

3.2.1 Gas residence time distribution

According to the method described above, we can obtain residence time distributions of gas in micro-fluidized beds under different experimental conditions. Fig. 3.6A shows the gas residence time distributions in a 15 mm diameter bed using five particles. It shows that when large particles are fluidized, the gas RTD curves are almost symmetrical with narrow widths and high peaks. In contrast, when small particles are used, they become asymmetrical with broader widths and lower peak heights. As described in the previous section, these results suggest that the extent of gas backmixing in the bed reduces when changing fluidized particles from Group A to Group B. This phenomenon, we believe, is due to the characteristics of gas–solid flow in micro-fluidized beds. As with that in large-scale fluidized beds, the gas backmixing in micro-fluidized beds is attributed to the downflow of particles. A downflow of gas

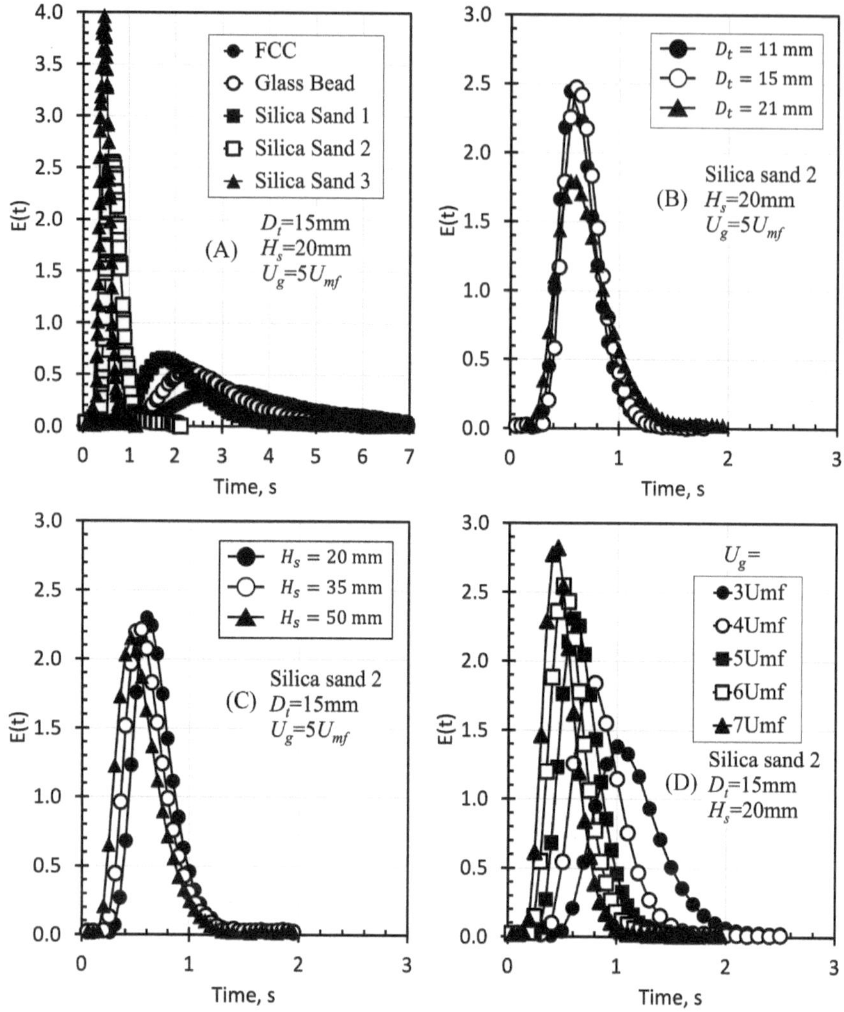

Figure 3.6 Gas residence time distributions in a micro-fluidized bed under various conditions. FCC: $d_p = 92\mu m$, $\rho_p = 1659 kg/m^3$; glass bead: $d_p = 85\mu m$, $\rho_p = 2450 kg/m^3$; silica sand 1: $d_p = 90\mu m$, $\rho_p = 2846 kg/m^3$; silica sand 2: $d_p = 155\mu m$, $\rho_p = 2846 kg/m^3$; silica sand 3: $d_p = 185\mu m$, $\rho_p = 2846 kg/m^3$ [7].

occurs when the velocity of descending solids exceeds the interstitial velocity of the gas in a dense phase. As discussed in Chapter 2, the wall effects lead to a lower bed voidage for Group B particles than for Group A particles. As a result, a fluidized bed with Group B particles would have a higher interstitial gas velocity (i.e., U_g/ε) in the dense phase than a bed with Group A particles at the same superficial gas velocity. Additionally, large Group B particles form a larger semi-cavity near the wall (i.e., a high voidage region), which results in a higher local upward gas velocity. This increased local upward gas velocity effectively suppresses the downflow of solids near the wall. In contrast, beds of Group A particles are featured with higher bed voidage and lower interstitial gas velocity. In addition, more bubbles form, coalesce, and break up in beds of fine particles. When bubbles erupt and burst at the bed surface, more particles entrained in the wake of bubbles fall back to the bed, creating a more intense circulation of solids. All these factors result in a greater extent of gas backmixing in beds of small particles. Therefore, if a close-to-plug flow is preferred, large particles should be used in micro-fluidized beds.

Fig. 3.6B shows the effect of bed diameter on residence time distributions for a Group B particle (the silica sand #2) under the conditions of $U_g = 5U_{mf}$ and $H_s = 20$ mm. It shows that the $E(t)$ curves obtained by the three beds (11, 15, and 21 mm in diameter) are similar, but the peak height $E(t)_{\max}$ is reduced when the bed diameter increases to 21 mm, suggesting a greater extent of gas backmixing in this diameter. The result indicates that the extent of gas backmixing would increase if the bed diameter is further increased. For the beds of 11 mm and 15 mm, no apparent differences in the $E(t)$ curves can be identified. As discussed before, the wall effect becomes insignificant when D_t/d_p approaches 150. For the 21 mm diameter bed with the silica sand particle used, D_t/d_p is 136, which is close to 150. For the beds of 11 and 15 mm in diameter, D_t/d_p is 71 and 97, respectively, so the reduced ratio D_t/d_p leads to an increase in the wall effect, which may reduce the downflow of particles, and thus the gas backmixing.

Fig. 3.6C presents the effect of static bed height on the gas residence time distribution in the micro-fluidized bed of 15 mm in diameter operated under the same gas velocity. It displays that the peak of $E(t)$ decreases with increasing static bed height, indicating that the gas backmixing increases with the static bed height. This result is expected because the downflow of particles is more pronounced in deeper fluidized beds.

The influence of gas velocity on the gas residence time distribution is elucidated in Fig. 3.6D. It indicates that with the gradual increase in gas velocity, the peak height of $E(t)$ curve increases and the spread reduces, approaching the typical distribution curve of plug flow. The result demonstrates that the gas backmixing reduces with increasing gas velocity. It is worthy to emphasize that the dependence of gas backmixing on gas velocity observed here is quite different from that reported in macro-fluidized beds. Usually, the extent of gas backmixing is proportional to gas velocity in large-scale bubbling fluidized beds [5,8,9] and independent or declining with gas velocity in large-scale dense high-velocity (e.g., turbulent) beds [10,11].

3.2.2 Axial gas dispersion coefficient

The determination of the axial gas dispersion coefficient is not straightforward. It needs, first, to choose the dispersion model appropriately and, second, to know the accurate bed voidage ε and height H at the corresponding operating conditions. The bed voidage is required to calculate the interstitial gas velocity ($u = U_g/\varepsilon$). Correctly obtaining actual bed height is also a challenge, so it is not uncommon to see that some researchers use the total column height (i.e., including freeboard or empty sections) in place of H in the dispersion number $D_{a,g}/uH$, or the Peclet number $uH/D_{a,g}$. Due to these uncertainties, it is advised that readers should pay attention to what values of H and u are used in the axial dispersion model when discussing or comparing the axial gas dispersion coefficients reported in the literature. Obviously, the accuracy and reliability of the results can be compromised if incorrect data of H and u are used. Nevertheless, if the tests are performed carefully, the values of $D_{a,g}$ obtained can shed light on gas backmixing in fluidized beds.

Figs. 3.7 and 3.8 show the effect of bed diameter on the axial gas dispersion coefficient $D_{a,g}$ in gas–solid fluidized beds. The data are collected from various literature studies conducted in considerably different operating conditions, such as particles and gas velocities [1,5,7–9,12–14]. For macro-fluidized beds, data cover fluidization regimes, including bubbling, slugging, turbulent, and fast fluidization. We do not mark these data separately corresponding to their specific conditions for simplicity. Nevertheless, the results indicate clear distinctions between micro- and macro-fluidized beds from the perspective of axial gas backmixing. The values of $D_{a,g}$ are in the order of magnitude of $10^{-4} \sim 10^{-2}$ m^2/s (mostly below 10^{-3} m^2/s) in micro-fluidized beds, which are significantly lower than the values of $D_{a,g}$ in macro-fluidized beds (greater than 10^{-2} m^2/s). It means that the value of $D_{a,g}$ in a micro-fluidized bed is lower than that in macro-fluidized beds by tens to hundreds of times. Therefore, we

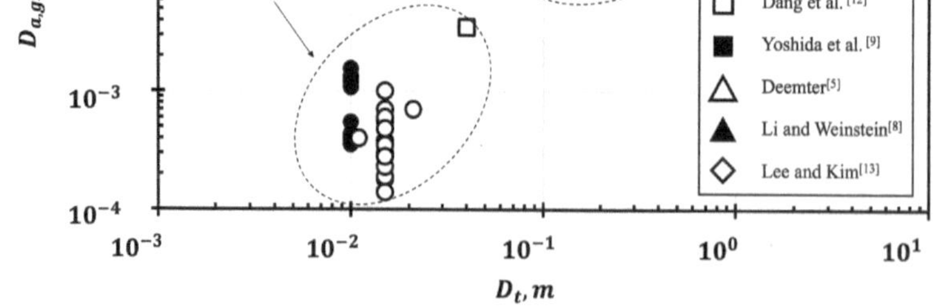

Figure 3.7 Variations of axial gas dispersion coefficient in micro- and macro-fluidized beds.

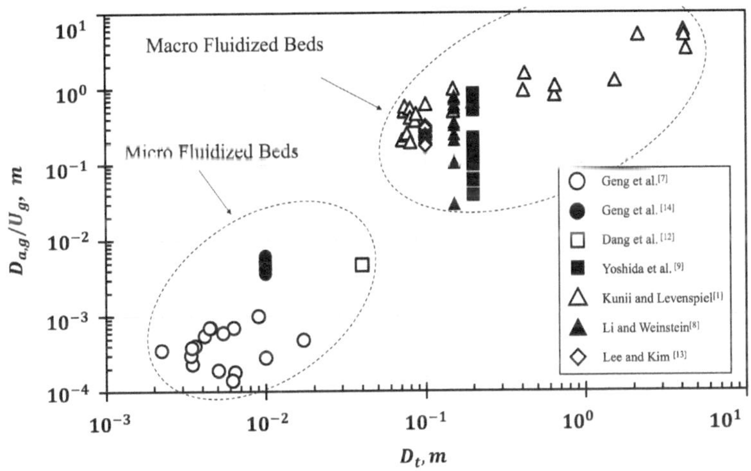

Figure 3.8 Variations of $D_{a,g}/U_g$ in micro- and macro-fluidized beds.

can conclude that the difference in $D_{a,g}$ makes a clear distinction between micro- and macro-fluidized beds. We will elaborate on this further in the next chapter.

3.2.3 Two-phase model analysis

The research results presented above demonstrate explicitly that the gas flow in micro-fluidized beds is characterized by significantly reduced gas backmixing compared to macro-fluidized beds. The axial gas backmixing can therefore be presented by the one-dimensional dispersion model. The one-dimensional dispersion model treats gas flow in a fluidized bed as a single phase with a certain degree of backmixing. In the one-dimensional dispersion model, the gas backmixing is characterized by a dispersion coefficient. The model is helpful for understanding the extent of gas backmixing in fluidized beds, but does not provide insights into how it is generated. Next, we present an analysis to explain the mechanism of gas backmixing in micro-fluidized beds based on a two-phase bubbling bed model.

A typical bubbling fluidized bed consists of two distinct phases—the bubble phase and the emulsion phase. Rising bubbles flow at a velocity U_b passing through the emulsion phase in which gas rises at an interstitial velocity, as conceptually shown in Fig. 3.9. Bubbles travel up by carrying a wake of solids behind them. When bubbles break up at the bed surface, most of the carried solids return to the bed and drift downward in the emulsion. This flow pattern sets up an internal circulation of solids in the bed. Depending on various factors, the downflowing solids can move slower or faster than the interstitial gas velocity. If the downflow velocity of emulsion solids is faster than the velocity of upflowing gas in the emulsion, a net downflow of emulsion gas (i.e., gas backmixing) occurs. Otherwise, when the downflow velocity of emulsion

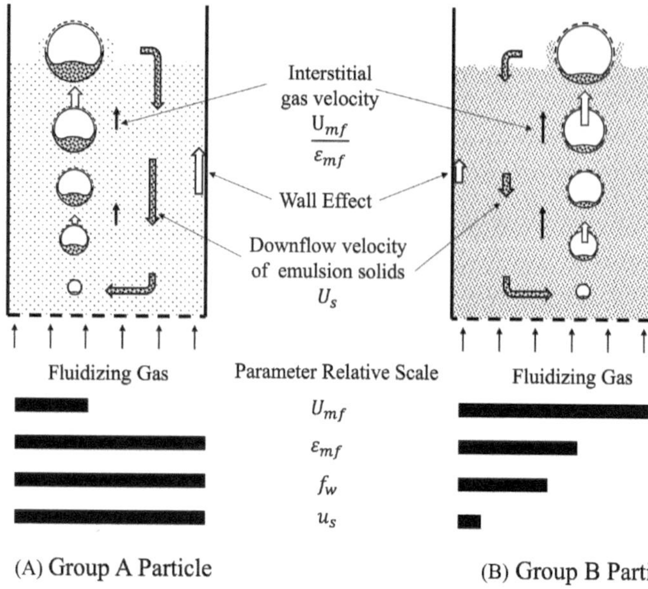

Figure 3.9 Illustrations of gas–solid flows in bubbling fluidized beds for typical Group A and B particles.

solids is slower than the velocity of upflowing gas in the emulsion, there is essentially no gas backmixing, though the gas flow can still have somewhat spread residence time distributions because of radial nonuniform bubble and gas flows. Therefore, the gas backmixing and dispersion are not precisely the same, although they are often used interchangeably.

According to the two-phase theory, the emulsion phase remains at the minimum fluidizing condition; hence, the interstitial velocity of the gas flowing through the emulsion is U_{mf}/ε_{mf}. Experiments in recent years have generated a good amount of data of U_{mf} and ε_{mf}, and accordingly, the empirical correlations of U_{mf} and ε_{mf} are also available [15]. These correlations make it possible to predict the interstitial velocity of the gas flowing through the emulsion as a function of bed diameter or ratio D_t/d_p for the given gas and solid properties. We write the downflow velocity of emulsion solids based on the two-phase model as follows:

$$u_s = \frac{f_w f_b U_b}{1 - f_b - f_w f_b} \tag{3.13}$$

where f_w is the ratio of wake volume to bubble volume and f_b is the fraction of the bed occupied by gas bubbles. The coefficient f_w can be estimated from the literature [2]. Usually, f_w is higher for Group A particles than for Group B particles. To calculate

f_b, the following equation can be derived based on the mass balance by assuming that the solid presence in gas bubbles is negligible: [1]

$$\varepsilon_f = f_b + (1 - f_b)\varepsilon_{mf}, \text{ or } f_b = \frac{\varepsilon_f - \varepsilon_{mf}}{1 - \varepsilon_{mf}} \tag{3.14}$$

For illustration, the bed voidage of a bubbling fluidized bed is calculated by the Richardson-Zaki approach [16]

$$\varepsilon_f = \left(\frac{U_g}{U_t}\right)^{1/n} \tag{3.15}$$

where the exponent n can be expressed as [17]

$$n = \left(4.4 + 18\frac{d_p}{D_t}\right)\mathrm{Re}_t^{-0.03}, \text{ where } \mathrm{Re}_t = \frac{\rho_g d_p U_t}{\mu_g} \tag{3.16}$$

where U_t is the terminal velocity of a particle, which can be estimated based on the correlations presented in the literature [1,11]. Wang and Fan [18] found that the bed expansion of the mini- and micro-channel fluidized beds could be presented by the Richardson-Zaki equation, although the exponent was slightly decreased with the channel size. Wang et al. [19]. confirmed that the bed expansion could be presented by the Richardson-Zaki equation for particles of 75 μm diameter in micro-bubbling and turbulent fluidized beds if $U_g < 0.11$ m/s. We apply this approach in our calculations. Assuming that the bubble velocity in bubbling beds follows the two-phase model, we can present the bubble velocity by

$$U_b = U_g - U_{mf} + U_{br} \tag{3.17}$$

The rising velocity of a single bubble is

$$U_{br} = 0.711(gd_b)^{1/2} \tag{3.18}$$

To estimate the bubble velocity U_b based on Eq. (3.17), we need the bubble size d_b. For this purpose, we take the Mori and Wen model: [20]

$$\frac{d_{bm} - d_b}{d_{bm} - d_{b0}} = e^{-0.3z/D_t} \tag{3.19}$$

where d_{b0} is the initial bubble size formed near the bottom of the bed, and d_{bm} is the maximum size of bubble expected in a very deep bed. This maximum size of the bubble is given by

$$d_{bm} = 0.374\left[\beta\frac{\pi}{4}D_t^2\left(U_g - U_{mf}\right)\right]^{0.4} \tag{3.20}$$

where β is 4 for large-scale fluidized beds. Wang and Fan [18] found that for mini- and micro-fluidized beds, β is a function of channel size. Based on the data reported by Wang and Fan [18], we correlate β as a function of bed diameter by

$$\beta = 0.35 + 3.65e^{-14.53/(1000D_t)} \tag{3.21}$$

The above equation approaches 4 when D_t is infinite (i.e., large-scale fluidized beds). The initial bubble size d_{b0} is dependent on the types and designs of gas distributors, so for simplicity and comparison purposes, we take $d_{b0} = 0$ for all beds. Now, for given operating conditions, we can estimate the bubble size as a function of bed height z and assume the mean bubble size

$$d_b = \frac{d_{b0} + d_{b,z=H_b}}{2} \tag{3.22}$$

Then, we calculate the rise velocity of emulsion gas through the bed

$$u_{ge} = \frac{U_{mf}}{\varepsilon_{mf}} - u_s = \frac{U_{mf}}{\varepsilon_{mf}} - \frac{f_w f_b U_b}{1 - f_b - f_w f_b} \tag{3.23}$$

Fig. 3.10 presents the predictions for a typical Group B particle based on the two-phase bubbling bed model described above. It shows, as expected, that increasing D_t/d_p decreases the interstitial gas velocity and also increases the downflow velocity of emulsion solids. Consequently, the rise velocity of emulsion gas through the bed is positive at low D_t/d_p and reverses to negative when D_t/d_p increases to reach a transition point (approx. 100–150). The reversal in the flow direction of emulsion gas means that gas backflows occur in beds. Therefore, we can see that the emulsion gas

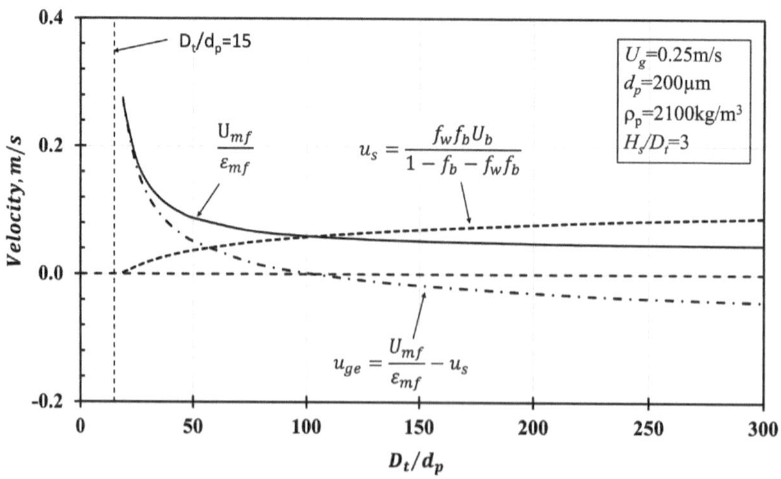

Figure 3.10 The two-phase model predictions for a typical Group B particle ($f_w = 0.2$).

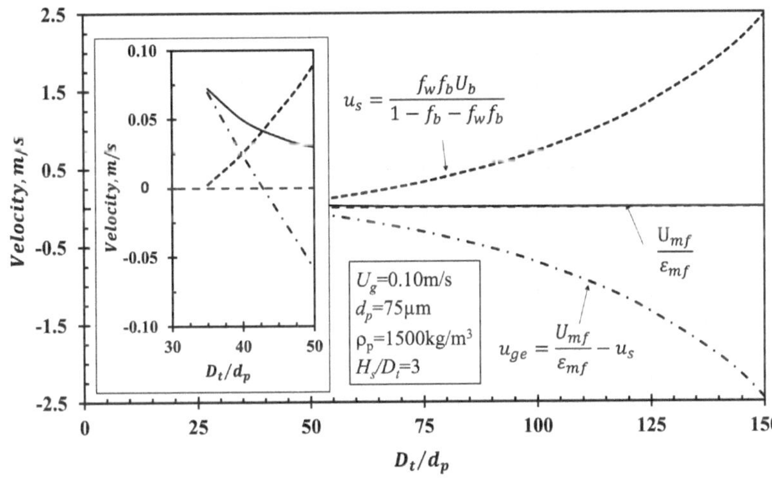

Figure 3.11 The two-phase model predictions for a typical Group A particle ($f_w = 0.45$).

moves upward and no gas backmixing occurs under the condition of small D_t/d_p. For large-scale beds, the gas backmixing becomes a serious concern as D_t/d_p increases.

For typical Group A particles, the predicted results are presented in Fig. 3.11. These fine particles have lower U_{mf} and higher ε_{mf}, leading to a lower interstitial gas velocity U_{mf}/ε_{mf} compared to that of Group B particles. Due to the high wake fraction of fine particles (here, we take $f_w = 0.45$ for a 75 μm particle based on the literature [2]), the solid circulation is significantly enhanced as more particles are carried up by bubble wakes, resulting in a remarkable increase in the downflow velocity of emulsion solids. Consequently, the rise velocity of emulsion gas through the bed is positive in a very small bed and quickly reverses to negative when the ratio D_t/d_p increases just over 45. From the above discussions, it can be concluded that the gas backmixing in the beds of Group A particles is greater than that in the beds of Group B particles. This is in good agreement with the experimental results [7].

Although the two-phase model can explain the gas mixing behavior and its variations with bed size and particle properties, the calculations show that the model predictions are highly sensitive to the bubble behavior parameters. Therefore, further work is still needed to develop a more comprehensive model to predict variations of gas mixing with operating parameters in micro-fluidized beds.

3.2.4 The criteria for plug flow of gas in micro-fluidized beds

It is clear that gas mixing in micro-fluidized beds is fundamentally different from that in macro-fluidized beds. Micro-fluidized beds feature a gas flow pattern that slightly deviates from ideal plug flow because of the wall effect that helps suppress the gas backmixing in beds, which is a severe issue of large-scale fluidized beds. However, a slight deviation from ideal plug flow or a small gas dispersion coefficient is still a term considered too qualitative to define the closeness of the gas flow to an ideal plug flow. Since a close-to-plug flow is critical for optimal operations of an MFB reactor, it is necessary to have a quantitative and easy-to-use criterion to determine whether a flow can be considered a close-to-plug flow in practical MFB operations.

Traditionally, researchers believe that a fluid is close-to-plug flow when the Peclet number ($Pe = uH_b/D_{a,g}$) exceeds 50, but other values such as 10, 20, or 100 have been mentioned [21]. As stated earlier, there are uncertainties related to using the Pe number to judge whether gas flow is close-to-plug flow. First, the Peclet number, according to its definition, does not only vary with the gas backmixing coefficient but also with the characteristic size. Therefore, even with the same gas backmixing or the same deviation from the ideal plug flow, the Peclet number can be different when the characteristic size (e.g., the distance between tracer gas injection and detection points) is different. For example, Geng et al. [7]. demonstrated that when the total bed height increased from 8 to 32 mm, the Peclet number varied from 31 to 71, while the gas backmixing coefficient changed merely from 0.00010 to 0.00017 m^2/s. Second, when the axial diffusion model is used to fit the experimental residence time distribution curve to estimate $D_{a,g}$, introduction of inappropriate H and u will result in errors in the estimated $D_{a,g}$. Third, the error in the estimated $D_{a,g}$ may be enlarged and further when the average residence time $\overline{\tau}$ is used in the axial dispersion model when gas flowing through the bed is nonuniform (e.g., channeling, bypassing). In fluidized beds, part of the gas flows through the particle bed in fast-rising bubbles, and the rest flows slowly through the emulsion. The bubble rise velocity varies with the bubble size and bed height, so the residence time of a bubble in the bed depends on where it forms and how it coalesces and breaks. This nature of gas flow in fluidized beds results in a mean residence time that may be very different from the space–time $\overline{\tau}$. Lastly, due to practical difficulties to obtain bed height H_b, the column height H_t is sometimes used in calculating Pe for simplicity. Thus, the computed Peclet number varies with H_t even if the heights of fluidized beds H_b and the values of $D_{a,g}$ are the same. For these reasons, we do not consider that it is appropriate to use the Peclet number as the criterion for defining the plug flow condition.

Again referring to Fig. 3.3, we can see that when $D_{a,g}/uH$ decreases to below 0.01, the one-dimensional dispersion model applies, the height of the residence time distribution increases, and the width of the distribution reduces. It is also seen that the peak height increases sharply as $D_{a,g}/uH$ decreases below 0.025 approximately, and correspondingly, the variation in the width of $E(\theta)$ becomes relatively small. This observation is further illustrated in Fig. 3.12 when plotting $E_{\theta,\max}$ as a function of $\sigma_\theta{}^2$. It shows that $E_{\theta,\max}$ increases with decreasing $\sigma_\theta{}^2$. When $\sigma_\theta{}^2$ is below about 4, the increase in $E_{\theta,\max}$ accelerates, especially when $\sigma_\theta{}^2$ is below about 0.25, $E_{\theta,\max}$ jumps up remarkably. It clearly shows that the residence time distribution approximates a

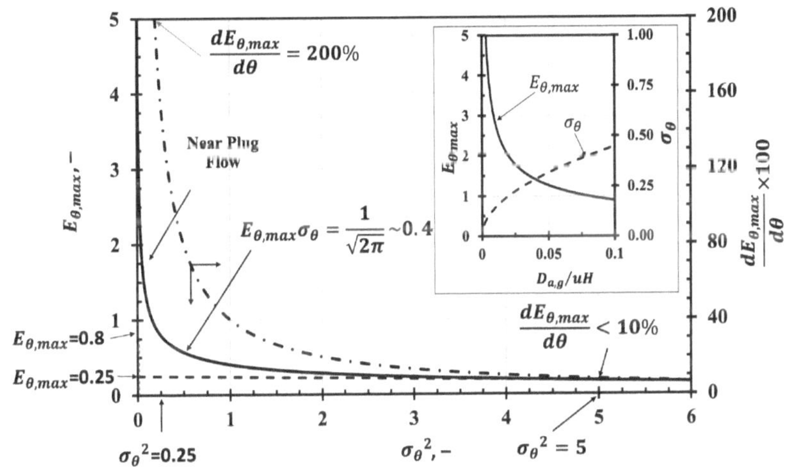

Figure 3.12 Relationship between $E_{\theta,\max}$ and σ_θ^2.

pulse with a very short time spread, that is, a characteristic of a close-to-plug flow. Therefore, we can suggest that when $\sigma_\theta^2 < 0.25$ or $E_{\theta,\max} > 0.8$ and $E_{\theta,\max}\sigma_\theta$ approximates 0.4 (i.e., Eq. (3.10)), an almost ideal plug flow is achieved, and the flow can be described by the one-dimensional dispersion model.

Geng et al. [7]. realized first that a criterion not based on the Pelect number is needed to describe gas flows in micro-fluidized beds. They argued that because the column height affected the Pelect number, a high Pelect number could come from a tall column even though the gas backmixing in a bed of particles remained almost unchanged. This realization led them to investigate relationships between $E(t)_h$ and σ_t^2 obtained from experiments with various particles, bed diameters, gas velocities, and static bed heights. The results are provided in Fig. 3.13A. In this chart, the experimental data of Jia [22], Boskovic and Loebbecke [23], and Adeosun and Lawal [24] are also presented. The solid black line represents Eq. (3.10). It is clear that all the data fit Eq. (3.10) satisfactorily except for FCC, glass beads, and silica sand 1. This means that the gas flows in the micro-fluidized beds with FCC, glass beads, and silica sand #1 particles cannot be modeled by Eqs. (3.7) and (3.8), which are based on the assumption of a small extent of dispersion. In contrast, the gas flows in the beds of the silica sand # 2 and #3 particles approximate plug flow.

This figure shows that when σ_t^2 is less than 0.25, $E(t)_h$ increases sharply with a further decrease in σ_t^2; when σ_t^2 is greater than 5.0 approximately, $E(t)_h$ reduces slightly. Based on their experimental data, Geng et al. [7]. proposed a criterion that a flow is defined as close-to-plug flow when σ_t^2 is below $0.25s^2$ or the peak height $E(t)_h$ is larger than 1.0/s. Under these conditions, the gas flow has a negligible backmixing in micro-fluidized beds. As shown in Fig. 3.13A, the gas flows in the beds of Group B

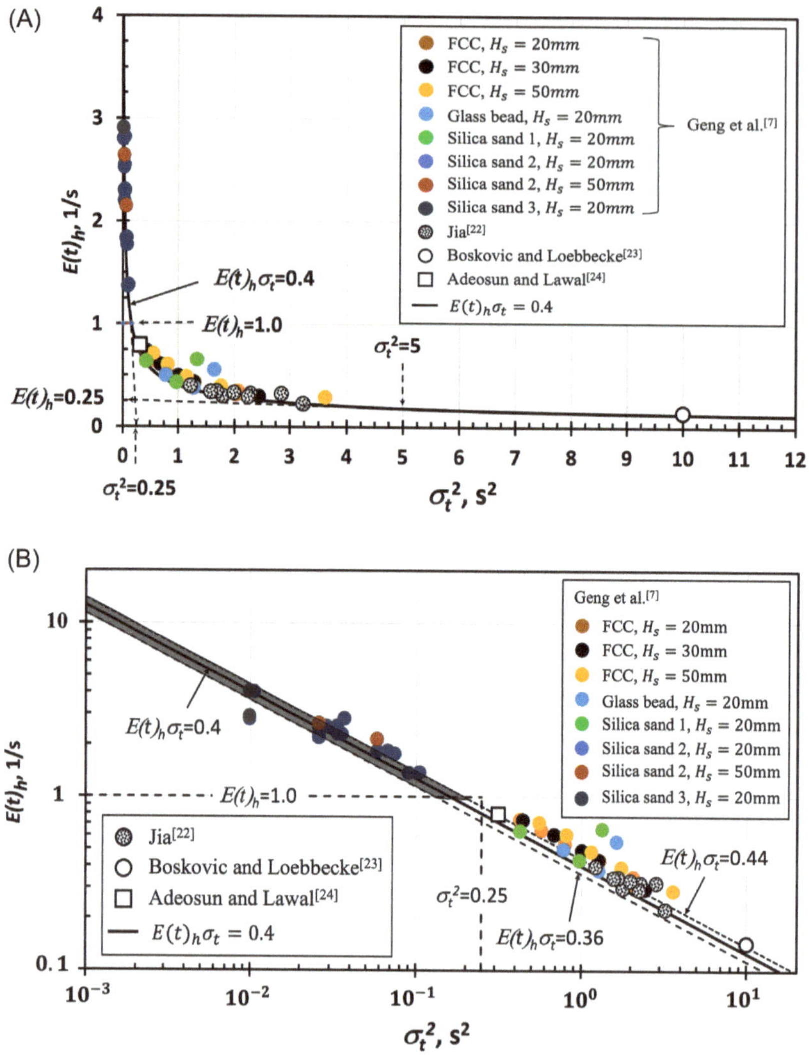

Figure 3.13 The relationship between $E(t)_h$ and $\sigma_t{}^2$. (A) Normal coordinates, (B) log-log coordinates.

particles (the silica sand 2 and #3) are close-to-plug flow. When $\sigma_t{}^2$ is above 5.0 or the peak height $E(t)_h$ is smaller than 0.25, an appreciable gas backmixing in micro-fluidized beds occurs. In between the above two cases, the gas flow has certain backmixing but is not significant, and the Group A particles (the FCC and the glass bead) shown in Fig. 3.13A belong to this region.

1. $E(t)_h$ versus $\sigma_t{}^2$,
2. Log–log relationship.

To examine the experimental data more explicitly, we replot the data of Fig. 3.13A in Fig. 3.13B in log–log coordinates. In addition to the solid line representing $E(t)_h\sigma_t = 0.40$, two broken lines showing Eq. (3.10) with +10% and −10% deviations (i.e., $E(t)_h\sigma_t = 0.40*1.1 = 0.44$ and $E(t)_h\sigma_t = 0.40*0.9 = 0.36$) are added. It is clear that for the data of FCC, glass beads, and silica sand 1 particles reported by Geng et al. [7] and the data reported by Jia [22], Boskovic and Loebbecke [23], and Adeosun and Lawal [24], all fall outside of the close-to-plug gas flow region (shaded area). For some experimental data, even though $E(t)_h\sigma_t$ is very close to 0.4, the corresponding values of $E(t)_h$ and σ_t^2 do not meet the plug flow requirements. In these experimental conditions, the gas flows in micro-fluidized beds will exhibit flow behavior that deviates largely from plug flow. The gas flows in micro-fluidized beds for the silica sand 2 and #3 particles, except for a few operating conditions (i.e., the static bed height $H_s = 20$ mm, too shallow to prevent channeling flow from occurring), meet the requirements for close-to-plug flow. In summary, the following criteria for a close-to-plug flow in micro-fluidized beds need to be satisfied:

1. The RTD curve is symmetric,
2. the RTD peak height $E(t)_h$ and variance σ_t approximately satisfy $E(t)_h\sigma_t = 0.4$, and
3. $E(t)_h$ is greater than 1.0 1/s or $\sigma_t^2 < 0.25$.

In practice, selecting the operation conditions (such as the use of Group B particles) and keeping $D_t/d_p = 10–150$ and $H_s/D_t = 2–4$ will provide good opportunities to approach plug flow of gas in micro-fluidized bed reactors.

3.3 Solid mixing

Solid mixing is an important issue in fluidized bed reactors, and it significantly influences heat transfer, gas–solid contact, and reaction. In a bubbling fluidized bed, rising bubbles promote particle mixing and transport particles upward in their wakes, while particles in the emulsion phase flow downward. As a result, solid mixing in fluidized beds is intense, confirmed theoretically and experimentally in large-scale fluidized beds. However, our knowledge of solid mixing in micro-fluidized beds is still limited. We, therefore, study this subject in this section based on the available theoretical modeling results. Interested readers are referred to pertinent references that may become available on the subject in the future.

3.3.1 Solid mixing simulation

Li and Ji [25] simulated the mixing behavior of particles in a micro-fluidized bed with a rectangular cross section (150 × 4 mm). As shown in Fig. 3.14, the micro-fluidized bed had an air distributor with a single air injecting orifice. At time $t = 0$ s, two kinds of colored particles (blue and black) were placed one above the other in the bed. Then, the fluidized gas was injected from a center orifice hole of the distributor plate

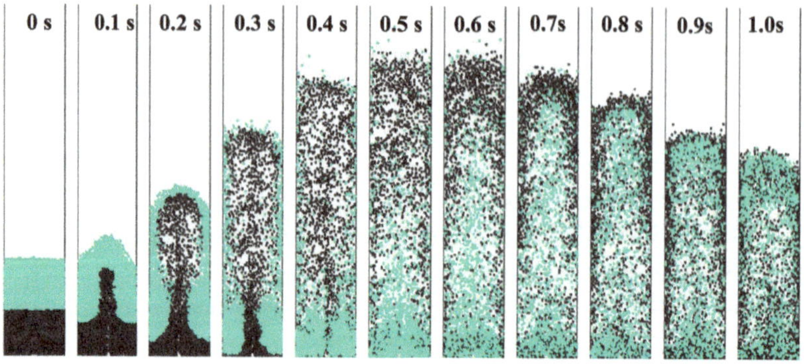

Figure 3.14 The simulation results of particle mixing in a micro-fluidized bed [25].

into the bed at an orifice velocity of 28 m/s. Under this condition, the model calculated the particle distributions at different times. It can be seen that the particles near the bottom nozzle are sucked into an upward gas stream by the fluidizing gas, carried by the gas into the bed top space, and then fall back to the bed. In the wall region, the low upward gas velocity leads to a drag force that is less than the weight of particles, and thus, the particles move downward along the wall. When particles fall back to the dense zone at the bed bottom, they slowly move downward and are sucked into the jet area again. Thus, a typical solids circulation is formed in the bed, featured by an upward-flowing center core surrounded by a downward-moving region near the wall. Because of the intensive internal circulation of particles, solid mixing in the micro-fluidized bed is intense and fast. It can be seen that the blue and black particles in the bed have been mixed very well in less than 1 s.

3.3.2 Particle feeding simulation

Yang et al. [26] investigated the particle mixing behavior in a micro-fluidized bed with different particle feeding methods. Fig. 3.15 shows five simulated structures. Case C places the injector tip in the bed center. Cases A and E place the injector tip near the sidewall. The injector tip is located at 1/4 bed diameter from the wall in the B and D cases. The injectors A and B are straight-inclined pipes. The injectors C, D, and E have an angled structure at their tip end, so the gas flow direction is perpendicular to the bed surface. During the simulations, 25–30 mg of sample particles were injected into the bed by flowing 3–5 mL gas in 20 MS. The sample particle size and density were the same as the bed materials. The SIMPLE coupled algorithm was used to solve the two-phase flow models. In the simulations, it was assumed that a plug flow condition was at the bed bottom and inside the injection pipe, and the ambient condition was at the bed exit.

In the cold model experiments (Fig. 3.16), sand particles (white color) of average diameter 200 μm and pulverized coal particles (black color) of 50 μm were used as the bed materials and the feeding sample particles, respectively. The volumetric flow of fluidizing gas was 1.4 L/min, and the experimental parameters were kept to be identical to those used in the simulations. A high-speed camera was used to record the

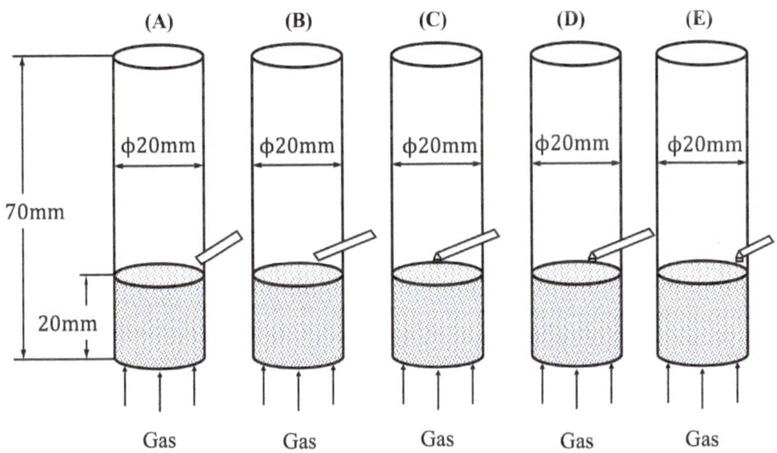

Figure 3.15 Five different feeding configurations simulated for a micro-fluidized bed [26].

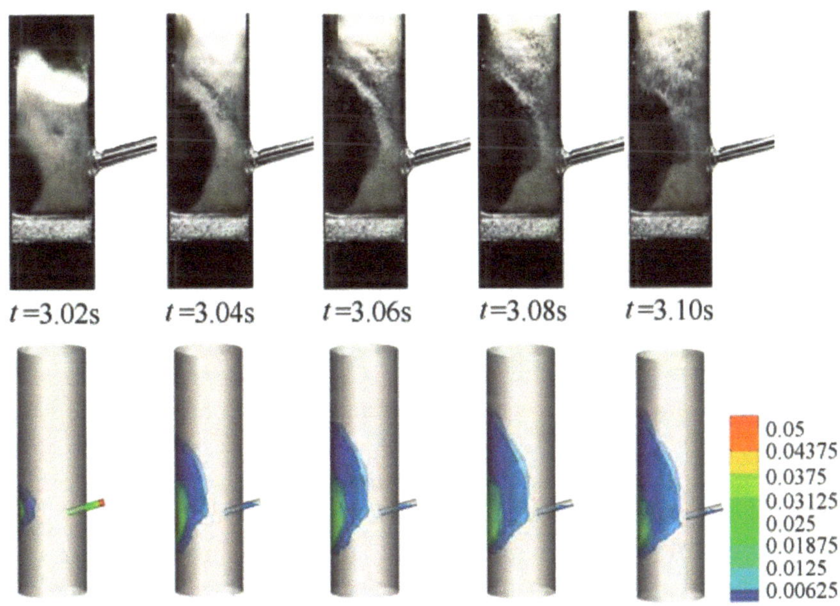

Figure 3.16 Comparison of experimental (the upper) and simulated (the lower) instantaneous particle concentration distributions after pulse feeding sample particles.

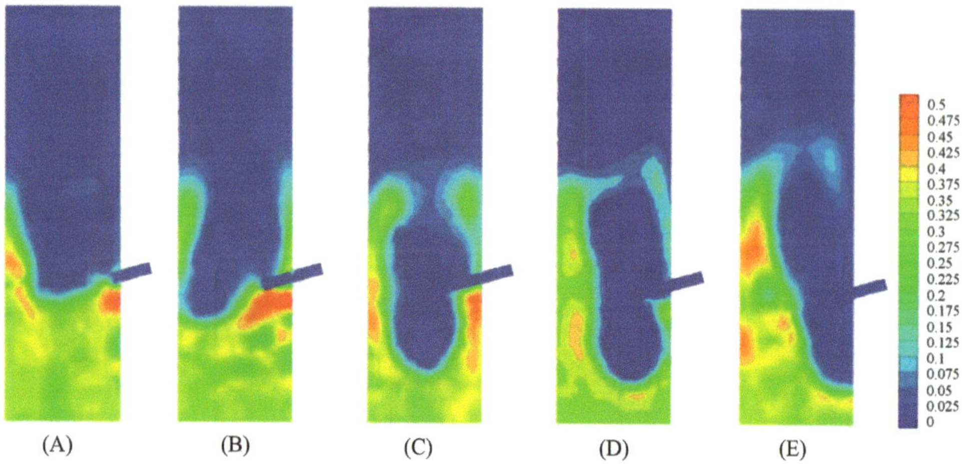

Figure 3.17 Instantaneous concentration distributions of bed material with different nozzle structures ($t = 3.04$ s). (A) Structure A, (B) structure B, (C) structure C, (D) structure D, (E) structure E.

fluidization patterns during the experiments. The camera speed was 200 frames per second, and the image resolution was 800×600.

Fig. 3.16 shows the instantaneous particle distributions for the feeding structure B. The upper images are taken by the high-speed camera, and the lower graphics are obtained by the simulations. The white and black particles in the images correspond to the bed materials and solid samples, respectively. It can be seen that the simulations and the experimental results are in good agreement.

The simulation results of the five sample feeding structures are shown in Fig. 3.17. Note that the sample injection is initiated at 3.04 s. It shows that the high-speed sample injection influences the fluidizing gas flow in the bed, and the influence varies with the feeding structure configuration. As can be seen in the figure, a large bubble forms instantly on the bed surface when the samples are injected by the nozzles C, D, and E. The bubble grows larger as it moves along the radial direction of the bed. The inclined structures A and B have certain influences on the bed material at the top area, but the effects diminish towards the wall.

To further evaluate the effect of sample feeding structure on solid mixing behavior, the center-of-gravity position (CGP) (Fig. 3.18) and the standard deviation of the bed particle concentration distribution for each simulation are calculated and shown in Fig. 3.19. It shows that the bed materials move upward immediately after the sample injections because of the formation of large bubbles. The perpendicular injections push the bed materials higher than the tilt injection. After injection, the disturbed beds of structures A and B return to stabilization quickly (within 0.1 s), but slowly (about 3.4 s) for structures C, D, and E.

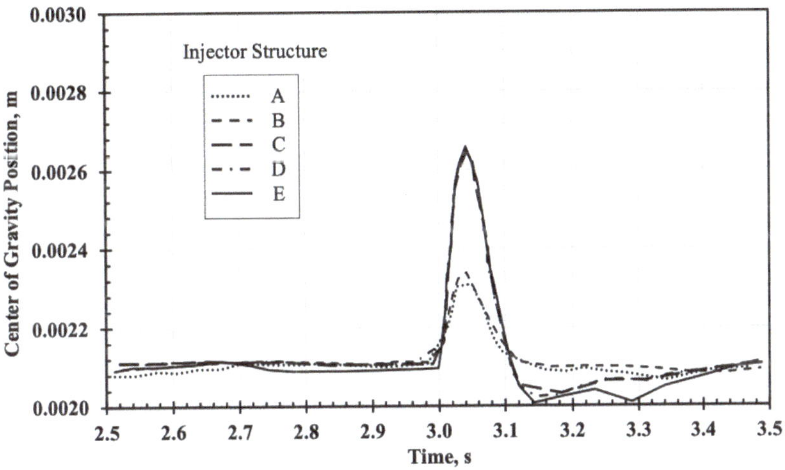

Figure 3.18 Variation of gravity center of bed material with time for different injector structures.

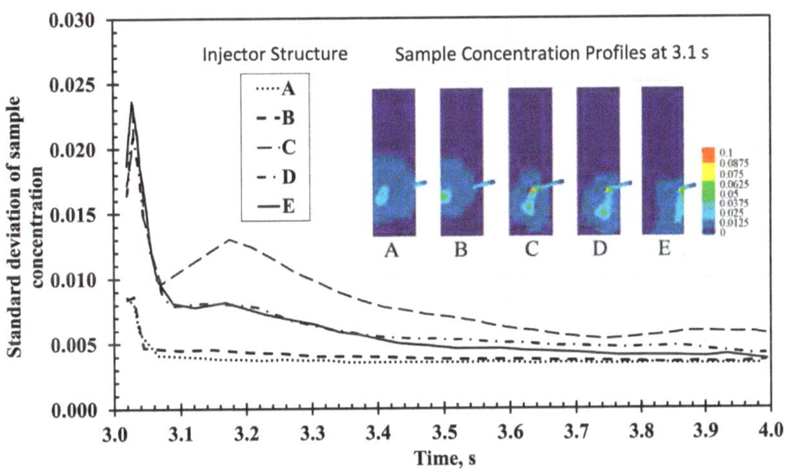

Figure 3.19 Variation of the standard deviation of sample concentration distribution for different injector structures.

Measured by the standard deviation of the entire bed particle concentration, the mixing uniformity between the bed material and the feeding sample particles is shown in Fig. 3.19. It shows that sample injections with the structures A and B produce the smallest standard deviation, indicating that the sample particles are mixed evenly with the bed material. In addition, the beds with structures A and B reach the minimum standard deviation in a shorter time than the beds with structures D and E. The bed mixing with nozzle C is far less ideal than the nozzles A and B.

The difference between structures A and B is the length of the nozzle inserted into the bed. It can be seen from Fig. 3.20 that if the nozzle is tilted too deep into the bed, the samples are more likely to collide directly with the opposite wall, which will affect the contact efficiency between the sample particles and the bed material.

Better mixing of particles in a micro-fluidized bed is one of the critical conditions to achieve superior reaction characteristics. It can be seen from this chapter that the study on gas and solid mixing in micro-fluidized beds is not sufficient because of the limitations of experimental techniques. The use of numerical simulation can be helpful to understand the mixing characteristics of gas and particles and provide guidance on the optimal design of micro-fluidized beds. However, as presented in this chapter, it must be noted that the numerical simulation results are sensitive to certain model parameters, which are determined empirically or even arbitrarily in some cases. Thus,

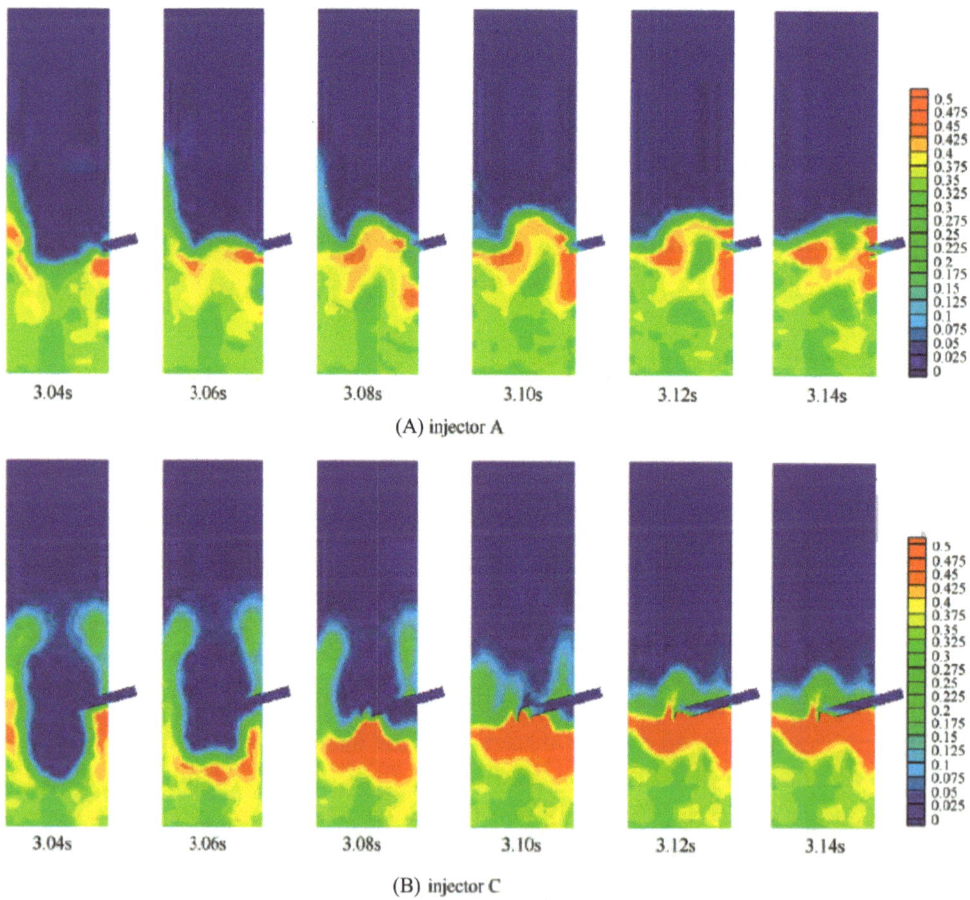

Figure 3.20 Sample concentrations after injection with the (A) and (B) feeding structures.

validating the model predictions is difficult in the absence of experimental data. To better understand the mechanism and behavior of gas–solid mixing in micro-fluidized beds, further experimental and modeling studies are required.

Abbreviation

MFB micro-fluidized bed
MS mass spectrometer
RTD residence time distribution

Nomenclature

$C(t)$ gas concentration at time t
$D_{a,g}$ axial gas dispersion coefficient (m^2/s)
d_b bubble size (m)
d_{b0} bubble size at $z = 0$ (m)
d_{bm} maximum stable bubble size (m)
d_p mean particle diameter (m)
$E(t)$ gas residence time distribution function (1/s)
$E(t)\ h$ peak height of $E(t)$ (1/s)
$E_{in}(t)$ gas residence time distribution at input location (1/s)
$E_{max}(t)$ maximum value of $E(t)$ (1/s)
$E_{out}(t)$ gas residence time distribution at the output location (1/s)
$E_{\theta,max}$ maximum value of dimensionless $E(\theta)$
f_b fraction of the bed occupied by gas bubbles
f_w ratio of wake volume to bubble volume
H bed height (m)
H_b height of the dense bed (m)
P_e Peclet number, $P_e = \mu H / D_{a,g}$
U constant velocity (m/s)
U_b bubble velocity (m/s)
U_{br} velocity of a single bubble (m/s)
U_g superficial gas velocity (m/s)
u_{ge} rise velocity of emulsion gas through the bed (m/s)
U_{mf} minimum fluidization velocity (m/s)
u_s downflow velocity of emulsion solids (m/s)
U_t terminal velocity of a single particle (m/s)
y axial coordinate (m)
z dimensionless axial coordinate

Greek letters

$\bar{t}$ mean residence time (s)
$\bar{\theta}$ dimensionless mean residence time
$\bar{\tau}$ mean residence time calculated by H/u
β a constant
ε average bed voidage
ε_f bed voidage of a bubbling fluidized bed calculated by the Richardson-Zaki equation

ε_{mf} average bed voidage at minimum fluidization condition
θ dimensionless time
ρ_p particle density (kg/m^3)
σ_t variance of the gas residence time distribution (s^2)
σ_θ dimensionless variance of the gas residence time distribution

References

[1] Kunii D, Levenspiel O. Fluidization engineering. 2nd ed. Boston: Butterworth—Heinemann; 1991.
[2] Grace JR. Hydrodynamics of gas fluidized beds Basu P. Fluidized bed boilers: design and application Pergamon Press Can Ltd 1984;13–30.
[3] Hu D, Zeng X, Wang F, et al. Comparison and analysis of gas backmixing in micro fluidized and spouted beds. In: The 11th Annual Conference and Cross-Strait Symposium on Particle Technology. Xiamen: Chinese Society of Particuology; 2020.
[4] Hu D, Geng S, Zeng X, et al. Gas back-mixing characteristics and the effects on gas-solid reaction behavior and activation energy characterization. CIESC Journal 2020;72(3):1354–63.
[5] Deemter JJ Van. Mixing patterns in large-scale fluidized beds; Grace JR, Matsen JM. Proceedings of the 1980 International Fluidization Conference. Henniker, New Hampshire, August 3–8,1980: Plenum Press, New York: 69–89.
[6] Levenspiel O. Chemical reaction engineering. 3rd ed. John Wiley & Sons, Inc; 1999.
[7] Geng S, Han Z, Yue J, et al. Conditioning micro fluidized bed for maximal approach of gas plug flow. Chemical Engineering Journal 2018;351:110–18.
[8] Li J, Weinstein H. An experimental comparison of gas backmixing. Chemical Engineering Science 1989;44(8):1697–705.
[9] Yoshida K, Kunii D, Levenspiel O. Axial dispersion of gas in bubbling fluidized beds. Industrial & Engineering Chemistry Fundamentals 1966;7:402–6.
[10] Bi HT, Ellis N, Abba IAA, et al. A state-of-the-art review of gas-solid turbulent fluidization. Chemical Engineering Science 2000;55(21):4789–825.
[11] Grace JR, Bi X, Ellis N. Essentials of fluidization technology. Essentials of fluidization technology, Wiley-VCH Verlag GmbH & Co. KGaA, Boschstr. 12, 69469 Weinheim, Germany; 2020.
[12] Dang TYNYN, Gallucci F, Van Sint Annaland M. Gas back-mixing study in a membrane-assisted micro-structured fluidized bed. Chemical Engineering Science 2014;108:194–202.
[13] Lee GS, Kim SD. Rise velocities of slugs and voids in slugging and turbulent fluidized beds. Korean Journal of Chemical Engineering 1989;6(1):15–22.
[14] Geng S, Yu J, Zhang J, et al. Gas back-mixing in micro fluidized beds. CIESC Journal 2013; 64(3):867–76.
[15] Han Z, Yue J, Geng S, et al. State-of-the-art hydrodynamics of gas-solid micro fluidized beds. Chemical Engineering Science 2021;232:116345.
[16] Richardson JF, Zaki WN. Sedimentation and fluidisation: part I. Chemical Engineering Research and Design 1954;75(3):S82–100.
[17] Davidson JF, Harrison D. Fluidization. Davidson JF, Clift R, Harrison D. 2nd ed. London and New York: Academic Press; 1971.
[18] Wang F, Fan LS. Gas-solid fluidization in mini- and micro-channels. Industrial & Engineering Chemistry Research 2011;50(8):4741–51.
[19] Wang J, Tan L, Van der Hoef MA, et al. From bubbling to turbulent fluidization: advanced onset of regime transition in micro-fluidized beds. Chemical Engineering Science 2011;66(9):2001–7.
[20] Mori S, Wen CY. Estimation of bubble diameter in gaseous fluidized-beds. AIChE Journal. American Institute of Chemical Engineers 1975;21(1):109–15.
[21] Deshmukh SARK, Laverman JA, Cents AHG, et al. Development of a membrane-assisted fluidized bed reactor - 1 - gas phase backmixing and bubble-to-emulsion phase mass transfer using tracer injection and ultrasound experiments. Industrial & Engineering Chemistry Research 2005;44:5955–65.

[22] Jia Z. Experimental and numerical investigation on hydrodynamics of micro fluidized bed. A master thesis submitted to Tianjin University; 2016.
[23] Boskovic D, Loebbecke S, Bošković D, et al. Modelling of the residence time distribution in micromixers. Chemical Engineering Journal 2008;135(Suppl. 1):S138–46.
[24] Adeosun J, Lawal A. Numerical and experimental studies of mixing characteristics in a T-junction microchannel using residence-time distribution. Chemical Engineering Science 2009;64:2422–32.
[25] Li B, Ji L. Numerical simulation of particle mixing in circulating fluidized bed with discrete element method. Proceedings of the CSEE 2012;32(20):42–8.
[26] Yang X, Liu Y, Yu J, et al. Numerical simulation of mixing characteristics of trace sample and bed material in micro fluidized bed reaction analyzer. CIESC Journal 2014;65(9):3323–30.

CHAPTER 4

Micro-fluidization regimes

Micro-fluidized beds are distinguished hydrodynamically from large-scale fluidized beds due to the wall effects, which become increasingly more pronounced as the bed-to-particle diameter ratio reduces [1−4]. The wall effects complicate the gas−solid flow behavior, including the $\Delta P_B \sim U_g$ relationship, gas backmixing, and solid mixing, as discussed in Chapters 2 and 3. The wall effects directly cause an increase in the particle-wall interactions and indirectly lead to significant variations in bed voidage and flow uniformity. It has been shown experimentally that the wall effects affect the fluidization regimes and their transitions, hence the gas and solid mixing, heat and mass transfer, and reactor performance. Therefore, a good understanding of these flow regimes is of great importance from the perspectives of multiphase flow science and practical industrial applications. This chapter discusses the fluidization regimes in micro-fluidized beds, with particular attention paid to identifying specific features that differentiate between micro- and macro-fluidized beds.

4.1 Experimental observations

Several research groups have investigated the flow patterns in micro-fluidized beds (e.g., references [4−6]) and reported that micro-fluidized beds exhibit unique features of flow regimes and their transitions compared to those typically observed in macro-fluidized beds. Employing high-speed cameras, Wang and Fan [4] and McDonough et al. [6] observed the gas-solid flow behavior in mini- and micro-fluidized beds with bed sizes ranging from 700 μm to 15 mm. During these experiments, high-speed videos of the fluidized beds were recorded to investigate the dynamic behavior of fluidization under different operating conditions. In addition to the visualization method, the gas-solid fluidization patterns or regimes are also identified and characterized by other experimental techniques, such as bed expansion and pressure drop measurements. To illustrate the progression of the flow regimes with increasing superficial gas velocity, Fig. 4.1 presents a representative $\Delta P_B \sim U_g$ relationship, accompanied by snapshots taken by a high-speed camera in a micro-fluidized bed of 5 × 5 mm using a 93 μm silica particle. As shown in Fig. 4.1, the bed remains stationary with a clear and flat top surface at low gas velocities. As the gas velocity increases, small voids appear in the area near the wall, and small gas bubbles are visible near the gas distributor but collapse quickly. Increasing the gas velocity further leads to more small gas bubbles visually observed in the bed, even though the bed has not yet been fluidized. The

Micro Fluidization
DOI: https://doi.org/10.1016/B978-0-443-18718-6.00004-8
© 2023 Chemical Industry Press Co., Ltd.
Published by Elsevier Inc., under an exclusive license with Chemical Industry Press Co., Ltd.

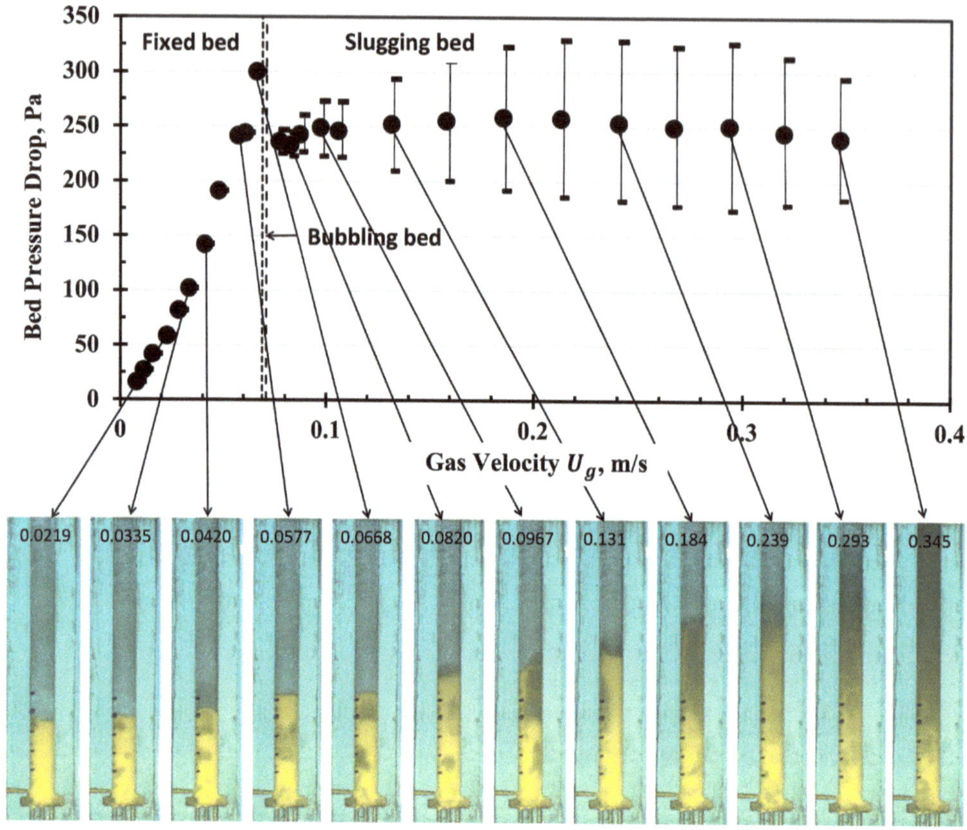

Figure 4.1 The gas-solid flow patterns recorded by a high-speed camera in a micro-fluidized bed of 5 × 5 mm with $H_s/D_t = 4$ [6].

formation of gas bubbles before the minimum fluidization is also confirmed by Liu et al. [3]. After reaching the minimum fluidization, the gas slugs form and grow quickly after the bed passes a narrow bubbling regime. The slugging fluidization regime is characterized by the increased fluctuations in the bed pressure drop. Further increasing superficial gas velocity, the bed enters a turbulent fluidization regime, in which breakups of gas bubbles become dominant. Thus, the standard deviation of pressure drop fluctuation starts to decline.

Visual observations and pressure drop analysis confirm that representative flow regimes in micro-fluidized beds include fixed bed, particulate fluidization, bubbling fluidization, slugging fluidization, and turbulent fluidization (Fig. 4.2). In principle, fast fluidization and pneumatic conveying can also be attained in micro-fluidized beds, but they have not been investigated yet. It is well understood that the characteristics and transitions of fluidization regimes in micro-beds are significantly affected by operating conditions and bed size. Next, we discuss these characteristics more specifically.

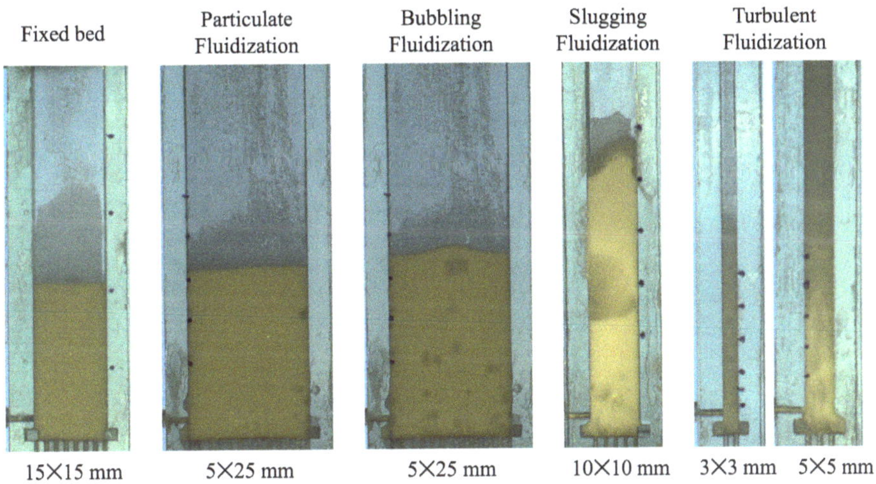

Figure 4.2 The gas-solid flow patterns snapshotted by a high-speed camera in micro-fluidized beds [6].

4.2 Fixed bed

A fixed bed is a fluid-particle contacting mode where a fluid (gas or liquid) passes through the void spaces between the stationary particles confined in a vessel. The bed pressure drop is approximately proportional to gas velocity in a gas—solid fixed bed. When the bed-to-particle diameter ratio D_t/d_p is large enough, particles can be considered to be in a randomly packed state, and the gas flow through the particle bed is relatively homogeneous. The $\Delta P_B \sim U_g$ relationship can therefore be represented satisfactorily by the classical Ergun equation [7,8]. However, these characteristics are affected significantly by the wall effects in micro-fixed beds. As discussed in Chapter 2, in addition to the wall frictional forces exerted directly on particles next to the wall, the wall confinement results in an increase in bed voidage and a nonhomogeneous gas flow across the bed cross section. These factors complicate the pressure drop predictions. Thus, the pressure drop in a wall-confined fixed bed can be higher or lower than the estimate of the Ergun equation [3,6,9—11], and the discrepancy depends on bed sizes and gas and particle properties.

Various correlations have been proposed in the literature to predict the fixed bed pressure drop under the conditions of wall confinement [10,12—15]. These correlations are commonly derived from the Ergun equation by incorporating wall effect parameters as follows:

$$\frac{\Delta P}{H} = A_E \frac{(1-\varepsilon)^2}{\varepsilon^3} \frac{\mu_g U_g}{{d_p}^2} + B_E \frac{1-\varepsilon}{\varepsilon^3} \frac{\rho_g {U_g}^2}{d_p} \tag{4.1}$$

where A_E is the viscous constant (=150 for the Ergun equation), B_E is the inertial constant (=1.75 for the Ergun equation), d_p is the particle diameter, U_g is the superficial gas velocity, H is the bed height, ΔP is the pressure drop, ε is the bed voidage, μ_g is the gas viscosity, and ρ_g is the gas density. Define

$$f_E = \frac{\Delta P}{H}\frac{\varepsilon^3}{1-\varepsilon}\frac{d_p}{\rho_g U_g^2}, \mathrm{Re}_g = \frac{\rho_g U_g d_p}{\mu_g(1-\varepsilon)} \tag{4.2}$$

Then, Eq. (4.1) can be expressed as

$$f_E = \frac{A_E}{\mathrm{Re}_g} + B_E \tag{4.3}$$

Table 4.1 summarizes various correlations available in the literature to evaluate A_E and B_E [14]. In this table, M is a function of the bed-to-particle diameter ratio, defined by

$$M = 1 + \frac{2}{3(1-\varepsilon)}\frac{1}{D_t/d_p} \tag{4.4}$$

Note that all the correlations are developed to account for the effect of the bed-to-particle diameter ratio. In practice, however, these correlations are all based on the data obtained in beds of millimeter particles rather than in beds of micrometer particles. Assessing the validity of these equations is problematic because it requires the bed voidage corresponding to the ratio D_t/d_p, which is not always available. In addition, the effects of oscillatory porosity and velocity variations with bed radius are not considered in all the models. Therefore, these correlations do not apply to micro-fluidized beds operating in fixed bed conditions, as shown in Fig. 4.3. In these calculations, the bed voidage data under the wall effect-free conditions are obtained by the experimentally measured bulk and particle densities (i.e., $\varepsilon = 1 - \rho_b/\rho_p$). It is clear that the predictions can be higher or lower than the experimentally measured values depending on the operating conditions. None of them can accurately predict the bed pressure drops in micro-fixed beds, although they all have the parameter D_t/d_p included in an attempt to correct the effect of the wall on the bed pressure drop. The reason is that the ratio D_t/d_p, a single parameter in these equations, cannot fully consider the changes in bed voidage and flow uniformity caused by wall constraints, as discussed in Chapter 2. Unfortunately, as of the writing of this book, we do not have enough data to distinguish between the effects of bed voidage variation and the flow uniformity. In the future, when the bed voidage data under the wall effects become available, attempts should be made to develop a more accurate and predictive correlation for estimating the bed pressure drop in micro-fluidized beds operating in fixed bed conditions.

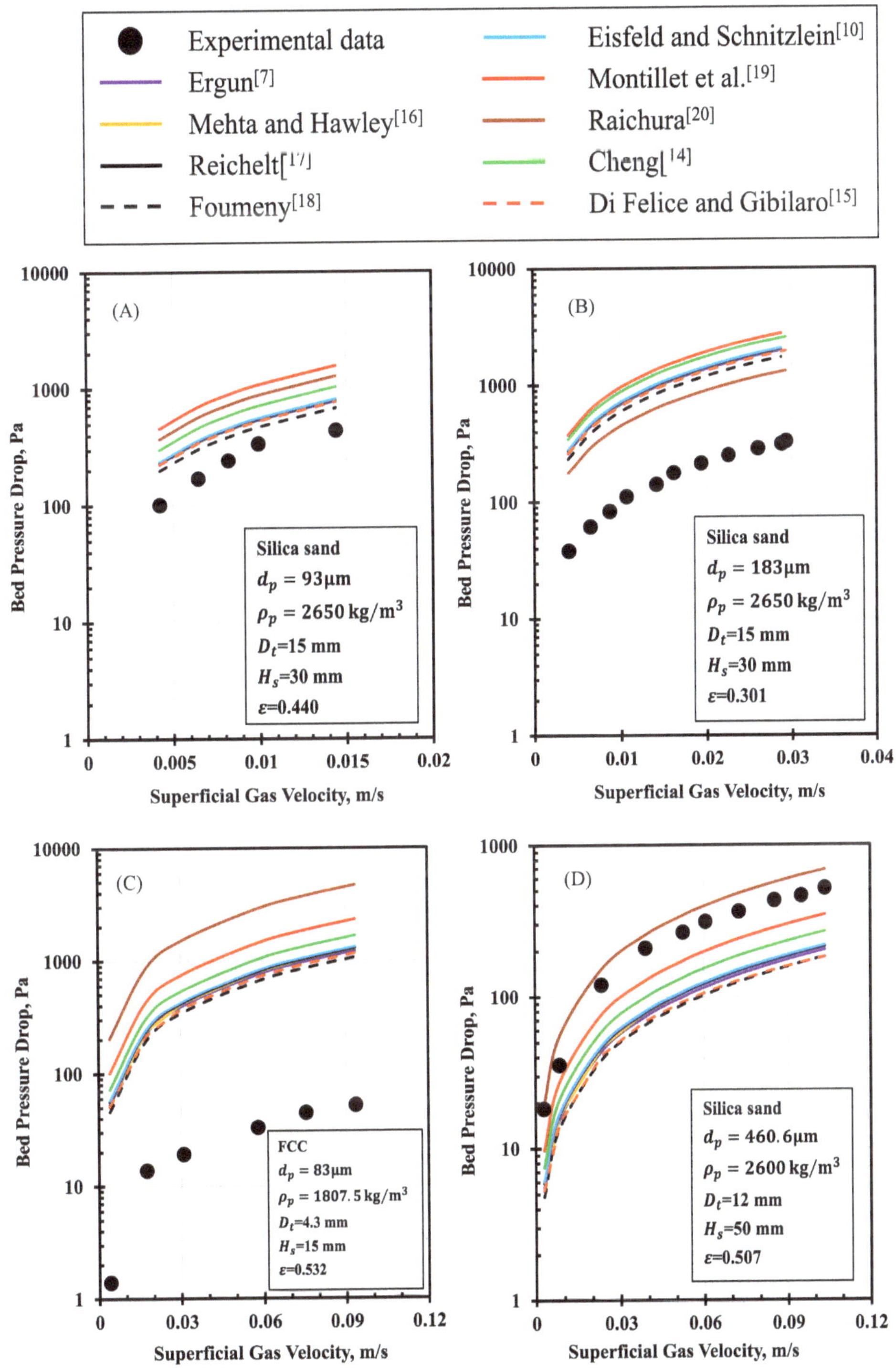

Figure 4.3 Comparisons between model-predicted and experimentally measured bed pressure drop data of (A,B) McDonough et al. [6], (C) Guo et al. [11], and (D) Liu et al. [3].

Table 4.1 Empirical formulas of the parameters A_E and B_E.

References	A_E	B_E	Note
Mehta and Hawley [16]	$150M^2$	$1.75M$	D_t/d_p:7–91
Reichelt [17]	$150M^2$	$\left[\frac{1.5}{(D_t/d_p)^2}+0.88\right]^{-2}M$	D_t/d_p:1.73–91
Foumeny et al. [18]	130	$\frac{D_t/d_p}{2.28+0.335D_t/d_p}$	D_t/d_p:3.23–23.8
Eisfeld and Schnitzlein [10]	$154M^2$	$\left[\frac{1.15}{(D_t/d_p)^2}+0.87\right]^{-2}M$	D_t/d_p:1.62–250
Montillet et al. [19]	$1000\alpha\left(D_t/d_p\right)^{0.2}\frac{1}{1-\varepsilon}$	$12\alpha\left(D_t/d_p\right)^{0.2}$	$\alpha = 0.061$ for dense packing and 0.05 for loose packing D_t/d_p:3.8–14.5
Raichura [20]	$103\left(\frac{\varepsilon}{1-\varepsilon}\right)^2\left(6(1-\varepsilon)+\frac{80}{D_t/d_p}\right)$	$2.8\frac{\varepsilon}{1-\varepsilon}\left(1-\frac{1.82}{D_t/d_p}\right)^2$	D_t/d_p:1.1–50.5
Cheng [14]	$185+17\frac{\varepsilon}{1-\varepsilon}\left(\frac{D_t/d_p}{D_t/d_p-1}\right)^2$	$1.3\left(\frac{1-\varepsilon}{\varepsilon}\right)^{1/3}+0.03\left(\frac{D_t/d_p}{D_t/d_p-1}\right)^2$	D_t/d_p:5–50
Di Felice and Gibilaro [15]	$\frac{150}{2.06-1.06\left(\frac{D_t/d_p-1}{D_t/d_p}\right)^2}$		D_t/d_p:1–40

4.3 Minimum fluidization velocity

In large-size or gas-solid macro-fluidized beds, fluidization is attained when the weight of solid particles is wholly supported by the upward drag force from a fluidizing gas. Mathematically, this can be determined by solving the force balance equation $\Delta P_E = mg/A$, where ΔP_E is the pressured drop caused by the drag force of gas, m is the mass of particles, A is the cross-sectional area of the bed, and g is the gravitational constant. Usually, ΔP_E can be reasonably estimated by the Ergun equation. The superficial gas velocity corresponding to $\Delta P_E = mg/A$ is the minimum fluidization velocity, U_{mf}.

In gas-solid micro-fluidized beds, U_{mf} is also experimentally determined by the $\Delta P_B \sim U_g$ relationship based on the descending gas velocity method, but the bed pressure drop is not necessarily equal to mg/A because the fluidized particles are supported by the drag force and the wall friction (see Chapter 2). Beyond U_{mf}, particles will be fully fluidized. Experimental research has demonstrated that the wall effects significantly delay the occurrence of complete fluidization.

4.3.1 Factors influencing U_{mf}

Fig. 4.4 shows the minimum fluidization velocity as a function of bed diameter for various particles and static bed heights [3]. It shows that decreasing bed diameter increases the minimum fluidization velocity when $D_t < 20$ mm. This result means that the onset of fluidization occurs with a delay for small diameter beds, while for beds of $D_t > 20$ mm, the effect of bed diameter on U_{mf} appears to be insignificant. The results

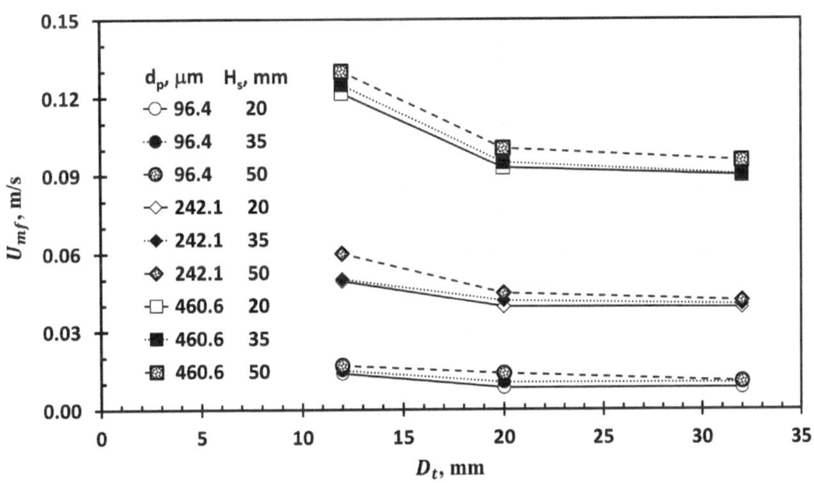

Figure 4.4 The minimum fluidization velocity as a function of bed diameter for different diameter particles and static bed heights. *Data from Liu et al. [3].*

also indicate that the static bed height has a relatively more significant influence on U_{mf} when $H_s > 35$ mm (i.e., $H_s/D_t > 2.9$).

Fig. 4.5 presents variations of U_{mf}/U_{mfm} with the bed diameter for a Group A particle (83 μm FCC) and a Group B particle (347 μm Glass beads), respectively. Here, U_{mfm} is the minimum fluidization velocity under the wall effect-free and homogenous flow conditions. For illustrative purposes, U_{mfm} used in Fig. 4.5 is estimated based on Wen and Yu's correlation using the same gas and solid properties as those used in the micro-fluidized bed experiments. It shows that the delay in fluidization becomes insignificant for the FCC particles (Fig. 4.5A) as the bed diameter increases. The ratio U_{mf}/U_{mfm} reaches 37 in the 4.3-mm-diameter bed (i.e., $D_t/d_p = 51.8$) and decreases to approximately 1 when the bed diameter is 25.5 mm (i.e., $D_t/d_p = 307$). For the Group B glass beads particles, as shown in Fig. 4.5B, the U_{mf}/U_{mfm} also decreases with D_t, but varies from 2.3 when $D_t = 4$ mm (i.e., $D_t/d_p = 11.5$) to approximately 1 when the bed diameter is 100 mm (i.e., $D_t/d_p = 288$).

The hysteresis of fluidization is indeed beneficial to micro-fluidized beds applications, especially when fine particles are fluidized. First, it permits beds of fine particles to be operated at higher gas velocities in micro-fluidized beds than in macro-fluidized beds, thus increasing the bed processing capacity. Otherwise, stable operation without significant particle entrainment is impossible at such high gas velocities. Second, it enhances the gas-solid contact efficiency, minimizes the gas backmixing, and promotes gas flow close to the plug pattern. These characteristics make gas-solid micro-fluidized beds fundamentally unique and advantageous over other gas-solid contact techniques, for example, gas-solid large-scale fluidized beds, especially when reaction analysis and fine chemical synthesis are concerned.

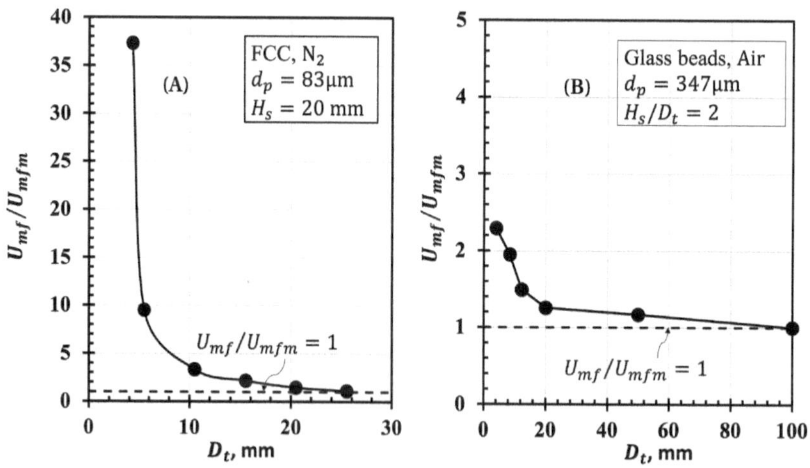

Figure 4.5 The ratio U_{mf}/U_{mfm} as a function of bed diameter for (A) Group A particles [11] and (B) Group B particles [5].

Using a gas mixture containing air, CO_2, and moisture as fluidizing gas, Prajapati et al. [21] investigated the effects of CO_2 concentration, particle diameter, and bed temperature on the minimum fluidization velocity in a 25-mm-diameter micro-fluidized bed. The results are shown in Fig. 4.6. It can be seen that for the identical particles, U_{mf} increases with an increase in CO_2 concentration and a decrease in bed temperature. Increasing particle diameter increases U_{mf} drastically. The results suggest that U_{mf} in micro-fluidized beds increases with increasing gas viscosity, which increases as the CO_2 concentration and bed temperature increase.

Han et al. [2] summarized the experimental results reported in the literature [3,5,6,11,21–24], as shown in Fig. 4.7. Note that Quan et al. [5] obtained the U_{mf} based on two sets of experimental data, that is, the variations in bed pressure drop and its standard deviation against the superficial gas velocity. The resultant data of U_{mf} are all plotted in the figure. In addition, the data reported by Wang and Fan [4] for mini-channel (channel size greater than 2 mm) and micro-channel beds (channel size smaller than 2 mm) fluidized are also presented. The results indicate that the bed diameter, the properties of gas (e.g., density and viscosity) and particles (e.g., diameter, density, shape), and the static bed height all have certain influences on U_{mf}. Due to the wall effect, U_{mf} increases remarkably for beds of smaller diameters or for D_t/d_p below about 150. The wall effect can increase U_{mf} by up to dozens of times compared to macro-fluidized beds for both Group A and B particles. When the ratio D_t/d_p approaches to about 10 for Group B particles and about 25 for Group A particles, U_{mf} increases sharply, indicating that particles are not likely to become fluidizable, which has been pointed out by McDonough et al. [6]. For mini- and micro-channel beds, U_{mf} also increases with reducing the ratio D_t/d_p [4], but the value is obviously lower than that of the millimeter- to centimeter-scale fluidized beds even under the same D_t/d_p ratios.

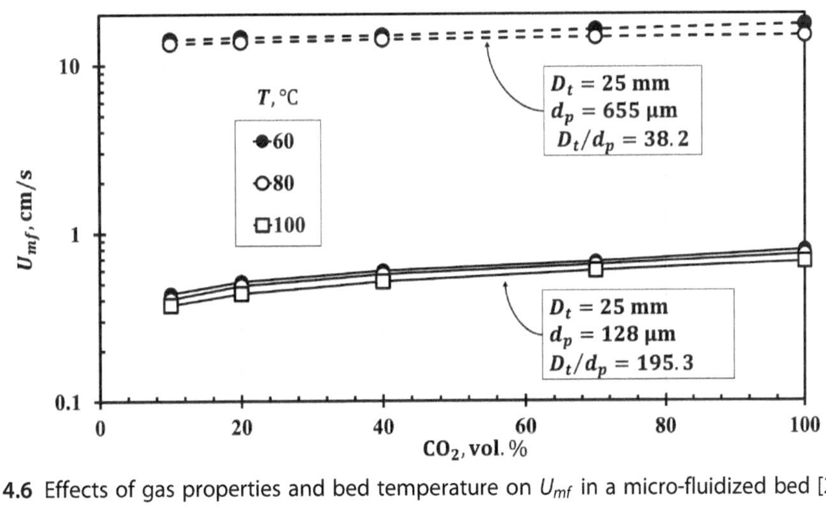

Figure 4.6 Effects of gas properties and bed temperature on U_{mf} in a micro-fluidized bed [21].

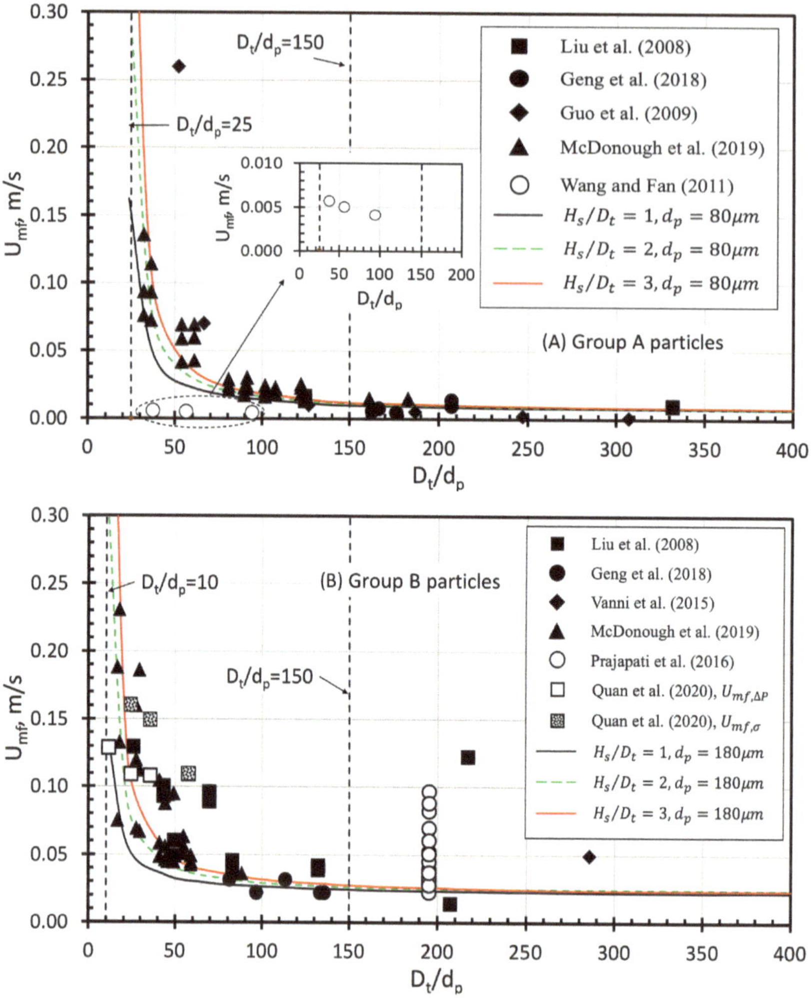

Figure 4.7 Variations of minimum fluidization velocities with D_t/d_p for (A) Group A particles and (B) Group B particles under various conditions [2].

This discrepancy may partly be due to the better radial uniformity in mini-channel beds, but further studies are needed to verify this.

4.3.2 Prediction of minimum fluidization velocity

Due to the wall effect, the correlations of minimum fluidization velocity commonly used for macro-fluidized beds are obviously not suitable for micro-fluidized beds. To develop a correlation of minimum fluidization velocity in micro-fluidized beds, Rao et al. [22] modified the Ergun equation by introducing Janssen's wall friction coefficient (the ratio of the normal stress exerted by the particles on the wall to the vertical normal stress). Based on this modified equation, the force balance under minimum fluidization conditions is given by

$$\left[1.75\frac{1-\varepsilon_{mf}}{\varepsilon_{mf}^{3}} - 120.4\tan\varphi\left(\frac{\phi d_p}{D_t}\right)\left(\frac{H_s}{D_t}\right)\right]\mathrm{Re}_{mf}^{2} + \left[150\frac{\left(1-\varepsilon_{mf}\right)^{2}}{\varepsilon_{mf}^{3}} - 2440\tan\varphi\left(\frac{\phi d_p}{D_t}\right)\left(\frac{H_s}{D_t}\right)\right]\mathrm{Re}_{mf} = \left(1-\varepsilon_{mf}\right)Ar \tag{4.5}$$

The above equation considers the wall effect by introducing the particle angle of internal friction φ and ratios D_t/d_p and H_s/D_t into the Ergun equation. The formula fits reasonably well with the experimental data of Liu et al. [3], except for particles smaller than 120 μm in a bed with an inner diameter of 16 mm. However, the use of this formula requires the value of friction angle between particles and wall, which is often unavailable because it is typically determined by specific rheology instruments.

Alternatively, Guo et al. [11] proposed an empirical correlation of U_{mf} based on their experimental data in micro-fluidized beds. The correlation reads

$$U_{mf} = \left[\frac{H_s}{d_p}e^{-6.312+242.272/(D_t/d_p)} + 1\right]\frac{7.169\times10^{-4}d_p^{1.82}\left(\rho_p-\rho_f\right)^{0.94}g}{\rho_g^{0.06}\mu_g^{0.88}} \tag{4.6}$$

Vanni et al. [23] found that the U_{mf} calculated by this empirical equation was much higher than the experimental data for their very dense particles ($\rho_p = 19{,}300$ kg/m^3), and the discrepancy tended to be more significant for beds of smaller diameters. They speculated that the relative weight assigned to the powder density in Eq. (4.6) was probably too high. Therefore, although it is convenient to use, the applicability of Eq. (4.6) is limited.

Han et al. [2] made efforts to develop a more generalized correlation to predict U_{mf} in micro-fluidized beds. The correlation is developed to predict the variations of U_{mf} with bed size and particle properties, represented mainly by the ratios D_t/d_p and H_s/D_t. The model assumes that the bed pressure drop can be presented by the Cheng equation (Table 4.1), and that at the minimum fluidization condition, the force balance can be written as

$$A_E\frac{\left(1-\varepsilon_{mf}\right)^{2}}{\varepsilon_{mf}^{3}}\frac{\mu_g U_{mf}}{\phi^{2}d_p^{2}} + B_E\frac{\left(1-\varepsilon_{mf}\right)}{\varepsilon_{mf}^{3}}\frac{\rho_g U_{mf}^{2}}{\phi d_p} = \left(\rho_p-\rho_g\right)g\left(1-\varepsilon_{mf}\right) \tag{4.7}$$

where A_E and B_E are related to D_t/d_p. Multiplying the two sides by $\frac{\rho_g(\phi d_p)^3}{\mu_g}$, Eq. (4.7) becomes

$$A_e\mathrm{Re}_{mf} + B_e\mathrm{Re}_{mf}{}^2 = Ar \tag{4.8}$$

where $\mathrm{Re}_{mf} = \frac{\rho_g U_{mf}\phi d_p}{\mu_g}$, $Ar = \frac{(\rho_p - \rho_g)\rho_g g(\phi d_p)^3}{\mu_g{}^2}$, $Ae = A_E\frac{1-\varepsilon_{mf}}{\varepsilon_{mf}{}^3}$, $Be = B_E\frac{1}{\varepsilon_{mf}{}^3}$. Solving Eq. (4.8) gives

$$U_{mf} = \mathrm{Re}_{mf}\frac{\mu_g}{\rho_f\phi d_p} = \frac{\mu_g}{\rho_f\phi d_p}\left[\frac{-A_e + \sqrt{A_e{}^2 + 4ArB_e}}{2B_e}\right] \tag{4.9}$$

It is clear that estimating U_{mf} requires the knowledge of ε_{mf}, in addition to properties of gas and solids as well as D_t/d_p. Note that ε_{mf} in micro-fluidized beds is also dependent on D_t/d_p (see Chapter 2). When no experimentally measured ε_{mf} is available, the correlations in Eqs. (2.10) and (2.11) presented in Chapter 2 can be employed. Fig. 4.8 compares the calculated U_{mf} with the experimental results. Note that the experimental data are obtained under considerably different conditions. Therefore, the calculation results can be considered to be reasonably well in agreement with the experimental data.

To further illustrate the effects of D_t/d_p and H_s/D_t on U_{mf}, Fig. 4.9 shows the predicted dimensionless gas velocities U^* $(=\mathrm{Re}/Ar^{\frac{1}{3}})$ versus dimensionless particle diameter $d_p{}^*(=\mathrm{Ar}^{1/3})$ for spherical Group B particles ($\phi = 1$). Here, $\mathrm{Re} = \rho_g d_p U_{mf}/\mu_g$, in which U_{mf} is calculated by Eq. (4.9) for micro-fluidized bed (i.e., $U_{mf}{}^*$) and by the

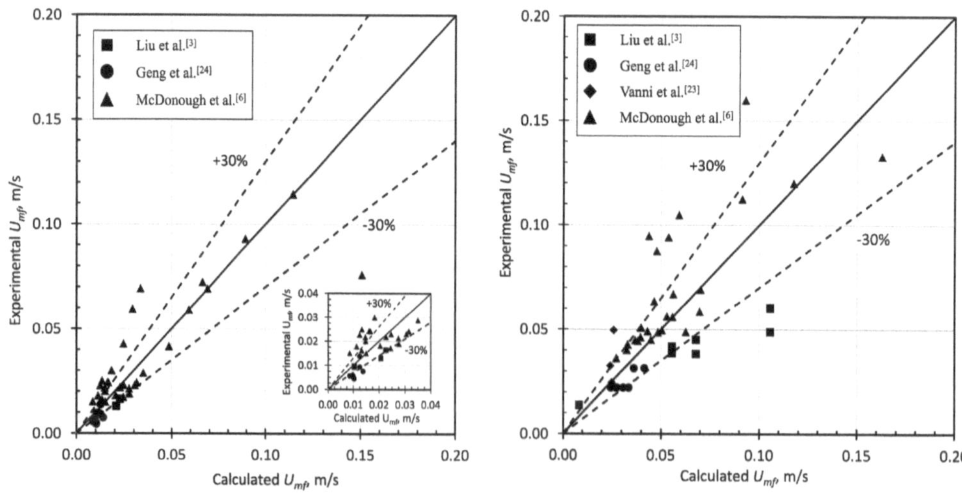

Figure 4.8 Comparisons of the calculated and experimental minimum fluidization velocities. Group A particles. Group B particles.

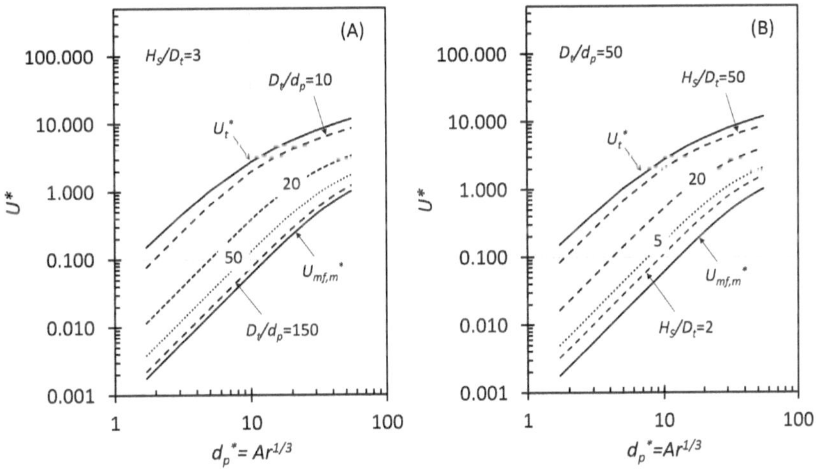

Figure 4.9 The predicted variations of the dimensionless gas velocities U^*versus d_p^* with (A) ratio D_t/d_p and (B) ratio H_s/D_t for spherical Group B particles.

Wen and Yu equation [25] for macro-fluidized beds (i.e., U_{mfm}^*). The dimensionless terminal velocity U_t^* is predicted by the Haider and Levenspiel equation [26]. In Fig. 4.9A, H_s/D_t is set as a constant of 3, and D_t/d_p is varied from 10 to 150. In Fig. 4.9B, H_s/D_t is changed from 2 to 50 with a constant D_t/d_p of 50. The results show that when $D_t/d_p > 150$, U_{mf}^* is very close to U_{mfm}^* in macro-fluidized beds and changes little with D_t/d_p and H_s/D_t. When D_t/d_p is below 150, U_{mf}^* increases as D_t/d_p is reduced. When D_t/d_p is close to 10 for Group B particles and about 25 for Group A particles, U_{mf}^* approaches to U_t^*. Further reducing D_t/d_p, Eq. (4.9) would yield a negative solution, implying that fluidization will not be possible under these circumstances. Fig. 4.9B also shows that when D_t/d_p is a constant, U_{mf}^* increases with increasing H_s/D_t. When H_s/D_t is increased to 50, the bed becomes too tall, so the wall effect becomes exceedingly strong that gas flow can no longer move particles. It is concluded that in micro-fluidized beds, the wall effect will become so prominent that fluidization becomes extremely difficult or even impossible when D_t/d_p is too small and/or H_s/D_t is too large. Therefore, from the perspective of practical operations, the ratios D_t/d_p and H_s/D_t are recommended to be 50–150 and 2–4, respectively.

4.4 Particulate fluidization

Following the minimum fluidization, particles start moving as the bed expands with increasing gas velocity. Although some small bubbles may appear at the walls or near the distributor, they cannot sustain and quickly collapse into a particle phase. The bed is essentially bubble-less and homogeneous across the cross section. This fluidization state is called "particulate fluidization." Particulate fluidization in gas-solid large-scale systems is typically observed only for Group A particles. In micro-fluidized beds, the particulate fluidization does occur for particles with the corresponding $Re_{mf} < 0.4$ [2].

Fig. 4.10A presents variations of overall bed voidage, dense phase voidage, and bubble phase fraction for an FCC particle in a 3-mm-diameter micro-fluidized bed [4]. The data are obtained by the bed collapsing method. Fig. 4.10B shows the ΔP and bed expansion for an FCC particle in a 66-mm fluidized bed [27]. These experimental methods identify the fixed bed, particulate, and bubbling fluidization regimes with increasing superficial gas velocity. In the particulate fluidization, the bed expands

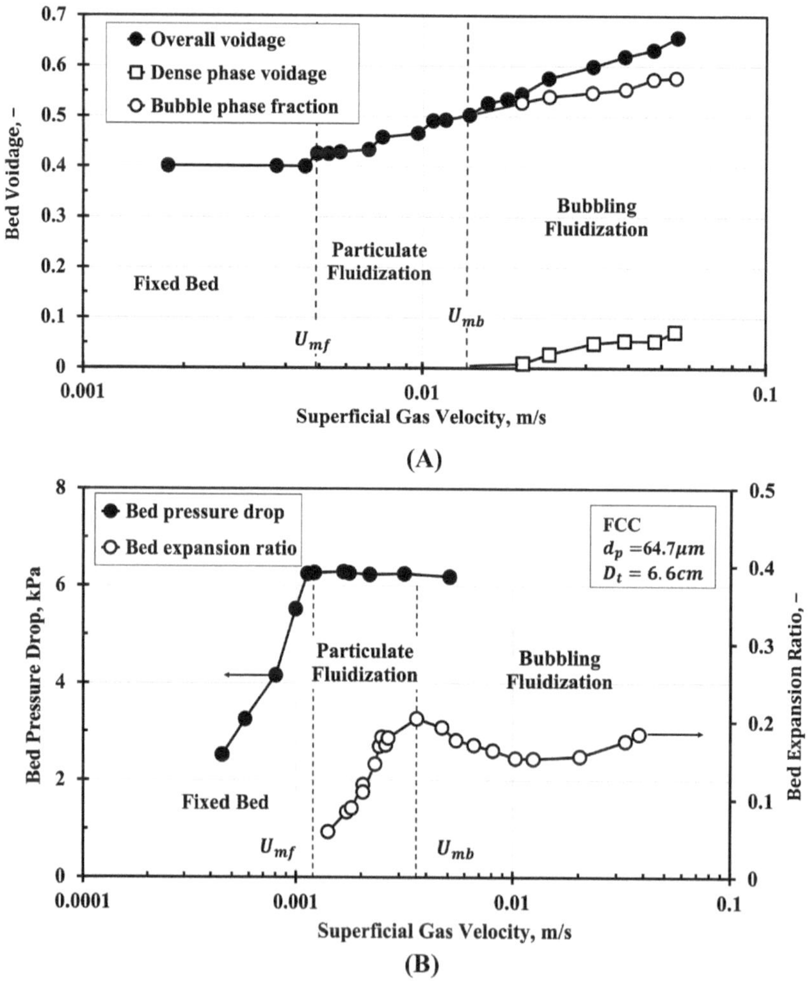

Figure 4.10 Fluidization regimes for FCC particles. (A) Bed voidage versus gas velocity for FCC particles in a 3-mm micro-fluidized bed [4]. (B) Bed pressure drop and expansion for FCC particles in a 66-mm fluidized bed [27].

smoothly with no bubbles observed. As the superficial gas velocity increases further to the minimum bubbling velocity U_{mb}, bubbles begin to form (i.e., the bubble fraction starts to increase from zero), and the bed height begins to decrease (i.e., the bed expansion starts declination after reaching a maximum). The bed voidage variation in particulate fluidization in micro fluidized beds where $U_{mf} < U_g < U_{mb}$ follows the Richardson-Zaki equation [28] given by

$$\frac{U_g}{U_t} = \varepsilon^n \tag{4.10}$$

The Richardson-Zaki exponent n is related to D_t/d_p by

$$n = \left(4.4 + \frac{18}{D_t/d_p}\right)\mathrm{Re}_t^{-0.03} \quad \mathrm{Re}_t = \frac{\rho_g d_p U_t}{\mu_g} \tag{4.11}$$

Wang and Fan [4] found that the Richardson-Zaki exponent slightly decreased with channel size, and the experimentally determined n was slightly higher than the prediction of the above equation.

4.5 Bubbling fluidized bed

4.5.1 The onset of bubbling fluidization

With a further increase in gas velocity, bubbles constantly form across the whole distributor and rise to the bed surface across the bed cross section. During the rising process, bubbles coalesce and break up frequently. Once stabilized, the bubble size remains relatively stable.

Direct visual observation is the simplest way to determine the onset of bubbling fluidization. In this way, the minimum bubbling velocity U_{mb} is determined by the superficial gas velocity at which the first bubble is visually observed at the bed surface [23]. The accuracy of this method is arguable, since it is rather subjective [3]. Therefore, the experimental methods shown in Fig. 4.10A and B can be used to estimate U_{mb}. Additionally, U_{mb} can also be determined as the superficial gas velocity at which the standard deviation of bed pressure drop starts to grow from zero [3,5,6], as shown in Fig. 4.11. Fig. 4.11 also shows that U_{mb} approximates U_{mf} for the 100-mm-diameter bed, but U_{mb} is slightly lower than U_{mf} for the 20-mm-diameter bed. This result suggests that gas bubbles appear before minimum fluidization in micro-fluidized beds for large particles, which has been confirmed by several researchers [3,6]. We will elaborate on this characteristic later.

Fig. 4.12 presents the variations of U_{mb} with the bed diameter for three kinds of particles at three static bed heights. The results indicate that for fluidized beds of $D_t > 20$ mm, U_{mb} is almost independent of D_t. When $D_t < 20$ mm, U_{mb} increases

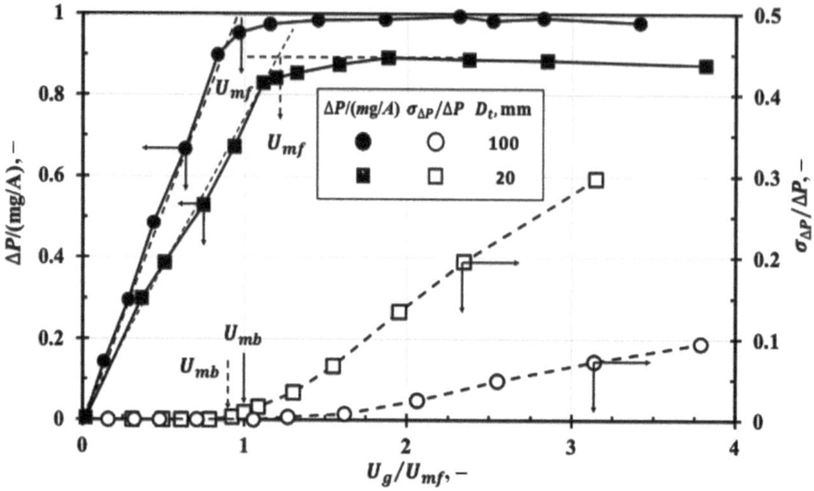

Figure 4.11 Variations of dimensionless pressure drop $\Delta P/(mg/A)$ and the corresponding standard deviation $\sigma_{\Delta P}/\Delta P$ with dimensionless gas velocity U_g/U_{mf}.

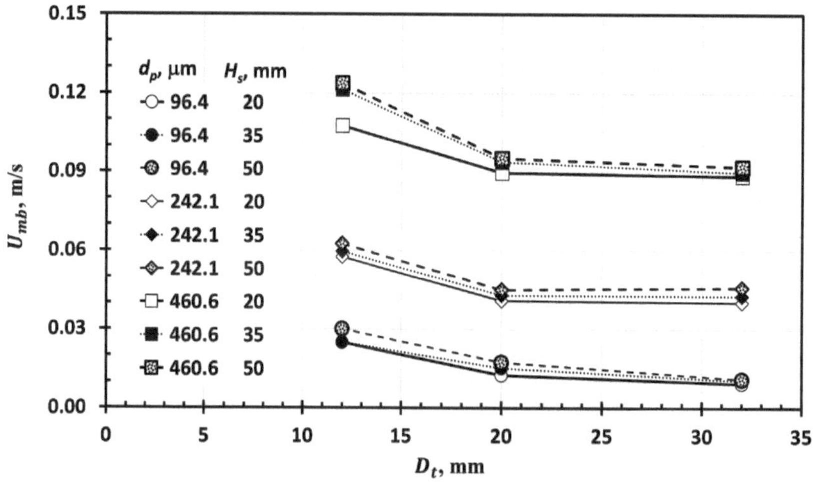

Figure 4.12 Variation of U_{mb} with D_t [3].

notably with decreasing D_t. For the same bed diameter, U_{mb} increases with increasing particle diameter and static bed height.

The experimental data of Liu et al. [3], McDonough et al. [6], Quan et al. [5], and Vanni et al. [23] are plotted in Fig. 4.13 to reveal the influence of the wall effect on U_{mb} in micro-fluidized beds. The results show that for the large ratios of D_t/d_p, U_{mb} is not influenced by D_t/d_p, but for the beds of D_t/d_p smaller than about 150, U_{mb} increases notably as ratio D_t/d_p decreases. Fig. 4.13 also includes the results of micro-channel

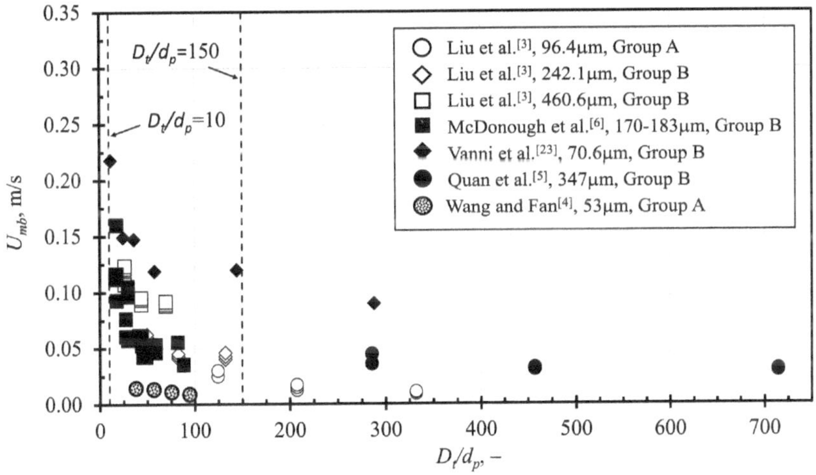

Figure 4.13 Variation of the minimum bubbling velocity with the D_t/d_p ratio.

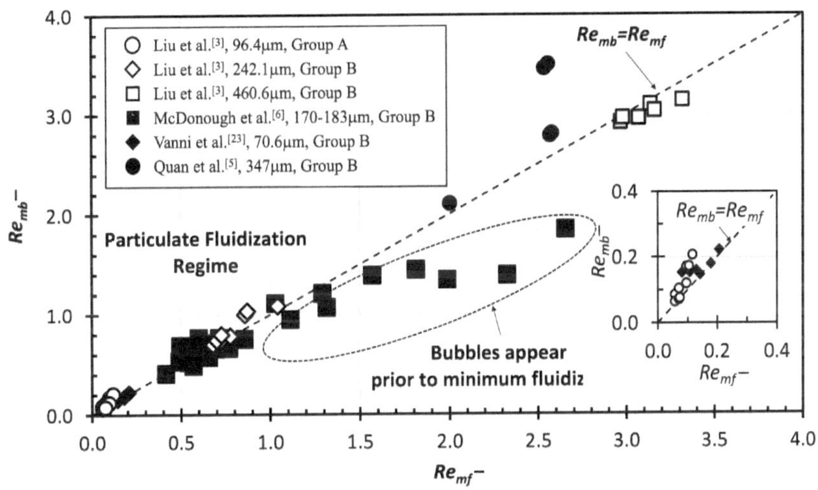

Figure 4.14 Relationship between Re_{mb} and Re_{mf} in micro-fluidized beds.

reactors from Wang and Fan [4], showing that in the mini- and micro-channel beds, the U_{mb} has a similar variation trend, but the value of U_{mb} is lower than that of micro-fluidized beds.

Fig. 4.14 further examines the relationship between U_{mb} and U_{mf} in terms of Reynold numbers Re_{mb} and Re_{mf}. It is clear that under the experimental conditions, both Re_{mb} and Re_{mf} are smaller than 4. When Re_{mf} is smaller than approximately 0.4, there is a very narrow particulate fluidization region where $Re_{mb} > Re_{mf}$ (see the insert). When Re_{mf} varies from approximately 0.4–3.0 corresponding to particles of

100–460 μm, most likely Re_{mb} is smaller than Re_{mf}; that is, gas bubbles will likely appear before the minimum fluidization. This result seems contrary to the consensus that Group B particles would have $U_{mb} \approx U_{mf}$ in large-scale fluidized beds. However, in micro-fluidized beds, it is true that U_{mb} can be slightly smaller than U_{mf}. In fact, McDonough et al. [6] observed with a high-speed camera that the bed pressure drop fluctuated slightly as gas bubbles formed near the wall and near the gas distributor prior to the minimum fluidization. When $Re_{mf} > 3$, for example, by increasing the particle size beyond 460 μm, Re_{mb} approximates Re_{mf}, suggesting that there will be neither particulate fluidization zone nor advance transition to bubbling zone for large particles; that is, the bubbling fluidization starts immediately after the bed is fluidized.

4.5.2 Prediction of minimum bubbling velocity

To develop a relatively generalized correlation for the prediction of U_{mb}, the experimental U_{mb} is plotted as a function of U_{mf} in Fig. 4.15. It shows that generally U_{mb} increases first proportionally and then slowly as U_{mf} increases. Based on the experimental data, the following empirical correlation is developed [2]:

$$\frac{U_{mb}}{U_{mf}} = 1.099 - 5.21\left(\frac{H_s}{D_t}\right)^{0.44}\left[e^{\frac{0.7}{D_t/d_p}} - 1\right] \tag{4.12}$$

For a total of 69 data points, Eq. (4.12) has a correlation coefficient of 0.96 and a relative average absolute error of 10.9%. From Eq. (4.12), the minimum bubbling fluidization velocity U_{mb} can be estimated once the minimum fluidization velocity U_{mf} is measured or estimated by Eq. (4.9).

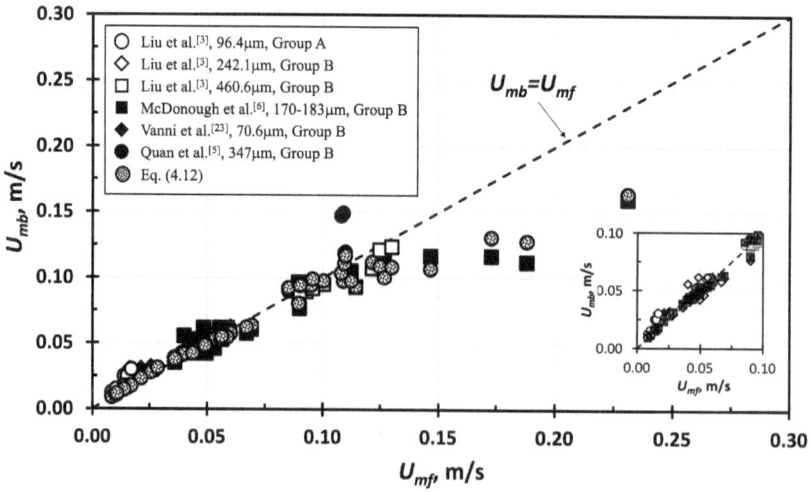

Figure 4.15 Relationship between U_{mb} and U_{mf}.

The minimum bubbling fluidization velocity can also be derived from the wave concept [4]. Consider that there are two types of wave velocity in a gas–solid fluidized bed: continuity wave velocity u_{cl} and dynamic wave velocity u_d, which can be expressed as

$$u_{cl} = U_g \frac{3-2\varepsilon}{\varepsilon} \qquad u_d = \sqrt{\frac{E}{\rho_p}} \tag{4.13}$$

where ε is the bed voidage, E is the elasticity modulus of the particle phase in the bed, and ρ_p is the particle density. In principle, a gas–solid fluidized bed is stable without a bubble formation if $u_{cl} < u_d$ but becomes unstable when a gas bubble is formed if $u_{cl} > u_d$. So, at the minimum bubbling fluidization condition $u_{cl} = u_d$, we have

$$U_{mb} = \frac{\varepsilon_{mb}}{3-2\varepsilon_{mb}} \sqrt{\frac{E}{\rho_p}} \tag{4.14}$$

where ε_{mb} is the bed voidage at the minimum bubbling fluidization condition. According to Wang and Fan [4], ε_{mb} decreases with increasing bed diameter. Therefore, the above equation indicates that U_{mb} decreases with D_t, which is consistent generally with the experimental results discussed above. However, using the above equation requires the knowledge of E and ε_{mb}, which may not be readily available for particles used for the investigation.

4.5.3 Bubble size

Similar to what occurs in macro-fluidized beds, small bubbles, once formed at the distributor, will grow larger as they rise toward the bed surface in micro-fluidized beds. Bubbles coalescence and breakup occur during the rising process. Eventually, a maximum stable bubble is reached when coalescence and breakup of bubbles attain equilibrium. The size of the maximum stable bubble d_{bm} for group A particles in a gas–solid fluidized bed is affected by the bed diameter, superficial gas velocity, and particle properties. The correlation of Mori and Wen [29] can be used to estimate the maximum stable bubble size

$$d_{bm} = 0.374\left[\beta A\left(U_g - U_{mf}\right)\right]^{0.4} \tag{4.15}$$

where A is the cross-sectional area of the bed, and β is the coefficient corresponding to the vertical distance between bubbles to avoid bubble coalescence. For gas–solid large-scale fluidized beds, $\beta = 0.4$ according to Mori and Wen [29], but Wang and Fan [4] found that β is affected by the bed size. Based on the experimentally measured

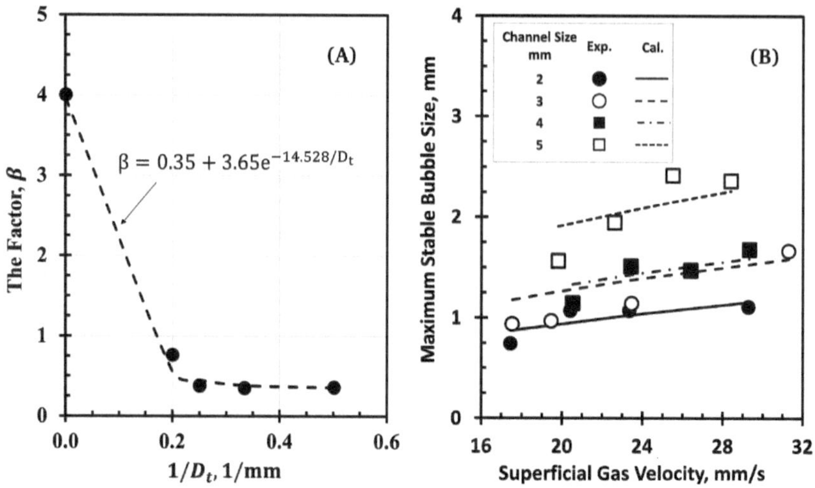

Figure 4.16 The coefficient β and the maximum stable bubble size in micro-fluidized beds. (A) The coefficient β and (B) the maximum stable bubble size in micro-fluidized beds.

maximum stable bubble size shown in Fig. 4.16B, the coefficient β is determined and given in Fig. 4.16A. It shows that β decreases from 4.0 to a roughly constant value of 0.37 when D_t is reduced or $1/D_t$ is increased. Based on the data shown in Fig. 4.16A, an empirical correlation for β is developed as follows:

$$\beta = 0.35 + 3.65e^{-14.528/D_t} \tag{4.16}$$

where D_t is in mm. Fig. 4.16B shows that the maximum stable bubble size increases with increasing gas velocity and bed size, and the calculated and experimental measured maximum stable bubble sizes are in good agreement.

4.6 Slugging fluidized bed

4.6.1 The onset of slugging fluidization

In gas–solid bubbling fluidized beds, increasing gas velocity further from the bubbling regime results in the continuous growth of bubbles in size and number due to the coalescence and breakup of bubbles. When the bubble size becomes comparable to the bed size or big enough to occupy a large portion of the bed cross section (i.e., $0.66D_t$), we define that the fluidization enters the slugging regime. In a slugging fluidization regime, the particle bed height varies greatly due to continuous geyser eruptions at the bed surface. When the slugging forms, the bubbles may have attained their maximum stable sizes (i.e., $d_{bm} \geq 0.66D_t$) or continue to grow. Since it takes time for a slug to grow up, the slugging fluidization can only be formed in a deeper bed; that is, a minimum bed height is needed to form the slugging fluidization. According to experimental data obtained in fluidized beds of four bed diameters (5−30 cm), Baeyens and Geldart [30] gave the height $H_s = 60D_t^{0.185}$ ($H_s$and $D_t$incm) at which complete slugging occurs. On this basis, we can estimate that slugging would take place when the ratio H_s/D_t reaches an apparently impossible large number 67 in a 20-mm-diameter fluidized bed. However, this estimated H_s/D_t ratio is too high to be practical and certainly not supported by the experiments reported on micro-fluidized beds in the literature. Slugs are observed in micro-fluidized beds of about 20 mm diameter even at significantly low H_s/D_t ratios (i.e., 1−5) [5,6]. Therefore, we are ensured that micro-fluidized beds are characteristically different from macro-fluidized beds.

When slugging occurs in fluidized beds, the bed pressure drop fluctuates considerably, characterized by an increased standard deviation and a high dominant frequency [5,6]. Fig. 4.17 shows typical power spectral density (PSD) functions for a micro-fluidized bed of 4 mm diameter. It shows that the magnitude of the pressure fluctuation increases drastically in slugging conditions, but the dominant frequency reduces with increasing gas velocity. At the full slugging, the magnitude of the pressure fluctuation is at its maximum. Beyond the full slugging, the magnitude of the pressure fluctuation reduces significantly as the gas velocity increases continuously. Quan et al. [5] found that a small bed diameter led to a high dominant frequency, while McDonough et al. [6] found that a large H_s/D_t ratio led to a small dominant frequency. Both researchers found that the dominant frequencies in the slugging regime were in the range of 5−12 Hz.

The onset of slugging fluidization can be determined by visualization or quantitatively analyzing bed pressure fluctuation signals measured experimentally. The minimum slugging velocity can be defined as the superficial gas velocity at which the standard deviation of bed pressure drop starts to fluctuate [31] or the dominant frequency of PSD reaches a maximum and becomes independent of U_g [30]. For micro-fluidized

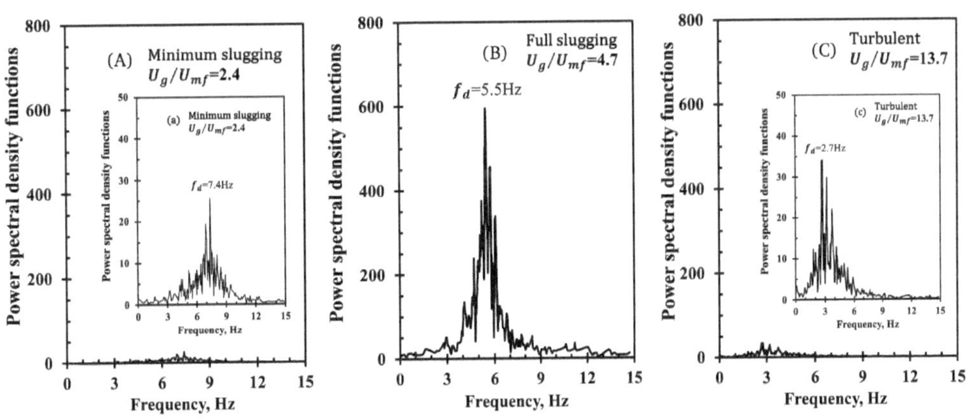

Figure 4.17 Power spectral density functions for a 4-mm-diameter micro-fluidized bed.

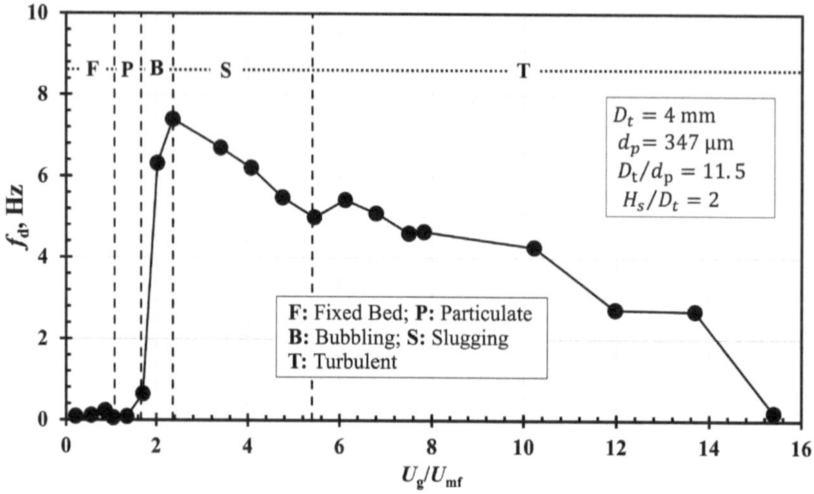

Figure 4.18 Variation of the dominant frequency with U_g/U_{mf} in a 4-mm-diameter micro-fluidized bed.

beds, McDonough et al. [6] observed that the dominant frequency might reach a maximum value and then decreased with increasing U_g due to the additional drag force induced by the walls, so they defined the gas velocities corresponding to the maximum and constant dominant frequencies as the transition to slugging and the full slugging velocities, respectively. A typical relationship between the dominant frequency of PSD and the gas velocity in a micro-fluidized bed is presented in Fig. 4.18. It shows that fluidization regimes in a micro-fluidized bed can be demarcated by the variation of the dominant frequency with the gas velocity.

4.6.2 Prediction of slugging velocity

Han et al. [2] summarized the influence of ratio D_t/d_p on the minimum slugging velocity U_{ms} in Fig. 4.19. The results indicate that when $D_t/d_p < 150$, U_{ms} increases with reducing D_t/d_p due to an increasingly pronounced wall effect. When D_t/d_p becomes larger than 150, U_{ms} is toward stabilizing, and further increasing D_t/d_p may lead to an increase in U_{ms} as commonly observed in macro-fluidized beds. It suggests that the observed variations of U_{ms} with D_t in micro- and macro-fluidized beds are opposite [32]. This phenomenon again demonstrates that the micro-fluidized bed has unique characteristics due to the wall effect, compared to the macro-fluidized bed with a negligible wall effect.

Fig. 4.20 correlates $Re_{ms}$with Re_{mf}. Although the data are relatively limited and scattering, a general trend can still be identified. Note that some data points are right on the line of $Re_{ms} = Re_{mf}$, suggesting that slugging occurs immediately as soon as beds are fluidized. These beds are operated at small D_t/d_p or high H_s/D_t ratios—in

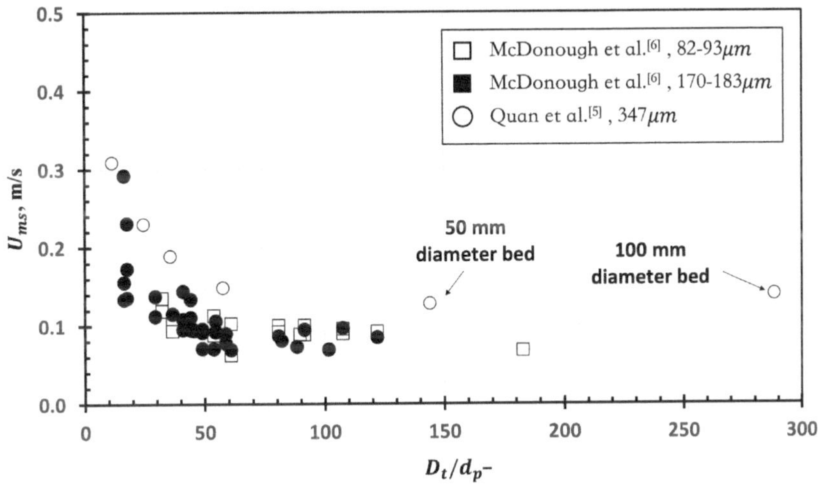

Figure 4.19 Influence of ratio D_t/d_p on the slugging fluidization velocity.

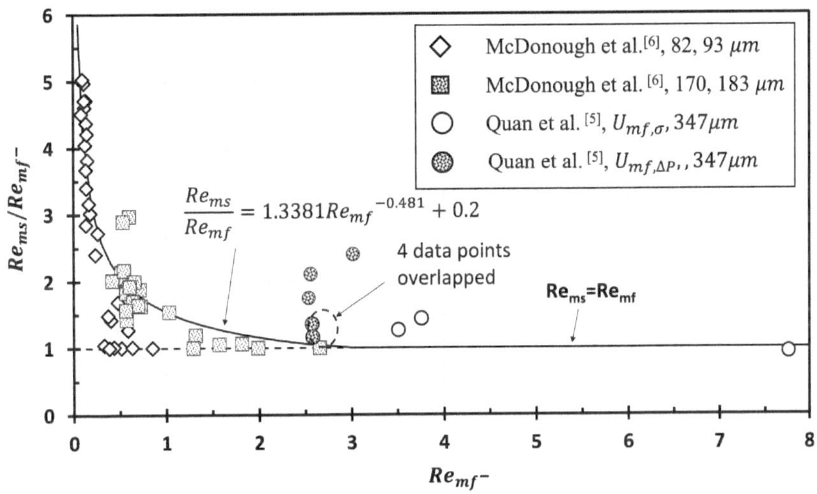

Figure 4.20 Relationship between Re_{ms} and Re_{mf}.

these beds, the wall effects are strong. As McDonough et al. [6] noted, the bubbling fluidization regime does not exist under some circumstances because of the easy formation and development of slugs in operating conditions such as $D_t/d_p < 50$ and $H_s/D_t > 3$. For other operating conditions, Re_{ms} is higher than Re_{mf}, and a distinctive bubbling fluidization regime exists before forming a slugging fluidization state. As shown in Fig. 4.20, when Re_{mf} is larger than about 2.0, no bubbling fluidization exists, and beds enter slugging operation directly once fluidized.

Fig. 4.20 also shows that the increase in Re_{ms}/Re_{mf} (equivalent to U_{ms}/U_{mf}) accelerates with decreasing Re_{mf} (equivalent to U_{mf}) progressively. As U_{mf} increases with decreasing bed diameter, a more significant difference between U_{ms} and U_{mf} in small beds is expected. This feature is obviously different from that found in macro-fluidized beds.

Based on the experimental data available, Han et al. [2] proposed the following empirical correlation:

$$Re_{ms}/Re_{mf} = 1.3381 Re_{mf}^{-0.481} + 0.2 \tag{4.17}$$

Compared with experimental data, Eq. (4.17) has a correlation coefficient of 0.898 and a relative absolute deviation of 22.8%. Although this correlation is not accurate enough, it helps estimate U_{ms} when no more accurate information is available.

4.7 Turbulent fluidized bed

4.7.1 The onset of turbulent fluidization

With increasing gas velocity further beyond the bubbling or slugging regime, the bubbling action becomes increasingly vigorous, and gradually, the bubble coalescence is taken over by breakup. As a result, the bed pressure fluctuations peak, decrease sharply or gradually, and then level off. In this case, the gas–solid flow transits to a state where more particles are embedded in the gas voids and more gas are inside the dense phase. At this moment, fluidization enters a typical turbulent fluidization regime.

Several experimental methods have been proposed to identify the onset velocity of turbulent fluidization (U_c), including visualization, bed expansion, voidage fluctuations, and pressure fluctuations [33]. For gas–solid mini- and micro-channel fluidized beds, Wang and Fan [4] reported that the U_c values determined by the methods of bed expansion, voidage fluctuation, and pressure fluctuation were different and ranged from 0.04 to 0.10 m/s using FCC particles. Dang et al. [34] also observed differences in U_c values determined by the standard deviations of solids concentration, absolute, and differential pressure fluctuations, as illustrated in Fig. 4.21. Nevertheless, given its simplicity, the time-series signal of pressure fluctuations is the most commonly measured and analyzed to determine U_c in gas–solid fluidization research.

Fig. 4.22 presents U_c as a function of D_t/d_p with the experimental data from various researchers. It shows that U_c generally decreases with reducing ratio D_t/d_p, indicating that the advance onset of turbulent fluidization in micro-fluidized beds occurs compared to macro-fluidized beds. This advancement is essentially the result of the wall effect—the wall friction force promotes the fragmentation of slugs and bubbles into small bubbles, which in turn increases the intensity of turbulence of the gas-solid flow. Consequently, the turbulent fluidization is attained at a small gas velocity.

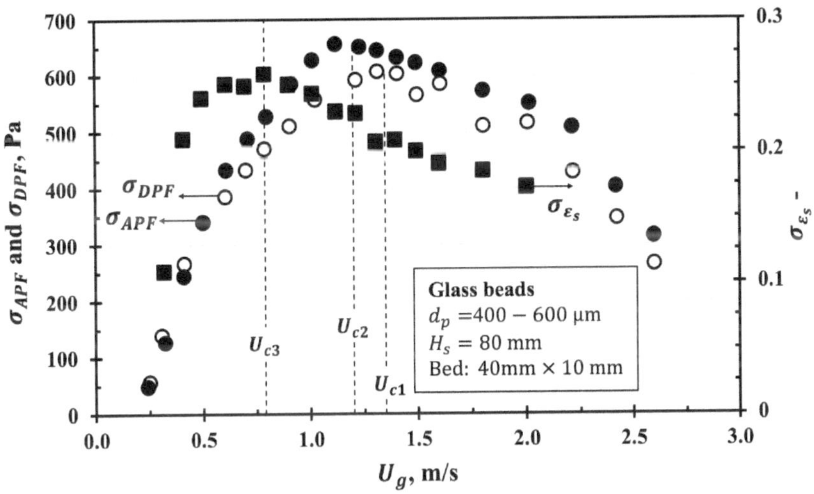

Figure 4.21 Determination of U_c based on the standard deviations of solid concentration, absolute and differential pressure fluctuations.

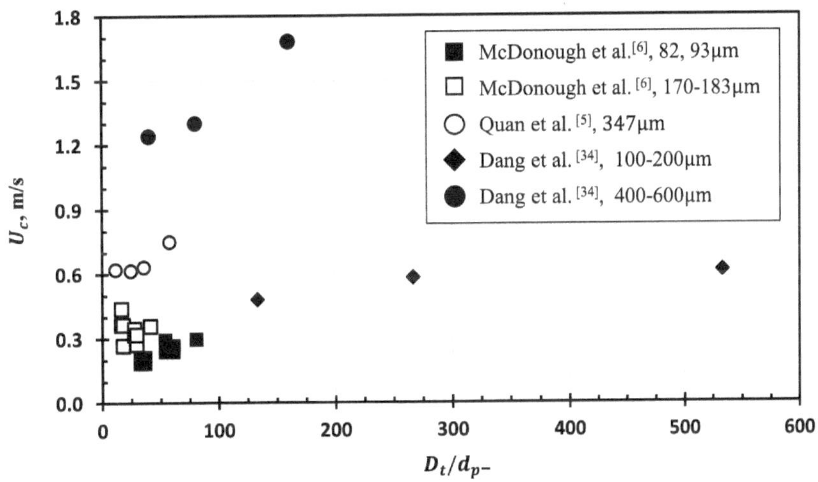

Figure 4.22 Variation of the transition velocity U_c with ratio D_t/d_p.

Operating micro-fluidized beds in the turbulent regime at significantly reduced gas velocities is a clear advantage over large-scale fluidized beds because the high intensity of turbulence greatly enhances mass and heat transfer.

Fig. 4.23 shows the local solid volume fraction and the corresponding standard deviation as a function of superficial gas velocity in three rectangular micro-fluidized beds [34]. It shows that the solid volume fraction decreases at a faster rate with

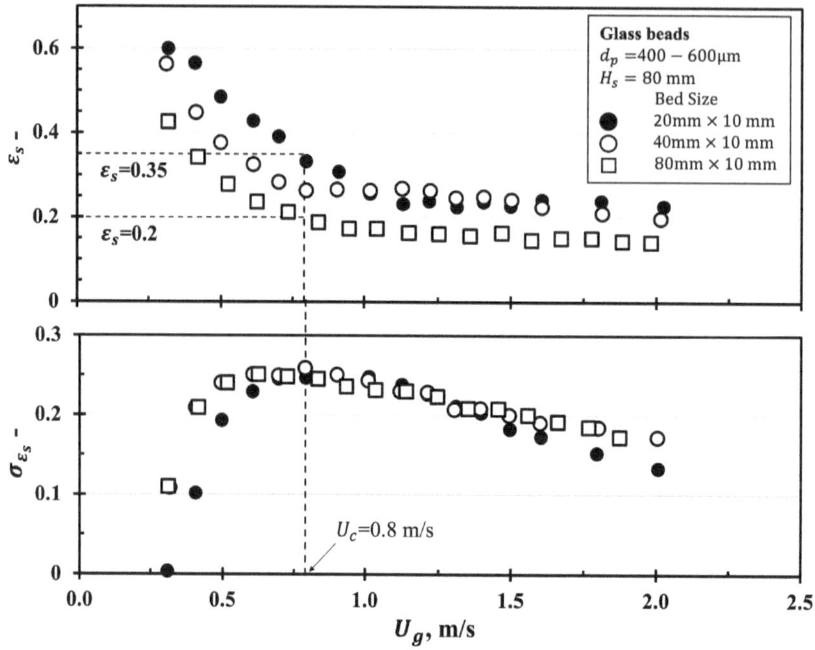

Figure 4.23 Local solid volume fraction (top) and its standard deviation (bottom) at 60 mm above the distributor as a function of superficial gas velocity for three micro-fluidized beds.

increasing gas velocity in the bubbling regime, and as the gas velocity approaches U_c, the change of the solid volume fraction becomes slower. As a result, the standard deviation reached its maximum value at U_c. The local solid volume fraction corresponding to U_c is between 0.20 and 0.35. Wang et al. [35] found that the solid volume fraction corresponding to U_c was 0.25 based on a numerical simulation. This solid volume fraction is slightly lower than the value of 0.35 that Bai et al. [36] concluded for gas–solid large-scale fluidized beds. Dang et al. [34] attributed this discrepancy to the difference between the local and bed average solid volume fractions used to develop these critical values. However, we believe that the lower critical solid volume fraction results from the wall effects. The wall effects provide additional forces to enhance the solid flow turbulence, thus making solid particles easier to attain the transition to turbulence.

4.7.2 Prediction of transition velocity U_c

Numerous correlations have been developed to predict U_c in macro-fluidized beds, which have been summarized by Bi et al. [33]. Almost all of these correlations, except those proposed by Cai et al. [37], do not consider bed diameter an influencing factor. However, in the equation of Cai et al. [37], the U_c is predicted to decrease with increasing bed diameter, contrary to the observations in micro-fluidized beds.

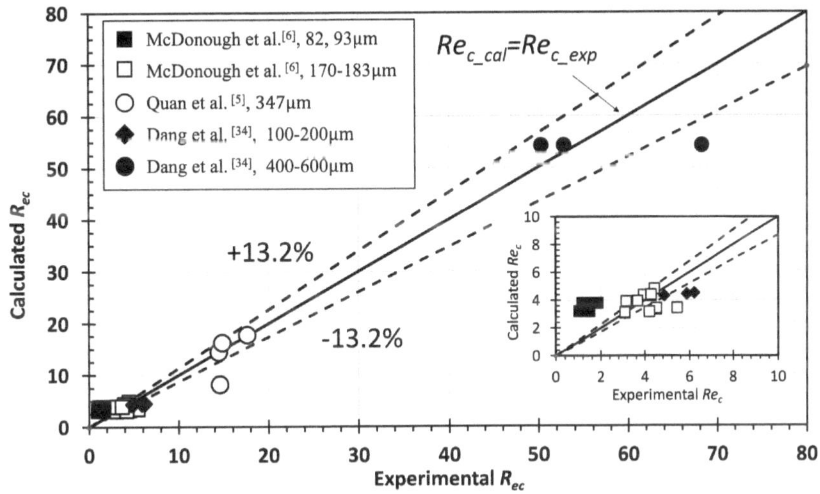

Figure 4.24 Comparison between the calculated and experimental Re_c.

Therefore, Han et al. [2] developed a new correlation of U_c in micro-fluidized beds based on the literature available data [5,6,34] as follows:

$$\mathrm{Re}_c = \left\{0.1477 + 0.8677\left(\frac{H_S}{D_t}\right)^{-0.068}\left[\exp\left(\frac{-1.3017}{D_t/d_p}\right) - 1\right]\right\}Ar^{0.5977} \tag{4.18}$$

The calculated and experimental values of Re_c are compared in Fig. 4.24. For a total of 36 data points, Eq. (4.18) has a correlation coefficient of 0.991and a relative absolute deviation of 13.2%. The ranges of parameters D_t/d_p, H_s/D_t, and Ar are 11.5–533, 2–4, and 42–19,996, respectively.

4.8 Distinction between micro- and macro-fluidized beds

As discussed, micro-fluidized beds differ characteristically from macro-fluidized beds in many fundamental aspects. These fundamental differences are due mainly to the wall effects of micro-fluidized beds, including pressure drop overshoot, pressure drop offset, delayed minimum, bubble and slugging fluidization, advance transition to turbulent fluidization, and increased bed voidage. Notably, the wall effect also leads to a close-to-plug gas flow in a micro-fluidized bed and essentially eliminates the backmixing of gas, thus providing a necessary condition for many chemical reactions to achieve the highest possible selectivity and product yield. Moreover, micro-fluidized beds open a new pathway for the development and scale-up of numerous chemical reactions. Table 4.2 summarizes these characteristics of micro-fluidized beds. Note that these

Table 4.2 Summary of hydrodynamic characteristics in micro-fluidized beds.

	Characteristics in micro-fluidized beds
Bed pressure drop	• Pressure drop overshoot before fluidization increases due to increased wall effects • Hysteresis occurs in ΔP versus U_g curves obtained from the ascending and descending gas velocity methods • Pressure drops in fixed beds deviate significantly from the estimates of the Ergun equation • Wall effects cause an increase in bed voidage and the flow nonuniformity • Wall frictional force can increase or decrease bed pressure drop depending on the solid circulation patterns, which are related to bed height and diameter • Wall effects offset part of the weight of particles to be supported by gas flows
Bed voidage	• Wall effects lead to an increase in bed voidage in both micro-fixed and fluidized beds
Minimum fluidization	• Wall effects lead to significant delay in minimum fluidization • U_{mf} increases with reducing D_t/d_p and increasing H_s/D_t
Bubbling fluidization	• U_{mb} increases with reducing D_t/d_p • U_{mb} can be slightly lower than U_{mf} for Group B particles. • The bubble size is small and breaks up more vigorously due to wall effects
Slugging fluidization	• U_{ms} increases with reducing D_t/d_p, opposite to the trend in macro-fluidized beds • Bed enters slugging regime without going through a bubbling regime (i.e., $U_{ms} \sim U_{mf}$) if $Re_{mf} > 2$ • Slugs form at H_s/D_t ratios significantly lower than predictions of Baeyens and Geldart [30]
Turbulent fluidization	• Wall effects promote gas–solids flow turbulence, thus the bed transitions to the turbulent regime at a reduced gas velocity and increased local voidage
Gas mixing	• Backmixing is small, and a close-to-plug flow is attainable when $D_t/d_p < 150$ • At conditions of close-to-plug flow, gas RTD parameters satisfy $E(t)_h \sigma_t \sim 0.4$
Solid mixing	• Fast solid mixing is achieved by solid internal circulation

characteristics are different from or opposite to those commonly encountered in macro-fluidized beds.

Analysis of hydrodynamic characteristics in micro-fluidized beds reveals that the wall effect becomes noticeable when D_t/d_p is less than about 150. The extensive experimental results of bed pressure drop and onset velocities of minimum, bubbling, slugging, and turbulent fluidization regimes depend strongly on D_t/d_p when

$D_t/d_p < 150$. On the other hand, when D_t/d_p is close to 10 for Group B particles or 25 for Group A particles, the minimum fluidization velocity, minimum bubbling fluidization velocity, slugging fluidization velocity, and the transition velocity to turbulent fluidization will converge. It suggests that in such low D_t/d_p ratios, the frictional force between particles and walls becomes so strong that particles will be non-fluidizable. Should the gas velocity be forcefully increased to overcome the wall confining force, the gas velocity would exceed the terminal velocity of particles, and stable fluidization would never be achievable.

The experimental data of σ_t^2 and $E(t)_h$ reported by Geng et al. [24] and Jia [38] are plotted against D_t/d_p in Fig. 4.25. Clearly, when D_t/d_p is lower than 150, the variance σ_t^2 is below 0.25, and the peak height $E(t)_h$ is above 1.0, indicating that a close-to-plug flow of gas is achieved according to the discussions in Chapter 3. When D_t/d_p is higher than 150, the variance σ_t^2 increases sharply, and the peak height $E(t)_h$ falls below 1.0, indicating that the extent of gas backmixing in fluidized beds increases.

Fig. 4.26 shows the variation of axial gas dispersion coefficient with D_t/d_p based on the literature data. Note that D_{ag} varies widely for fluidized beds, depending on the operating conditions covering bubbling, slugging, and turbulent flow regimes using various particles.

Nevertheless, the results show that D_{ag} usually is higher than 0.01 m^2/s when $D_t/d_p > 150$ and smaller than 0.001 m^2/s when $D_t/d_p < 150$. Therefore, we can conclude that gas flow in micro-fluidized beds will exhibit a close-to-plug flow behavior when D_t/d_p is below 150, and the gas backmixing would become more pronounced when $D_t/d_p > 150$.

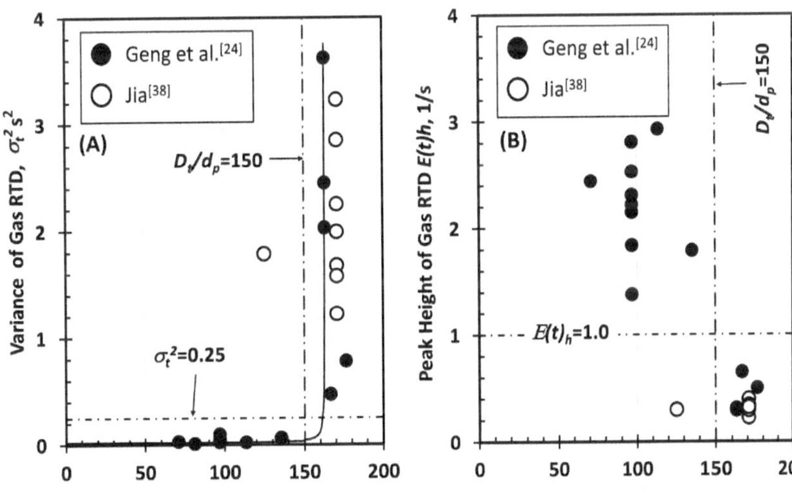

Figure 4.25 The dependence of σ_t^2 and $E(t)_h$ on the ratio D_t/d_p. (A) $\sigma_t^2 \sim D_t/d_p$. (B) $E(t)_h \sim D_t/d_p$.

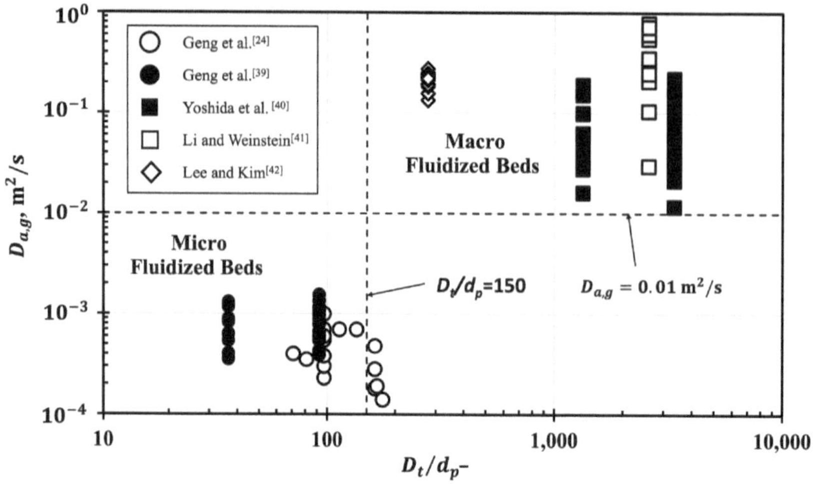

Figure 4.26 The dependence of D_{ag} on the ratio D_t/d_p based on the literature data [24,39–42].

In summary, micro-fluidized beds differentiate characteristically from macro-fluidized beds due to the wall effects. Micro-fluidized beds can be quantitatively defined if the condition $D_t/d_p = 10-150$ (group B particles) or 25–150 (group A particles) is satisfied.

4.9 Fluidization regime map for micro-fluidized beds

As discussed, micro-fluidized beds can be operated in flow regimes, including fixed bed, particulate, bubbling, slugging, and turbulent fluidization, depending on gas velocity and other operating parameters. Several fluidization regime maps have been proposed for specific micro-beds [4–6].

Fig. 4.27A shows the relationship of various fluidization regimes for FCC particles in mini- and micro-channels. It indicates that the transition of the fixed bed to the particulate regime, bubbling regime, slugging regime, and then to the turbulent regime will occur sequentially as superficial gas velocity increases continuously in the channel of above 2 mm. For the channels of 700 μm and 1 mm, the particulate fluidization is attained at low gas velocities after bubbles and slugs form and disappear after a certain time. With an increase in the gas velocity, the bubbling/slugging fluidization to particulate fluidization transition disappears, and the bed enters the slugging regime directly from the fixed bed. Fig. 4.27B shows the variation of fluidization regime with channel size and superficial gas velocity quantitatively.

Fig. 4.28 is a fluidization regime map for Group B particles [5]. It shows a narrow particulate fluidization regime (pseudo-fluidization in Fig. 4.28A) for small diameter beds before the beds enter the bubbling and slugging regimes as the gas velocity increases. For large diameter beds, no particulate fluidization regime exists, and the

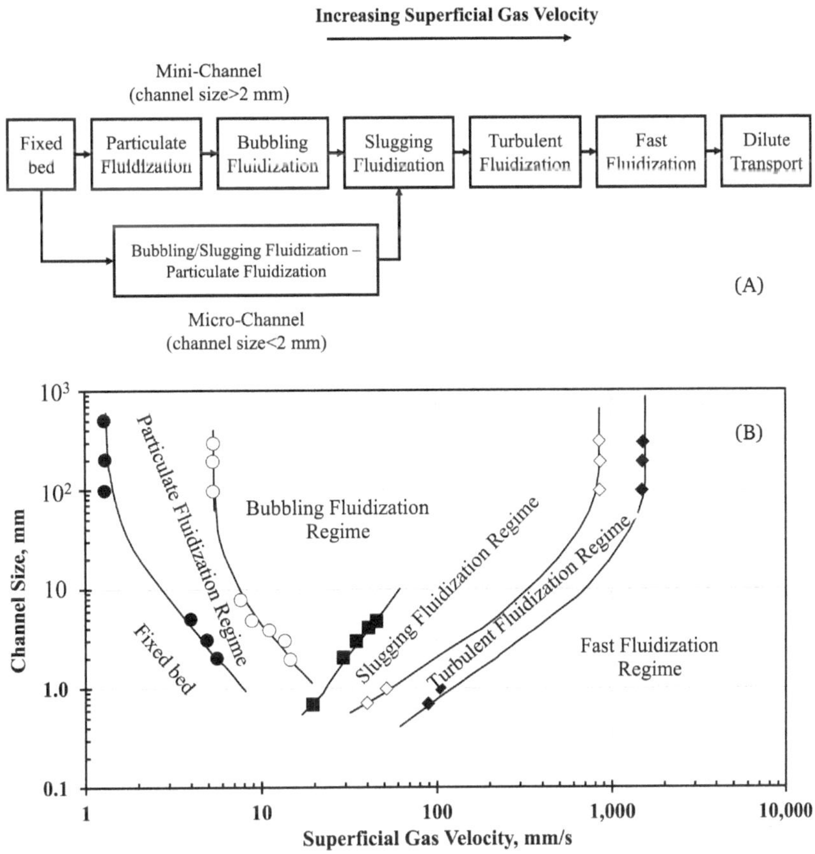

Figure 4.27 A fluidization map for FCC particles in mini- and micro-channel fluidized beds [4]. (A) The relationship of various fluidization regimes, (B) variation of fluidization regime with channel size and superficial gas velocity.

bed enters a slugging regime after a very narrow bubbling regime with increasing gas velocity. Fig. 4.28B explicitly indicates the dependence of fluidization regime on bed diameter in the range of 4—100 mm.

McDonough et al. [6] developed a set of fluidization maps for both Group A and B particles based on the analysis of bed pressure drop data, as shown in Fig. 4.29. For Group A particles, no particulate fluidization is observed for all H_s/D_t ratios. The minimum fluidization velocity decreases with increasing D_t/d_p and stabilizes when D_t/d_p is greater than 75. U_{ms} decreases with an increase in D_t/d_p. There is no clear regime of bubbling fluidization in micro-fluidized beds under the conditions of $H_s/D_t \geq 3$ and $D_t/d_p \leq 50$. In the cases of $H_s/D_t = 2$ and $D_t/d_p \geq 90$ as well as $H_s/D_t \geq 3$ and $D_t/d_p \geq 75$, there is a transition zone between the bubbling and slugging fluidization. In this transition zone, the gas bubble size reaches the slugging condition (i.e., $\geq 0.66D_t$). For Group B particles, bubbles are formed

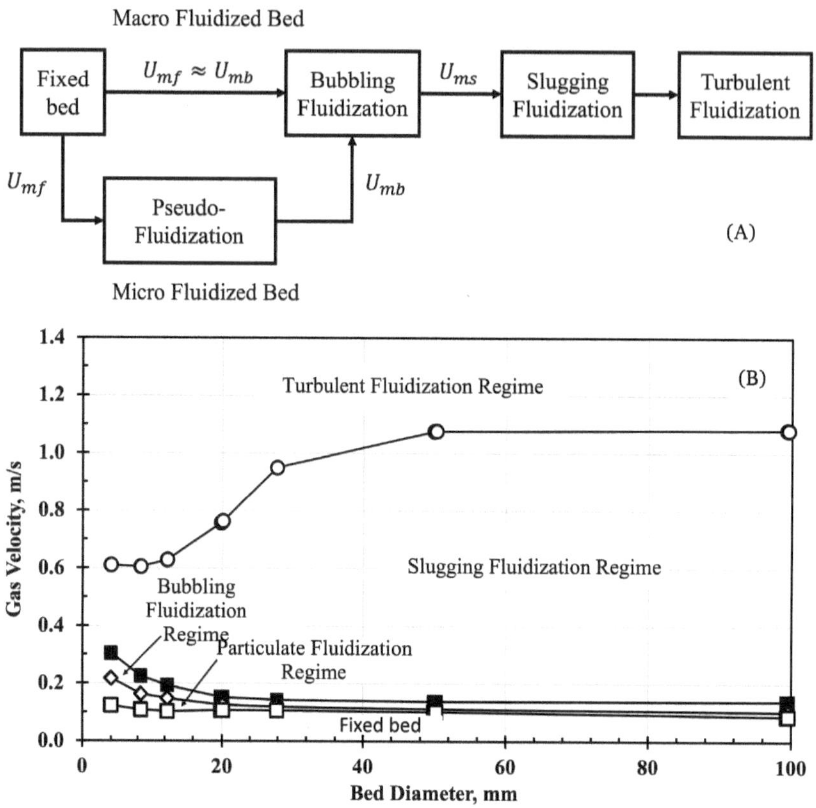

Figure 4.28 A fluidization map for class B particles in micro- and macro-fluidized beds [5]. (A) The relationship of various fluidization regimes, (B) variation of fluidization regime with bed diameter and superficial gas velocity.

before minimum fluidization at small D_t/d_p (<30) and high H_s/D_t (≥ 3) ratios. Compared with Group A particles, Group B particles have a small bubble fluidization region.

The flow regime maps presented above are very useful for better understanding the fluidization regime transitions in micro-fluidized beds, but their applicability is limited to their test conditions. Therefore, Han et al. [2] developed generalized flow regime maps, as shown in Fig. 4.30.

For Group A particles (Fig. 4.30A), there are significant delays in transition from the fixed bed to the minimum fluidization regime as D_t/d_p decreases, which can be well judged by the increased difference between the minimum fluidization velocities of micro-fluidized beds (U_{mf}) and macro-fluidized beds (U_{mfm}). When D_t/d_p is greater than 75, the minimum bubbling velocity is slightly greater than the minimum fluidization velocity, but when D_t/d_p is less than 75, the relative magnitudes of the two fluidization velocities are opposite. This fact leads to the almost nonexistence of a particulate fluidization regime. Apparently, this is contradictory to the observations of Wang and Fan [4]

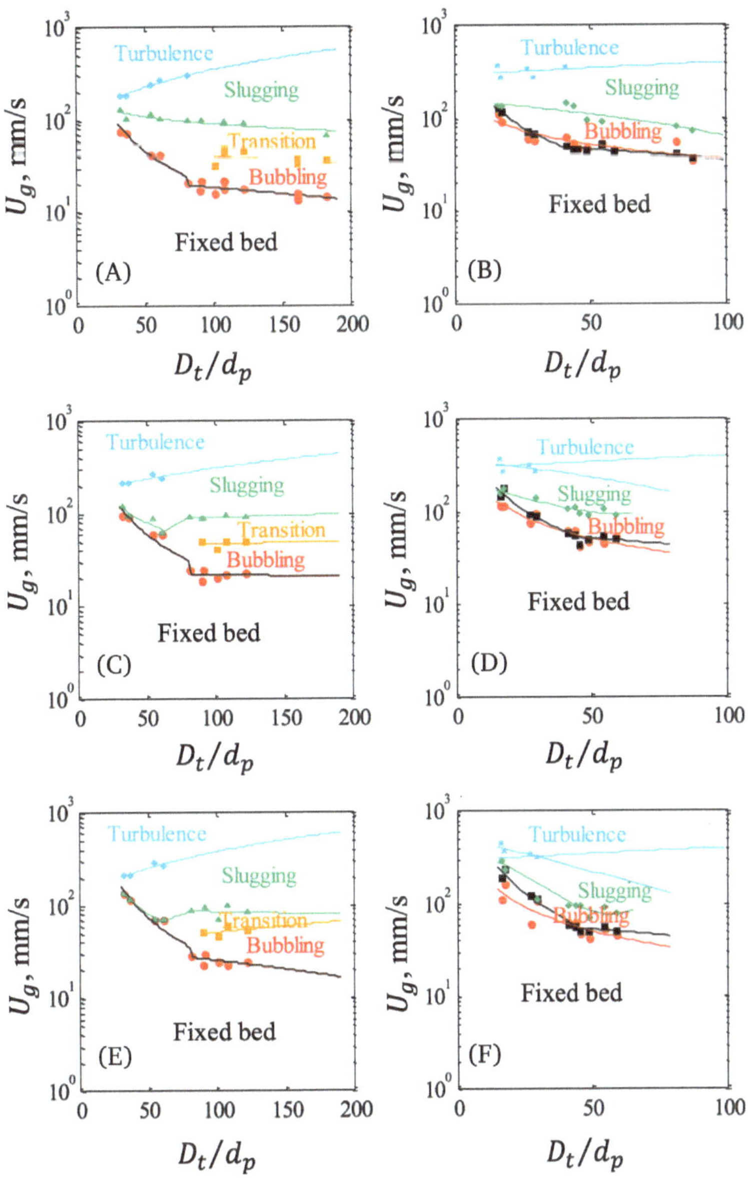

Figure 4.29 Flow regime maps of micro-fluidized beds (left (A, C, E): diameters 82 and 93 μm Group A particles; right (B, D, F): diameters 180 and 183 μm Group B particles; first row (A, B): $H_s/D_t = 2$; second row (C, D): $H_s/D_t = 3$; third row (E, F): $H_s/D_t = 4$) [6].

for FCC particles in mini- and micro-channel fluidized beds with channel sizes of 2–5 mm, but it is consistent with the report of McDonough et al. [6] who found that for Group A particles there was no particulate fluidization in micro-fluidized beds under all

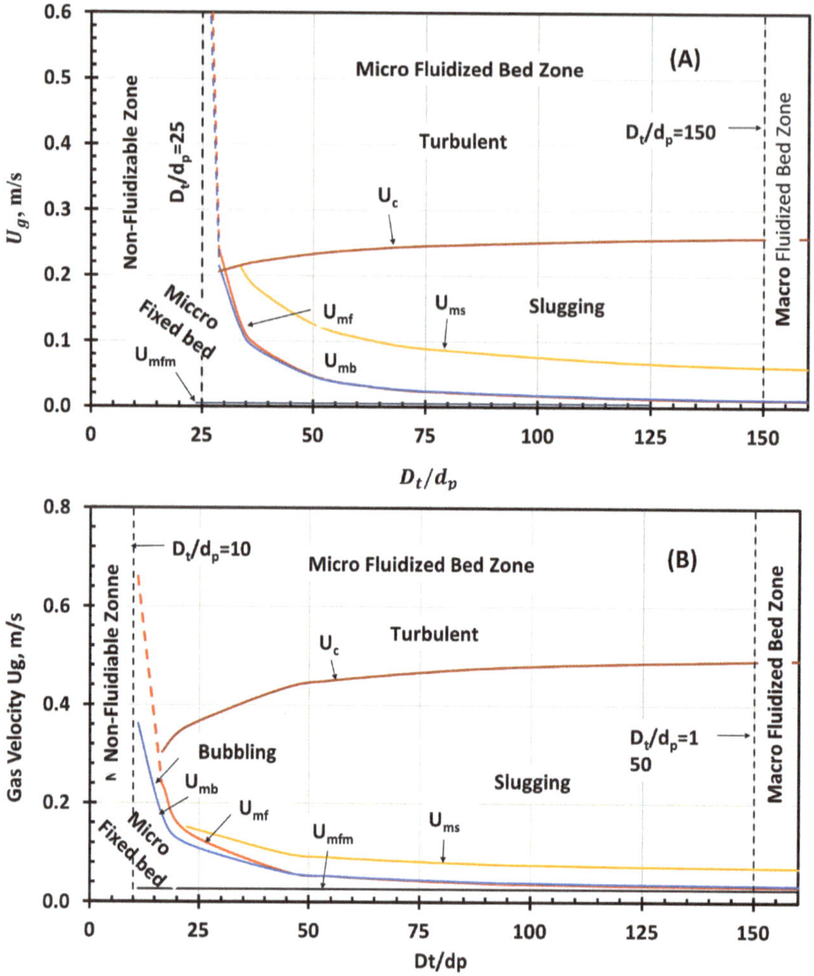

Figure 4.30 A fluidization regime map for gas–solid micro-fluidized beds. (A) Group A particles, 80 μm, 2475 kg/m^3, $H_s/D_t = 3$; (B) Group B particles, 180 μm, 2475 kg/m^3, $H_s/D_t = 3$.

H_s/D_t tested. When D_t/d_p ratio is reduced to approximately 25, the velocities U_{mf}, U_{mb}, and U_{ms} appear to merge and then increase sharply, indicating that a stable fluidization state is impossible to realize under this condition because the wall friction becomes so strong that the gas can no longer pass through the highly compacted particles. Under all D_t/d_p investigated, there is a broader bubbling fluidization regime, although the regime becomes relatively narrower when D_t/d_p is reduced to approximately 25. Under the conditions of $D_t/d_p < 35$, the transition to a turbulent fluidization regime directly from the bubbling regime may occur; otherwise, the gas–solid flow enters slugging fluidization and then to turbulent fluidization if the superficial gas velocity continuously increases. From Fig. 4.30A, it can be seen that the predicted U_{ms}/U_{mf} ratios vary between 2 and 6. Therefore, for Group A particles, the operating gas velocity U_g is recommended to be 2–6 times the minimum fluidization velocity to attain a stable fluidization condition in micro-fluidized beds. For Group B particles (Fig. 4.30B), the delay of the onset of fluidization is relatively reduced compared to that for Group A particles in micro-fluidized beds. Under the conditions of $H_s/D_t \geq 3$ and $D_t/d_p \leq 50$, gas bubbles form prior to minimum fluidization, resulting in a narrow and homogenous transition regime (i.e., particulate fluidization regime). As $D_t/d_p > 50$, the bubbling fluidization regime starts immediately as the bed is fluidized and then moves to the slugging fluidization regime as the gas velocity U_g is increased to 2–4 times the minimum fluidization velocity ($U_g/U_{mf} = 2–4$). When the ratio D_t/d_p approaches 10, fluidization becomes difficult because the wall effect is so strong.

Based on the correlations presented above, more generalized flow regime maps are produced for Group A (Fig. 4.31A) and Group B (Fig. 4.31B) particles. In these charts, the dimensionless velocities U^* $(=\mathrm{Re}/Ar^{\frac{1}{3}} = \mathrm{Re}/d_p{}^*)$ are plotted against the dimensionless particle size $d_p{}^*$ $(d_p{}^* = Ar^{1/3})$ in logarithm coordinates. It shows again that there is a more significant delay in minimum fluidization for Group A particles than for Group B particles. For both A and B particles, the minimum bubbling velocities are very close to the minimum fluidization velocities in all ranges of $d_p{}^*$. As $d_p{}^*$ increases, all the transition velocities increase. The bubbling fluidization regime becomes narrow as $d_p{}^*$ increases. For Group B particles, bubbling fluidization may not occur when $d_p{}^*$ is greater than about 14; in this case, the bed may transfer from fixed bed to slugging

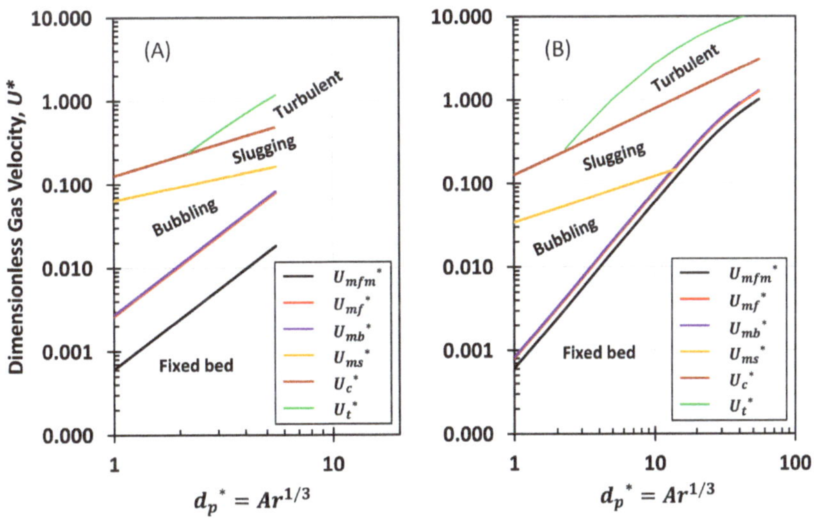

Figure 4.31 Generalized fluidization regime maps for gas–solid micro-fluidized beds ($D_t/d_p = 50$, $H_s/D_t = 3$). (A) Group A particles, (B) group B particles.

fluidization directly. Moreover, when $d_p{}^*$ is smaller than about 2, stable turbulent fluidization may not be achieved, since the gas velocity may have exceeded the terminal velocity of particles and particles are likely to be carried out of the bed.

In summary, gas–solid micro-fluidized beds can be defined based on the following conditions: $D_t/d_p = 25-150$ for Group A particles and $D_t/d_p = 10-150$ for Group B particles. To achieve stable fluidization, the recommended operating U_g/U_{mf} is 2–6 for Group A particles and 2–4 for Group B particles. To maintain stable fluidization, $H_s/D_t = 2-4$ is recommended. Micro-fluidized beds operated under these conditions have been well tested. The experimental results have confirmed that operating micro-fluidized beds under these conditions will yield the most desirable hydrodynamic characteristics for reaction analysis, that is, excellent heat/mass transfer, uniform flow and temperature distributions, and close-to-plug flow of gas [24,43].

Nomenclature

A	cross-sectional area of the bed (m^2)
A_e	modified viscous constant
A_E	viscous constant
Ar	Archimedes number
B_e	modified inertial constant
B_E	inertial constant
d_{bm}	size of the maximum stable bubble (m)
d_p	particle diameter (m)
$d_p{}^*$	dimensionless particle size
D_t	bed diameter (m)
E	elasticity modulus of the particle phase
$E(t)_h$	peak height of $E(t)$
f_E	friction coefficient
g	acceleration of gravity (m/s^2)
H	bed height (m)
H_S	static bed height (m)
M	a function of the bed-to-particle diameter ratio
m	mass of particles (kg)
n	Richardson-Zaki exponent
Re	Reynolds number
Re_c	Reynolds number based on Uc
Re_g	Reynolds number based on Ug
Re_{mb}	Reynolds number based on Umb
Re_{mf}	Reynolds number based on Umf
Re_{ms}	Reynolds number based on Ums
Re_t	Reynolds number based on Ut
U^*	dimensionless velocities
U_c	onset velocity of turbulent fluidization (m/s)
U_g	superficial gas velocity (m/s)
U_{mb}	minimum bubbling velocity (m/s)

U_{mf} minimum fluidization velocity (m/s)
U_{mf}^* dimensionless minimum fluidization velocity
U_{mfm} minimum fluidization velocity under the wall effect-free and homogenous flow conditions (m/s)
U_{mfm}^* dimensionless minimum fluidization velocity under the wall effect-free and homogenous flow conditions
U_{ms} minimum slugging velocity (m/s)
u_{cl} continuity wave velocity (m/s)
u_d dynamic wave velocity (m/s)
U_t terminal velocity (m/s)
U_t^* dimensionless terminal velocity
ΔP bed pressure drop (Pa)
ΔP_E the pressured drop caused by the Ergun equation or modified Ergun equation (Pa)

Greek letters

β coefficient corresponding to the vertical distance between bubbles
ε bed voidage
ε_{mb} bed voidage at the minimum bubbling condition
ε_{mf} bed voidage at the minimum fluidization condition
μ_g gas viscosity (Pa · s)
ρ_b bulk density (kg/m^3)
ρ_g gas density (kg/m^3)
ρ_p particle density (kg/m^3)
σ_t^2 variance of $E(t)$ (s^2)
φ particle angle of internal friction (degrees)
Φ particle sphericity

References

[1] Zivkovic V, Biggs MJ. On importance of surface forces in a microfluidic fluidized bed. Chemical Engineering Science 2015;26:143–9.
[2] Han Z, Yue J, Geng S, et al. State-of-the-art hydrodynamics of gas-solid micro fluidized beds. Chemical Engineering Science 2021;232:116345.
[3] Liu X, Xu G, Gao S. Micro fluidized beds: wall effect and operability. Chemical Engineering Journal 2008;137(2):302–7.
[4] Wang F, Fan LS. Gas-solid fluidization in mini- and micro-channels. Industrial and Engineering Chemistry Research 2011;50(8):4741–51.
[5] Quan H, Fatah N, Hu C. Diagnosis of hydrodynamic regimes from large to micro-fluidized beds. Chemical Engineering Journal 2020;391.
[6] Mcdonough JR, Law R, Reay DA, et al. Fluidization in small-scale gas-solid 3D-printed fluidized beds. Chemical Engineering Science 2019;200:294–309.
[7] Ergun S. Fluid flow through packed columns. Chemical Engineering Progress 1952;48:1179–84.
[8] Ergun S, Orning AA. Fluid flow through randomly packed columns and fluidized beds. Industrial and Engineering Chemistry 1949;41(6):1179–84.
[9] Winterberg M, Tsotsas E. Impact of tube-to-particle-diameter ratio on pressure drop in packed beds. AIChE Journal 2000;46(5):1084–8.
[10] Eisfeld B, Schnitzlein K. The infuence of confining walls on the pressure drop in packed beds. Chemical Engineering Science 2001;56(14):4321–9.
[11] Guo Q, Xu Y, Yue X. Fluidization characteristics in micro-fluidized beds of various inner diameters. Chemical Engineering and Technology 2009;32(12):1992–9.

[12] Tian FY, Huang LF, Fan LW, et al. Wall effects on the pressure drop in packed beds of irregularly shaped sintered ore particles. Powder Technology 2016;301:1284–93.
[13] Navvab Kashani M, Elekaei H, Zivkovic V, et al. Explicit numerical simulation-based study of the hydrodynamics of micro-packed beds. Chemical Engineering Science 2016;145:71–9.
[14] Cheng NS. Wall effect on pressure drop in packed beds. Powder Technology 2011;210:261–6.
[15] Di Felice R, Gibilaro LG. Wall effects for the pressure drop in fixed beds. Chemical Engineering Science 2004;59:3037–40.
[16] Mehta D, Hawley MC. Wall effect in packed columns. Industrial and Engineering Chemistry Process Design and Development 1969;8(2):280–2.
[17] Reichelt W. Calculation of pressure-drop in spherical and cylindrical packings for single-phase flow. Chemie Ingenieur Technik 1972;44(18):1068–71.
[18] Foumeny EA, Benyahia F, Castro JAA, et al. Correlations of pressure drop in packed beds taking into account the effect of confining wall. International Journal of Heat and Mass Transfer 1993;36 (2):536–40.
[19] Montillet A, Akkari E, Comiti J. About a correlating equation for predicting pressure drops through packed beds of spheres in a large range of Reynolds numbers. Chemical Engineering and Processing 2007;46(4):329–33.
[20] Raichura R. Pressure drop and heat transfer in packed beds with small tube-to-particle diameter ratio. Experimental Heat Transfer 1999;12(4):309–27.
[21] Prajapati A, Renganathan T, Krishnaiah K. Kinetic studies of CO_2 capture using K_2CO_3/activated carbon in fluidized bed reactor. Energy and Fuels 2016;30(12):10758–69.
[22] Rao A, Curtis JS, Hancock BC, et al. The effect of column diameter and bed height on minimum fluidization velocity. AIChE Journal 2010;56(9):2304–11.
[23] Vanni F, Caussat B, Ablitzer C, et al. Effects of reducing the reactor diameter on the fluidization of a very dense powder. Powder Technology 2015;277:268–74.
[24] Geng S, Han Z, Yue J, et al. Conditioning micro fluidized bed for maximal approach of gas plug flow. Chemical Engineering Journal 2018;351:110–18.
[25] Wen CY, Yu YH. A generalized method for predicting the minimum fluidization velocity. AIChE Journal 1966;12(3):610–12.
[26] Haider A, Levenspiel O. Drag coefficient and terminal velocity of spherical and nonspherical particles. Powder Technology 1989;58(1):63–70.
[27] Kunii D, Levenspiel O. Fluidization engineering. 2nd ed. Boston: Butterworth—Heinemann; 1991.
[28] Richardson JF, Zaki WN. Sedimentation and fluidisation: part I. Chemical Engineering Research and Design 1954;75(3):S82–100.
[29] Mori S, Wen CY. Estimation of bubble diameter in gaseous fluidized-beds. AIChE Journal 1975;21 (1):109–15.
[30] Baeyens J, Geldart D. An investigation into slugging fluidized beds. Chemical Engineering Science 1974;29:255–65.
[31] Makkawi YT, Wright PC. Fluidization regimes in a conventional fluidized bed characterized by means of electrical capacitance tomography. Chemical Engineering Science 2002;57(13):2411–37.
[32] Wu C, Cheng Y. Grace JR, Bi X, Ellis N. Essentials of Fluidization Technology Downer reactors, 12. Wiley-VCH Verlag GmbH & Co. KGaA, Boschstr; 2020. p. 69469. Weinheim, Germany: 499–530.
[33] Bi HT, Ellis N, Abba IAA, et al. A state-of-the-art review of gas-solid turbulent fluidization. Chemical Engineering Science 2000;55(21):4789–825.
[34] Dang NTY, Gallucci F, van Sint Annaland M. An experimental investigation on the onset from bubbling to turbulent fluidization regime in micro-structured fluidized beds. Powder Technology 2014;256:166–74.
[35] Wang J, Tan L, van der Hoef MA, et al. From bubbling to turbulent fluidization: advanced onset of regime transition in micro-fluidized beds. Chemical Engineering Science 2011;66(9):2001–7.
[36] Bai D, Shibuya E, Nakagawa N, et al. Characterization of gas fluidization regimes using pressure fluctuations. Powder Technology 1996;87(2):105–11.

[37] Cai P, Jin Y, Yu Z-Q, et al. Mechanism of flow regime transition from bubbling to turbulent fluidization. AIChE Journal 1990;36(955−956).
[38] Jia Z. Experimental and Numerical Investigation on Hydrodynamics of Micro Fluidized Bed. A mater thesis submitted to Tianjin University; 2016.
[39] Geng S, Yu J, Zhang J, et al. Gas back-mixing in micro fluidized beds. CIESC Journal 2013;64 (3):867−76
[40] Yoshida K, Kunii D, Levenspiel O. Axial dispersion of gas in bubbling fluidized beds. Industrial and Engineering Chemistry Fundamentals 1966;7:402−6.
[41] Li J, Weinstein H. An experimental comparison of gas backmixing. Chemical Engineering Science 1989;44(8):1697−705.
[42] Lee GS, Kim SD. Rise velocities of slugs and voids in slugging and turbulent fluidized beds. Korean Journal of Chemical Engineering 1989;6(1):15−22.
[43] Wang F, Zeng X, Geng S, et al. Distinctive hydrodynamics of a micro fluidized bed and its application to gas-solid reaction analysis. Energy and Fuels 2018;32(4):4096−106.

CHAPTER 5

Hydrodynamic modeling of micro-fluidized beds

Computational fluid dynamics (CFD) modeling is a powerful analytical tool for understanding the characteristics of flow dynamics, heat and mass transfer, and reaction performance of multiphase systems. It is particularly useful for investigating micro-fluidized beds where some conventional experimental techniques (e.g., optical fiber probes, capacitance probes, radiative probes) are less applicable because of limited accessibility, space limitation, or interference effects. For example, the distributions of gas and particle velocities and the solid concentration in both the bed axial and radial directions are difficult to obtain by experimental methods in micro-fluidized beds. Under these circumstances, hydrodynamic modeling is of great use to provide spatial and temporal information and to gain insights into complex phenomena that are not readily measurable in space-constrained micro-fluidized beds. However, only a few CFD simulations for micro-fluidized beds are performed and published at the time of writing. This chapter summarizes the CFD methods and the available modeling results of gas-solid micro-fluidized beds.

5.1 CFD modeling approaches

Several modeling approaches are available to investigate complex fluid dynamics in gas-solid fluidized beds [1]. Fig. 5.1 presents the most commonly used simulation approaches with corresponding resolution characteristics (i.e., simulated levels of details and accuracies) and computational easiness (i.e., computational time requirements and costs).

As shown in Fig. 5.1, two general methods are used to describe the motion of particles in gas-solid fluidized beds. The first is the Eulerian method. This method does not directly follow the motion of particles. Instead, it monitors variations of flow parameters at the spatial points in the flow field with time. In this way, the motion of the whole fluid is obtained by combining information from enough space points. The second is the Lagrangian method. By tracking the motion of each and every fluid particle, this method observes the changes in the motion parameters (position coordinate, velocity, acceleration, etc.) when the particle travels from one point to another in the flow field. The entire flow behavior is obtained by summarizing the changes in the motion parameters of all the fluid particles. In practice, some hybrid methods are also

Micro Fluidization
DOI: https://doi.org/10.1016/B978-0-443-18718-6.00005-X
© 2023 Chemical Industry Press Co., Ltd.
Published by Elsevier Inc., under an exclusive license with Chemical Industry Press Co., Ltd.

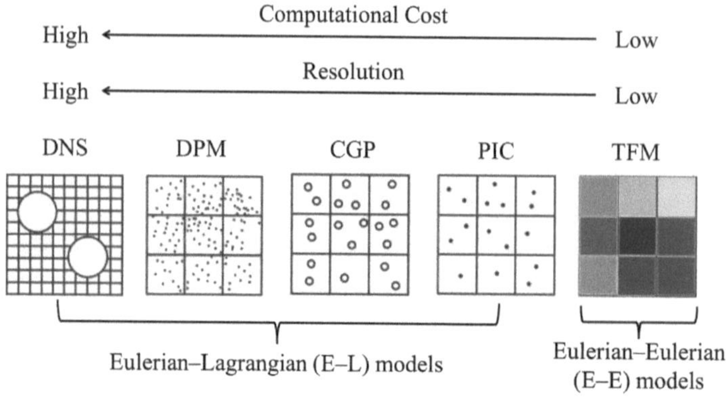

Figure 5.1 Different CFD models for simulations of gas-solid flows. *DNS*, direct numerical simulation; *DPM*, discrete particle method; *CGP*, coarse-grained particle; *PIC*, particle in cell; *TFM*, two-fluid method [1].

used. In these cases, the gas phase and particle phase are modeled by the Eulerian and Lagrangian methods, respectively. These methods are called Euler—Lagrangian (EL) models. In the EL models, the interface between gas and solid phases is calculated by an average area bounded by a number of particle trajectories. Hence, a large number of particle trajectories are used to obtain meaningful hydrodynamic properties of the continuous phase. The Eulerian—Eulerian (EE) models, based on the interpenetrating continuum assumption, use the Eulerian approach to deal with gas and solid phases. In this approach, the trajectories of particles are obtained at a hypothetical level rather than at a physical level compared to the EL models. The EE models can be applied to multiphase flow processes containing a large volume fraction of solid particles. For simulations of gas-solid systems, the particle flow can be described by one of the following four major methods (Fig. 5.1):

- **Direct numerical simulation (DNS) method**. This model directly solves the governing Navier—Stokes equations of the gas-solid flow with exact boundary conditions imposed on each particle surface to obtain complete three-dimensional time-dependent velocity and pressure fields. The model does not require additional closure equations because the number of equations equals the number of unknowns. The DNS approach offers high spatial and temporal resolutions, but is limited to applications with low Reynolds numbers and small-size computational domains. Otherwise, the instantaneous range of scales in turbulent flows increases rapidly with an increase in Reynolds number, and the computational cost would be too high to be affordable.
- **Discrete particle method (DPM)**. The DPM, or the discrete element method (DEM), calculates the movements of individual particles by considering all particle—particle and particle—geometry interactions. Individual particles are simulated and tracked by solving the governing equations based on Newton's second law of

motion combined with specific time-stepping algorithms. In this method, interactions between particles are assumed to exist only between those in contact with each other. It is featured by the inclusion of rotational degrees of freedom and stateful contact for even complicated geometries (including polyhedral). The DPM requires the computing power to increase as the simulating number of particles increases. Therefore, its applicability to solid flows involving a large number of particles is very challenging and costly. So, the DPM or DEM is more suitable for micro-solid flow systems.

- **Coarse-grained particle (CGP) method**. The CGP model tracks particles by the so-called coarse grains containing some real particles. In each coarse grain, the particles are identical: they all have the same properties, including density, diameter, temperature, and velocity. In this way, the model does not calculate the collision between particles inside the coarse grains, thus reducing the computational time and cost, but sacrificing the simulation accuracy.
- **Particle-in-cell (PIC) method**. The PIC simulates individual particles but tracks a group of particles as a point without computation of collision forces, favoring a simpler continuum solid pressure model. Without resolving the interparticle contact, this method provides superior performance in the computational speed, but with the compromise of accuracy in the calculation results.
- **Two-fluid method (TFM)**. The TFM considers both the particulate and the gas phases to be continuous fluids, fully interpenetrating continua coupled with an interaction term. This model is the most popular one because it does not need to calculate the motion of individual or grouped particles and thus requires low computation power, but it has a low resolution.

Despite extensive research and applications of the CFD simulations for gas-solid fluidized beds, only TFM and CFD-DPM (or CFD-DEM) are considered suitable for micro-fluidized beds when the accuracy of simulation results and the cost of computation are taken into consideration. In both the TFM and the CFD-DPM, closure models are needed for the gas-solid interactions in the momentum and energy equations. For gas-solid flow in fluidized beds of practical interest, the mass ratio of solid to gas is far greater than unity so that the drag becomes the dominant mechanism to characterize the momentum exchange and drive the solid flow. Therefore, selecting an appropriate drag coefficient correlation is vital because different drag models can lead to remarkably different results. The following descriptions will focus on the applications of TFM and CFD-DEM in micro-fluidized beds.

5.2 Two-fluid method

The TFM takes the gas and solid particles as two continuous fluid phases in the Eulerian frame. Each phase is governed by a set of conservation equations of mass, momentum, and energy, and a drag coefficient links the continuous gas and solid–fluid phases. To describe the dependence of the averaged field of one phase on the other, we have interaction terms included in these balance equations. Thus, we have six differential field equations with three interfacial transfer conditions to govern the macroscopic two-phase flow systems. Moreover, since the solid flow is treated as a continuous fluid phase, it is necessary to introduce a stress tensor and a viscosity to describe the rheology of the particulate phase. Among various constitutive relations for the solid phase, the kinetic theory of granular flow, derived from the analogy with the kinetic theory of gas, is the most widely applied theory to describe the rheology of particulate phases. It is essential to carefully specify the wall boundary conditions when solving the TFM equations. As explicitly described in Chapter 2, a wall region with high voidage in gas-solid micro-fluidized beds presents a challenge for the two-fluid modeling because it is not a continuum in the region.

Liu et al. [2] and Liu et al. [3] assessed the applicability of the TFM in micro-fluidized beds using typical Group A and Group B particles. The studies indicate that the wall boundary conditions have remarkable influences on the predicted hydrodynamic characteristics in micro-fluidized beds.

5.2.1 TFM formulation

In the Eulerian approach, gas and solid are treated as fully interpenetrating continua. The mass conservation equations for each phase are written as follows:

$$\frac{\partial \varepsilon_g \rho_g}{\partial t} + \nabla \cdot \left(\varepsilon_g \rho_g \overrightarrow{u_g}\right) = -\gamma_{gs} \tag{5.1}$$

$$\frac{\partial \varepsilon_s \rho_p}{\partial t} + \nabla \cdot \left(\varepsilon_s \rho_p \overrightarrow{u_p}\right) = \gamma_{gs} \tag{5.2}$$

$$\varepsilon_g + \varepsilon_s = 1 \tag{5.3}$$

where: ρ_g: gas density ρ_p: solid particle density $\overrightarrow{u_g}$: mean gas velocity vector $\overrightarrow{u_p}$: mean solid particle velocity vector ε_g: gas-phase volume fraction ε_s: solid-phase volume fraction γ_{gs}: rate of mass transfer from the gas phase to the solid phase due to chemical reactions (combustion, gasification, etc.) or physical processes, such as evaporation, adsorption, and desorption. When there is no mass transfer, $\gamma_{gs} = 0$.

Note that the above governing equations should be closed using averaged field variables based on either time average, space average, or ensemble average approaches.

The momentum equations for the gas and particulate phases are expressed by the Navier–Stokes equation as the following:

$$\frac{\partial \varepsilon_g \rho_g \overrightarrow{u_g}}{\partial t} + \nabla \cdot \left(\varepsilon_g \rho_g \overrightarrow{u_g}\, \overrightarrow{u_g} \right) = -\varepsilon_g \nabla P + \nabla \cdot \left(\varepsilon_g \tau_g \right) + \varepsilon_g \rho_g \vec{g} - \beta_{gs} \left(\overrightarrow{u_g} - \overrightarrow{u_p} \right) - m_{gs} \tag{5.4}$$

$$\frac{\partial \varepsilon_s \rho_p \overrightarrow{u_p}}{\partial t} + \nabla \cdot \left(\varepsilon_s \rho_p \overrightarrow{u_p}\, \overrightarrow{u_p} \right) = -\varepsilon_s \nabla P - \nabla P_s + \nabla \cdot \left(\varepsilon_s \tau_s \right) + \varepsilon_s \rho_p \vec{g} + \beta_{gs} \left(\overrightarrow{u_g} - \overrightarrow{u_p} \right) + m_{gs} \tag{5.5}$$

where τ_g and τ_s are the shear stress tensors of gas and solid phases, β_{gs} is the drag coefficient between gas and solid phases, $\vec{g}$ is the gravitational acceleration vector, and m_{gs} is the momentum transfer due to the mass transfer between phases. If there is no momentum transfer between the two phases, $m_{gs} = 0$. For the gas phase, the shear stress tensor can be simply calculated by

$$\tau_g = \mu_g \left(\nabla \overrightarrow{u_g} + \overrightarrow{u_g}^{T} \right) - \frac{2}{3} \mu_g \nabla \cdot \left(\overrightarrow{u_g} \right) \bar{I} \tag{5.6}$$

where $\bar{I}$ is the identity matrix. The kinetic theory of granular flow (KTGF) is used to close the momentum equations for the solid phase. The stresses exerted on particles due to transient and instantaneous collisions are referred to as solid-phase kinetic and collisional stresses, which depend on the magnitude of the particle velocity fluctuations, characterized by the so-called granular temperature, Θ_s. Based on the KTGF [4], the granular temperature can be estimated by the following transport equation:

$$\frac{3}{2} \left[\frac{\partial}{\partial t} \left(\varepsilon_s \rho_p \Theta_s \right) + \nabla \cdot \left(\varepsilon_s \rho_p \overrightarrow{u_s} \Theta_s \right) \right] = \left(-P_s \bar{I} + \varepsilon_s \tau_s \right) \nabla \overrightarrow{u_s} + \nabla \cdot \left(\kappa_s \nabla \Theta_s \right) - \gamma - 3\beta_{gs} \Theta_s \tag{5.7}$$

where the conductivity of the fluctuating energy κ_s and the collisional energy dissipation γ are calculated as follows:

$$\kappa_s = \frac{150 \rho_p d_p \sqrt{\Theta_s \pi}}{384 (1 + e_{pp}) \chi_r} \left[1 + \frac{6}{5} \varepsilon_s \chi_r \left(1 + e_{pp} \right) \right]^2 + 2 \rho_p \varepsilon_s^{2} d_p \left(1 + e_{pp} \right) \chi_r \sqrt{\frac{\Theta_s}{\pi}} \tag{5.8}$$

$$\gamma = \frac{12 (1 - e_{pp}^{2}) \chi_r}{d_p \sqrt{\pi}} \rho_p \varepsilon_s^{2} d_p \Theta_s^{3/2} \tag{5.9}$$

In the above equations, e_{pp} is the restitution coefficient for particle–particle interaction, representing the collision elasticity between particles. A restitution coefficient of completely elastic collisions is one and that of completely inelastic collisions is zero. It should be mentioned that the value of the coefficient influences the kinetic and collisional behavior of particles, but no theoretical methods for its estimation. χ_r is the radial distribution function, which can be expressed as

$$\chi_r = \left[1 - \left(\frac{\varepsilon_s}{\varepsilon_{s,\max}}\right)^{1/3}\right]^{-1} \tag{5.10}$$

where $\varepsilon_{s,\max}$ is the maximum volume fraction of solid particles. Now, combining equations from (5.7) to (5.10), the granular temperature, Θ_s, can be estimated. Then, the solid stress tensor τ_s and solid pressure P_s can be calculated based on the following correlations:

$$\tau_s = \mu_s\left(\nabla\vec{u_s} + \vec{u_s}^T\right) + \left(\lambda_s - \frac{2}{3}\mu_s\right)\nabla\cdot(\vec{u_s})\bar{I} \tag{5.11}$$

$$P_s = \left(\varepsilon_s\rho_p + 2\rho_p\varepsilon_s^2\chi_r\right)\Theta_s \tag{5.12}$$

where the solid-phase bulk viscosity λ_s and the solid-phase shear viscosity μ_s are calculated by the following equations:

$$\lambda_s = \frac{4}{3}\rho_p\varepsilon_s d_p\left(1 + e_{pp}\right)\chi_r\sqrt{\frac{\Theta_s}{\pi}} \tag{5.13}$$

$$\mu_s = \frac{4}{5}\rho_p\varepsilon_s d_p\left(1 + e_{pp}\right)\chi_r\sqrt{\frac{\Theta_s}{\pi}} + \frac{10\rho_p d_p\sqrt{\Theta_s\pi}}{96\varepsilon_s\left(1 + e_{pp}\right)\chi_r}\left[1 + \frac{4}{5}\varepsilon_s\left(1 + e_{pp}\right)\chi_r\right]^2 + \frac{\rho_p \sin\phi}{2\sqrt{I_{2D}}} \tag{5.14}$$

where ϕ is the specularity coefficient, an empirical parameter ranging from 0 to 1 to characterize the nature of the particle–wall collision. A value of 0 and 1 denotes a perfect specular and fully diffusive collision, respectively. Again, note that there are no theoretical methods for estimating the specularity coefficient.

In the TFM, the gas phase is coupled with the particle phase through an interphase momentum exchange, generally expressed by a drag coefficient, β_{gs}. Currently, there are two types of drag models in the literature [5,6]. The first is homogeneous drag models developed by assuming a constant solid concentration in the control volume.

The most popular homogeneous drag model developed by Gidaspow [4] is given below:

$$\beta_{gs} = \begin{cases} 150\frac{\varepsilon_s{}^2\mu_g}{\varepsilon_g d_p{}^2} + 1.75\frac{\rho_g\varepsilon_s\left|\vec{u_g} - \vec{u_p}\right|}{d_p}, \varepsilon_g \leq 0.8 \\ \frac{3}{4}C_d\frac{\rho_g\varepsilon_g\varepsilon_s\left|\vec{u_g} - \vec{u_p}\right|}{d_p}\varepsilon_g{}^{-2.65}, \varepsilon_g > 0.8 \end{cases} \quad (5.15)$$

$$\text{where } C_d = \begin{cases} \frac{24}{\mathrm{Re}_p}\left[1 + 0.15\mathrm{Re}_p{}^{0.687}\right], \mathrm{Re}_p = \frac{\rho_g\varepsilon_g\left|\vec{u_g} - \vec{u_p}\right|d_p}{\mu_g} < 1000 \\ 0.44, \mathrm{Re}_p \geq 1000 \end{cases}$$

where C_d is the drag coefficient. The second is heterogeneous or structure-dependent drag models, which are developed to account for small-scale structures such as clusters and bubbles in the control volume. The typical heterogeneous drag models are based on the energy minimization multiscale (EMMS) model [7–12] or coarse grid filtering of fine grid simulation [13,14]. These models are often complex and case-dependent, so we will not explain them further. Since many drag models are reported in the literature and their detailed descriptions are beyond the scope of this book, we encourage readers interested in this area to consult the original literature for details of model development and application [5,7,9,10,12,15–19].

Now, we have Eqs. (5.6)–(5.15) to form the constitutive closure equations for the basic TFM (i.e., Eqs. (5.1)–(5.5)). Note that alternatives to the constitutive equations are also available in the literature [6,20–24]. Doubtlessly, a successful prediction of the hydrodynamics of gas-solid micro-fluidized beds relies on the introduced constitutive equations and values of the involved empirical parameters (e.g., the specularity coefficient and the restitution coefficients for particle–particle and particle–wall interactions). It is worth pointing out that despite being significant, the selection of drag models is purely empirical, subjective, and even speculative. The discussion below will illustrate this phenomenon.

5.2.2 TFM simulations and validations

Liu et al. [2] studied the fluidization behavior of Geldart A particles in a gas-solid micro-fluidized bed by the Eulerian–Eulerian numerical simulation. The simulated system is based on the experimental setup of Wang and Fan [25], which consists of a rectangular channel of 3 mm wide and 12.1 mm high and a disengagement section of 6 mm wide and 8.8 mm high. The particles have an average diameter of 53 μm with a density of

1400 kg/m^3. In all the simulation calculations, it is assumed that the gas velocity at the bottom inlet is uniform and the outlet at the top of the bed is atmospheric pressure.

Given the importance of wall effects, the wall boundary conditions need to be specified very carefully in numerical simulations of gas-solid two-phase flows in the wall-confining micro-fluidized beds. Liu et al. [2] defined the wall boundary conditions using the Johnson and Jackson approach [26] as follows:

$$\frac{\vec{U}_{sw} = \frac{6\mu_s\varepsilon_{s,\max}}{\sqrt{3}\pi\phi\rho_p\varepsilon_s\chi_r\sqrt{\Theta_s}}\partial\vec{U}_{sw}}{\partial\vec{n}}$$

$$\frac{\Theta_s = \frac{\kappa_{\Theta_s}\Theta_s}{\gamma_w}\frac{\partial\Theta_s}{\partial\vec{n}} + \sqrt{3}\pi\phi\rho_p\varepsilon_s\chi_r\vec{U}_{sw}^{\,2}\Theta_s^{3/2}}{6\varepsilon_{s,\max}\gamma_w} \tag{5.16}$$

$$\gamma_w = \frac{\sqrt{3}\pi(1 - e_{pw}^{\,2})\rho_p\varepsilon_s\chi_r\Theta_s^{3/2}}{4\varepsilon_{s,\max}}$$

where $\vec{U}_{sw}$ represents the particle slip velocity parallel to the wall, $\vec{n}$ is the unit vector normal to the wall, and e_{pw} is the restitution coefficient for particle–wall collision. In general, the coefficients of particle–wall collision differ from those of particle–particle collision, both of which affect the motion of fluidized particles by affecting the energy lost during particles collision with confining walls and with each other.

Liu et al. [2] first conducted numerical simulations using the Gidaspow drag model [4] to describe the gas-solid interactions with the simulation conditions presented above. The results reveal that the TFM incorporating the Gidaspow drag model into the simulations fails to predict the bed expansion and the minimum bubbling velocity even with grid resolution up to one particle diameter. Bubbles are predicted to appear at a superficial gas velocity of ~6 mm/s based on the simulations, as shown in Fig. 5.2A, which is far smaller than the experimentally observed minimum bubbling velocity of 13.5 mm/s. The results show that the model prediction in the bed expansion is too large and in the onset of bubbling is too early.

The predicted over-bed expansion shown in Fig. 5.2A is due to the excessive drag force induced by the Gidaspow drag model. To correct this issue, McKeen and Pugsley [19] proposed an alternative drag model as follows:

$$\beta_{gs} = \begin{cases} C_{mp}\left[\dfrac{17.3}{\mathrm{Re}_p} + 0.036\right]\dfrac{\rho_g\left|\vec{u_g} - \vec{u_p}\right|}{d_p}\varepsilon_g^{1.8}\varepsilon_s \\ \mathrm{Re}_p = \dfrac{\rho_g\varepsilon_g\left|\vec{u_g} - \vec{u_p}\right|d_p}{\mu_g} \end{cases} \tag{5.17}$$

This empirical model requires the scaling factor C_{mp} to be specified for every specific application. The values of C_{mp} ranging from 0.001 to 0.3 have been used in the literature to simulate Geldart A particles in bubbling/turbulent fluidized beds [2,27]. Therefore, determining an appropriate C_{mp} value may require the trial-and-error approach in practice for specific simulations. Taking $C_{mp} = 0.2$, the predicted minimum bubbling velocity is approximately 12–14 mm/s (Fig. 5.2B), which is in good agreement with the experimental value of 13.5 mm/s. Fig. 5.2C further shows that the predicted variation of bed solid volume fractions with the superficial gas velocity is in good agreement with the experimental data.

Fig. 5.3 displays the instantaneous voidage profiles in a 3 × 3 mm gas-solid micro-fluidized bed simulated at $U_g = 25$ and 55 mm/s. It shows that gas slugging occurs

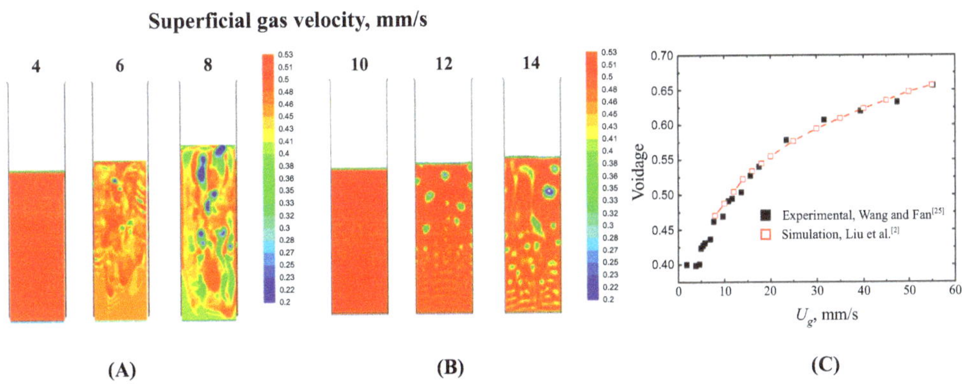

Figure 5.2 The simulated bed solid volume fraction as a function of superficial gas velocity compared with experimental data with $\phi = 1$. (A) Gidaspow drag model. (B) McKeen and Pugsley drag model. (C) Average bed voidage versus U_g.

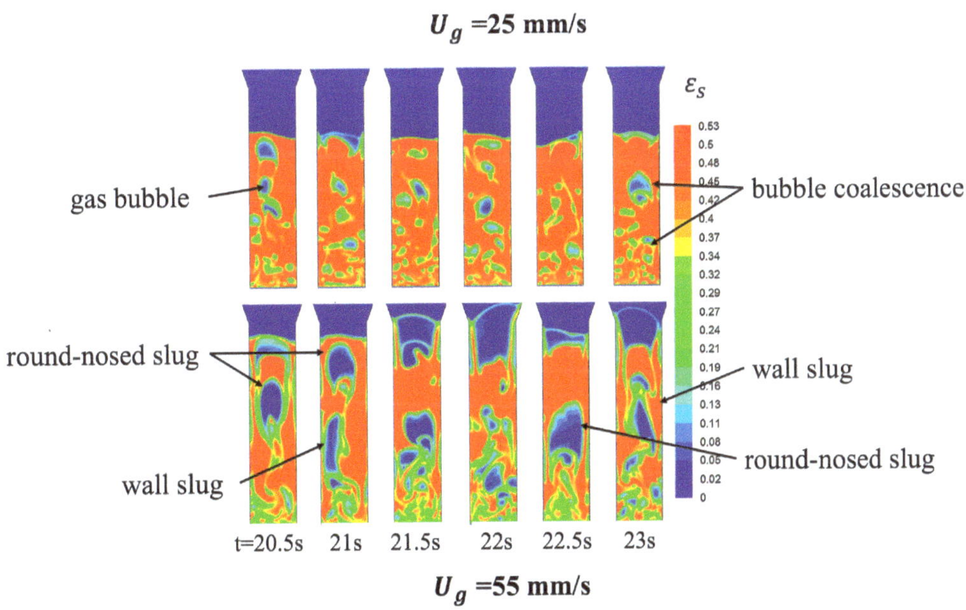

Figure 5.3 The simulated instantaneous voidage profiles [2].

when the superficial gas velocity is between 25 and 35 mm/s, which agrees well with the experimental result. In addition, the simulations predict the existence of round-nosed slugs and wall slugs, consistent with the experimental observations.

Piemjaiswang et al. [28] compared the Gidaspow model with the EMMS drag model using computational fluid dynamics simulation to study the effect of design parameters on system mixing in a micro-fluidized bed reactor. Fig. 5.4 shows that the Gidaspow drag model again overestimates the bed expansion. In comparison, the EMMS drag model considers the local gas-solid flow heterogeneity and is more rational in predicting the gas-solid hydrodynamics in fluidized beds.

The EMMS drag model is expressed as

$$\beta_{gs} = \begin{cases} 150\dfrac{\varepsilon_s^2\mu_g}{\varepsilon_g d_p^2} + 1.75\dfrac{\varepsilon_s\rho_g|\vec{u_g} - \vec{u_p}|}{d_p} & \varepsilon_g < 0.74 \\ \dfrac{3}{4}\dfrac{\varepsilon_s\varepsilon_g\rho_g|\vec{u_g} - \vec{u_p}|}{d_p}C_d\omega & \varepsilon_g \geq 0.74 \end{cases}$$
$$\omega = \begin{cases} -0.576 + \dfrac{0.0214}{4(\varepsilon_g - 0.7463)^2 + 0.0044} & 0.74 \leq \varepsilon_g \leq 0.82 \\ -0.0101 + \dfrac{0.0038}{4(\varepsilon_g - 0.7789)^2 + 0.0040} & 0.82 < \varepsilon_g \leq 0.97 \\ -31.8295 + 32.8295\varepsilon_g & \varepsilon_g > 0.97 \end{cases} \quad (5.18)$$

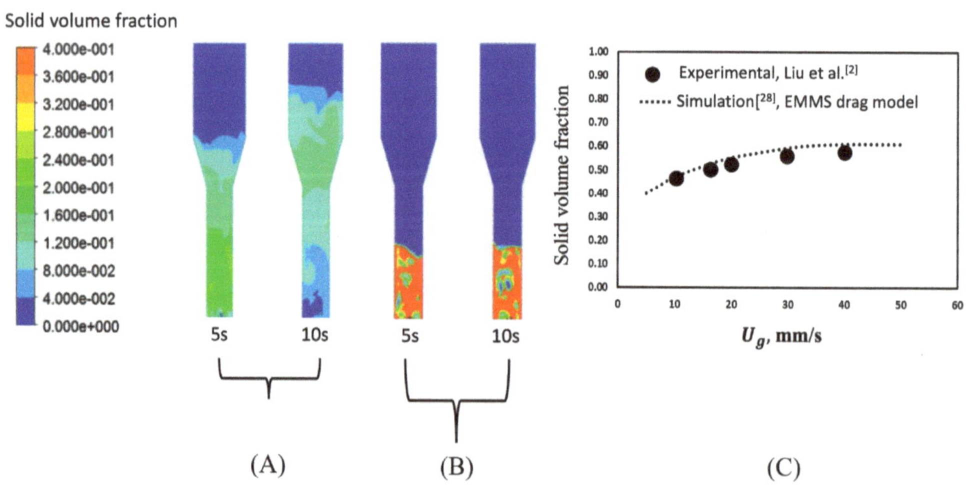

Figure 5.4 The simulated time-averaged solid volume fraction as a function of superficial gas velocity compared with experimental data. (A) Gidaspow drag model. (B) EMMS drag model. (C) average solid volume fraction versus U_g.

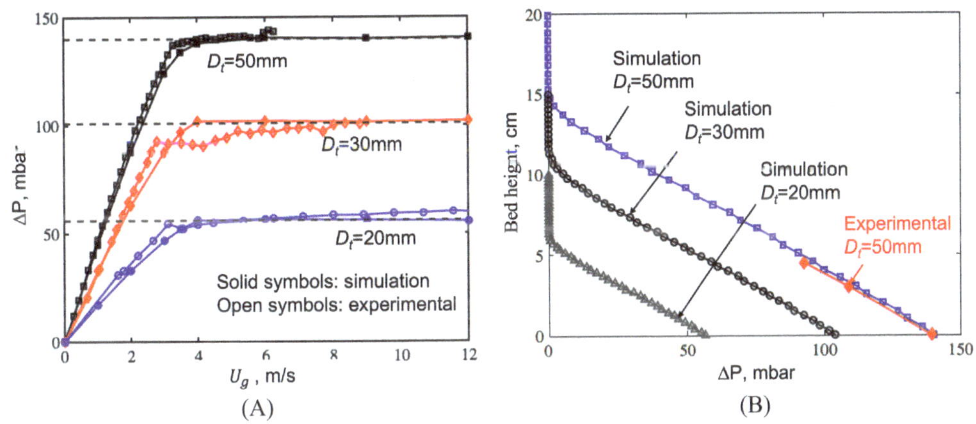

Figure 5.5 Comparison between experimental measurements and numerical results [29]. (A) The bed pressure drop versus superficial gas velocity U_g curves. (B) Time-averaged axial profiles of gas pressure ($U_g = 12$ cm/s).

where ω is a correction factor. Ansart et al. [29] simulated a micro-fluidized bed of dense Tungsten particles ($\rho_p = 19300$ kg/m^3) using an Eulerian n-fluid approach. The Wen and Yu drag model [30] (the second equation in Eq. (5.15)) is used as the drag model. Fig. 5.5 shows a comparison between the simulated and experimental-measured bed pressure drop in the three steel columns [29]. In Fig. 5.5A, the horizontal dashed lines correspond to the bed weight of particles per column surface area. In Fig. 5.5B, P_{ref} is the reference pressure above the dense fluidized bed. A good agreement can be observed between the model predictions and the experimental data. Fig. 5.5A also indicates that the numerical simulations predict an increase in minimum fluidization velocity with decreasing bed diameter, a trend consistent with the experimental observations.

5.2.3 TFM predicted MFB hydrodynamics

Fig. 5.6A plots the predicted axial profiles of the time-averaged solid volume fraction in a 3×3 mm gas-solid micro-fluidized bed. It shows that increasing the superficial gas velocity leads to a decrease in the solid volume fraction and an increase in the bed height. The model-predicted bed solid volume fraction and height are seemingly independent of the specularity coefficient ϕ. However, the influence of the specularity coefficient ϕ on the lateral distribution of solid volume fraction is significant. As shown in Fig. 5.6B, the free-slip boundary wall condition ($\phi = 0$) results in higher solid concentrations near the wall and at the center than the no-slip boundary wall condition ($\phi = 1$). A dense annular zone near the wall is seen under the free-slip boundary wall condition, but not under the no-slip boundary wall condition. Fig. 5.7 shows the predicted profiles of time-averaged gas and solid axial velocities under conditions of $\phi = 0$ and $\phi = 1$. It clearly shows that the radial distributions of time-averaged gas and solid axial velocities are strongly affected by the value of ϕ. Particularly, distributions of time-averaged gas and solid axial velocities in the region near the wall are completely different when ϕ is changed from 0 to 1.

Jia [31] simulated a 2.8 × 6 mm gas-solid micro-fluidized bed using the TFM with varying the specularity coefficient from 0 to 1. The simulated lateral profiles of solid volume fraction and axial velocity are shown in Fig. 5.8. The results show that the specularity coefficient has a pronounced impact on the predicted flow structure. Depending on the bed height, the influence of the specularity on the solid volume fraction and axial velocity appears differently.

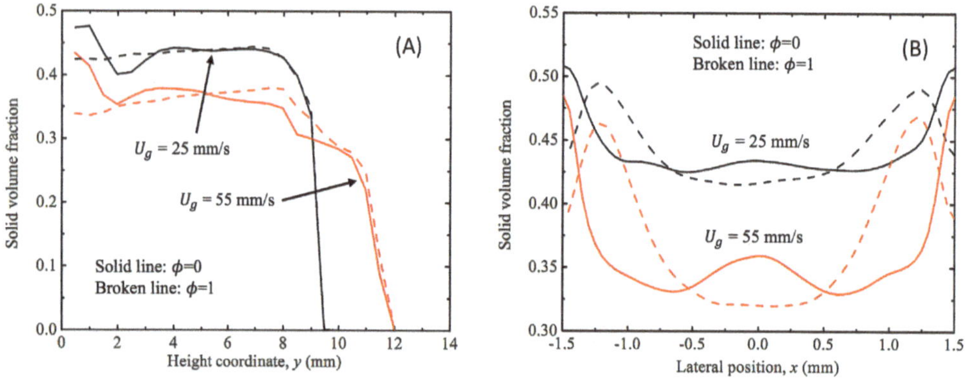

Figure 5.6 The simulated (A) axial and (B) lateral profiles of time-averaged solid volume fraction of a Group A particle in a 3 × 3 mm gas-solid micro-fluidized bed [2].

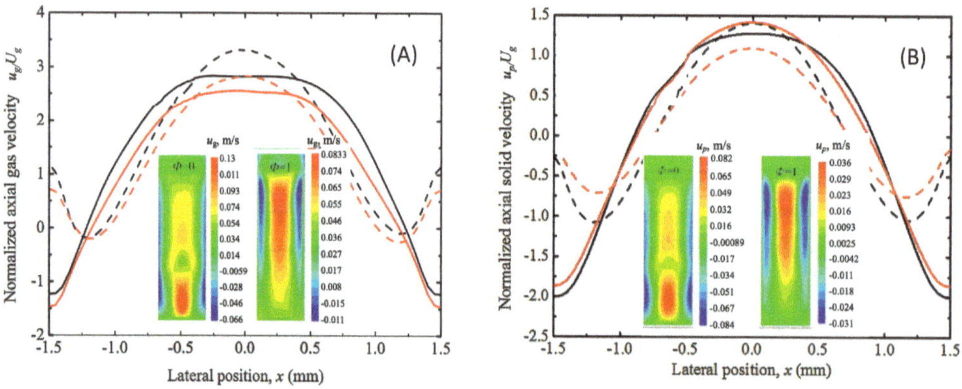

Figure 5.7 The simulated lateral profiles of time-averaged (A) gas and (B) solid axial velocities of a Group A particle in a 3 × 3 mm gas-solid micro-fluidized bed. Solid line: $\phi = 0$, broken line: $\phi = 1$; black line: $U_g = 25$ mm/s, red line: $U_g = 55$ mm/s. Insets provide snapshots of the time-averaged gas-solid axial velocities at $U_g = 25$ mm/s [2].

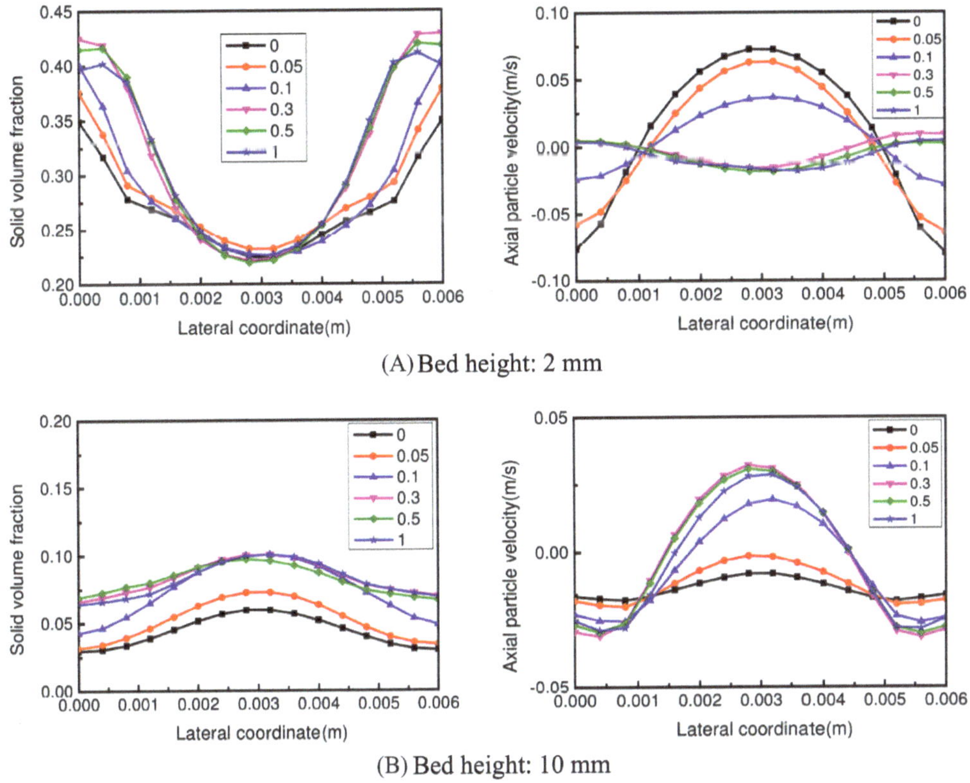

Figure 5.8 The lateral profiles of time-averaged gas and solid axial velocities in a 2.8 × 6 mm gas-solid micro-fluidized bed simulated by the TFM with different specularity coefficients at $U_g = 120$ mm/s and at bed heights of (A) 2 mm and (B) 10 mm [31].

Fig. 5.9 shows the TFM simulation results of three micro-fluidized beds of the inner diameter of 20, 30, and 50 mm with a Group B particle [29]. In these simulations, the no-slip wall boundary condition ($\phi = 1$) is used for the gas and the particles. As shown in Fig. 5.9A, the time-averaged solid volume fraction is the smallest at the center of the bed and increases along the radial direction toward the wall until it reaches the maximum value and then decreases to a lower value at the wall. This result confirms the previous discussions on the effect of confining walls on the solid volume fraction presented in Chapter 2; that is, the smaller the bed diameter, the higher the bed voidage and the flow nonuniformity. Fig. 5.9B shows that the radial distribution of the time-averaged vertical component of particle velocity features a positive maximum at the center and a negative minimum near the wall. The results show that the maximum particle velocity and the solid flux at the center of the bed decrease with reducing bed diameter, since the small beds limit the growth and flow of bubbles.

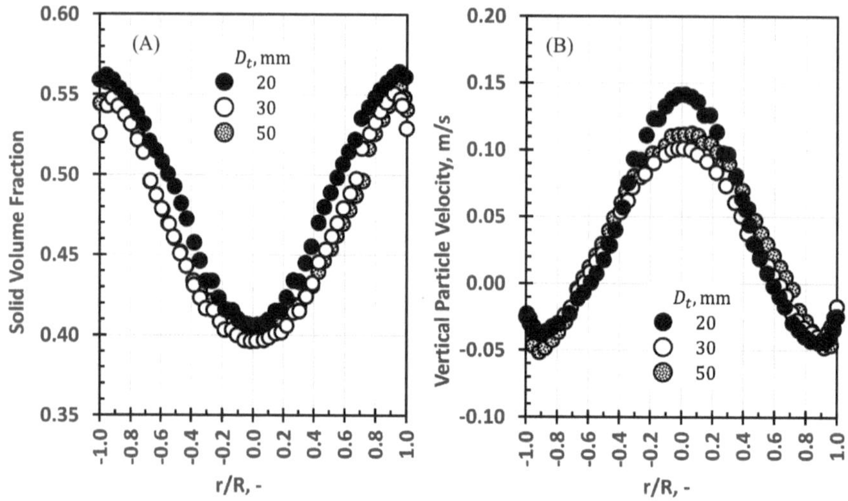

Figure 5.9 Effect of column diameter on the time-averaged (A) radial profiles of solid volume fraction and (B) vertical particle velocity at 2/3 of the bed height and U_g = 12 cm/s [29].

Fig. 5.10 shows the simulated gas residence time distributions in a gas-solid micro-fluidized bed of diameter 15 mm using the TFM. The particles used in the simulations are 88 μm in diameter and 1659 kg/m^3 in density. The experimental RTD is also plotted for comparison. It shows that the gas flow exhibits some extent of backmixing characterized by somewhat asymmetrical shapes and tails of the RTD curves, as expected for the Group A particles (see Chapter 3). The simulated RTD curves generally agree with the experimental curve, but the accuracy needs further improvement. Nevertheless, the simulations show that the drag models (Fig. 5.10A), the restitution coefficient (Fig. 5.10B), and the specularity coefficient (Fig. 5.10C) have noticeable impacts on the simulation results, although they have little effects on the calculated mean residence time and variance.

In summary, the TFM can simulate the hydrodynamics in gas-solid micro-fluidized beds. The challenge for a successful simulation is to set up the closure equations and the associated parameters properly. The presentation above demonstrates that in order to make simulations of a model reasonable and accurate, it is critical to select an appropriate drag model and correctly specify the parameters, such as the specularity coefficients. In fact, ga-solid fluidization is a highly complex system, wherein some factors, such as the scale-dependent wall effects, gas-solid interactions, and interparticle forces, have not been fully understood. In practice, almost all drag models contain case-dependent empirical parameters, and there is no theoretical, universally applicable drag model available yet. For micro-fluidized beds, lacking direct experimental results makes it difficult to validate the adopted closure equations and parameters. These issues need to be addressed in the future as the TFM continues to be developed as a major modeling approach for micro-fluidized bed applications.

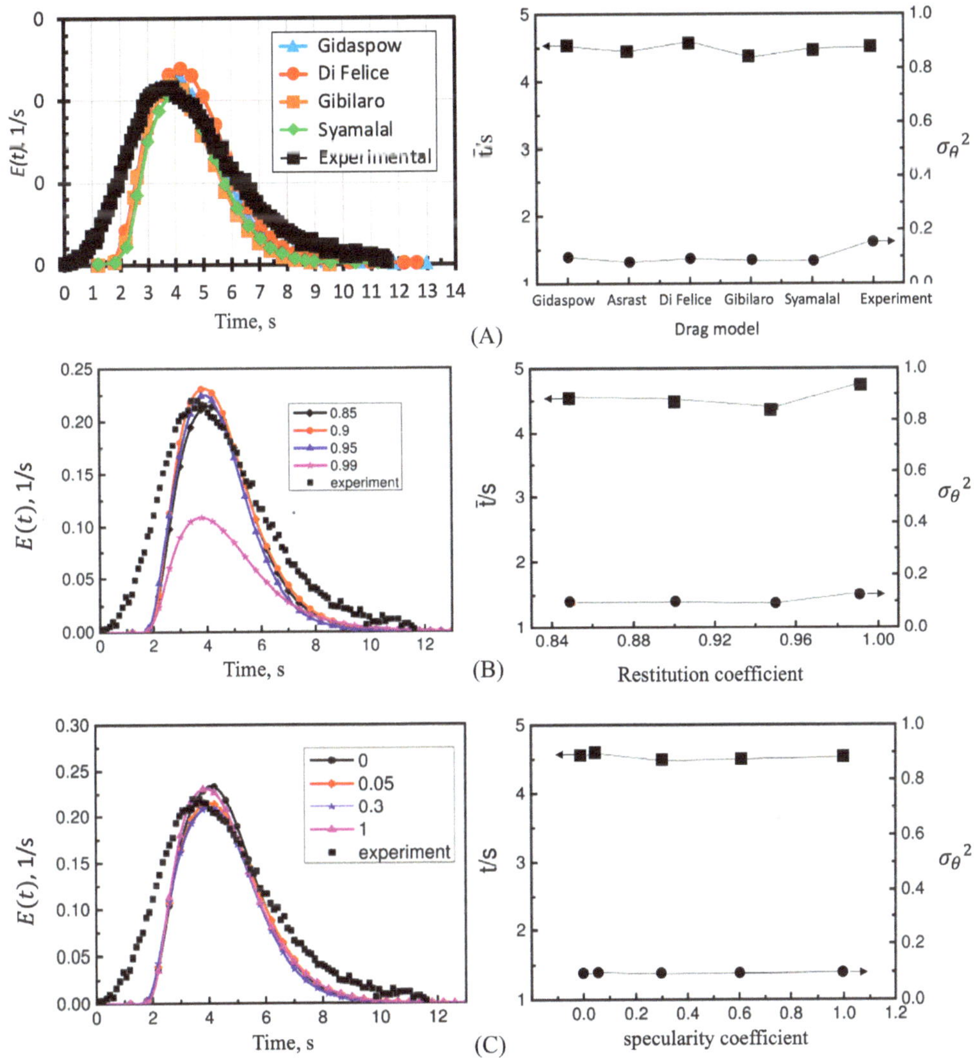

Figure 5.10 The simulated gas RTD curves in a gas-solid micro-fluidized bed of diameter 15 mm with a particle of diameter 88 μm by the TFM with (A) different drag models (Gidaspow [4], Di Felice [32], Gibilaro [33], Syamalal [34]), (B) different restitution coefficients, and (C) different specularity coefficients, at $U_g = 15$ mm/s [31].

5.3 The discrete element method

The DEM is a popular CFD modeling approach to tracking the motion of individual particles. The DEM simulations take all particle–particle and particle–geometry interactions into consideration. The DEM has been used to model gas-solid micro-fluidized beds by several researchers [3,23,35–38]. Wang et al. [38] modeled a membrane-assisted micro-fluidized bed reactor with the coupled CFD-DEM approach. The study found that the onset velocity of turbulent fluidization was significantly reduced in micro-fluidized beds compared to large beds. Li and Ji [37] employed CFD-DEM to investigate solid mixing behavior in a two-dimensional (4 × 150 mm) micro-fluidized bed. The simulation results reveal that the solid mixing in the micro-fluidized bed is intensive. Tan et al. [23] used a discrete particle model to investigate the effect of gas permeation on the hydrodynamic characteristics of a membrane-assisted micro-fluidized bed. The simulations indicate that gas permeation may have an adverse impact on bed expansion. In addition, the formation of an increased solid holdup zone near the membrane wall, which is observed in the case of gas extraction, may increase the bed-to-membrane mass transfer resistance. Galvin and Benyahia [39] studied the effects of van der Waals-type cohesive forces on fluidization and defluidization of Group A powders in a micro-fluidized bed by a numerical method. The results demonstrate that the interparticle cohesive forces are crucial to maintaining close contact between particles. As a result, there is a noticeable spike in pressure drop and hysteresis around the minimum fluidization. Xu et al. [35] investigated the effects of bed diameter and static bed height on the fluidization behavior of a Geldart B particle in micro-fluidized beds. The predictions of the CFD-DEM model agree qualitatively well with the experimental observations in terms of pressure overshoot and delay of the minimum fluidization. Liu et al. [3] studied the bed hydrodynamics using CFD-DEM and used the results as a benchmark to investigate the applicability of TFM for simulating micro-fluidized beds. Recently, Finn et al. [36] employed CFD-DEM to investigate flue gas CO_2 capture by micro-encapsulated solvents in a 24-mm-diameter micro-fluidized bed. The model consists of mass, momentum, energy, and CO_2 absorption/desorption equations. The validity of the model is confirmed by comparing the model-predicted and experimental-measured CO_2 absorption rates and gas temperature evolutions.

The main challenge in adequately accounting for interparticle forces (IPFs) in hydrodynamic correlations lies in the unavailability of a reliable technique for quantifying IPFs in a fluidized bed, particularly at elevated temperatures.

5.3.1 Model formulation

In the CFD-DEM approach, the particulate phase is represented by a finite number of spherical particles. Assume that a single particle has a diameter of d_p, a density of ρ_p, a volume of $v_p = (\pi/6)d_p{}^3$, and a mass of $m_p = (\pi/6)d_p{}^3\rho_p$. Let $\boldsymbol{x_p}$ denote the position of a particle and F_c denote the net contact force due to particle contact with walls and other particles. The motion of each particle i is described by Newton's laws of motion as follows:

$$\frac{d\boldsymbol{x}_{p,i}}{dt} = \overrightarrow{u_{p,i}} \tag{5.19}$$

$$m_{p,i}\frac{d\overrightarrow{u_{p,i}}}{dt} = m_{p,i}g - v_p\nabla P + v_p\frac{\beta_{gs}}{\varepsilon_s}\left(\overrightarrow{u_g} - \overrightarrow{u_{p,i}}\right) + F_{c,i} \tag{5.20}$$

Assuming that the particle angular moment of inertia $\boldsymbol{\Gamma}_i$ is generated solely by torque from particle–particle and particle–wall collisions, the rotational motion of the particle i is given by

$$I_{p,i}\frac{d\boldsymbol{\omega}_i}{dt} = \boldsymbol{\Gamma}_i \tag{5.21}$$

where $I_p = m_p d_p^{\,2}/10$ is the moment of inertia, and $\boldsymbol{\omega}$ is the angular velocity.

In calculating particle–particle and particle–wall interactions, two types of collision models are widely used, namely, the hard-sphere and the soft-sphere models. The hard-sphere model assumes that the particles interact through instantaneous binary collisions, and momentum conservation is applied to calculate the velocities after a particle collision. On the other hand, the soft-sphere model calculates the contact force between colliding particles based on the deformation of colliding particles using a combination of spring and dashpot. The soft-sphere model is typically used for modeling dense gas-solid fluidized beds.

With the soft-sphere model, the total contact force on the particle i is calculated by the sum of contact forces with adjacent particles in normal and tangential directions according to appropriate models. It is also worth noting that interparticle forces can be caused by a variety of factors, for example, capillary forces, electrostatic forces, and van der Waals interactions. In Eq. (5.20), these interparticle forces may also be included [39,40]. Readers are encouraged to consult with the cited and other relevant literature for details [3,23,35,41,42].

5.3.2 DEM simulations and validations

Fig. 5.11 shows the variations of model-predicted minimum fluidization velocity U_{mf} with the ratio D_t/d_p. It shows that U_{mf} increases with decreasing D_t/d_p and increasing particle–wall frictional coefficient μ_{pw}. In general, the variation trends of U_{mf} predicted by the DEM with the ratio D_t/d_p are qualitatively consistent with the experimental observations (see Chapter 4). However, the DEM predictions are quantitatively lower than the results calculated based on the correlation proposed by Rao et al. [43], but higher than the results calculated based on the correlation

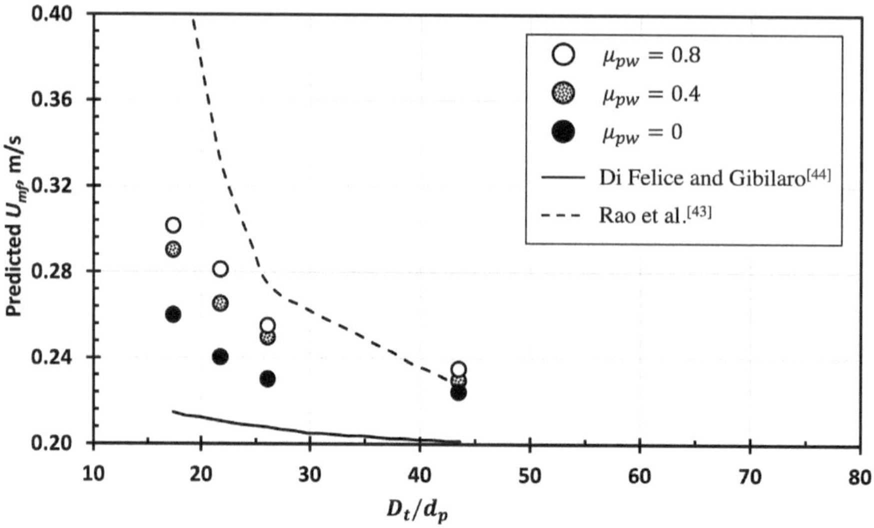

Figure 5.11 Influence of bed inner diameter and particle–wall frictional coefficient on the minimum fluidization velocity predicted by CFD-DEM simulations [35].

developed by Di Felice and Gibilaro [44]. Liu et al. [3] explained that the discrepancies might be partially attributable to the different bed voidage values and the two empirical correlations used in the model. Furthermore, the DEM predicts that U_{mf} increases with the increase of wall friction for the same-diameter beds. This is reasonable because, as discussed in Chapter 2, the confining wall can also lead to nonuniform flow distributions and bed voidage variations, both of which affect the gas-solid flow in micro-fluidized beds.

Fig. 5.12 plots the variations of the standard deviation of the bed pressure drop as a function of superficial gas velocity in 10- and 20-mm-diameter micro-fluidized beds. As indicated in the figure, the gas velocity corresponding to the maximum standard deviation is the onset velocity of turbulent fluidization U_c. It shows that U_c ranges from 0.85 to 1.15 m/s when $D_t = 10$, and from 1.5 to 2.0 m/s when $D_t = 20$ mm [3]. The result agrees with the experimental observations that an advanced onset of turbulent fluidization occurs due to the decrease in bed diameter, as discussed in Chapter 4.

5.3.3 DEM predicted MFB hydrodynamics

Fig. 5.13 presents the bed pressure drop overshoots predicted by DEM under various values of D_t/d_p and H_s/D_t. It shows that pressure drop overshoots are close to zero when $D_t/d_p > 150$ and $H_s/D_t < 2$. The simulation results show that uneven lateral distributions of gas and solid flows may occur if the bed is too shallow. Therefore, to ensure a uniform gas-solid flow and minimized influence of the wall effect, it is

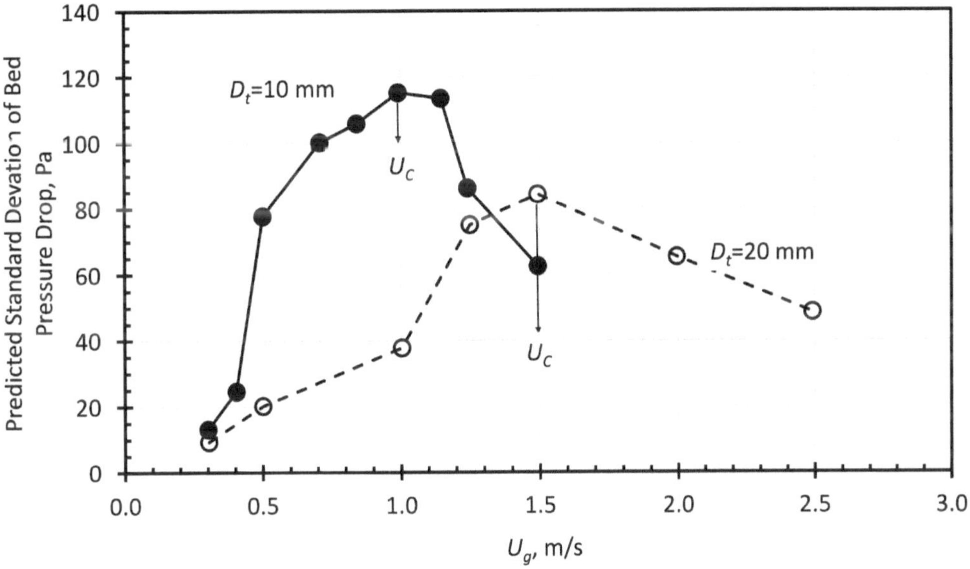

Figure 5.12 The CFD-DEM predicted standard deviations of bed pressure as a function of superficial gas velocity in two micro-fluidized beds of $D_t = 10$ and 20 mm. ($d_p = 460\ \mu m$, $\rho_p = 2600\ kg/m^3$) [3].

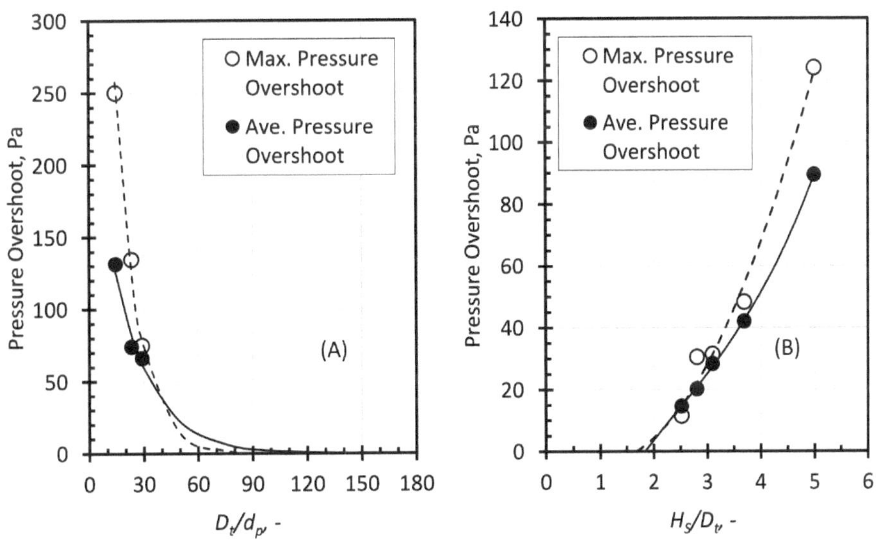

Figure 5.13 The predicted variations of the pressure drop overshoot with ratios (A) D_t/d_p and (B) H_s/D_t [35].

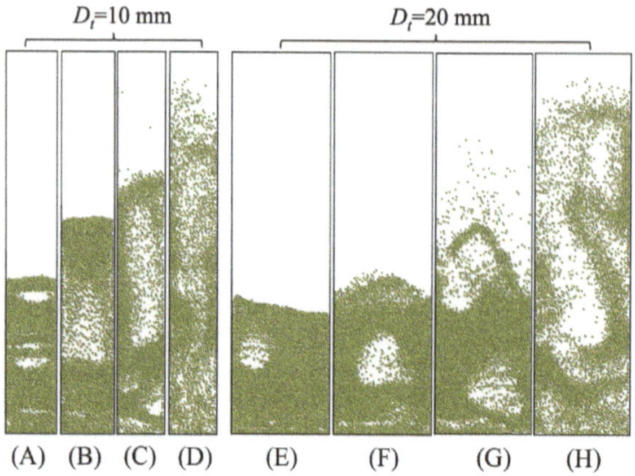

Figure 5.14 CFD-DEM simulated flow patterns at different superficial gas velocities. (A, E) $U_g = 0.3$ m/s; (B, F) $U_g = 0.5$ m/s; (C, G) $U_g = 1.0$ m/s; (D, H) $U_g = 1.5$ m/s. $d_p = 460$ μm, $\rho_p = 2600kg/m^3$, $U_{mb} = 0.265$ m/s [3].

recommended to use a fluidized bed with a diameter of 50–150 times the particle diameter and a static bed height of 2–4 times the bed diameter.

For the gas-solid micro-fluidized beds of 10 and 20 mm in diameter, the flow structures simulated by the CFD-DEM are illustrated in Fig. 5.14 [3]. We can see that gas bubbles form almost immediately after minimum fluidization, as shown in Fig. 5.14A and E. Increasing gas velocity leads to growth in the bubble size. When the bubble size becomes substantially large to be comparable to the bed diameter, the bed enters the slugging regime, as shown in Fig. 5.14(B and F) and (C and G). When the superficial gas velocity continues to increase to 1.5 m/s, as shown in Fig. 5.14D and H, stable gas slugs cannot be sustained due to their vigorous breakup, so the bed enters the turbulent regime. According to the simulation results, it can be confirmed that reducing the bed size will lead to serious slugging fluidization, featured by the increased bed pressure fluctuation and the advanced transition to turbulent fluidization in micro-fluidized beds, which is consistent with the experimental results presented in Chapter 4.

Fig. 5.15 shows the radial distributions of solid volume fraction and axial velocity calculated by the CFD-DFM [3] in the micro-fluidized beds of 10 and 20 mm in diameter. The results show that the solid volume fraction is high near the vicinity of the walls where the particles flow downward. By comparison, the solid volume fraction in the central region of the bed where the particles flow upward is relatively low. It indicates a characteristic core-annular gas-solid flow structure. Fig. 5.15 also shows that the smaller diameter beds, especially with higher static beds, have a substantial wall effect characterized by increased uniformity in the distributions of solid volume fraction and axial velocity.

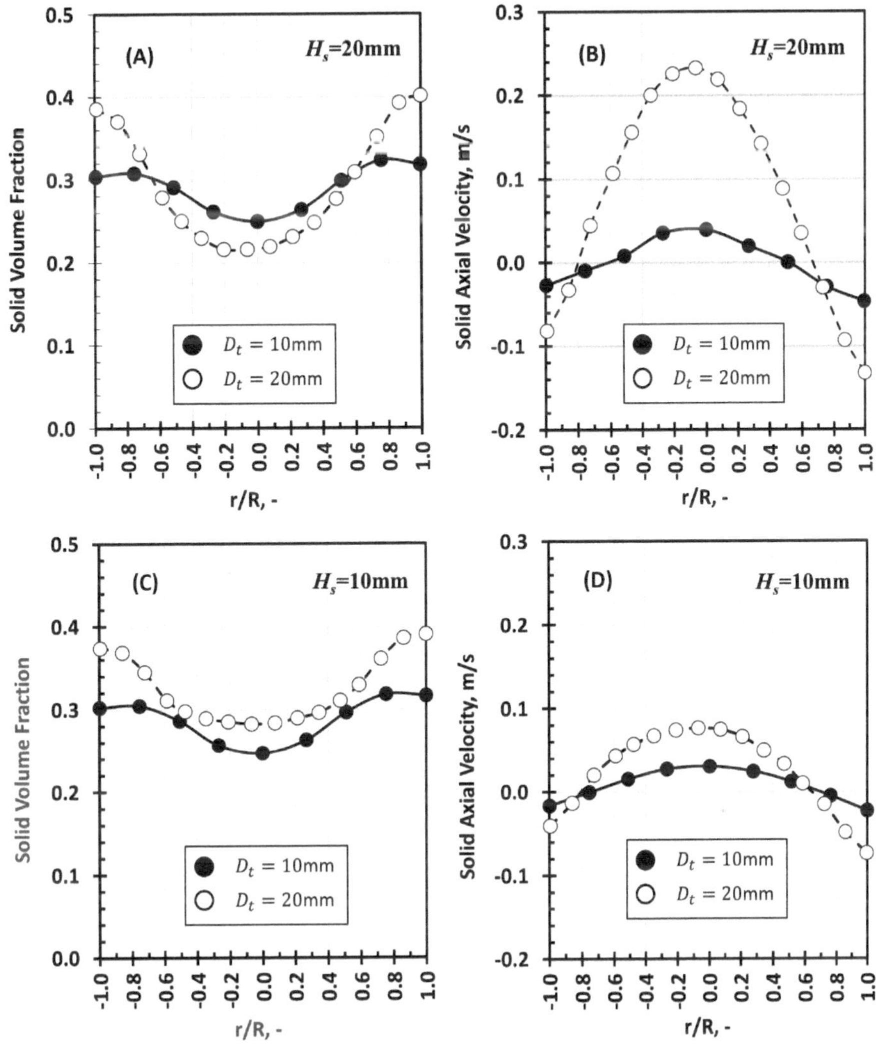

Figure 5.15 The radial distribution of solid volume fraction (A) $H_s = 20$ mm, (C) $H_s = 10$ mm and axial profile of particle velocity, (B) $H_s = 20$ mm, (D) $H_s = 10$ mm. predicted by the CFD-DEM ($d_p = 460\mu m$, $\rho_p = 2600 kg/m^3$) [3].

As mentioned early, numerical simulations are powerful to provide valuable insights into the gas-solid flow characteristics, which are hardly obtainable by experimental techniques in some circumstances. For illustration purposes, the CFD-DEM simulations of an 8-mm-diameter gas-solid micro-fluidized bed are shown in Fig. 5.16. In these simulations, the particle–wall friction coefficient μ_{pw} varies from 0.0 to 0.4 while the bed diameter is kept unchanged. The numerical modeling makes it possible to separate the effect of wall friction and that of the bed size on bed hydrodynamics. In Fig. 5.16A, the pressure drops of the bed with varying particle—wall frictional coefficient μ_{pw} are presented. It shows that when $\mu_{pw} = 0$, no noticeable pressure overshoot is predicted, which is consistent with the discussions previously in Chapter 2. As μ_{pw} increases, both the pressure overshoot and average bed pressure drop increase, as indicated in Fig. 5.16B. The increase in simulated bed pressure drop suggests that the wall friction adds the resistance for the fluidizing gas to overcome in fluidization. However, this prediction contradicts some experimental results that the increased wall effect results in pressure offsets [45,46]. Nevertheless, the simulations suggest, doubtlessly, that the wall effects complicate the hydrodynamics of micro-fluidized beds.

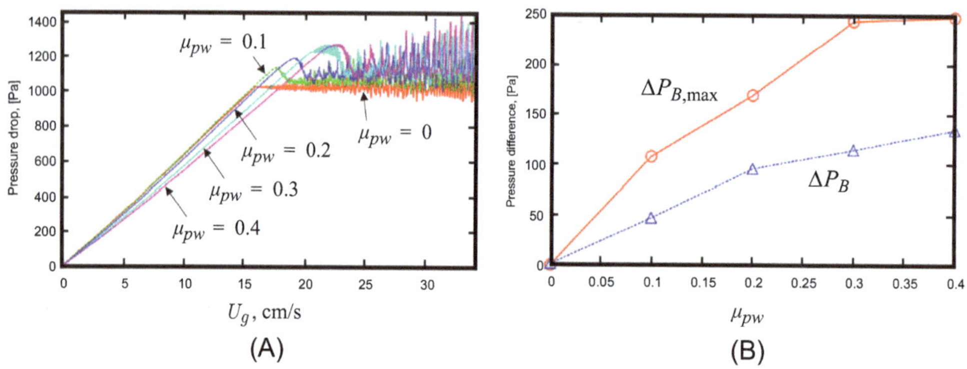

Figure 5.16 CFD-DEM simulation results in a gas-solid micro-fluidized bed of 8 mm in diameter. (A) Pressure drop, (B) pressure overshoot and average bed pressure drop. [35].

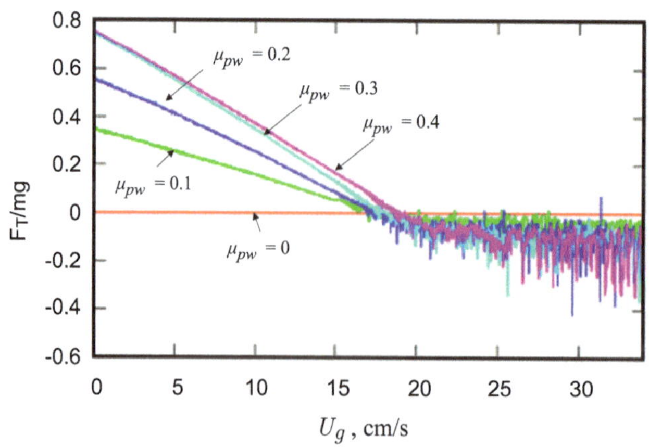

Figure 5.17 The tangential force exerted on the particles by the wall in a micro-fluidized bed of 8 mm in diameter [35].

Fig. 5.17 shows the total tangential forces between the particles and the wall estimated by the CFD-DEM under different particle–wall friction coefficients. It indicates that the tangential force is positive in the fixed bed ($U_g < 15–20$ cm/s), suggesting that the weight of the particles is partially supported by the wall friction in the fixed bed. The tangential force turns negative when the bed is fluidized, indicating that particles move upward during fluidization. The tangential force in both the fixed and fluidized beds increases with increasing wall friction coefficient, but the increase diminishes when $\mu_{pw} > 0.3$.

The effect of particle–wall friction on the minimum fluidization velocity and the corresponding bed solid volume fraction are predicted by the CFD-DEM, as shown in Fig. 5.18. The predicted solid volume fraction decreases from 0.634 to 0.614 as μ_{pw} increases from 0 to 0.4. The results suggest that the wall friction supports the weight of the particles in the fixed bed, and the bed voidage increases accordingly. When there is no particle–wall friction ($\mu_{pw} = 0$), the minimum fluidization velocity predicted by the CFD-DEM agrees well with that calculated by the Ergun equation, which is only applicable to the wall friction-free conditions. However, when the particle–wall friction exists ($\mu_{pw} > 0$), the minimum fluidization velocity predicted by the CFD-DEM is larger than that calculated by the Ergun equation[47], and the difference becomes significant as the particle–wall friction coefficient increases. Therefore, the particle–wall friction delays the minimum fluidization, consistent with the previous discussions in Chapter 4.

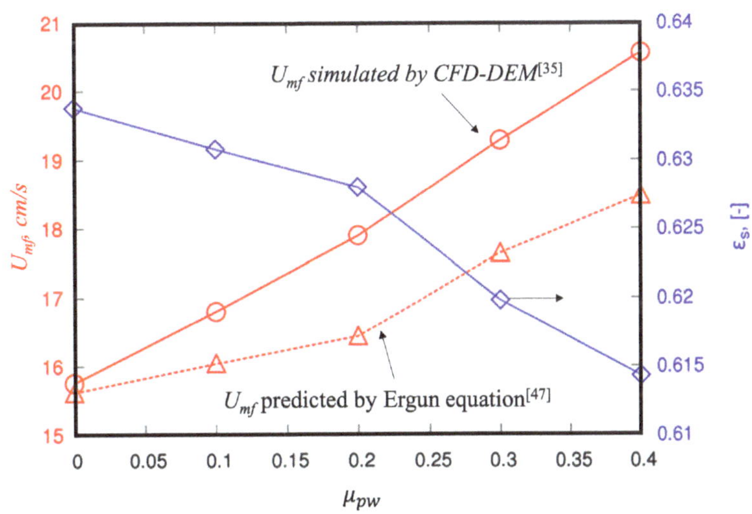

Figure 5.18 Minimum fluidization velocity and solid volume fraction under different particle–wall friction coefficients [35].

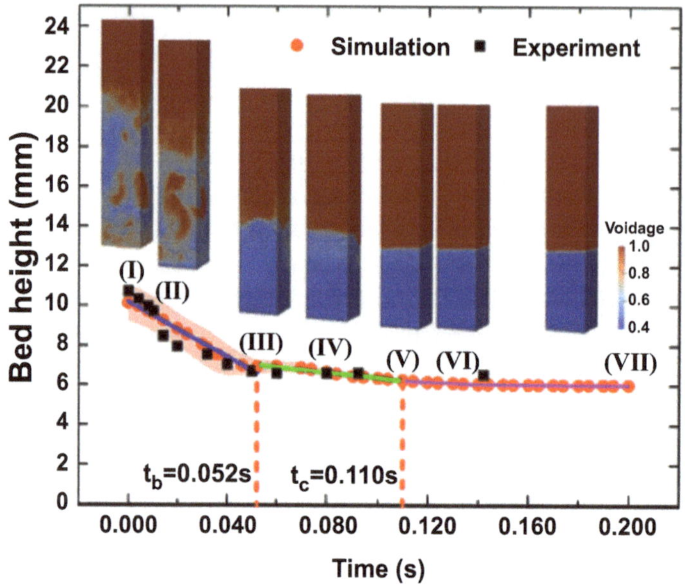

Figure 5.19 Comparison of bed collapse curves obtained from CFD-DEM simulation [48] and experiment of Wang and Fan [25]. Typical snapshots of simulation with a superficial gas velocity of 55 mm/s. (I) $t = 0$ s; (II) stage 1: bubble escape; (III) $t = 0.052$ s; (IV) stage 2: hindered sedimentation; (V) $t = 0.110$ s; (VI) stage 3: solid consolidation; (VII) $t = 0.200$ s.

In recent work, Li et al. [48] simulated the bed collapsing behavior of Geldart A particles in a $3 \times 3 \times 27$ mm gas-solid micro-fluidized bed using the CFD-DEM modeling. Gidaspow's model was used to quantify particle–gas interphase drag force. The simulated bed collapsing behavior is shown in Fig. 5.19. Because bubbles flow vigorously inside the bed and break up at the bed surface, the instantaneous voidage changes momentarily. Furthermore, the initial state setups also lead to a significant difference in the calculated axial bed voidage profiles. Therefore, the average results of 500 simulations with different initial conditions are presented in Fig. 5.19. Generally, the bed collapse curve obtained from the CFD-DEM simulation agrees well with the experimental measurement. The model successfully predicts the three stages of the bed collapsing process: 1: bubble escape stage (II); 2: hindered sedimentation stage (IV); and 3: solid consolidation stage (VI).

5.4 A brief discussion and future perspective

This chapter briefly presents the most commonly used TFM and CFD-DEM modeling with exemplary applications. It demonstrates that these two models can provide an incredible amount of information on gas-solid flow structures in micro-fluidized beds (e.g., gas-solid interaction, particle–wall interaction, solid volume fraction, particle velocity, and flow regime transitions). Doubtlessly, the modeling can be used as a complementary tool to help interpret experimental findings, guide experimental designs, improve measurement techniques, and testify new technological concepts. Also, the CFD modeling can help reveal some physical and chemical phenomena that are hard to obtain experimentally due to practical constraints in bed geometric size and/or experimental techniques. Therefore, it is reasonable to expect that CFD modeling will continue to play an increasingly important role in the research of micro-fluidized beds in the future.

This chapter and the cited literature demonstrate that the CFD modeling requires inputs of some empirical parameters, but in some cases, no direct experimental data are available to validate these empirical parameters. It is a huge challenge and an opportunity for CFD modeling in this research area.

Abbreviation

CFD computational fluid dynamics
CGP coarse-grained particle
DNS direct numerical simulation
DPM discrete particle method
IPF interparticle force
MFB micro-fluidized bed
PIC particle in cell
TFM two-fluid method

Nomenclature

$\overrightarrow{u_g}$ mean gas velocity vector (m/s)
$\overrightarrow{u_p}$ mean solid particle velocity vector (m/s)
$\vec{U}_{sw}$ particle slip velocity parallel to the wall (m/s)
$\vec{g}$ gravitational acceleration vector (m/s^2)
I_p moment of inertia
$\vec{I}$ identity matrix
$\vec{n}$ unit vector normal to the wall
U_c onset velocity to turbulent fluidization (m/s)
$\boldsymbol{C_d}$ drag coefficient
$\boldsymbol{C_{mp}}$ scaling factor
$\boldsymbol{d_p}$ particle diameter (m)
$\boldsymbol{e_{pp}}$ restitution coefficient for particle—particle interaction
$\boldsymbol{e_{pw}}$ restitution coefficient for particle—wall collision
$\boldsymbol{F_c}$ net contact force due to particle contact with walls and other particles (N)
$\boldsymbol{m_{gs}}$ momentum transfer
$\boldsymbol{m_p}$ mass of a particle
$\boldsymbol{P}$ pressure of gas phase (pa)
$\boldsymbol{P_s}$ pressure of particle phase (pa)
$\boldsymbol{Re_p}$ Reynolds number based on particle velocity

Greek letters

x_p position of a particle
Γ_i particle angular inertia moment
μ_{pw} particle–wall frictional coefficient
κ_s conductivity of the fluctuating energy
χ_γ radial distribution function
β_{gs} drag coefficient between gas phase and solid phase
γ collisional energy dissipation
γ_{gs} mass transfer rate from the gas phase to the solid phase
ε_g gas-phase volume fraction
ε_s solid-phase volume fraction
$\varepsilon_{s,max}$ maximum volume fraction of solid particles
Θ_s granular temperature
λ_s solid-phase bulk viscosity (pa.s)
λ_s solid-phase shear viscosity (pa.s)
ν_p volume of a particle (m^3)
ρ_g gas density (kg/m^3)
ρ_p particle density (kg/m^3)
τ_g shear stress tensors of gas phase
τ_s shear stress tensors of solid phase
ω angular velocity, or correction factor
ϕ specularity coefficient

References

[1] Grace J.R., Bi X., Ellis N. Essentials of fluidization technology, Wiley-VCH Verlag GmbH & Co. KGaA, Boschstr. 12, 69469 Weinheim, Germany; 2020.
[2] Liu X, Zhu C, Geng S, et al. Two-fluid modeling of Geldart A particles in gas-solid micro-fluidized beds. Particuology 2015;21:118–27.
[3] Liu X, Su J, Qian Y, et al. Comparison of two-fluid and discrete particle modeling of gas-particle flows in micro fluidized beds. Powder Technology 2018;338:79–86.
[4] Gidaspow D. Multiphase flow and fluidization: continuum and kinetic theory descriptions with applications. New York: Academic Press, Inc; 1994.
[5] Qi H, Dai Q, Chen C. The key scientific problems in the eulerian modeling of large-scale multi-phase flows- drag model. Mechanics in Engineering 2014;36(3):269–77.
[6] Loha C, Gu S, De Wilde J, et al. Advances in mathematical modeling of fluidized bed gasification. Renewable and Sustainable Energy Reviews 2014;40:688–715.
[7] Nikolopoulos A, Papafotiou D, Nikolopoulos N, et al. An advanced EMMS scheme for the prediction of drag coefficient under a 1.2 MWth CFBC isothermal flow-Part I: numerical formulation. Chemical Engineering Science 2010;65(13):4080–8.
[8] Wang W, Li J. Simulation of gas–solid two-phase flow by a multi-scale CFD approach – of the EMMS model to the sub-grid level. Chemical Engineering Science 2007;62(1–2):208–31.
[9] Tian Y, Lu B, Li F, et al. A steady-state EMMS drag model for fluidized beds. Chemical Engineering Science 2020;219:115616.
[10] He M, Zhao B, Wang J. A unified EMMS-based constitutive law for heterogeneous gas-solid flow in CFB risers. Chemical Engineering Science 2020;225:115797.
[11] Yang N, Wang W, Ge W, et al. CFD simulation of concurrent-up gas–solid flow in circulating fluidized beds with structure-dependent drag coefficient. Chemical Engineering Journal 2003;96(1):71–80.
[12] Wang W, Lu B, Geng J, et al. Mesoscale drag modeling: a critical review. Current Opinion in Chemical Engineering 2020;29:96–103.

[13] Zhu L-T, Liu Y-X, Tang J-X, et al. A material-property-dependent sub-grid drag model for coarse-grained simulation of 3D large-scale CFB risers. Chemical Engineering Science 2019;204:228–45.
[14] Chen X, Song N, Jiang M, et al. Theoretical and numerical analysis of key sub-grid quantities' effect on filtered Eulerian drag force. Powder Technology 2020;372:15–31.
[15] Pang B, Wang S, Lu H. A modified drag model for power-law fluid-particle flow used in computational fluid dynamics simulation. Advanced Powder Technology 2021;32(4):1207–18.
[16] Zhao L, Chen X, Zhou Q. Inhomogeneous drag models for gas-solid suspensions based on sub-grid quantities. Powder Technology 2021;385:170–84.
[17] Jiang Y, Chen X, Kolehmainen J, et al. Development of data-driven filtered drag model for industrial-scale fluidized beds. Chemical Engineering Science 2021;230:116235.
[18] Gao X, Wu C, Cheng YWEI, et al. Experimental and numerical investigation of solid behavior in a gas-solid turbulent fluidized bed. Powder Technology 2012;228:1–13.
[19] McKeen T, Pugsley T. Simulation and experimental validation of a freely bubbling bed of FCC catalyst. Powder Technology 2003;129(1):139–52.
[20] Haider A, Levenspiel O. Drag coefficient and terminal velocity of spherical and nonspherical particles. Powder Technology 1989;58(1):63–70.
[21] Johnson PC, Nott PR, Jackson R. Frictional–collisional equations of motion for particulate flows and their application to chutes. Journal of Fluid Mechanics 1990;210:501–36.
[22] Bougamra A, Huilin L. Modeling of chemical looping combustion of methane using a Ni-based oxygen carrier. Energy & Fuels 2014;28:3420–9.
[23] Tan L, Roghair I, van Sint Annaland M. Simulation study on the effect of gas permeation on the hydrodynamic characteristics of membrane-assisted micro fluidized beds. Applied Mathematical Modelling 2014;38(17–18):4291–307.
[24] Ghadirian E, Arastoopour H. Numerical analysis of frictional behavior of dense gas–solid systems. Particuology 2017;32:178–90.
[25] Wang F, Fan LS. Gas-solid fluidization in mini- and micro-channels. Industrial and Engineering Chemistry Research 2011;50(8):4741–51.
[26] Johnson PC, Jackson R. Frictional–collisional constitutive relations for granular materials, with application to plane shearing. Journal of Fluid Mechanics 1987;176:67–93.
[27] Zivkovic V, Biggs MJ, Alwahabi ZT. Experimental study of a liquid fluidization in a microfluidic channel. AIChE Journal 2013;59(2):361–4.
[28] Piemjaiswang R, Charoenchaipet J, Saelau T, et al. Effect of design parameters on system mixing for a micro fluidized bed reactor using computational fluid dynamics simulation. Brazilian Journal of Chemical Engineering 2020;(0123456789).
[29] Ansart R, Vanni F, Caussat B, et al. Effects of reducing the reactor diameter on the dense gas–solid fluidization of very heavy particles: 3D numerical simulations. Chemical Engineering Research and Design 2017;117:575–83.
[30] Wen CY, Yu YH. A generalized method for predicting the minimum fluidization velocity. AIChE Journal 1966;12(3):610–12.
[31] Jia Z. Experimental and numerical investigation on hydrodynamics of micro-fluidized bed. A mater thesis submitted to Tianjin University; 2016.
[32] Felice RD. The voidage function for fluid-particle interaction systems. International Journal of Multiphase Flow 1994;20(1):153–9.
[33] Gibilaro LG, Di Felice R, Waldram SP, et al. Generalized friction factor and drag coefficient correlations for fluid-particle interactions. Chemical Engineering Science 1985;40(10):1817–23.
[34] Syamlal M., Rogers W.A., O'Brien T. MFIX documentation: theory guide, technical note DOE/METC-95/1013; 1993.
[35] Xu Y, Li T, Musser J, et al. CFD-DEM modeling the effect of column size and bed height on minimum fluidization velocity in micro fluidized beds with Geldart B particles. Powder Technology 2017;318:321–8.
[36] Finn JR, Galvin JE, Hornbostel K. CFD investigation of CO_2 absorption/desorption by a fluidized bed of micro-encapsulated solvents. Chemical Engineering Science 2020;6 Article 100050.

[37] Li B, Ji L. Numerical simulation of particle mixing in circulating fluidized bed with discrete element method. Proceedings of the CSEE 2012;32(20):42–8.
[38] Wang J, Tan L, van der Hoef MA, et al. From bubbling to turbulent fluidization: advanced onset of regime transition in micro-fluidized beds. Chemical Engineering Science 2011;66 (9):2001–7.
[39] Galvin JE, Benyahia S. The effect of cohesive forces on the fluidization of aeratable powders. AIChE Journal 2014;60(2):473–84.
[40] Ye M, van der Hoef MA, Kuipers JAM. From discrete particle model to a continuous model of Geldart a particles. Chemical Engineering Research and Design 2005;83(7 A):833–43.
[41] Cundall PA, Strack ODL. A discrete numerical model for granular assemblies. Geotechnique 1979;29(1):47–65.
[42] Wu C, Cheng Y. Downer reactors Grace JR, Bi X, Ellis N. Essentials of Fluidization Technology Wiley-VCH Verlag GmbH & Co. KGaA, Boschstr 2020;12:499–530 69469 Weinheim, Germany.
[43] Rao A, Curtis JS, Hancock BC, et al. The effect of column diameter and bed height on minimum fluidization velocity. AIChE Journal 2010;56(9):2304–11.
[44] Di Felice R, Gibilaro LG. Wall effects for the pressure drop in fixed beds. Chemical Engineering Science 2004;59:3037–40.
[45] Mcdonough JR, Law R, Reay DA, et al. Fluidization in small-scale gas-solid 3D-printed fluidized beds. Chemical Engineering Science 2019;200:294–309.
[46] Guo Q, Xu Y, Yue X. Fluidization characteristics in micro-fluidized beds of various inner diameters. Chemical Engineering and Technology 2009;32(12):1992–9.
[47] Ergun S. Fluid flow through packed columns. Chemical Engineering Progress 1952;48:1179–84.
[48] Li S, Zhao P, Xu J, et al. Direct comparison of CFD-DEM simulation and experimental measurement of Geldart A particles in a micro-fluidized bed. Chemical Engineering Science 2021;242 Article 116725.

CHAPTER 6

Microreactors for thermal analysis of gas−solid thermochemical reactions

We devoted the preceding chapters to the fundamentals of micro-fluidization. From this chapter to Chapter 10, we will focus on the applications of gas−solid micro-fluidized beds. As of the writing of this book, thermal analysis of thermochemical reactions is the only commercial application of the micro-fluidized bed technology. This chapter will provide a brief background regarding thermal analysis and commonly used microreactors before we present the micro-fluidized bed reaction analyzers and their applications in detail in later chapters.

6.1 Thermal analysis approaches

6.1.1 Thermochemical reaction pathways

Gas−solid thermochemical reactions are widely employed in many industrial processes to produce essential and value-added products such as chemicals, energy, materials, foods, and pharmaceutical products. These reactions are heat-activated or heat-driven. Typical examples of such reactions include extractions of metals from their raw ores, pyrolysis/gasification/combustion of coal, biomass, and other carbonaceous fuels, thermal cracking, decomposition, reduction, numerous heterogeneous gas catalytic reactions, etc. For the purpose of illustration, the thermochemical conversion pathways of typical carbonaceous materials are presented in Fig. 6.1. It shows that thermochemical conversions are very diverse and complex. Some reactions may occur independently or as an intermediate step of multistage reactions. For example, pyrolysis, activated by heating the fuel in the absence or insufficiency of oxygen to a preferred temperature, has been used to directly produce oil, gas, and char. To produce value-added products, the pyrolysis products can also be further processed or refined by various catalytic or noncatalytic processes. Besides, pyrolysis is also an intermediary reaction step that gasification and combustion must go through. Therefore, the reaction performance of gasification and combustion depends greatly on the characteristics of the pyrolysis-produced char (such as specific surface area and pore structure) and the reaction behavior of char particles [1−3]. Due to the diversity and complexity of gas−solid reactions, it is vital to correctly understand the reaction performance, mechanism, and kinetics for operating the reactor efficiently and safely.

Micro Fluidization
DOI: https://doi.org/10.1016/B978-0-443-18718-6.00006-1
© 2023 Chemical Industry Press Co., Ltd.
Published by Elsevier Inc., under an exclusive license with Chemical Industry Press Co., Ltd.

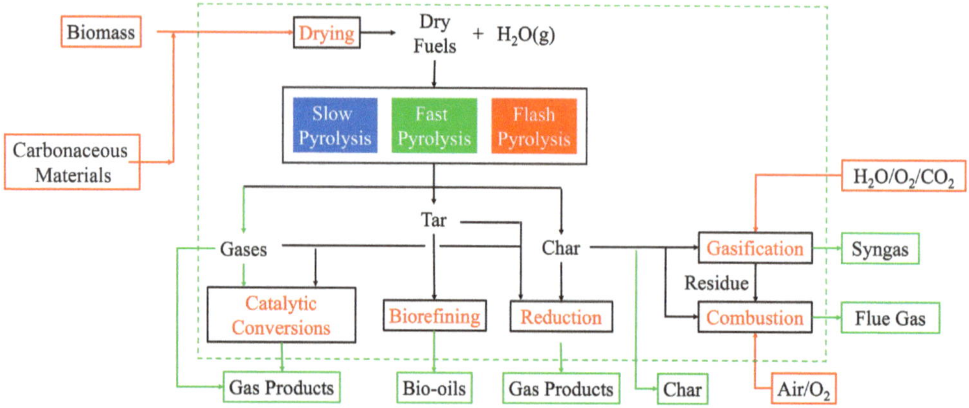

Figure 6.1 The typical pathways of gas–solid thermochemical conversions.

6.1.2 General requirements for thermal analysis approaches

A variety of thermal analysis techniques have been developed and are available to characterize gas–solid reactions. Usually, these techniques measure the temperature-dependent changes in specific physiochemical properties of a substance when it is subjected to one particular program-controlled temperature variation process. In general, thermal analysis characterizes one or more of the following properties:

1. the global reaction kinetic behavior by measuring weight loss and heat flow with appropriate thermal analysis techniques, for example, TGA, DTA, or DSC;
2. the profiles and yields of the product gas evolved (i.e., evolved gas analysis, EGA) by coupling the thermal reactor with basic analytical techniques, such as gas chromatography (GC), Fourier-transform infrared spectroscopy (FTIR), mass spectrometry (MS), and gas chromatography and mass spectrometry (GC-MS);
3. the reaction mechanism and intrinsic kinetics.

For gas–solid reactions shown in Fig. 6.1, the following factors need to be carefully considered when selecting an appropriate analytic technique to ensure that the required information can be obtained.

1. Gas flow condition

   A close-to-plug flow of gas is an essential requirement for accurately detecting and measuring the sequence and dynamic release of the product gas from sample particles. The close-to-plug flow of gas prevents gases produced at different reaction times from mixing and minimizes or eliminates the possible secondary reactions among primary products. Experimental research has confirmed that the deviation from the plug flow increases the activation energy of the reaction [4]. In practice, we highly recommend that the effect of the gas flow rate is examined to ensure that the close-to-plug flow conditions are attained in the adopted thermal analysis system. The criterion for a close-to-plug gas flow has been discussed and defined in Chapter 3.

2. Mass/heat transfer effect

   Mass/heat transfer effects occur inside and outside the sample particles loaded in the thermal analytical reactor. Reactor designs, in principle, have no direct impact on the particle's internal transport behavior, but can affect the external mass/heat transfer appreciably during thermal analysis [5]. Generally, internal mass/heat transfer effects can be significantly reduced by using sample particles of small sizes, such as <100 μm [6–8]. The external transport effect may be significant when sample particles are loaded in a reactor in the stacked or packed form, especially in a sample cup [9]. In practical situations, experimental results should be examined by some simple approaches, for example, by observing the kinetic data corresponding to different sample loadings and gas velocities, to confirm that the external transport effect is negligible [10]. A negligible external transport effect can be assumed if the kinetic data are not affected by these operating parameters.
3. Sample heating rate

   Sample heating rate is a crucial operating parameter that affects the yields and compositions of products produced by biomass thermochemical reactions [11]. Investigations of fast reactions (e.g., pyrolysis and combustion) require a heating rate as fast as over 10^{3}°C/s or even 10^{4}°C/s, but a lower heating rate may be acceptable for slow reactions (e.g., gasification). Therefore, it is essential to carefully select reactor configuration and operating conditions by considering the characteristics of reaction kinetics and production preference.
4. Temperature and reactant distributions

   Temperature and reactant distributions in a reaction zone also significantly impact the accuracy of reaction kinetic data. A nonuniform temperature distribution may flaw the results of fast reactions even though the heating rate is high enough. Particular attention should be paid to reactors where sample particles are loaded in a cup or stacked between heating elements because these reactors may potentially have a maximum temperature deviation of several hundreds of degrees of Celsius from the preferred reaction temperature [12,13]. When a nonuniform temperature distribution occurs, the reactions of the test sample particles do not take place at the same rate. Thus, the reliability of the test results is significantly compromised.
5. Ability to prepare nascent intermediary product

   As stated previously, the gasification or combustion of char particles is the rate-controlling step of the global biomass/coal gasification or combustion process [1,14,15]. Investigations on char gasification and combustion processes are therefore of practical significance for the design and operation of gasifiers and boilers. For this purpose, the following char particles have been prepared and investigated:

   - ***Slowly cooled char:*** This is the most commonly used char by researchers [1,16]. The char particles are prepared in a separate pyrolysis operation and are then slowly or naturally cooled and stored before being used for the gasification tests. According to Wang et al. [1], the char prepared in this way has the smallest total surface area and the largest average pore diameter compared to char prepared by other methods. Besides, during cooling and reheating processes, secondary reactions may take place, leading to changes in the structural, microcrystalline, and physicochemical properties of char [1,2,14]. For both gasification [1] and combustion [2], the slowly cooled char shows the lowest reactivity and the highest activation energy.

- ***Rapidly cooled char:*** This char is prepared by using a dual reactor system (e.g., Fig. 7.4A), in which the solid fuel is pyrolyzed in the first reactor under an inert gas atmosphere (e.g., Ar or N_2). When the pyrolysis is complete, the produced hot char is quickly blown into the second reactor held at ambient temperature to be rapidly cooled [2]. Experiments have shown that the reactivity of the rapidly cooled char is increased compared with that of the slowly cooled char, but still lower than that of the ex/in situ hot char [2].
- ***Ex situ hot char:*** This char is also prepared in a dual reactor system (e.g., Fig. 7.4A), in which the solid fuel is pyrolyzed in the pyrolysis reactor and then blown into another reactor held at preset gasification or combustion temperature [2]. According to Fang et al. [2], the combustion rate of the ex situ hot char is 1.1−1.2 times that of the rapidly cooled char and 1.5−1.6 times that of the slowly cooled char.
- ***In situ hot char:*** The so-called in situ hot char is prepared in a reactor where the pyrolysis is conducted at the same temperature, but in different atmospheres, as the successive gasification or combustion. The reactor atmosphere changes by switching an inert gas (Ar) for pyrolysis to an oxidant (O_2, CO_2, etc.) for gasification or combustion [17].
- ***In situ nascent char:*** This char is produced in situ at a reactor where the pyrolysis is conducted at the same temperature and atmosphere as the successive gasification or combustion [1]. The preparation and reaction of this char mimic the process that occurs in real gasifiers or combustors, and therefore, the in situ nascent char particles provide true reaction reactivity because they are not affected by any other artificial treatment.

It is evident that correct char reaction performance can only be obtained by using in situ nascent char. Doing so requires performing fuel pyrolysis and char gasification/combustion reactions at the same temperature and atmosphere in the same reactor without interruption.

6. Liquid feeding and atomization

With few exceptions, the tar produced by pyrolysis needs to be further refined through thermal cracking, reforming, or other conversion routes in order to produce lighter hydrocarbon fuels [18−21]. In the case of gasification, the tar in the syngas is also required to be converted to hydrocarbon gases because it can cause troublesome operational issues, such as fouling the surface of pipelines, filters, turbines, and engines. Early research on tar conversions, such as steam reforming, dry reforming, thermal cracking, hydrocracking, or water-gas reaction, is generally conducted by using either model compound or tar condensate collected from pyrolysis test. In this way, the liquid sample needs to be reheated and vaporized, but keep in mind that some side reactions are likely to occur during this process. Ideally, the characterization of the tar conversion process should be conducted in a reaction environment that closely mimics that of tar in actual reactors, but this is a challenge for almost all of the standard thermal analysis techniques. Loading a liquid sample before heating or dropping it into a microreactor with a sample cup does not guarantee the correct reaction kinetics to be obtained, and in this case, online feeding of liquid reactants using a liquid injector or syringe pump is recommended [22].

7. Online sampling and characterization of solid particles

Pyrolysis gases often need to be further transformed catalytically into high-quality products. In the search for the best catalyst formulation, the most challenging task is to formulate the catalyst and determine the catalyst performance, especially the deactivation mechanism of the catalyst. To discover the deactivation mechanism of a catalyst, most of the research so far has relied on the method of post-experimental analysis, which is not only time-consuming and costly but also likely to produce uncertain or even misleading information. It is, therefore, ideal if the characterization of the catalyst could be carried out in real time during the reaction process. Except for fluidized bed reactors capable of solids sampling in real time, no other analytical techniques are currently able to achieve this objective (see Chapter 7).

8. Change of reacting atmosphere during reactions

Being able to change or adjust reacting atmospheres smoothly and quickly during a reaction is very useful for experimental investigation of many chemical reactions occurring in series, such as alternative oxidation and reduction of oxygen carriers (e.g., oxidation of metal and reduction of the metal oxides), catalytic reaction (in which catalysts deactivate due to carbon deposition), and catalyst regeneration (i.e., oxidation or gasification of the coke deposited on the catalyst). Having this capability would significantly improve the efficiency of reaction research, saving time and cost of the research.

In principle, any reactor can be operated in a gas-switching manner, that is, changing the reacting atmosphere from one to another. However, two factors need to be met to do it quickly and smoothly. First, a close-to-plug flow condition needs to prevail inside the reactor, and it ensures that after gas switching, the gas backmixing does not occur between the outgoing and incoming gas streams. Second, the total gas storage volume (including reactor and all pipelines from the gas inlet to the gas analyzer) needs to be small enough compared to the gas volumetric flow rate; thus, the average gas replacement time will be greatly shortened.

6.2 Microreactors for thermal analysis

6.2.1 General approaches and requirements

There are two traditional methods for testing and analyzing reaction kinetics: integral and differential methods. The integral measurement method is used to measure the gas penetration (release) curve of the experimental samples in a fixed-bed reactor. The curve is then fitted based on assumed reaction kinetics to obtain the reaction kinetics parameters. The integral method is susceptible to the complexity of the heat and mass transfer characteristics of the reaction and the conditions of the reactor (such as the reactant concentration and temperature distributions in the particle interior and the bed axial and radial coordinates). By comparison, the differential method is based on the test of a small amount of sample (i.e., the reaction zone can be regarded as a differential unit in an integral reactor), so the concentration and temperature of reactants in the reaction volume can be taken as constants. Therefore, the differential method is more straightforward than the integral method in determining the kinetic parameters and reaction mechanism of a chemical reaction. Based on the accuracy and operability of experimental data, the differential method is also, in most cases, superior and suitable for the study of gas-solid reaction kinetics and reaction mechanism.

6.2.2 Classification of microreactors

As stated earlier, gas-solid reactions, whether catalytic or noncatalytic, can have very different characteristics and special requirements in the design or selection of microreactors to investigate the reaction mechanism and kinetics. An in-depth analysis of the most popular micro gas–solid reactors available in the literature shows that these reactors can be classified based on the heating and gas-solid contacting methods. As shown in Fig. 6.2, there are four principal methods to heat the solid particles (note not the reactor), namely,

1. **Furnace heating**: Furnace heating is the most popular and widely used heating method. Solid particles in a microreactor are heated by a furnace powered either electrically or inductively. Depending on the gas-solid contact mode, we classify furnace heating reactors into three types: (1) solid particles are packed in a sample crucible (e.g., TGA) or a cup (e.g., microtubular reactors) where heat transfer is dominated by conduction between the wall and particles and between particles. A carrier or reacting gas gets into and out of packed particles through diffusion. As a result, this type of reactor has low heating rates (typically $<10°C/s$). (2) Solid particles are in a typical fixed bed state. As gas flows between particles, the heat transfer rate increases compared to diffusion-controlled reactors. The "drop-in" method can be employed to load the solid particles to increase the heating rate further for this type of reactor, in which the reactor is first brought to a preset temperature, and then, the solid particles are dropped into the reactor with a cup or basket (i.e., the fixed bed reactor,

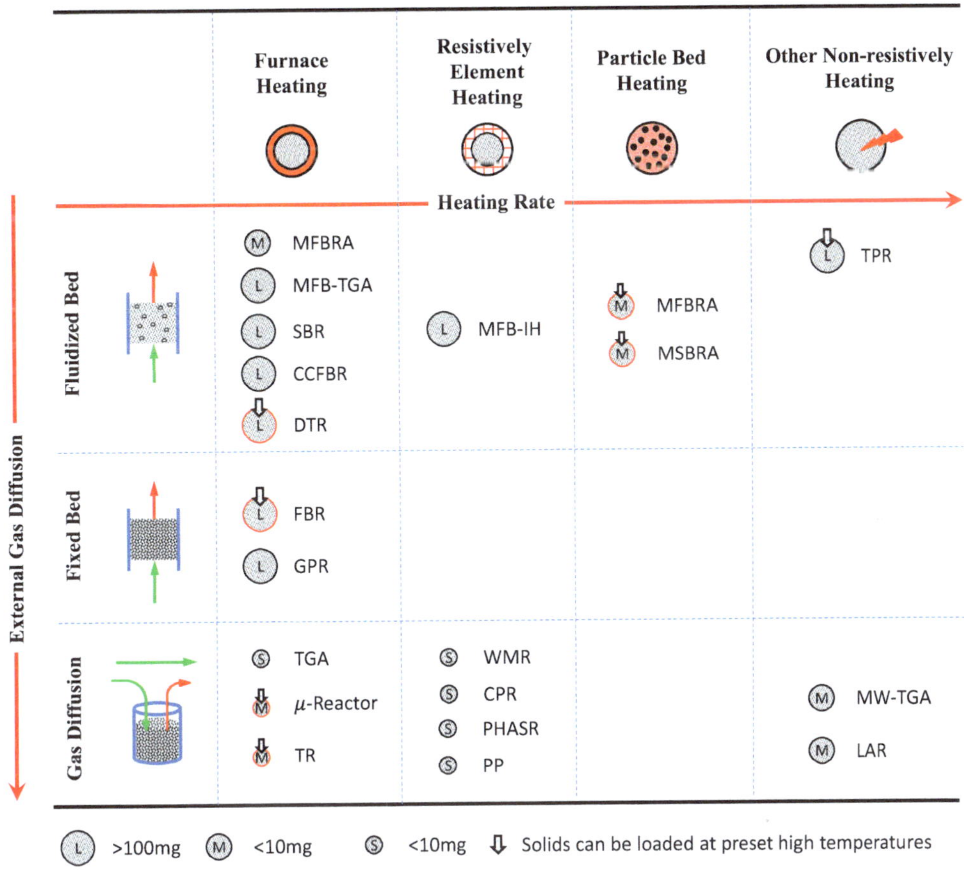

Figure 6.2 Classification of micro gas–solid reactors used for thermal analysis. *DTR*, drop-tube reactor; *MFB-IH*, inducting heating micro-fluidized bed; *MFB-TGA*, micro-fluidized bed thermogravimetric analyzer; *CCFBR*, catalyst cell fluidized bed reactor; *FBR*, fixed bed reactor; *GPR*, gas-pulsed reactor; *TGA*, thermogravimetric analyzer; *μ-reactor*, microtubular reactor; *TR*, tandem microtubular reactor; *WMR*, wire mesh reactor; *CPR*, Curie point reactor; *PHASR*, pulse-heated analysis of solid reaction reactor; *PP*, Pyroprobe; *MFBRA*, micro-fluidized bed reaction analyzer; *MSBRA*, micro-spouted bed reaction analyzer; *SBR*, spouted bed reactor; *MW-TGA*, microwave thermogravimetric analyzer; *LAR*, laser ablation reactor; *TPR*, thermal plasma reactor. Circle symbol size: L: large solid sample amount (usually larger than 100 mg); M: medium solid sample amount (i.e., 10–100 mg); S: smaller solid sample amount (i.e., <10 mg). The *red* circle symbol with a downward arrow means that the solid sample particles can be added to a preheated reactor; the *black* circle symbol presents that the solid particles are loaded before heating the reactor.

microtubular reactor, and the tandem reactor). In this way, the sample heating rate increases approximately to the order of 100°C/s, and the operation can be considered close to isothermal [23,24]. (3) Solid particles are fluidized to take advantage of fluidization, for example, high mass/heat transfer and uniform temperature distribution.

In this way, the heating rate in micro-fluidized bed reactors is easy to control, but the maximum heating rate is limited by the furnace, which usually is lower than approximately 100°C/s.

2. **Element heating**: Solid particles are intimately in contact with certain heat-generation elements on which solid particles are placed. The heat-generation elements can be electric resistors powered by electricity (e.g., wire mesh reactors, Pyroprobe) or certain magnetic materials (e.g., ferromagnetic metals) activated by high-frequency magnets (i.e., Curie point reactors). The element heating reactors are mainly employed for investigations of fast pyrolysis. Normally, a small amount of solid samples (usually less than 1 mg) is enclosed in or deposited on heat-generation elements. A carrier or reacting gas flows over the solid particles, transporting the pyrolysis products to the downstream analyzer such as GC and MS for analysis. Element-heating reactors can attain heating rates in the order of 10^2-10^{3}°C/s, and sometimes up to 10^{4}°C/s. However, using element-heating microreactors for the evolved gas analysis (EGA) would be impractical due to the small gas flow. Also, collecting and analyzing particle samples during the reaction process are difficult, if not impossible, to achieve. Therefore, the element-heating microreactors are more applicable to pyrolysis rather than other gas–solid reactions. Note that element-heating microreactors may suffer from low accuracy of temperature measurements and poor reproducibility because of nonuniform temperature distribution over the sample placement area. Sample placement in terms of thickness, area, and distance to the heating element also affects the temperature of the sample particles. Besides, the heating elements may not suit corrosive and/or pollutant environments, since damaged components may be hard or costly to replace.
3. **Particle bed heating**: Particle bed heating is realized by pulse or continuous feeding of the sample particles into a fluidized bed reactor, in which inert bed materials are maintained at a preset temperature by a furnace. The fluidized bed materials serve as a very stable thermal bath to heat the sample particles. When the amount of sample particles is small enough compared to the bed materials, the addition of sample particles causes negligible disturbance to the bed temperature, ensuring the operation stability. Particle bed heating microreactor takes advantage of fluidized beds, providing a very fast heating rate ($>10^{4}$°C/s) and solid mixing with significantly improved temperature uniformity. The high heating rate and excellent solid mixing allow precise control of reaction temperature and maximally eliminate mass/heat transfer inhibition to gas–solid reactions.
4. **Non-resistively heating**: This includes unconventional non-resistively heating methods such as microwave, laser, and plasma. This type of reactor has been designed customarily or specifically for particular reaction research purposes, but has scarce applications due to various practical constraints (e.g., materials, applications).

Fig. 6.2 illustrates that microreactors with diffusion-controlled gas-solid contacts are more suitable for the characterization of gas-solid reactions with slow reaction rates or inherently rate controlled by mass transfer, although small quantities of solid samples can be used to minimize the diffusion effect. However, using a too-small sample amount presents a challenge in producing reproducible and reliable results, especially when complex and/or heterogeneous feedstocks are characterized. In fixed beds and fluidized beds, the solid sample amount can be increased to get better reproducibility, but the temperature and solid distribution uniformity within the reactor may become an issue. Therefore, it is necessary to understand the characteristics and limitations of the reactor when selecting the reactor to characterize a gas—solid reaction. The following sections discuss the major features of the reactors listed in Fig. 6.2.

6.3 Furnace heating microreactors

6.3.1 Micro-fixed bed reactor

A micro-fixed bed reactor is a cylindrical tube heated by a well-controlled furnace with appropriate heating elements. Micro-fixed beds have been widely used to investigate heterogeneous gas-solid reactions. Due to the simplicity and low cost, micro-fixed bed systems are often customized to various lengths and diameters and engineered for various pressures, temperatures, and construction materials. Fig. 6.3A schematically shows a micro-fixed bed reactor that is operated non-isothermally when solid particles are pre-loaded and then heated to the desired reaction temperature. This type of operation is suitable for some reactions, for example, catalyst screening and performance evaluation or slow gas—solid reactions, but not for the fast reactions

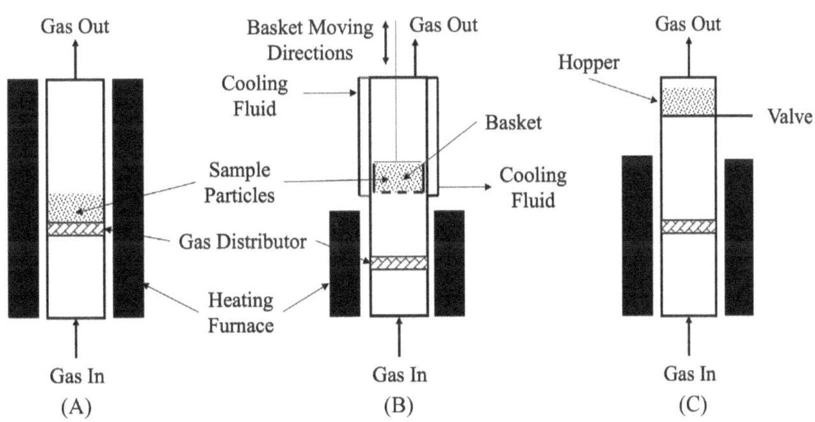

Figure 6.3 Schematics of micro-fixed bed reactors. (A) A fixed bed reactor with pre-loaded particles, (B) a fixed bed with particles pre-loaded in a drop-in basket, and (C) a fixed bed with particles pre-loaded in a hopper above a valve.

(e.g., biomass pyrolysis), since denaturation, degradation, or thermosetting of samples and evaporation of volatile and semi-volatile compounds in samples may occur during the slow-paced heating period. The sample particles can be placed in a basket [23,24] (Fig. 6.3B) or a hopper [25] (Fig. 6.3C) that is held at an ambient temperature to reduce the chances of evaporation, degradation, or thermosetting. The micro-fixed bed is preheated to the desired temperature. Then, the solid samples are dropped into the reactor and quickly heated to the reaction temperature to undergo the gas–solid reactions. Since the heating rate is relatively fast, the reaction can be reasonably considered to be taken place isothermally, especially for gas-solid reactions with reaction rates not very fast. However, when micro-fixed bed reactors are applied to such reactions that have very fast reaction rates or are strong exothermic or endothermic or deal with solid particles which are sensitive to pretesting thermal exposure even for a significantly short time, it is recommended to evaluate the applicability of the reactors before experiments.

6.3.2 Gas-pulsed microreactor

The so-called gas-pulsed microreactor is a particular micro-fixed bed reactor operated with the reactant gas injected into the reactor in pulses. The pulse injection can be applied at least to one gas stream while the other gas stream flows through the bed continually. Two gas streams can also be alternatively injected in pulses. Gas-pulsed microreactors have been used to study the kinetics of gas-solid reactions [26–28]. Fig. 6.4 shows a typical gas-pulsed microreactor system, in which a six-port valve realizes the pulse injections of the reactant gas and carrier gas. In this exemplary application of the pulsed microreactor, the Cu oxidation by air and CuO reduction by methane were characterized [28]. In the experiment, a fixed amount of solid CuO was placed in the reactor maintained at a preset temperature, and then, alternate cycles

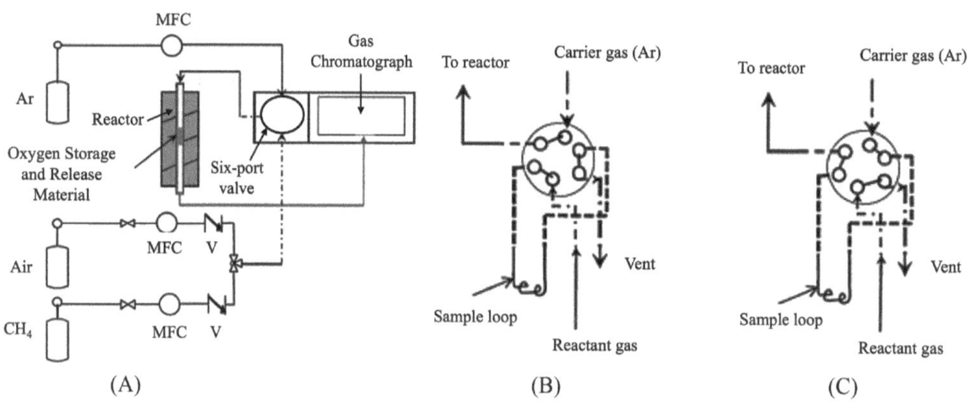

Figure 6.4 The pulsed microreactor system [28]. (A) Pulsed microreactor system. (B) Normal operation. (C) Pulse injection.

of oxidation and reduction were conducted. During each process, several pulses of the reactant gas and the carrier gas were introduced into the reactor, and the product gas after each pulse injection was analyzed. When there was no change in the product gas composition, the reactor changed operation to the next reaction cycle (e.g., oxidation to reduction or vice versa). The gas-pulsed microreactor is very useful for understanding kinetic phenomena at a microscopic level, but the data analysis involved is mathematically complex.

6.3.3 Thermogravimetric analyzer

A thermogravimetric analyzer (TGA) is a precise microbalance connected to a sample pan or crucible inside an electric furnace. TGAs have been commercially available and widely used to measure physical and chemical characteristic properties of sample materials by monitoring the weight change with time under a computer-controlled temperature program (i.e., heating, cooling, isothermal hold, or a combination of them) [29,30]. Sample particles can be loaded from the top, bottom, or side of the reactor before testing, as shown in Fig. 6.5. Traditional TGA cannot characterize reactions with no mass change (e.g., melting, crystallization, glass transition, solid-solid transition, phase transition, and polymorphic transformation), but modern TGA has been upgraded by integrating with differential scanning calorimetry (DSC) or differential thermal analysis (DTA), so these transformations can be characterized. TGA-DSC and TGA-DTA monitor both heat flow and mass change, and they can identify staged reactions and the corresponding temperature ranges. Also, coupling TGA and gas analyzers (e.g., MS, FTIR) can determine global reaction and gas product generation kinetics. [5,9,31].

However, TGAs have some noticeable limitations [9]. First, its heating rate is generally limited to below 100°C/min. Second, the external transport effect on the performance of reactions is significant because sample particles loaded in a sample crucible or cup are in the packed state, as conceptually illustrated in Fig. 6.6A. To reduce the

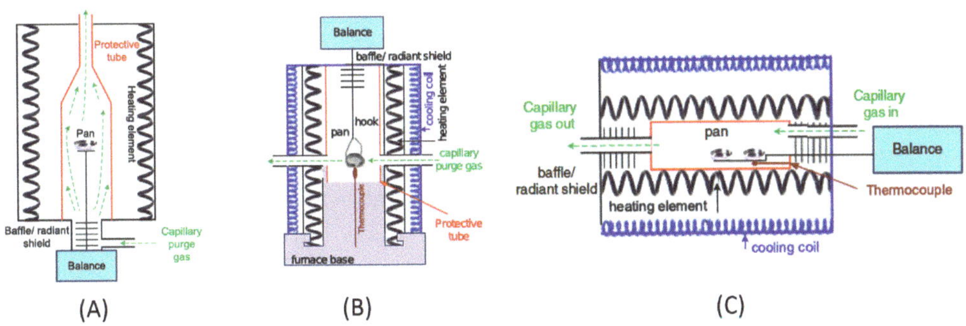

Figure 6.5 Three sample loading methods of TGAs. (A) Top loading. (B) Bottom loading/hanging down. (C) Side loading.

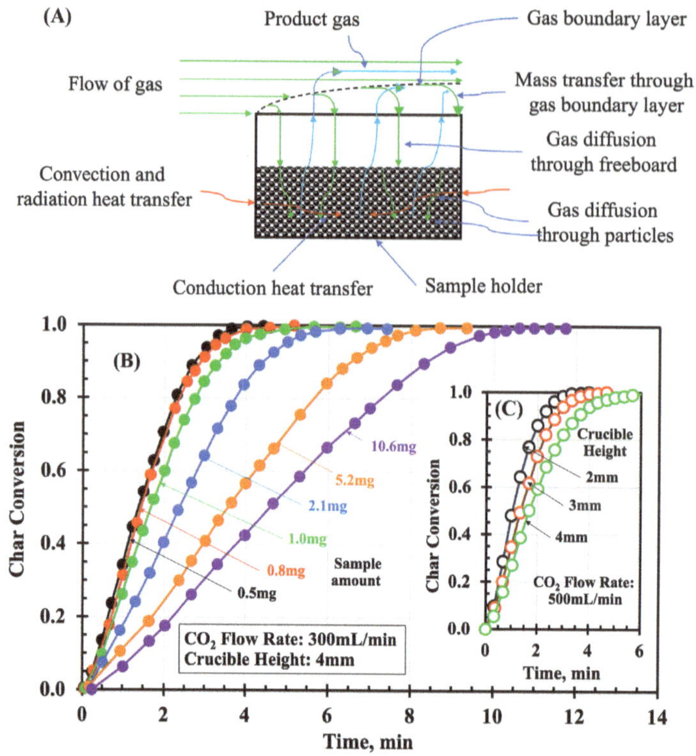

Figure 6.6 (A) Illustration of the transport process around and inside a sample holder and (B,C) the experimental results showing the effect of mass transport on CO_2 gasification of char particles in a TGA [10].

transport effect, Zeng et al. [10] recommended that the sample amount in TGA should not be higher than 1 mg (Fig. 6.6B), and the crucible height should not be over 2 mm (Fig. 6.6C). Besides, we suggest that the carrier gas flow rate should be higher than 300 mL/min, and the sample particle size should be below 100 μm when using TGA for reaction characterizations.

6.3.4 The single and tandem μ-reactors

Fig. 6.7 shows a single microtubular reactor (Fig. 6.7A) and a tandem microreactor (Fig. 6.7B), designated as μ-reactor by Frontier Laboratories Ltd. [32]. The single microtubular reactor is preheated by an electric furnace and held at the predetermined temperature (40°C–900°C), and sample particles are loaded in a reaction-inactive sample cup. The sample cup is dropped into the reactor once its temperature is stabilized. In this way, sample particles are heated to the preset temperature at a heating rate of up to 10^{4}°C/s and then undergo volatilization reactions. The tandem microreactor, an extension of the single μ-reactor, is designed to carry out fuel volatilization and the

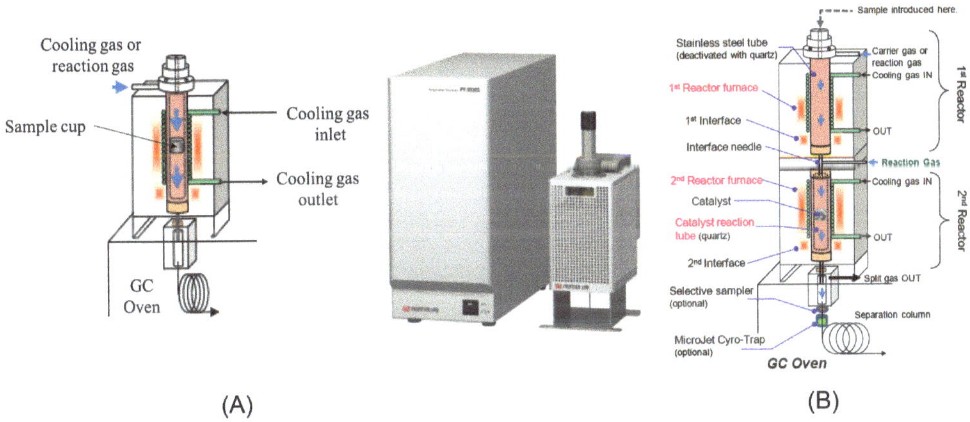

Figure 6.7 Schematics of the single and tandem μ-reactors. (A) Single μ-reactor, and (B) tandem μ-reactor.

volatiles conversion reaction kinetics in a single reactor. This microreactor has been extensively used to investigate gas-phase reaction kinetics of the volatiles released from various solid and liquid fuels. The tandem microreactor consists of two vertically connected microreactors. The first microreactor (same as the single μ-reactor) performs pyrolysis or volatilization of the sample particles in the dropped cup. The second microreactor is used for gas-phase catalytic reactions of the volatiles released in the first reactor. The temperatures of the two reactors are controlled independently at their respective preferred temperatures. A small flow of carrier gas sends the products exited from the microreactor to the GC/MS system for chemical analysis. Note that a close-to-plug flow of gas is not guaranteed because of the low carrier gas flow rate. When the microreactor is used as a pyrolyzer, the system is called Py-GC/MS, which has various designs available today (e.g., Shimadzu, SGE, Frontier Lab, CDS Analytical, Gerstel-Py, Advance Riko). There are two major types of pyrolysis units, the continuous mode and the pulse mode, to be chosen depending on the actual circumstances. Note that the use of Py-GC/MS is quite time-consuming, and the reproducibility of the results may be poor.

Note that temperature transition sections inevitably exist in microtubular reactors before and after the stable-temperature zone [32]. This can become a concern for heat-sensitive samples and fast reactions (e.g., fast and flash pyrolysis reactions). As shown in Fig. 6.8, despite a slight variation in the stable-temperature zone within a 40 mm length of the catalyst bed, the temperature in the entrance section varies in the range of few hundreds of degrees Celsius. Consider sample particles falling into this section to reach the reaction zone of stable temperature. In this entrance section, sample particles are exposed to a varying thermal environment via which some physical and chemical changes may have occurred. These unintentional consequences will

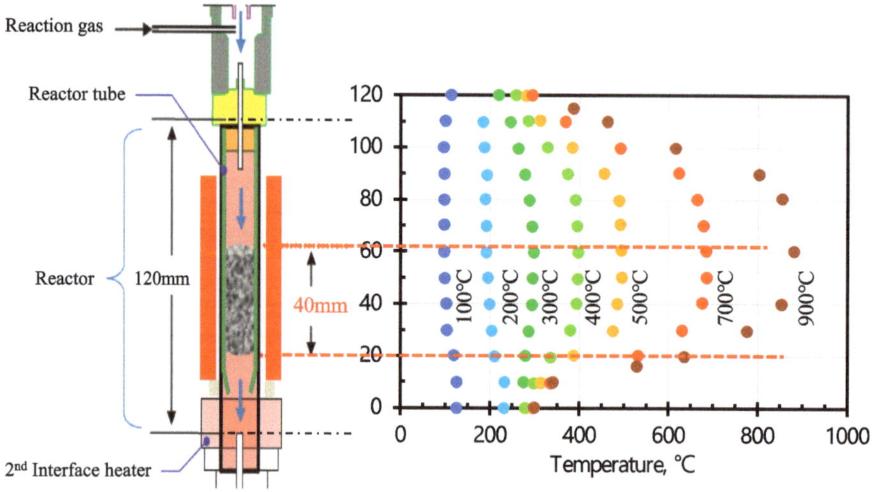

Figure 6.8 Axial temperature profiles along the microreactor of the μ-reactor.

certainly flaw the experimental results because the reaction performance is not produced entirely at the reported reaction temperature. Therefore, this fact needs to be considered for investigations of thermally sensitive reactions.

6.3.5 Drop-tube reactor

A drop-tube reactor is a kind of downer reactor and is sometimes called a laminar entrained-flow reactor [33]. The reactor is simply a long vertical concentric tube heated by an electric furnace. Solid particles are fed into the reactor from the top and then fall with or without a reacting gas flow. Solid particles can be fed mechanically or pneumatically [34–36], as schematically shown in Figs. 6.9A and B, respectively. Solid residues are collected at the reactor bottom, and gas products are sent to an analyzer. In most cases, the furnace is installed with multiple independently controlled heaters to obtain a preferred axial temperature distribution (Fig. 6.9B). However, since a temperature gradient in the entrance section always exists [34,35], sample particles may need to be fed through a vertical pipe with a water cooling jacket to minimize their exposure to undesired thermal conditions, as shown in Fig. 6.9C [37]. In addition, extra efforts are also needed to disperse the cohesive solid particles before and during the reaction processes. Fig. 6.9D shows a micro-drop-tube reactor with an air-cooled particle injector tube and the oil-cooled product sampling tube. The distance between the cooling tubes before and after the reaction zone can be adjusted to vary the length of the pyrolysis zone, thus the pyrolysis reaction time, and therefore this reactor is suitable for investigations of the effect of reaction time on pyrolysis reactions [38].

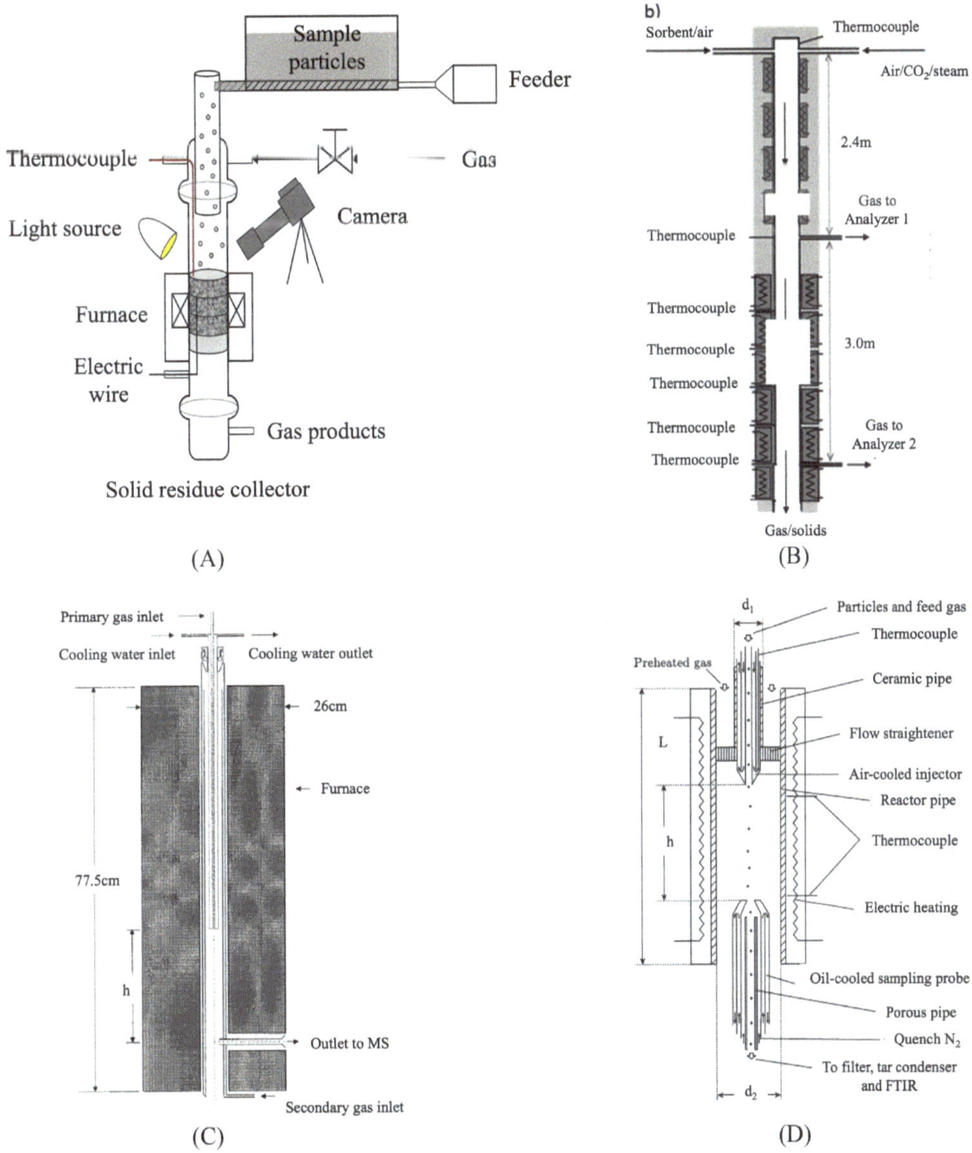

Figure 6.9 A schematic illustration of typical drop-tube reactors. (A) Mechanical feeding; (B) pneumatical feeding; (C) feeding with a cooling water jacket; and (D) air-cooled feeding and oil-cooled sampling.

6.3.6 Catalyst cell fluidized bed reactor

The catalyst cell fluidized bed reactors are developed based on a commercial CCR1000 catalyst cell reactor from Linkam Scientific Instruments (Figs. 6.10A and B) to investigate

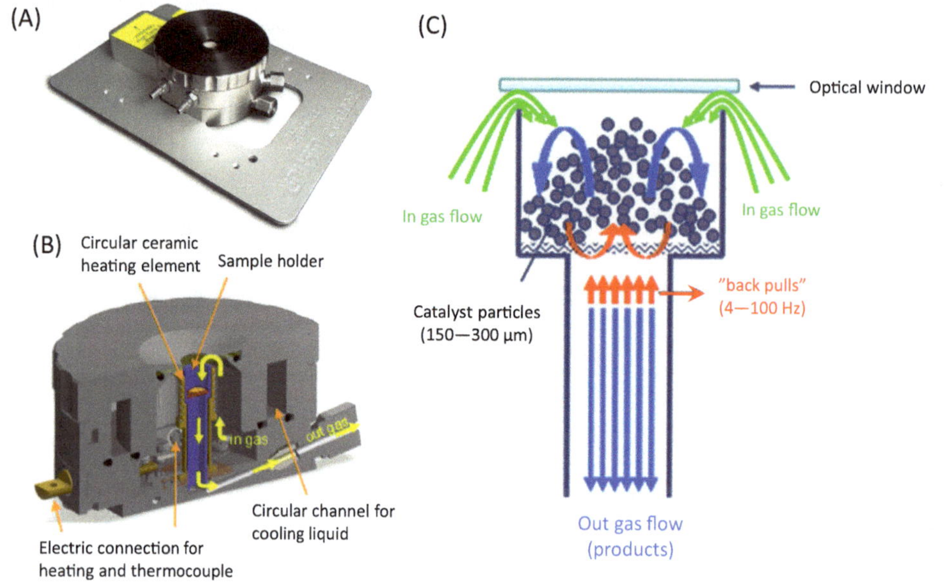

Figure 6.10 A micro-fluidized bed based on the Linkam CCR1000 microreactor [39]. (A) A photograph of the CCR1000 microreactor, (B) a cutaway view of the CCR1000 microreactor, and (C) the principle of fluidized bed of particles.

heterogeneous catalytic reactions [39,40] under operando conditions. Homogeneously fluidizing particles allow heterogeneous catalytic reactions to be monitored by operando Raman spectroscopy without the drawback of laser heating or damage. As shown in Fig. 6.10C, sample particles are supported in a ceramic sample holder on a ceramic fiber filter, and a miniature device is placed downstream of the reactor. As the reaction gas flows down through the particles, the miniature device is activated in a controlled manner, generating pressure oscillations to push the gas flow upward and downward in pulses of 40–100 Hz frequency. In such a way, the particles are homogeneously fluidized, so heterogeneous catalytic reactions can take place without laser-induced heating and damage while high laser power is still used. A ceramic heater heats the reactor at a heating rate of up to 200°C/min. The maximum working temperature is 1000°C.

6.4 Resistively heated microreactors

6.4.1 Wire mesh reactor

A wire mesh reactor (WMR) is made of electrically conducting wires powered by an electric circuit or magnetic field. The wires generate heat in situ to heat sample particles placed on and between them [41–43]. In practice, there are various WMR design variations, such as heated grid reactors (HGRs) [12] and screen heater reactors (SHRs)

[44], each having different geometries and configurations. Typically, a small amount of sample particles (e.g., <10 mg) is sandwiched between two layers of wire meshes. A control system controls the heating rate and the stable-temperature holding time. Fig. 6.11 shows photographs of wire meshes, sample particles placed on wire meshes, and a WMR operated at 1000°C [45]. By selecting the wire materials properly, wire mesh reactors can operate at temperatures up to 2000°C [46].

For WMR, the selection of mesh material deserves careful consideration according to the melting and caking propensities of the test material. In addition, mesh apertures should be appropriately designed to accommodate the particle size of test materials. A particle size of around 100 μm is generally recommended so that it is small enough to minimize the mass transfer effect but still big enough to facilitate the design of mesh apertures. Another challenge for WMR is related to the measurement and control of reaction temperature when a thermocouple is used. Considering that the thermocouple tip size is significantly smaller than the area occupied by sample particles, the measured temperature by the thermocouple corresponds only to the local spot where the thermocouple tip is placed. In this way, the measured temperature is likely not representative because the temperature is usually ununiformly distributed over the sample placement area [12,13], as shown in Fig. 6.12, where the maximum temperature deviation can amount to several hundreds of degrees of Celsius. Fig. 6.13 further indicates that the sample loading location and size affect the temperature distribution significantly. Note that the average error caused by temperature measurement is only 1°C–5°C [12], so the nonuniform temperature distribution must be caused by the heat transfer limitation between heat-generation wires and particles and the uneven distribution of biomass particles on the heated grid. Due to this nonuniform temperature

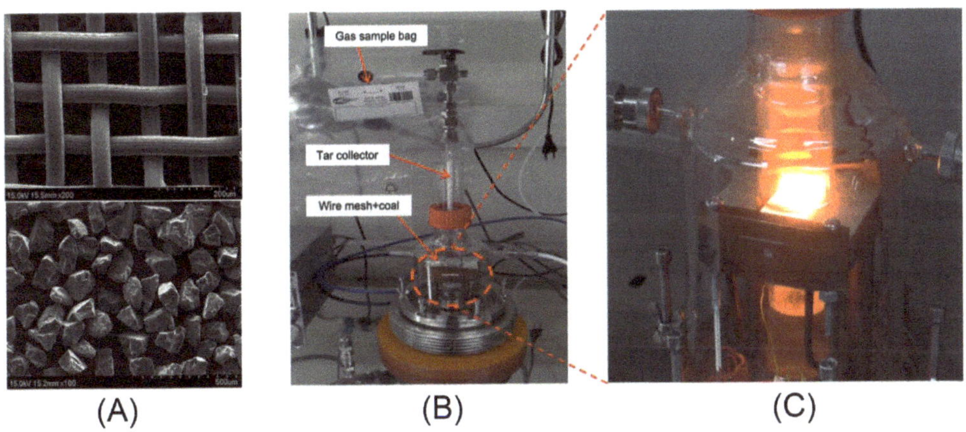

Figure 6.11 (A) Sample particles placed on the wire mesh, (B) a photograph of wire mesh reactor setup, and (C) a view of the operating wire mesh reactor [45].

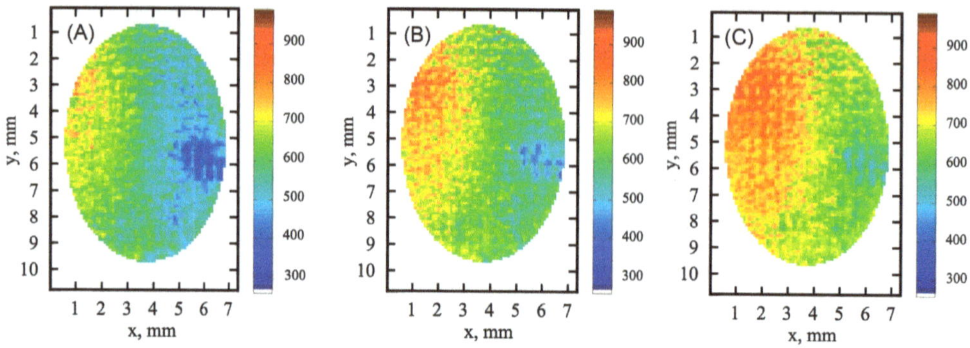

Figure 6.12 Two-dimensional images of the measured temperature distributions of (A) 400°C, (B) 500°C, and (C) 600°C on the grid of a mesh reactor.

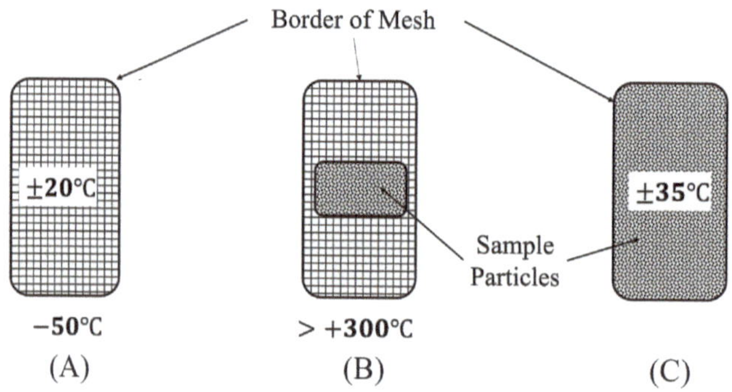

Figure 6.13 Temperature distribution over mesh after a holding time of 1 s. The temperature at the center is 500°C, and the numbers indicate the temperature differences from the center temperature [13]. (A) No sample particles loaded. (B) Sample particles loaded in the center location. (C) Sample particles fully covered the meshes.

distribution, the release of volatiles from biomass pyrolysis is not at the same rate in the area between the grids, greatly affecting the reliability of the test results.

6.4.2 Curie point reactor

A Curie point reactor is a high-frequency induction-heating pyrolyzer [47,48]. Solid or liquid samples are wrapped in a pyrofoil made of the selected ferromagnetic material and then placed into the sample tube (Fig. 6.14A). Each pyrofoil has one fixed Curie temperature. Japan Analytical Industry has developed 21 different Curie point pyrofoils with the Curie temperature in the range of 160°C–1040°C (i.e., 160, 170, 220, 235, 255,280, 315, 333, 358, 386, 423, 445, 485, 500, 590, 650, 670, 740, 764, 920, 1040) [48]. Curie point reactors have two operational models available commercially, that is, portable (Fig. 6.14A) and automatic models (Fig. 6.14B) [48]. The two models are

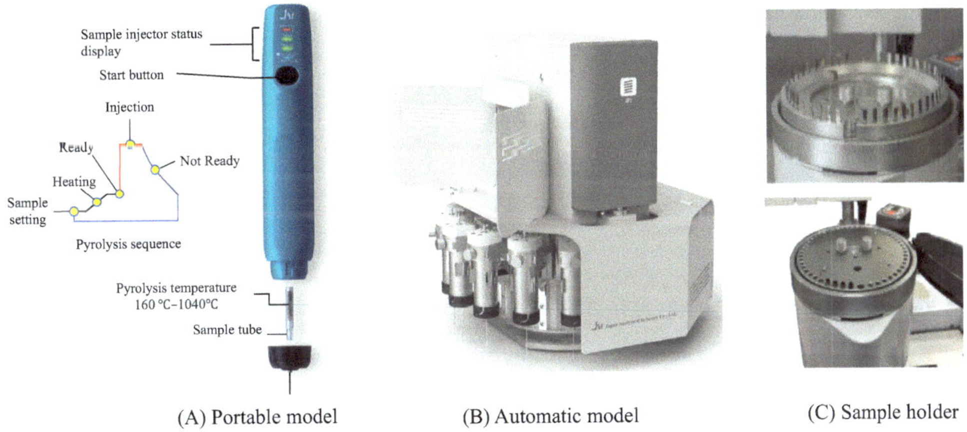

Figure 6.14 Portable and automatic Curie point reactors [48]. (A) Portable model. (B) Automatic model. (C) Sample holder.

basically the same in principle, but the automatic model is equipped with an automatic sample feeder (Fig. 6.14C). For the Curie point reactor, the heating rate can be up to 5000°C/s (i.e., the heating time < 0.2 s), but the operating temperature is inflexible because it depends on the specific Curie point temperature of the selected material. A small pulse of carrier gas is injected into the reactor to sweep away the product gas during operations, in which a constant plug flow of gas is not guaranteed. Since only a small pulse of carrier gas is injected, the pyrolysis reaction is seriously limited by mass transport and thus takes a long time to complete (a few seconds up to a few hours).

6.4.3 Pulse-heated analysis of solid reaction reactor

The pulse-heated analysis of solid reaction (PHASR) reactor is a laboratory self-made analytic equipment to measure the kinetics of high-temperature fast reactions [49]. As shown in Fig. 6.15, it has a top resistive heating section with 1000 Hz feedback control (Fig. 6.15A) and a bottom cooling section with a high-velocity cooling fluid (Fig. 6.15B). The heating and cooling sections are joined together with a PTFE gasket. A thin layer of solid sample particles (10 – 50 μm in thickness, 50 – 350 μg in weight) is sandwiched between the two sections (Fig. 6.15C and D). In this way, the microreactor heats and cools feedstocks in pulses on a millisecond timescale. The PHASR is designed to have volatile products produced, quenched, and swept away quickly so that secondary reactions or the condensation of volatile products can be maximally minimized.

6.4.4 Microprobe reactor

A microprobe reactor is one typical Py-GC/MS analyzer, as shown in Fig. 6.16A. It has a horizontal tubular reactor heated resistively [50]. When the reactor is stabilized at

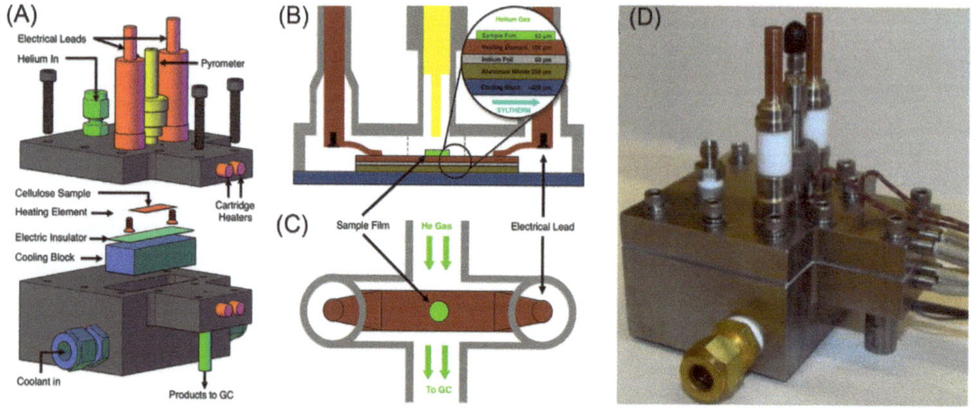

Figure 6.15 The pulse-heated analysis of solid reaction (PHASR) reactor. (A) an exploded diagram CAD model of the PHASR reactor with an upper resistive heating section and a lower high velocity coolant section, (B) A side cross-sectional view of the PHASR reactor with two electrical leads and the coolant, (C) A top cross-sectional view of the PHASR reactor showing the heating element with a thin-film sample loaded on the center and helium sweep gas constantly flowing across the center, and (D) A photograph of the laboratory-scale PHASR reactor. [49].

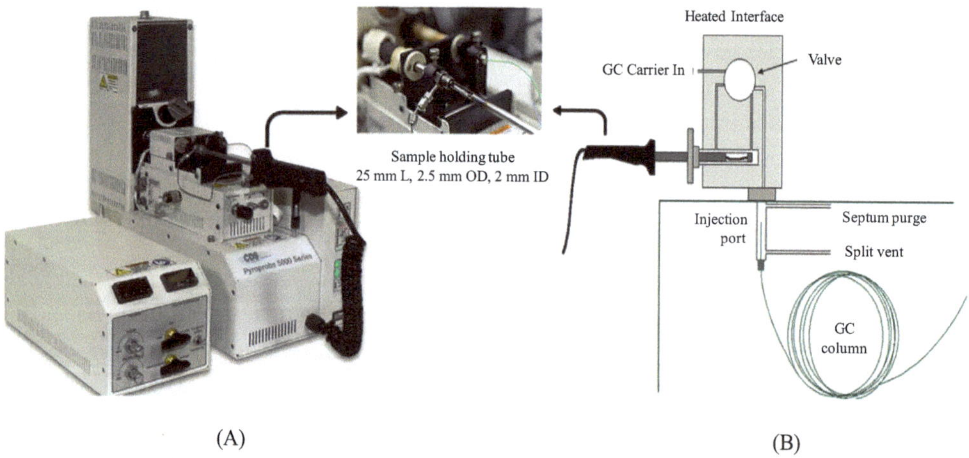

Figure 6.16 Schematic of a Pyroprobe combined with GC-MS system. (A) a photograph of the Pyroprobe analyzer, and (B) an illustration of the principle of the Pyroprobe [50].

a preset temperature (maximum up to 1400°C), the samples placed in a quartz holder are quickly inserted into the heated interface (manifold), as shown in Fig. 6.16B. There are three models with a maximum heating rate of 1000°C/min, 1000°C/s, and 20,000°C/s, respectively. The rapid heating rate activates bond fragmentation within the macromolecular structure, producing a set of low molecular weight chemical components and macromolecule types (e.g., lignin, cellulose, hemicelluloses). The mixture of products is swept into the GC and GC-MS analytical column by injecting a pulse of a small amount of carrier gas. Accordingly, the volatiles take a long time to leave the reactor, creating secondary reactions and condensation possibilities. In addition, the microprobe-type reactors cannot analyze liquid samples. Note that this instrument has been designed to be capable of operating at pressures up to 3.4 MPa.

6.5 Particle bed heating microreactors

6.5.1 Micro-spouted bed reactor

The micro-spouted bed (or spouted fluidized bed) reactor is an expansion of micro-fluidized bed reactors [51]. The principle of spouted bed is illustrated in Fig. 6.17A. The reactor has a conical bottom section designed to help fluidize wet, coarse, nonuniformly sized, and sticky particles. Gas is injected vertically through a small opening at the base center with a high enough velocity so that the resulting high-speed jet causes large, wet, or sticky particles to rise in the center core. After rising above the bed surface, the particles fall back and move slowly down in the outer annulus, and as a result, a cyclic pattern of solids movement is formed.

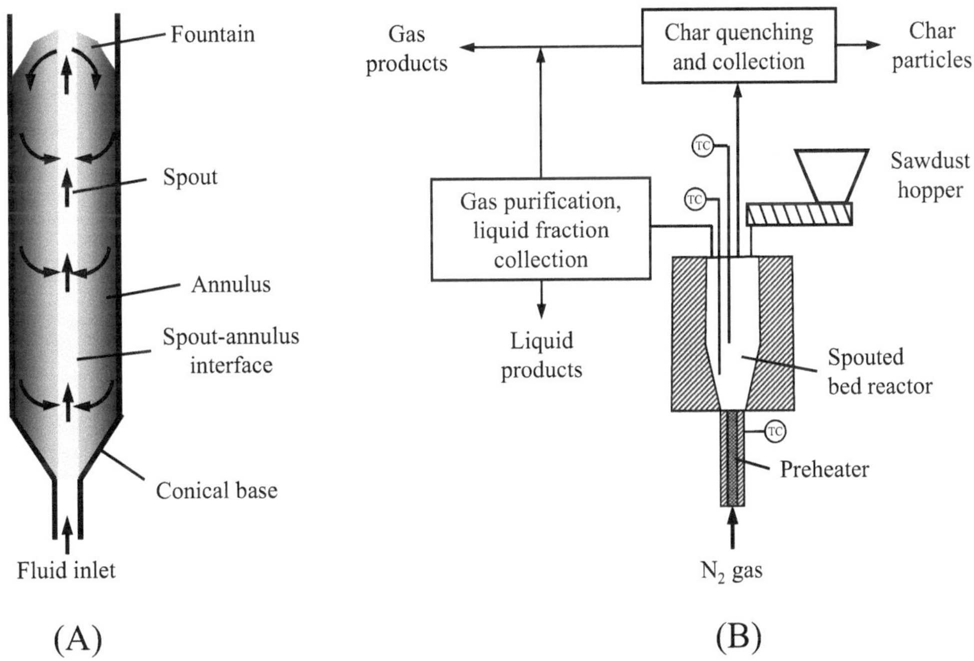

Figure 6.17 The spouted bed principle and (B) its application for the characterization of biomass pyrolysis. (A) an illustration of spouted bed concept, and (B) a spouted bed system used for biomass pyrolysis.

Fig. 6.17B shows a typical reactor system for the characterization of biomass pyrolysis reactions [52]. Experiments show that under similar operating conditions, the gas backmixing is lower in micro-spouted beds than in micro-fluidized beds [51] because the upward flow of gas is concentrated in the central region, and the resulting flow velocity is high.

6.5.2 Micro-fluidized bed reactor

The micro-fluidized bed reaction analyzer (MFBRA) is an advanced catalytic and noncatalytic reaction analysis system [53–55]. MFBRA has shown unique advantages and application characteristics compared to other thermal analysis reactors described in this chapter, and we will elaborate on this advanced reactor in Chapters 7–10.

6.6 Other non-resistively heating microreactors

6.6.1 Microwave microreactor

The microwave microreactor heats samples with microwave power. The reactor consists of an insulated, transparent-to-microwave transmission quartz vessel, and an infrared thermopile, as shown in Fig. 6.18. Neutral microwave receptors, such as silicon carbide, are added to heat nonpolar samples. The microwave input power controls the heating rate; however, maintaining linear heating rates and measuring sample temperatures accurately are challenging [30]. In the design shown in Fig. 6.18, an optical filter is required to measure the interior of the bed temperature, but inserting an optical filter limits the operable temperature range of the instrument, which is extremely undesirable. Since the reliable temperature measurement with a combination of silicon and quartz filters can

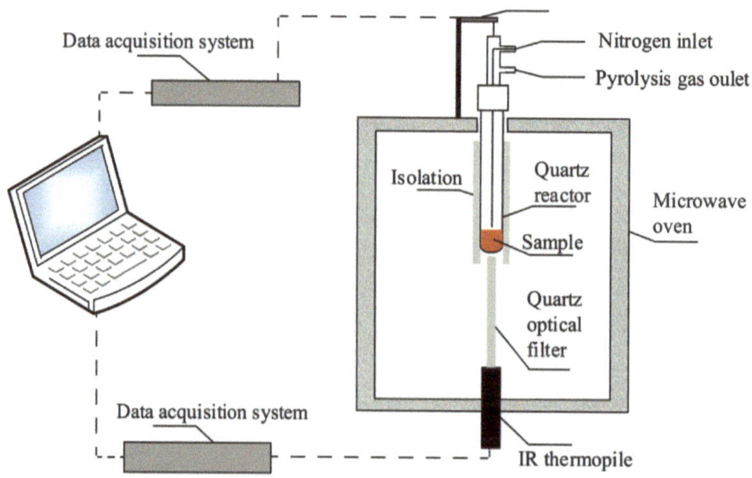

Figure 6.18 Schematic of a microwave reactor [56].

only be achieved above 300°C, the MW-TGA is not suitable for characterizations of reactions below this temperature.

6.6.2 Laser ablation reactor

Fig. 6.19A shows a laser ablation/pulsed sample introduction/mass spectrometry platform that integrates pyrolysis and/or laser ablation with resonance-enhanced multiphoton ionization reflectron time-of-flight mass spectrometry. The laser beam has a pulse duration of 8 ns and can be focused on an area of 0.03 mm^2. The short duration ensures very fast heating rates to limit secondary reactions, and the small spot area permits investigations of localized samples.

Fig. 6.19B is a hot-stage pyrolysis reactor. Biomass materials are placed in a circular copper sample holder and securely attached to stainless steel washers. The sample holder sits above the heating element and is heated to the preset temperature at a heating rate of ~100°C/min, and then, pyrolysis and laser ablation tests are conducted.

6.6.3 Thermal plasma reactor

A thermal plasma reactor can realize ultra-high-temperature reactions with a heating rate of 10^4–10^{6}°C/s [57,58]. Fig. 6.20 shows a schematic of the laboratory-scale thermal plasma reactor system used for rapid coal pyrolysis. The system consists of a plasma torch, mixing section, reaction chamber, quench device, gas-solid separator, feeding subsystem, and measurement/analysis subsystem. The plasma torch has a maximum power input of 10 kW. The electrodes are protected by a cooling system to reduce electrode wear at the expense of a reduced torch thermal efficiency. The cathode, made of thoriated tungsten, is embedded in a water-cooled copper holder located at the center of the torch. The hollow cathode has an inner diameter of 7 mm, and the

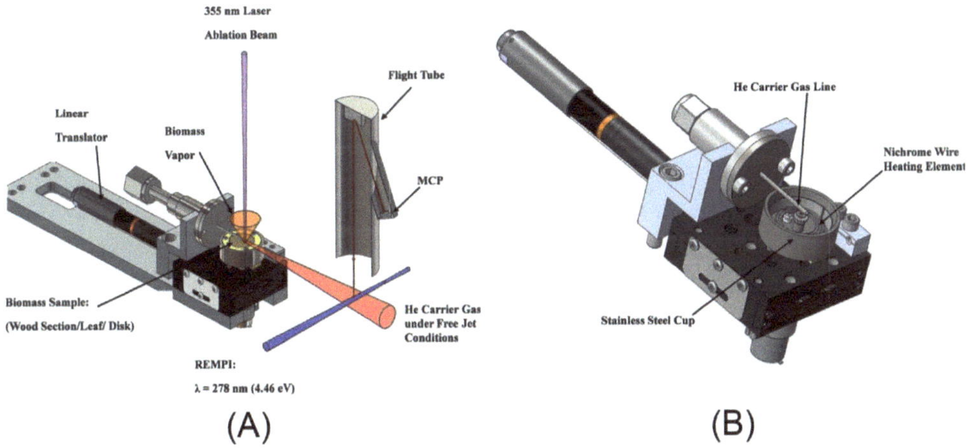

Figure 6.19 (A) A diagram of the pyrolysis/laser ablation photoionization mass spectrometer and (B) a hot-stage pyrolysis reactor [59].

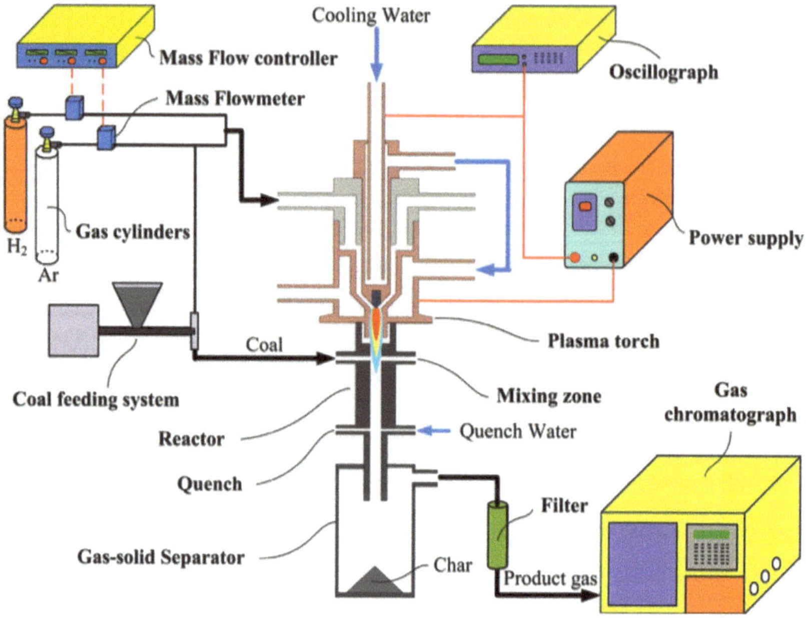

Figure 6.20 Schematic drawing of the laboratory-scale thermal plasma reactor for coal pyrolysis [58].

button-type cathode has an outer diameter of 6 mm. The insulation gas injection is made of Teflon. The water-cooled copper anode in a nozzle shape is used. The inner diameter of the nozzle is 6 mm.

6.7 Remarks

Thermal analysis is of paramount importance in decoding the mechanism and quantifying the kinetics of gas–solid reactions to ensure the successful development and operation of the thermochemical reaction processes. Over the years, various microreactors have been developed to meet requirements for characterizing a wide array of gas-solid reactions involving varying operating conditions (pressure, temperature, atmosphere, etc.), diversified feedstocks, and products. These reactors can behave quite differently because of different heating methods and rates, solid sample amounts and placement methods, and gas–solid contact efficiencies in the reactors. Aiming to provide general guidance on the selection and operation of the microreactor to satisfy one's practical needs, an overview of these reactors has been presented in this chapter.

Despite the relatively vast availability and significant advancements in the field of thermal measurement and analysis, some areas require further studies:

- Most of the available microreactors today for thermal analysis are built with solid particles stacked in a sample holder. Generally, these microreactors suffer from severe mass transfer limitations. The conduction-dominated heat transfer is likely to lead to a nonuniform temperature distribution over the area of solid sample placement additionally. These characteristics may not be so critical for slow reactions, but would have a more significant impact on the performance of fast reactions. Therefore, it is necessary to develop techniques continuously to minimize the diffusion effect and improve the uniformity of temperature distribution in microreactors stacked with solid particles.

- Fast reactions (e.g., pyrolysis and combustion) have been extensively studied for many years. However, there are controversies over the product profiles, product release behavior, pyrolysis mechanisms, and kinetics, mainly caused by inappropriate reactant heating and product cooling capabilities. Therefore, research on new technologies and instruments capable of fast heating and cooling needs to be pursued.
- It is critical to attain a gas plug flow through a microreactor and the pipeline that transport the product gas from the reactor to the gas analyzer to determine the evolution behavior of the dynamic reaction product. In the future, measures to make gas flow in microreactors close to the plug flow of gas need to be further explored.
- There is a broad array of multistaged gas–solid reactions. The mechanism and kinetics of each step reaction often need to be investigated. For example, the gasification of char particles, which is the controlling step of the solid fuel gasification process, has been extensively studied in the literature using various microreactors with ex situ prepared char particles. Due to inevitable changes in the physicochemical properties of char particles caused by the ex situ preparation process (e.g., cooling and heating), there are considerable discrepancies in the reported results. Therefore, further research is needed to develop microreactors that can produce nascent solid samples and conduct the subsequent reaction analysis in situ for multistaged complex reactions.
- Sometimes, it is highly desirable to have direct information about what is happening to the solid catalysts or reactants during the reaction process, rather than speculating about what happens by characterizing the posttest solid samples. Therefore, research is needed to develop a technology that can collect solid samples online or monitor and analyze solid samples in real time in microreactors.
- Some reactions may be highly exothermic/endothermic, explosive or occur under high-pressure, high-temperature, corrosive, or highly oxidizing environments. Few microreactors can currently operate under such challenging or harsh conditions. There is, therefore, a crying need for solving this issue.
- The analytical methods and instruments of the products are of equal importance to thermal measurement and analysis. Notably, two requirements for gas analyzers are needed: (1) being able to detect and analyze the gas products in a sufficiently increased sampling frequency (e.g., >100 Hz) and (2) being able to detect, identify,

and quantify the full spectrum of the reaction products, especially the high-molecular components.

In conclusion, there has been a diversified thermal measurement and analysis instruments portfolio. However, a great deal of research and development is still required in many areas to advance the technology further. Under this background, a more advanced thermal analysis technique, that is, a micro-fluidized bed reaction analyzer (MFBRA), has been developed. The principles, features, and applications will be presented in the following chapters.

Abbreviation

CCFBR	catalyst cell fluidized bed reactor
CPR	Curie point reactor
DSC	differential scanning calorimetry
DTA	differential thermal analysis
DTR	drop-tube reactor
EGA	evolved gas analysis
FBR	fixed bed reactor
FTIR	Fourier-transform infrared spectroscopy
GC	gas chromatography
GPR	gas-pulsed reactor
HGR	heated grid reactor
LAR	laser ablation reactor
MFB	micro-fluidized bed
MFB-IH	induction heating micro-fluidized bed
MFB-TDA	micro-fluidized bed thermogravimetric analyzer
MFBRA	micro-fluidized bed reaction analyzer
MS	mass spectrometry
MSBRA	micro-spouted bed reaction analyzer
MW-TGA	microwave thermogravimetric analyze
PHASR	pulse-heated analysis of solid reaction reactor
PP	Pyroprobe
SBR	spouted bed reactor
SHR	screen heater reactor
TGA	thermogravimetric analyzer
TPR	thermal plasma reactor
TR	tandem microtubular reactor
WMR	wire mesh reactor

References

[1] Wang F, Zeng X, Shao R, et al. Isothermal gasification of in situ/ex situ coal char with CO_2 in a micro fluidized bed reaction analyzer. Energy and Fuels 2015;29(8):4795–802.

[2] Fang Y, Luo G, Chen C, et al. Combustion kinetics of in-situ char and cold char in micro-fluidized bed. Journal of Combustion Science and Technology 2016;22(2):148–54.

[3] Mahinpey N, Gomez A. Review of gasification fundamentals and new findings: reactors, feedstock, and kinetic studies. Chemical Engineering Science 2016;148:14–31.

[4] Hu D, Geng S, Zeng X, et al. Gas back-mixing characteristics and the effects on gas-solid reaction behavior and activation energy characterization. CIESC Journal 2020;72(3):1354–63.
[5] Barr MR, Volpe R, Kandiyoti R. Influence of reactor design on product distributions from biomass pyrolysis. ACS Sustainable Chemistry and Engineering 2019;7:13734–45.
[6] Yu J, Yao C, Zeng X, et al. Biomass pyrolysis in a micro-fluidized bed reactor: characterization and kinetics. Chemical Engineering Journal 2011;168(2):839–47.
[7] Chen H, Zheng Z, Shi W. Investigation on the kinetics of iron ore fines reduction by CO in a micro-fluidized bed. Procedia Engineering 2015;102:1726–35.
[8] SriBala G, Carstensen HH, Van Geem KM, et al. Measuring biomass fast pyrolysis kinetics: state of the art. Wiley Interdisciplinary Reviews: Energy and Environment 2019;8:e326.
[9] Nowak B, Karlström O, Backman P, et al. Mass transfer limitation in thermogravimetry of biomass gasification. Journal of Thermal Analysis and Calorimetry 2013;11:183–92.
[10] Zeng X, Wang F, Wang Y, et al. Characterization of char gasification in a micro fluidized bed reaction analyzer. Energy and Fuels 2014;28(3):1838–45.
[11] Zanzi R, Sjöström K, Björnbom E. Rapid high-temperature pyrolysis of biomass in a free-fall reactor. Fuel 1996;75(5):545–50.
[12] Prins MJ, Lindén J, Li ZS, et al. Visualization of biomass pyrolysis and temperature imaging in a heated-grid reactor. Energy and Fuels 2009;23(2):993–1006.
[13] Hoekstra E, van Swaaij WPM, Kersten SRA, et al. Fast pyrolysis in a novel wire-mesh reactor: design and initial results. Chemical Engineering Journal 2012;191:45–58.
[14] Guo Y, Zhao Y, Gao D, et al. Kinetics of steam gasification of in-situ chars in a micro fluidized bed. International Journal of Hydrogen Energy 2016;41(34):15187–98.
[15] Zhang Z, Li Z, Cai N. Comparison and analysis of different models of pulverized coal char combustion. Journal of Combustion Science and Technology 2014;20(5):393–400.
[16] Zhang R, Wang QH, Luo ZY, et al. Competition and inhibition effects during coal char gasification in the mixture of H2O and CO2. Energy and Fuels 2013;27(9):5107–15.
[17] Guo Y, Zhao Y, Meng S, et al. Development of a multistage in situ reaction analyzer based on a micro fluidized bed and its suitability for rapid gas – solid reactions. Energy and Fuels 2016;30:6021–33.
[18] García-Labiano F, Gayán P, de Diego LF, et al. Tar abatement in a fixed bed catalytic filter candle during biomass gasification in a dual fluidized bed. Applied Catalysis B: Environmental 2016;188:198–206.
[19] de Diego LF, García-Labiano F, Gayán P, et al. Tar abatement for clean syngas production during biomass gasification in a dual fluidized bed. Fuel Processing Technology 2016;152:116–23.
[20] Gai C, Dong Y, Lv Z, et al. Pyrolysis behavior and kinetic study of phenol as tar model compound in micro fluidized bed reactor. International Journal of Hydrogen Energy 2015;40(25):7956–64.
[21] Zeng X, Wang F, Sun Y, et al. Characteristics of tar abatement by thermal cracking and char catalytic reforming in a fluidized bed two-stage reactor. Fuel 2018;231:18–25.
[22] Gai C, Dong Y, Fan P, et al. Kinetic study on thermal decomposition of toluene in a micro fluidized bed reactor. Energy Conversion and Management 2015;106:721–7.
[23] Encinar JM, Beltrán FJ, Ramiro A, et al. Catalyzed pyrolysis of grape and olive bagasse. Influence of catalyst type and chemical treatment. Industrial and Engineering Chemistry Research 1997;36(10):4176–83.
[24] Encinar JM, Beltrán FJ, González JF, et al. Pyrolysis of maize, sunflower, grape and tobacco residues. Journal of Chemical Technology and Biotechnology 1997;70(4):400–10.
[25] Zhang Y, Zheng Y. Co-gasification of coal and biomass in a fixed bed reactor with separate and mixed bed configurations. Fuel 2016;183:132–8.
[26] Sárkány J, Bartók M, Gonzalez RD. The modification of CO adlayers on Pt SiO2 catalysts by preadsorbed oxygen: an infrared and pulse microreactor study. Journal of Catalysis 1983;81(2):347–57.
[27] Benn FR, Dwyer J, Esfahani AAA. The catalytic isomerisation of cyclopropane over NaCoX zeolites using a pulsed microreactor. Inorganica Chimica Acta 1978;31:101–4.
[28] Deshpande A, Krishnaswamy S, Ponnani K. Pulsed micro-reactor: an alternative to estimating kinetic parameters of non-catalytic gas-solid reactions. Chemical Engineering Research and Design 2017;117:382–93.
[29] Wunderlich B. Chapter 7 Thermogravimetry. Thermal Analysis. Academic Press: 371–416.

[30] Saadatkhah N, Carillo Garcia A, Ackermann S, et al. Experimental methods in chemical engineering: thermogravimetric analysis—TGA. Canadian Journal of Chemical Engineering 2020;98:34–43.
[31] Samih S, Chaouki J. Development of a fluidized bed thermogravimetric analyzer. AIChE Journal 2015;61:84–9.
[32] Frontier Laboratories, Rapid screening of catalysts. Frontier Laboratories, 2020. http://www.frontier-lab.com/.
[33] Wu C, Cheng Y. Downer reactors Grace JR, Bi X, Ellis N. Essentials of Fluidization Technology Boschstr. 12, 69469 Weinheim, Germany: Wiley-VCH Verlag GmbH & Co. KGaA; 2020. p. 499–530.
[34] Fan DQ, Sohn HY, Elzohiery M. Analysis of the reduction rate of hematite concentrate particles in the solid state by h2 or co in a drop-tube reactor through CFD modeling. Metallurgical and Materials Transactions B: Process Metallurgy and Materials Processing Science 2017;48(5):2677–84.
[35] Turrado S, Arias B, Fernández JR, et al. Carbonation of fine CaO particles in a drop Tube reactor. Industrial and Engineering Chemistry Research 2018;57(40):13372–80.
[36] Do HS, Bunman Y, Gao S, et al. Reduction of NO by biomass pyrolysis products in an experimental drop-tube. Energy and Fuels 2017;31(4):4499–506.
[37] Brown AL, Dayton DC, Nimlos MR, et al. Design and characterization of an entrained flow reactor for the study of biomass pyrolysis chemistry at high heating rates. Energy and Fuels 2001;15(5):1276–85.
[38] Ontyd C, Pielsticker S, Yildiz C, et al. Experimental determination of walnut shell pyrolysis kinetics in N2 and CO2 via thermogravimetric analysis, fluidized bed and drop tube reactors. Fuel 2021;287 (September 2020):1–12.
[39] Beato P, Schachtl E, Barbera K, et al. Operando Raman spectroscopy applying novel fluidized bed micro-reactor technology. Catalysis Today 2013;205:128–33.
[40] Linkam Scientific Instruments. CCR1000 - catalysis cell reactor. Product Brochure, 2020: 1–4. https://www.linkam.co.uk/ccr1000.
[41] Zhang Y, Niu Y, Zou H, et al. Characteristics of biomass fast pyrolysis in a wire-mesh reactor. Fuel 2017;200:225–35.
[42] Damartzis T, Ioannidis G, Zabaniotou A. Simulating the behavior of a wire mesh reactor for olive kernel fast pyrolysis. Chemical Engineering Journal 2008;136(2–3):320–30.
[43] Trubetskaya A, Jensen PA, Jensen AD, et al. Influence of fast pyrolysis conditions on yield and structural transformation of biomass chars. Fuel Processing Technology 2015;140:205–14.
[44] Zhou S, Garcia-Perez M, Pecha B, et al. Effect of the fast pyrolysis temperature on the primary and secondary products of lignin. Energy and Fuels 2013;27(10):5867–77.
[45] Ra HW, Seo MW, Yoon SJ, et al. Devolatilization characteristics of high volatile coal in a wire mesh reactor. Korean Journal of Chemical Engineering 2014;31(9):1570–6.
[46] Zeng C, Chen L, Liu G, et al. Advances in the development of wire mesh reactor for coal gasification studies. Review of Scientific Instruments 2008;79 Article 084102.
[47] Fan H, He K. Fast pyrolysis of sewage sludge in a curie-point pyrolyzer: the case of sludge in the city of Shanghai, China. Energy and Fuels 2016;30:1020–6.
[48] Japan Analytical Industry Co. Ltd. Curie Point Pyrolyzer, 2020. https://www.jai.co.jp/english/index.html.
[49] Krumm C, Pfaendtner J, Dauenhauer PJ. Millisecond pulsed films unify the mechanisms of cellulose fragmentation. Chemistry of Materials 2016;28:3108–14.
[50] Wu S, Lv G, Lou R. Applications of chromatography hyphenated techniques in the field of lignin pyrolysis Reza D. Applications of Gas Chromatography 2012;41–64.
[51] Hu D, Zeng X, Wang F, et al. Comparison and analysis of gas backmixing in micro fluidized and spouted beds. In: The 11th Annual Conference and Cross-Strait Symposium on Particle Technology. Xiamen: Chinese Society of Particuology; 2020.
[52] Aguado R, Olazar M, San José MJ, et al. Pyrolysis of sawdust in a conical spouted bed reactor. Yields and product composition. Industrial and Engineering Chemistry Research 2000;39 (6):1925–33.
[53] Yu J, Yue J, Liu Z, et al. Kinetics and mechanism of solid reactions in a micro fluidized bed reactor. AIChE Journal 2010;56(11):2905–12.

[54] Wang F, Zeng X, Geng S, et al. Distinctive hydrodynamics of a micro fluidized bed and its application to gas-solid reaction analysis. Energy and Fuels 2018;32(4):4096–106.
[55] Yu J, Zeng X, Zhang J, et al. Isothermal differential characteristics of gas-solid reaction in micro-fluidized bed reactor. Fuel 2013;103:29–36.
[56] Leclerc P, Doucet J, Chaouki J. Development of a microwave thermogravimetric analyzer and its application on polystyrene microwave pyrolysis kinetics. Journal of Analytical and Applied Pyrolysis 2018;130:209–15.
[57] Cheng Y, Yan BH, Cao CX, et al. Experimental investigation on coal devolatilization at high temperatures with different heating rates. Fuel 2014;117(Part b):1215–22.
[58] Yan B, Xu P, Guo CY, et al. Experimental study on coal pyrolysis to acetylene in thermal plasma reactors. Chemical Engineering Journal 2012;207–208:109–16.
[59] Mukarakate C, Scheer AM, Robichaud DJ, et al. Laser ablation with resonance-enhanced multiphoton ionization time-of-flight mass spectrometry for determining aromatic lignin volatilization products from biomass. Review of Scientific Instruments 2011;82(3):1–10.

CHAPTER 7

System of micro-fluidized bed reaction analysis

This chapter presents an advanced thermochemical reaction analyzer based on micro-fluidized bed technology. The research and development of the micro-fluidized bed reaction analyzer (MFBRA) were initiated by the Institute of Process Engineering (IPE) of the Chinese Academy of Sciences (CAS) in 2005 and collaborated on by a team of scientists and engineers across China sponsored by the Chinese Academy of Sciences (CAS), the Ministry of Science and Technology of the People's Republic of China, and the National Natural Science Foundation of China [1,2]. The MFBRA incorporates a micro-fluidized bed reactor, an online pulse feeder capable of quick feeding milligrams of fine solid reactants to the reactor, and a real-time analyzer for monitoring time-dependent variations in gas products. It can analyze complex gas—solids reactions both isothermally and non-isothermally with a substantially minimized gas diffusional effect. As a fluidized bed, the micro-fluidized bed exhibits advantages of high mass and heat transfer rates, uniform temperature distribution, excellent mixing between gas and solids as well as between bed and sample particles, and isothermal/non-isothermal differential operation ability. As a microchemical reactor, it is convenient to operate with low cost, high efficiency, high reliability, and exceptionally enhanced safety. Today, the MFBRA has been systematically and comprehensively researched [3—6] and developed into a series of standardized reaction analytical instruments. In the past 10 years, the MFBRA has been successfully applied to characterizations of a variety of gas-solids reactions.

7.1 System configurations

7.1.1 System configurations of micro-fluidized bed reaction analyzer

A typical flow diagram and a photograph of a micro-fluidized bed reaction analyzer are separately shown in Fig. 7.1A and B. The core of the analyzer is the micro-fluidized bed reactor (reactor tube) supported by auxiliary subsystems, including an electric heating furnace, a sample feeding, a reacting gas supplying and a product gas cleaning device, online measurement, process control, and data acquisition components. The micro-fluidized bed reactor is fabricated with appropriate materials (e.g., stainless steel, glass, ceramics, alundum) depending on the requirements of operating

Micro Fluidization
DOI: https://doi.org/10.1016/B978-0-443-18718-6.00007-3
© 2023 Chemical Industry Press Co., Ltd.
Published by Elsevier Inc., under an exclusive license with Chemical Industry Press Co., Ltd.

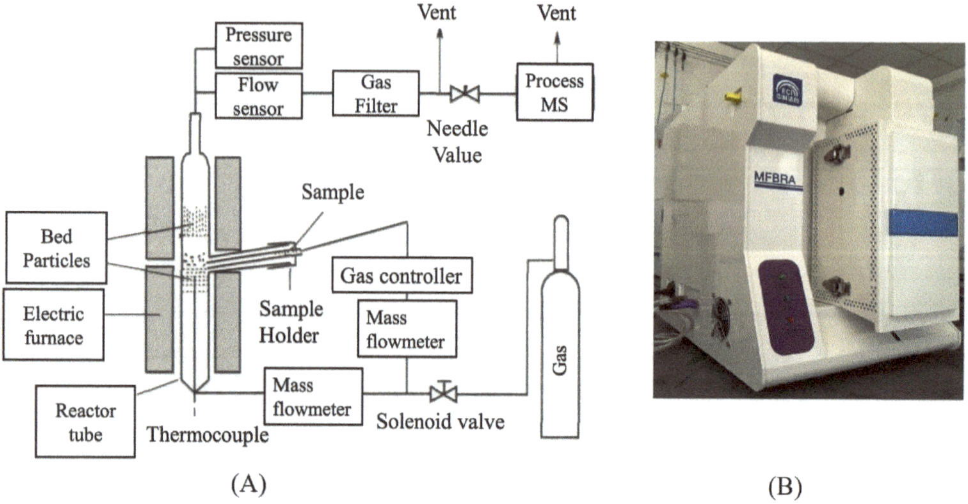

Figure 7.1 A typical flow diagram of a micro-fluidized bed reaction analyzer and an MFBRA photograph. (A) A typical MFBRA system. (B) A typical MFBRA.

temperature, pressure, and atmosphere. The micro-fluidized bed reactor is installed in an electrically powered heating furnace, and appropriate heating elements are used to provide the capability to achieve desired reaction temperatures [7]. The heating hardware and software are designed to maintain a stable temperature in the reactor with an allowable temperature fluctuation of as low as $\pm 1^{\circ}C$ [8]. For temperature measurement and sample particle feeding, openings on the side or top of the reactor can be designed to accommodate thermocouples and sample input ports. Given the limited space available, a bare thermocouple with a fine wire is immersed in the bed to measure and control the reaction temperature. Due to a small bed volume, intense gas—solid mixing, and a controllable rapid heating rate, the micro-fluidized bed can be considered a differential reactor [6,9].

For isothermal differential reaction analysis, a solid sample (coal, biomass, char, oil shale, mineral powder, etc.) is rapidly pulse-injected into the micro-fluidized bed held at the preferred temperature. Once in the bed, the sample particles are quickly mixed with the bed material and heated to the bed temperature. A reacting or inert fluidizing gas is preconditioned and measured to obtain the required gas composition, temperature, and pressure before flowing into the reactor. The stream of product gas flows out of the reactor quickly in a close-to-plug flow condition. When required, the product gas is cleaned through dust filters and condensers to have particles, tar, water, and any other containments removed before being sent to a quick responding and accurate process analyzer for quantification. The gas analyzer that can be used includes analytical instruments such as process mass spectrometer (Ms), photoionization mass spectrometer (SPI-Ms), gas detection sensors, and/or process IR gas analyzer. The analyzer should be capable of responding and recording the data with a sufficiently high frequency (e.g., >50 Hz) and accuracy. The operational and performance parameters, such as bed temperature, pressure, reaction/fluidizing gas composition, and flow rate, are monitored, controlled, and recorded in real time by the process control and data acquisition system. Operating automatically with a small amount of sample particles, the analyzer provides opportunities to conduct experiments and reaction data analysis quickly, reproducibly, efficiently, and cost-effectively.

Fig. 7.2 shows an alternative flow diagram of the MFBRA system [8,10], consisting of a gas supply system (GSS), a gas flow switching system (GFSS), a reaction system (RS), a sample feeding system (SFS), and an online gas analysis system (OGAS). This MFBRA system allows feeding samples and in situ intermediates to undergo stepwise reactions at different temperatures and/or gas reaction atmospheres, either isothermally or non-isothermally.

- *The gas supply system (GSS)* consists of several precise mass flow meters and controllers. A gas mixer is also included to prepare a gas mixture with the desired gas composition. Moreover, an appropriate preconditioning system is integrated to adjust the reactant gas to meet the targeted moisture content, pressure, and temperature.
- *The gas flow switching system (GFSS)* consists of several electrically controlled solenoid valves and pressure transducers. A back-pressure controller modulates the differential pressure between the two gas flows to reduce the transient instability caused by gas switching during the operation. The GFSS switches the reacting gas from one to another smoothly and quickly to meet the requirements of a series of reactions in, respectively, different reaction environments.

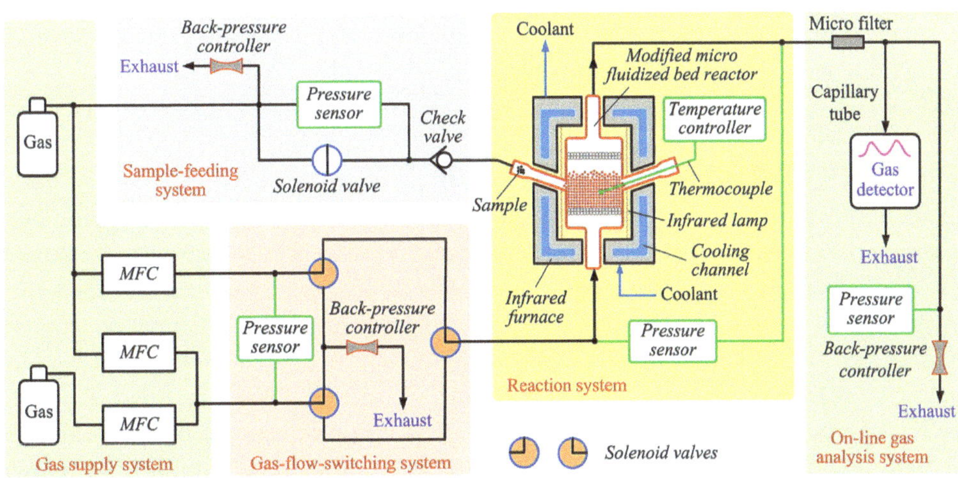

Figure 7.2 A schematic diagram of multistage micro-fluidized bed reaction analyzer [8].

- *The reaction system (RS)* consists of a micro-fluidized bed reactor enclosed in a furnace. The furnace is capable of both heating and cooling. The temperature can reach 1200°C−1600°C with fluctuations of $< \pm 1°C$ with an electrical or infrared heating furnace. The maximum linear heating rate gets up to 400°C/min. The furnace can be cooled at a rate of up to −650°C/min with enclosed cooling channels. The reactor can be made of quartz, stainless steel, alumina, or other appropriate materials. According to the reaction requirements, it can be designed with a gas inlet section, a gas preheating section, a reaction zone, and an exit section. Besides, small diameter tubes can be attached to either side of the reactor or to the inlet/outlet to install thermocouples, solid feeding, or discharging ports. Inert materials can be packed in the lower part of the reactor to intensify the gas preheating. Porous sintered plates are installed at the inlet and outlet of the reaction zone to homogenize the fluidizing gas and prevent fine particles from escaping the reaction zone. The inner diameter of the reaction zone usually is 10 − 25 mm. A bare thermocouple with a wire diameter of 0.5 − 0.8 mm is immersed in the fluidized bed to measure and control the reaction temperature. The reactor inlet and outlet use reduced diameter tubes to accelerate the gas flow, providing a close-to-plug and quick flow condition.
- *The sample feeding system (SFS)* consists of a solid sample holder, a check valve, and a solenoid valve to activate the supply of compressed gas to inject sample particles into the fluidized bed. The inlet pressure is optimized so that solid particles can be injected into the bed at a pulse of 10−20 Ms while maintaining a minimized disturbance of the pulse injection to the stable operations of the bed.
- *The online gas analysis system (OGAS)* consists of a microfilter, a capillary sampling tube, an online process mass spectrometer, an absolute pressure sensor, and a back-pressure controller. The OGAS detection frequency is 5−100 Hz for scanning a single mass number m/z. The gas concentration can be quantitatively determined by using an appropriate gas analytical instrument.

7.1.2 Micro-fluidized bed reactor design

Micro-fluidized bed reactors can be standardized or customized to meet specific purposes. Fig. 7.3A shows an MFB reactor configuration most commonly employed. Fig. 7.3B depicts an alternative MFB design in which inert particles are filled in a lower stage to preheat and pre-distribute the fluidizing gas. Fig. 7.3C presents another MFB design, in which an upper stage can be served as a fixed or fluidized bed for multistage reactions or simply as a particle filter. A micro-fluidized bed may also have lower and upper stages, as illustrated in Fig. 7.3D. Furthermore, the injection of sample particles and insertion of thermocouples can be from the top of the bed, as shown in Fig. 7.3E.

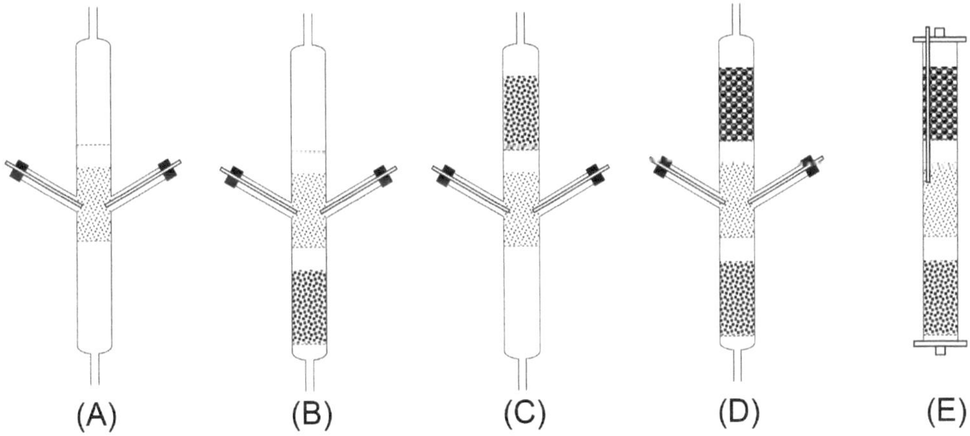

Figure 7.3 Exemplary design options for micro-fluidized bed reactors. (A) The standard design, (B) an alternative design with a lower stage to preheat and pre-distribute the fluidizing gas, (C) an alternative design with an upper stage fixed or fluidized bed for multistage reactions or particle filtration, (D) an alternative design with both lower and upper stages, and (E) an alternative design with the injection of sample particles from the top of the reactor.

The micro-fluidized bed can also be combined with other types of microreactor, such as a micro-fixed bed or micro-drop tube, to form a multistage reaction analysis system. Each reactor may operate at different temperatures and atmospheres according to specific reaction requirements. Fig. 7.4A shows a dual reactor that integrates a micro-drop tube reactor used for fuel pyrolysis and a micro-fluidized bed reactor used for combustion of nascent char produced in the micro-drop tube reactor [11]. In the in situ char combustion tests, the drop tube and micro-fluidized bed reactors are preheated to their predetermined temperatures. Then, the preferred gases flow into the two reactors separately at specified compositions and flow rates. Once the system is stabilized, a small sample of raw coal particles is carried into the drop tube reactor by a carrier gas to undergo pyrolysis reactions. After the char particles are separated from the volatiles at the elbow section, a pulsed air blows the hot char particles into the micro-fluidized bed for isothermal combustion. Note that this dual reactor system can also produce rapidly cooled char particles—the hot char particles produced by the micro-drop tube reactor are blown to the micro-fluidized bed maintained at ambient temperature. Rapid cooling of the char particles is realized by quickly mixing and contacting the low-temperature fluidized bed particles at a cooling rate of up to 10^4-10^5°C/s.

Fig. 7.4B shows a dual reactor with a micro-fluidized bed reactor and a horizontal tubular reactor [12]. In these reactors, the primary tar is generated by pyrolysis in MFBR, and secondary reactions occur in the tubular reactor.

Many high-temperature gas–solids reactions take place through multiple parallel or continuous sub-reactions. To accurately understand each sub-reaction's kinetic

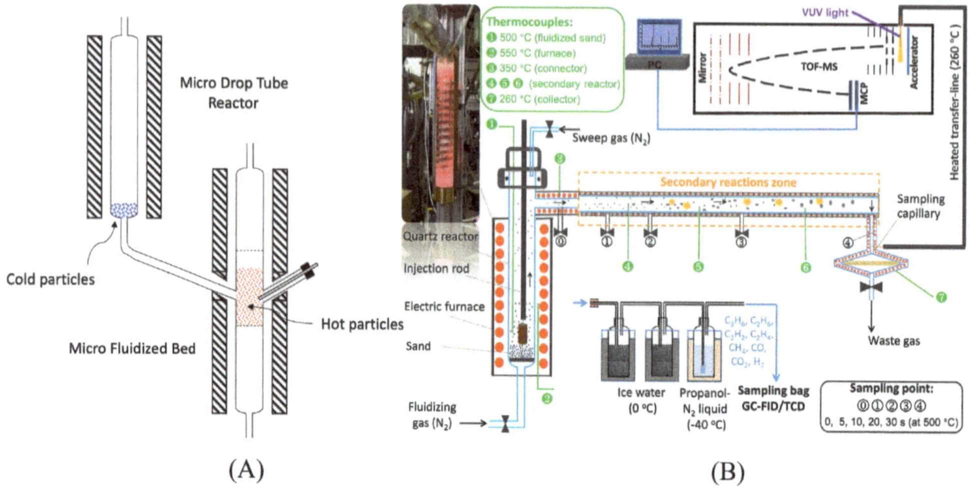

Figure 7.4 (A) A micro-fluidized bed reactor integrated with a micro-drop tube reactor [11] and (B) a micro-fluidized bed integrated with a horizontal tubular reactor [12].

behavior, it would be ideal to prepare in situ nascent reactants of each reaction under the real reaction conditions and then conduct the reaction. However, the traditional thermal analysis reactor is unable to meet this requirement. Conventional experimental methods usually prepare the intermediary reactants offsite separately under conditions different from those used in the actual reaction. The ex situ samples generally have different physical, structural, and chemical characteristics than the in situ nascent reactants. For example, to study the characteristics of char combustion and gasification processes, the char particles are first prepared by pyrolysis, cooled down, and stored separately. The prepared sample char particles are heated again to the temperature of the gasification or combustion. The reaction results obtained in this way are different from those produced by in situ samples in actual reactors. These problems can be overcome by the improved micro-fluidized bed reactors, which have rapid heating and cooling capabilities [8] (see Fig. 8.2).

7.1.3 Solid sample feeding method

In isothermal differential reaction analysis, solid sample particles of 10–50 mg can be fed to the micro-fluidized bed reactor by a compressed gas pulse in a period as short as less than 10 Ms, as shown in Figs. 7.1 and 7.2. Experiences show that the influence of pulse feeding method on the stability of bed operation is negligible when the mass ratio of bed material to sample particles is appropriate. The bed temperature fluctuation caused by pulses can be controlled below ± 5°C. The sample particles injected into the bed are quickly mixed with and heated by the bed materials to the bed temperature in less than 0.1 s.

To ensure stable MFBRA operations, it is necessary to optimize the structure of the sample feeding tube, the feeding position, and the compressed gas pressure and flow rate through experiments or numerical simulations. Yang et al. [13] investigated the solids mixing behavior by considering different fitting tube configurations, and the results are presented in Chapter 3.

In addition, solid samples can be fed to the reactor from its top, as schematically shown in Fig. 7.5. This configuration makes it easy to fabricate the furnace and operate the reactor under high pressures.

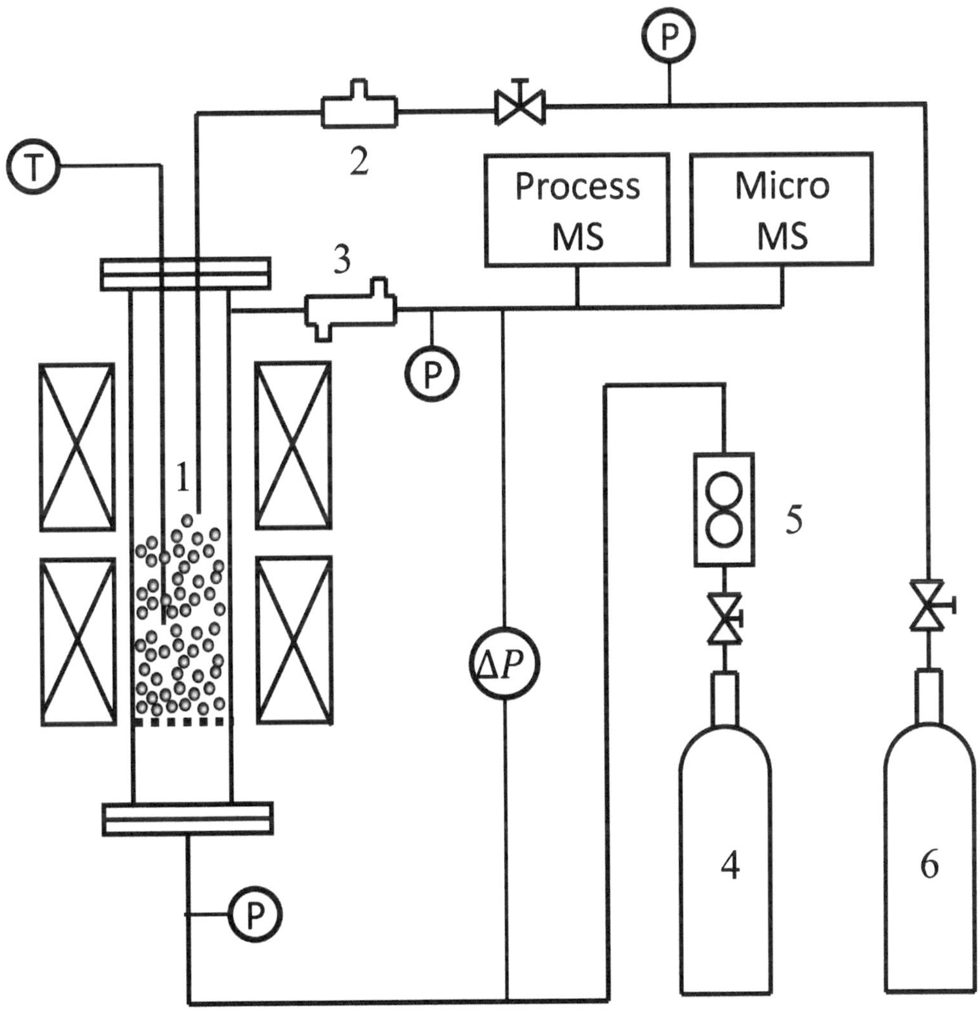

Figure 7.5 A schematic diagram of micro-fluidized bed reaction analyzer featured by solid sample feeding from the reactor top. 1. micro-fluidized bed, 2. particle sample holder, 3. product gas cleaning, 4. fluidizing gas, 5. flowmeter, 6. high-pressure gas.

7.1.4 Liquid sample feeding method

Micro-fluidized bed reaction analyzer also offers the capability to feed liquid reactant samples such as oil, tar, or other liquids. A schematic flow diagram and a photograph of the micro-fluidized bed reaction analyzer are depicted in Fig. 7.6A and B. The reactor is equipped with an automatic-working liquid injector, also called a syringe pump [14]. It is developed to feed liquid-phase reactant samples into the micro-fluidized bed reactor in an injection time as short as 0.1 s. Typical examples of reactions with liquid feedstock, including waste oil, tar, and various model hydrocarbons, have been tested successfully in micro-fluidized beds [14–16].

7.1.5 Online gas sampling and analysis

Gas products produced by micro-fluidized beds can be sampled in real time and analyzed online or offline. Fig. 7.7 shows a gas sampling system. A precision temperature controller is used to control the capillary temperature within a variation of $\pm 0.2°C$ to ensure a quick response with satisfactory reproducibility and stability, which is critical for achieving desirable stabilities of sampled gas flow and vacuum of an Ms chamber. The sampling stability can be improved further by precisely regulated absolute pressure together with a high-precision back-pressure controller (± 0.02 kPa). A dilution gas can also be introduced into the gas sampling line to improve the analysis stability and accuracy while extending the analyzer's lifetime. This dilution is especially necessary for the case where gas concentration changes in a relatively large span.

Detecting and quantifying the gas-phase products produced by a micro-fluidized bed reactor require an appropriate gas analytical instrument. A variety of gas analyzers can be used according to the requirements to analyze the product gas. An electron impact ionization (EI) mass spectrometry (Ms) can be used for permeant gases. The EI mass spectra have a drawback of producing overlapping ionization fragments and thus

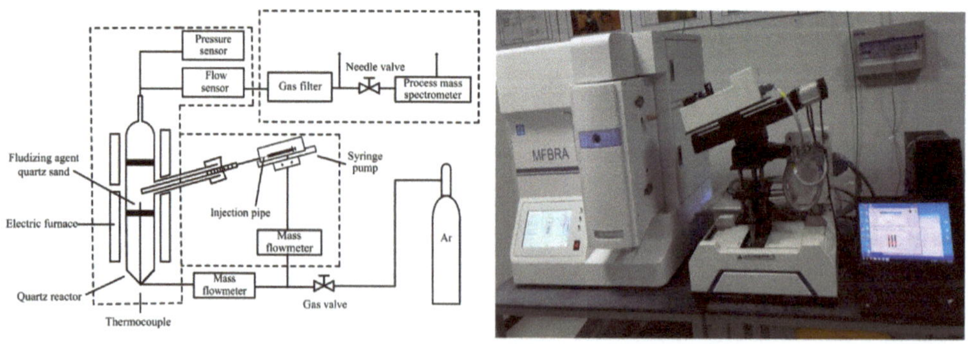

(A) (B)

Figure 7.6 A schematic diagram (A) and photograph (B) of micro-fluidized bed reaction analyzer with liquid sample injector [14].

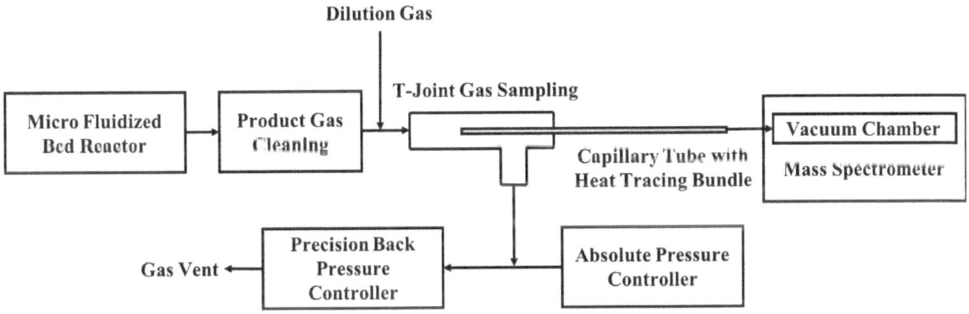

Figure 7.7 A schematic diagram of gas sampling system used in micro-fluidized bed reaction analyzers [17].

are unsuitable for analyzing gases of high molecules. In this case, some soft ionization techniques, including chemical ionization (CI), electrospray ionization (ESI), matrix-assisted laser desorption/ionization (MALDI), and photon ionization (PI), can be used [18,19].

7.1.6 Online particle sampling

Traditionally, tests of catalyst reduction, performance evaluation, deactivation and regeneration, and constituents screening are carried out by using lab-scale fixed-bed reactors. However, the catalysts packed in fixed-bed reactors are not uniformly distributed; therefore, the catalysts in different locations experience different fluid flow rates, temperatures, and reactant concentrations. Characteristics of catalysts in one place do not represent those of the entire bed. In addition, traditional methods can only characterize the catalyst samples collected after experiments. Although the state of catalysts collected post-experiments is related to the reaction process, the physiochemical properties of the collected catalysts may have experienced some changes during sample cooling, collection, and storing operations. Therefore, relying on the properties of the catalysts based on the post-experimental samples to draw any insights into the catalyst behavior during the reaction is sometimes uncertain or even skeptical. It is also difficult to decode the mechanism and identify the key factors that change the reaction performance (e.g., catalyst deactivation with time).

Micro-fluidized beds offer good mixing of solid particles, and thus, the catalysts in the beds are uniformly distributed. The state of particles at one location is the same as that anywhere else in the bed; therefore it can represent all the catalysts in the bed. Micro-fluidized beds with the capability of online particle sampling can be used to collect and characterize solid materials in the bed, either solid reactants or catalysts, with non-interruption and negligible interference to the reaction.

Fig. 7.8 shows two micro-fluidized bed reaction analyzers with an online sampling of bed particles [20,21]. The online particle sampling system consists of a particle

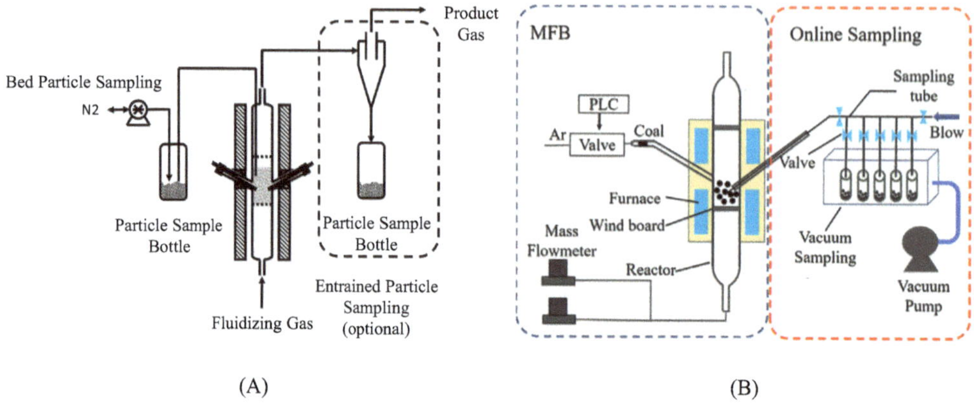

Figure 7.8 Illustrations of the sampling methods of bed and entrained particles [20,21]. (A) online particle sampling from the reactor top and (B) online particle sampling from the reactor side.

suction tube inserted into the bed from the top (Fig. 7.8A) or side (Fig. 7.8B), an inert gas purge line, and a differential pressure measurement. Independent suction and purge systems are used to collect particle samples and purge the sampling line and the bottle. In the sampling process, an inert gas purging tube is constantly in operation to ensure that the catalyst particles are sucked out of the bed, and a three-way valve is used to regulate and control the flow of suction gas so that the sampled particles are not left in the sampling tube, sampling bottle, and other places. In this way, the particles can be kept at their original reaction state. Each sample amount is generally controlled to be less than 1.0 g to reduce the impact of sampling on the reaction process, and the sampling time for each sample is shorter than 1.0 s. The samples are then analyzed. Variations in their physical and chemical properties can be examined as a function of reaction time and related to the corresponding gas-phase behavior. This method helps identify the key factors that directly lead to changes in reaction performance.

7.1.7 Change of reaction atmosphere

Changing or adjusting reacting atmospheres smoothly and quickly during reactions is very useful for experimental investigations of many chemical reactions occurring in series, such as alternative oxidation and reduction of oxygen carriers for chemical looping reactions, and catalytic reaction and regeneration of the deactivated catalysts. This capability can significantly save time and cost of experimental research because the consecutive reactions can be characterized under their required atmospheres in one experiment operation without interruption. For fast chemical reactions, such as combustion, it is particularly crucial to change the reaction atmosphere quickly and effectively so that the accuracy and reliability of the experimental results can be assured. In principle, all reactors can be operated in a gas-switching manner to change

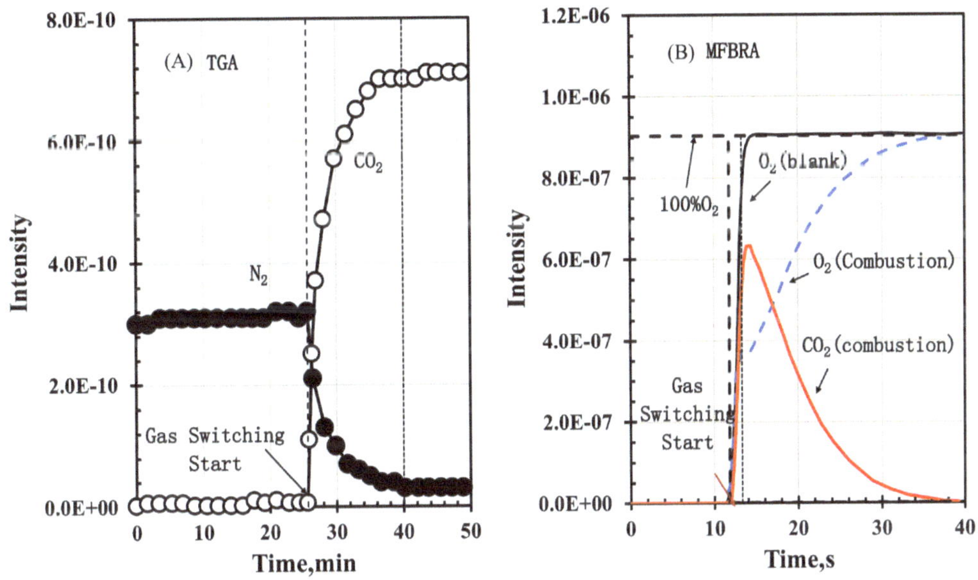

Figure 7.9 Gas-switching experiments conducted in (A) a TGA [22] and (B) an MFBRA [8].

the reacting atmosphere, but the switching time can vary significantly from reactor to reactor. To do gas switching quickly and smoothly, several conditions need to be met. First, a close-to-plug flow condition must prevail inside the reactor and along the pipeline connecting the reactor and the analytical detector. Second, the total gas storage volume (including reactor and all pipelines from the gas inlet to the gas analyzer) needs to be small enough compared to the gas volumetric flow rate. As a result, the average gas replacement time will be significantly shortened. Additionally, the mass transfer through solid particles needs to be negligible. Fig. 7.9A shows an example of gas switching in a thermogravimetric analyzer [22]. It shows that completely replacing N_2 with CO_2 takes tens of minutes due to gas diffusion inhibition in the TGA. In comparison, the MFBRA takes only a few seconds to complete the gas replacement, as shown in Fig. 7.9B [8].

Steam is used as a reactant or reaction environment for reactions such as steam gasification [23–28] and steam reforming [29]. For various reasons, the steam supply rate needs to be alternated to create desired reaction conditions during reactions. Several alternative approaches can accomplish this requirement. First, the steam can be generated by an external steam generator [26–28], as shown in Fig. 7.10A. This method is simple and does not need to change the reactor configuration, but requires more auxiliary components (three-way valve, steam generator, etc.). The second steam generation method is shown in Fig. 7.10B. In this approach, water is premixed with the fluidizing gas and enters the MFB reactor to contact the hot packing materials for the generation of steam [23]. However, this method may experience pipe clogging or flooding when high water

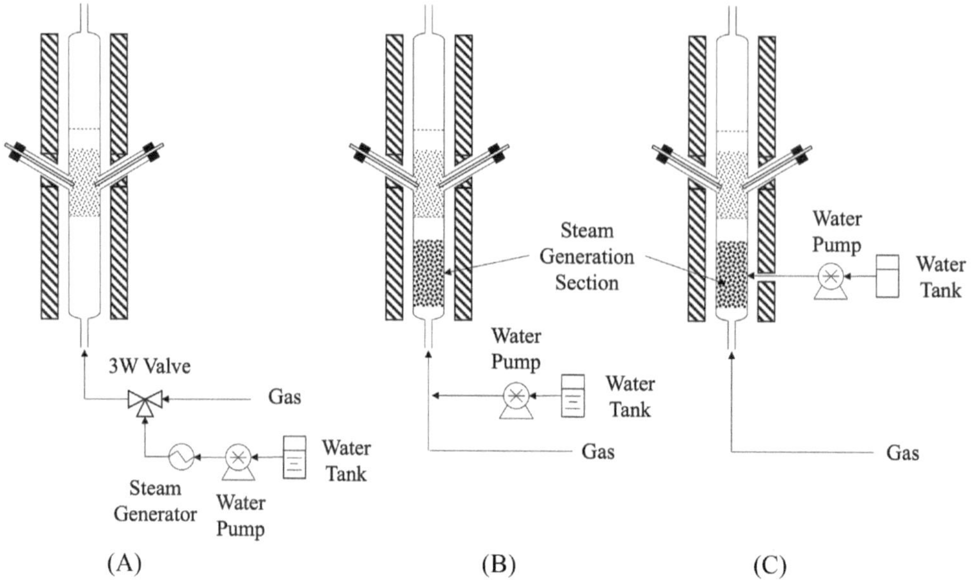

Figure 7.10 Methods to supply steam for micro-fluidized beds. (A) An external steam generator, (B) a packed bed steam generator with water supplied to the fluidizing gas, and (C) A packed bed steam generator with water supplied to the packed bed.

supply rates are required. The third method supplies water directly to the vaporizing section of MFB, as shown in Fig. 7.10C [30]. In this manner, the steam environment can be changed instantly without interference in the system's operation stability.

7.2 Kinetic data analysis

7.2.1 Data acquisition

As shown in Fig. 7.11, a user-friendly interface makes it easy to operate the micro-fluidized bed reaction analyzer. The system records the data of pressure, temperature, and signal intensities of gas species by the gas analyzer (i.e., Ms). The control system stores the experimental data for detailed analysis, and the data can also be downloaded for offline analysis.

7.2.2 Data processing

Fig. 7.12A shows a typical time-dependent release curve of the product gas by a micro-fluidized bed reaction analyzer. From this result, the cumulative production amounts of the gas component i released from time t_0 to time t and the reaction end t_f can be calculated by the following integral equations:

$$m_i(t) = \frac{M_i}{V_0} \int_{t_0}^{t} F_0(t) \cdot C_i(t)dt \tag{7.1}$$

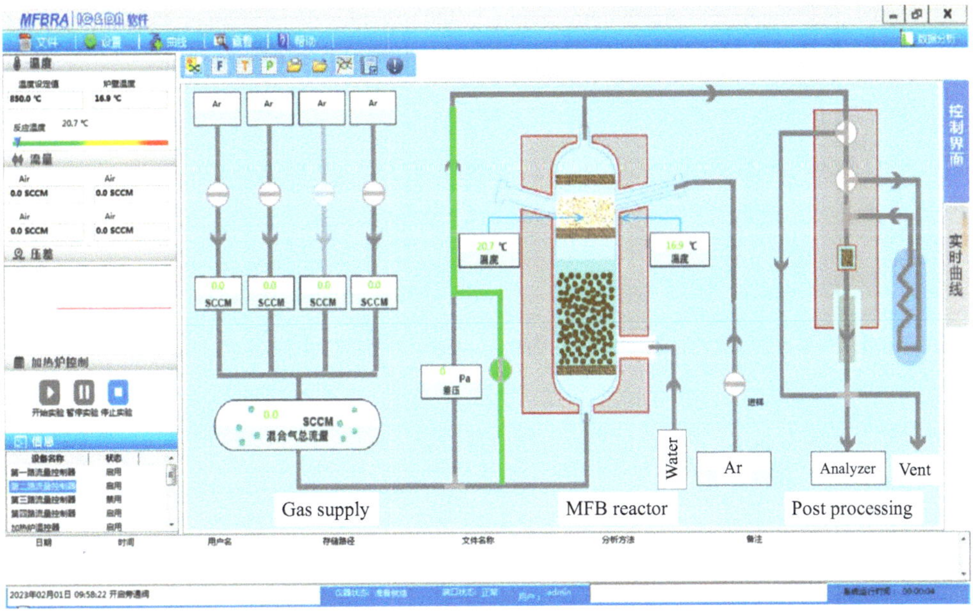

Figure 7.11 A snapshot of the user-friendly interface of the micro-fluidized bed reaction analyzer for operation and data acquisition.

$$m_{i,f} = \frac{M_i}{V_0} \int_{t_0}^{t_f} F_0(t) \cdot C_i(t) dt \tag{7.2}$$

where C_i is the volumetric concentration of the gas component i, F_0 is the total volumetric flow rate under the standard conditions, M_i is the molecular weight of the gas component i, and V_0 is the moles per standard volume of gas (22.4 L/mole). The relationship between the volumetric gas concentration C_i and the corresponding signal intensity I_i can be calibrated experimentally.

Combining Eqs. (7.1) and (7.2), the relative conversions of gas component i and the total gas mixture can be defined as follows:

$$x_i = \frac{m_i(t)}{m_{i,f}} \tag{7.3}$$

$$x_o = \frac{\sum_i m_i(t)}{\sum_i m_{i,f}} \tag{7.4}$$

Note that "relative conversion," referred to as "conversion," is used for simplicity in the following sections. Based on Eq. (7.3), the conversion is calculated as a function of time, as shown in Fig. 7.12B.

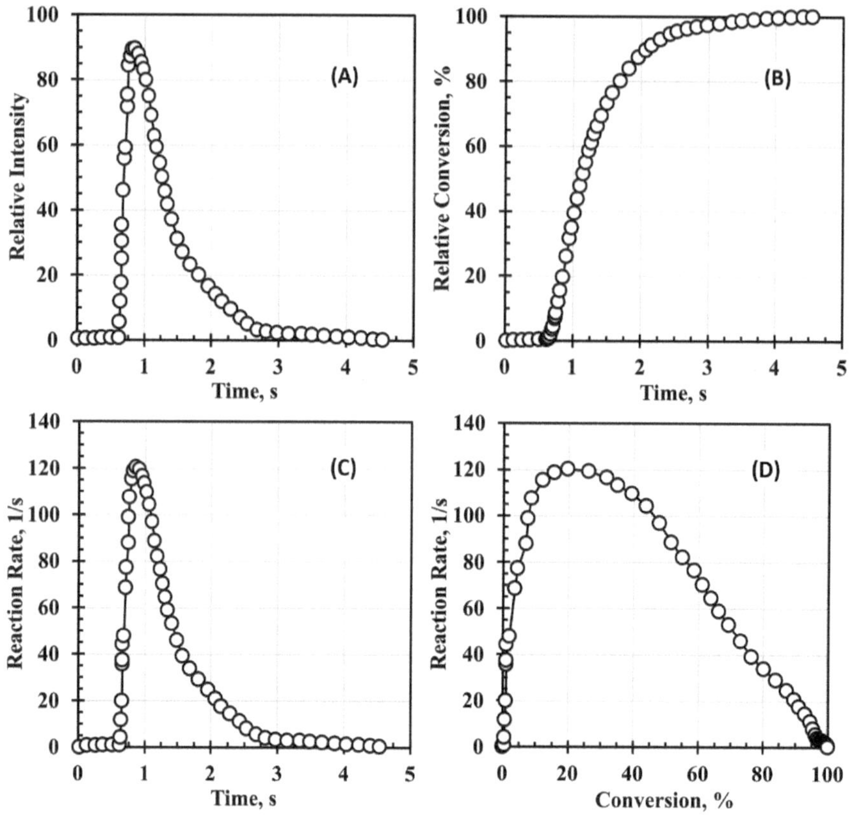

Figure 7.12 MFBRA data processing procedure. (A) Raw data, (B) relative conversion versus time, (C) reaction rate versus time, and (D) reaction rate versus conversion.

The chemical reaction rates of the gas component i and the total gas mixture can be estimated by

$$r_i = \frac{1}{m_{i,f}}\frac{dm_i(t)}{dt} = \frac{dx_i}{dt} \approx \frac{\Delta x_i}{\Delta t} \tag{7.5}$$

$$r_o = \frac{1}{\sum_i m_{i,f}}\frac{d\sum_i m_i(t)}{dt} = \frac{dx_o}{dt} \approx \frac{\Delta x_o}{\Delta \mathrm{t}} \tag{7.6}$$

Fig. 7.12C shows the reaction rate as a function of time. Furthermore, the reaction rate can be expressed as a function of the conversion, as shown in Fig. 7.12D.

Now from Eqs. (7.5) and (7.6), the gas production kinetics can be estimated based on the appropriate reaction kinetic models and analysis approaches [31–33].

7.2.3 Kinetic modeling

For a typical gas–solid reaction, the reaction rate can be expressed as

$$\frac{dx}{dt} = k(T) \cdot f(x) \tag{7.7}$$

where $k(T)$ is the reaction rate constant, and $f(x)$ is the reaction model function. The reaction rate constant is related to the reaction temperature by the Arrhenius equation

$$k(T) = A \cdot \exp\left(-\frac{E_a}{RT}\right) \tag{7.8}$$

where A is the pre-exponential factor, E_a is the activation energy, R is the gas constant (8.314 J/(mole · K)), and T is the reaction temperature. Substituting Eq. (7.8) into (7.7) gives

$$\frac{dx}{dt} = A \cdot \exp\left(-\frac{E_a}{RT}\right) \cdot f(x) \tag{7.9}$$

In Eq. (7.9), the reaction rate dx/dt is expressed as functions of temperature T and the conversion x. To determine the reaction kinetic parameters A and E_a, various numerical approaches have been employed [34–46]. These approaches, including their principles and analytical or numerical procedures, have been well documented in the literature, and hence detailed descriptions are omitted here. Interested readers are referred to the original literature.

7.3 New developments in MFBRA

Continuous research and development have been in progress to enhance the capabilities and expand applications of micro-fluidized bed reaction analyzers. Some of the new developments are presented below.

7.3.1 MFB thermogravimetric analyzer

As presented early, the MFBRA is initially designed to decode the reaction mechanisms and kinetics based on monitoring and analyzing the time-dependent product gas profiles. In addition, today's MFBRA works with solid samples in the range of 5–50 mg. These features allow the MFBRA to generate close-to-intrinsic kinetics of gas production in a close-to-ideal differential isothermal condition. However, in practice, there are two concerns that need to be noted. First, except for reactions of one step or one product, the kinetics of the product gas reflects part of the reaction kinetic behavior, but not the whole. For complex reactions, gas analyzers may be unable to receive or analyze the whole spectrum of the products, and therefore the global-reaction kinetic behavior cannot be obtained. Second, the representability of test samples and the reproducibility and reliability of experimental results may be surrendered by using small samples when characterizing highly heterogeneous and complex feedstocks.

Given these circumstances, combining a micro-fluidized bed reactor and a thermogravimetric analyzer is rational and meaningful to take full advantage of the two technologies. Samih et al. [47] were the first to report a concept of MFB-TGA, which was designed to operate at atmospheric pressure and temperatures from ambient to 1200°C with a maximum sample loading of 50 g (Fig. 7.13A). In an MFB-TGA, the entire fluidized bed reactor is placed on a load cell to measure the reactor weight. It is well understood that when fluidizing gas flows through the bed of particles, it causes an additional buoyance force or the pseudo-variation of the weight. Therefore, to obtain the real weight loss resulting from the reaction, it is necessary to subtract the

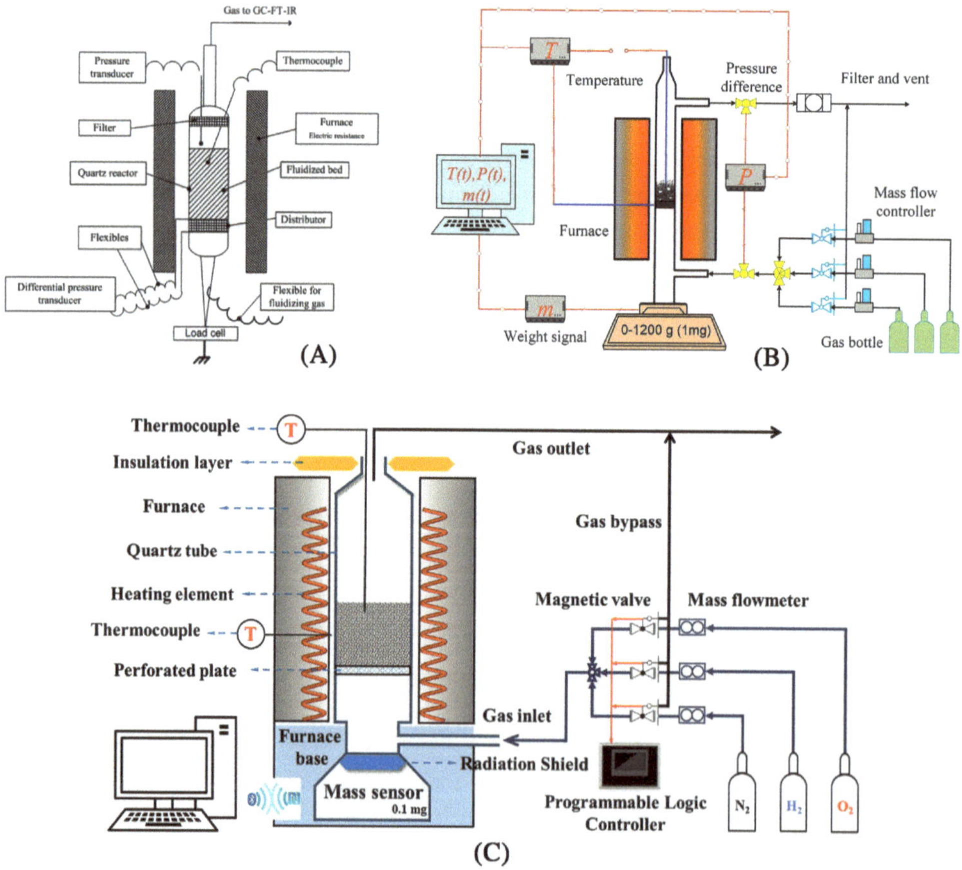

Figure 7.13 Schematic diagrams of micro-fluidized bed thermogravimetric analyzers. (A) The concept of Samih et al. [47], (B) the concept of Li et al. [53], and (C) the concept of Yan et al. [55].

pseudo-variation of the weight from the total weight loss measured by the load cell. Based on this MFB-TGA, characterizations of $Ca(OH)_2$ decomposition [48], coal pyrolysis [49], coal gasification [50], and synthesis of spinel lithium titanate [51] have been successfully accomplished.

Fig. 7.13B is a micro-fluidized bed thermogravimetric analysis (MFB-TGA) system proposed by Li et al. [52,53]. The design is, in principle, similar to the one shown in Fig. 7.13B, but the measurement range is extended to 1200 g with a precision of 1 mg using a high-precision weight transducer. Using this MFB-TGA, kinetics experiments [53,54] have been conducted with milligrams (up to 50 mg) of solid sample particles, and about 10 grams of bed materials.

Recently, Yan et al. [55] proposed an improved design of MFB-TGA, as shown in Fig. 7.13C. The fluidized bed reactor is placed on a high-precision electronic balance with a precision of 0.1 mg and a measuring range of 0−500 g, without contact with any part of the furnace. A layer of thermal insulation is installed on the reactor top to reduce the heat exchange between the reactor and the external environment. Furthermore, a flow guiding ring is installed in the upper bed section to minimize the influence of gas disturbance caused by inner temperature gradients on mass measurement. The mass signal is transmitted to the computer wirelessly via Bluetooth for recording and analysis at a sampling frequency of 10 Hz. Using this MFB-TGA, Yan et al. [55] investigated the performance of chemical looping combustion with manganese ore as the oxygen carrier.

The advantage of MFB-TGA is obvious. It measures the change of solid mass and the release of product gas simultaneously, providing a more comprehensive understanding of reaction kinetic performance. However, to obtain the actual change in the mass of the sample, it is necessary to determine the effects of buoyancy, disturbances caused by changes in fluidization gas velocity and composition, reaction temperature, and pressure drop through the gas distributor on the measured total weight of the reactor. Due to the instability of the bed at high superficial gas velocities, the MFB-TGA is unsuitable for characterizing reactions under vigorously bubbling and turbulent fluidized bed conditions [52]. For future studies, this issue remains to be resolved.

7.3.2 Induction heating MFB

In an MFBRA, sample particles are heated by hotbed materials, but the reactor body is heated by an electric furnace. Therefore, the reactor temperature may not be heated or cooled fast enough in practical operations, preventing experiments from quick startup, termination, and timely change in reaction temperature during experiments. In addition, a temperature gradient between bed particles and the wall surface may exist, leading to unwanted parasitic thermal reactions near the reactor wall. To overcome these problems, a rapidly heating MFB, i.e., induction heating MFB, has been developed [56,57].

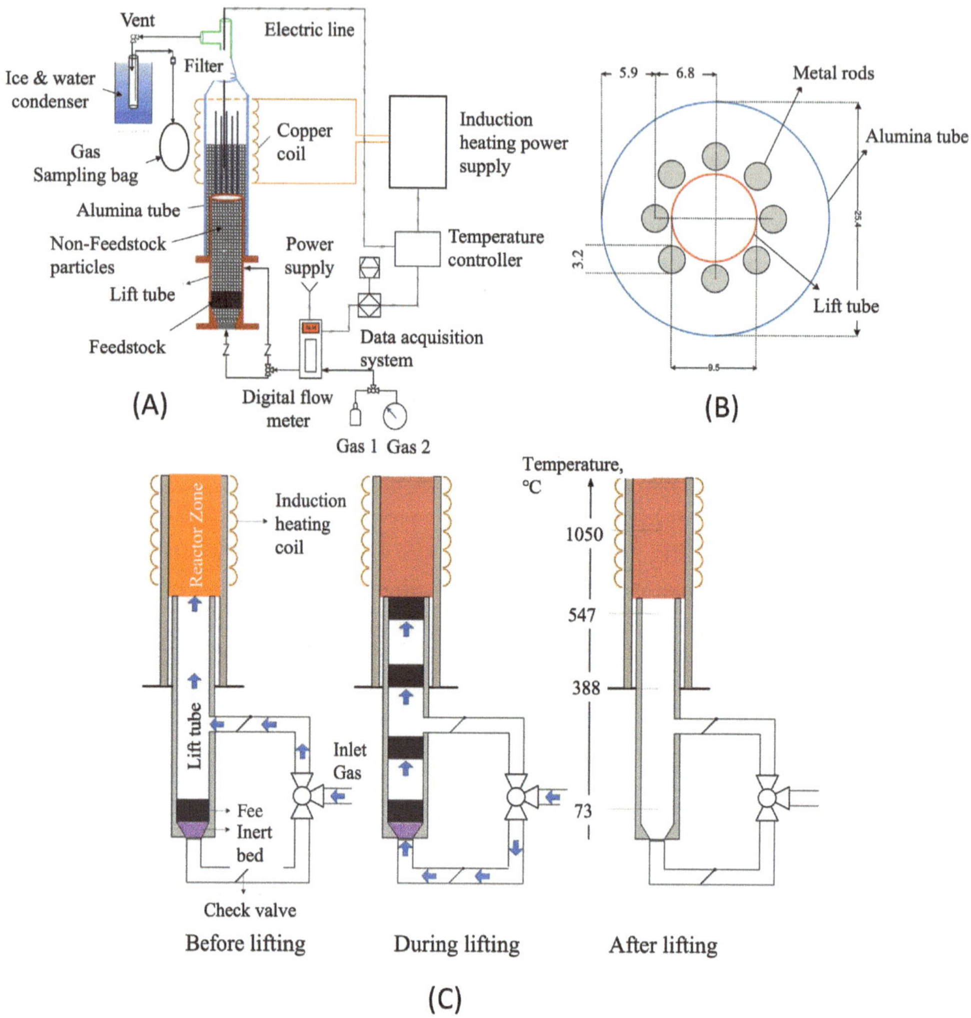

Figure 7.14 Schematic of an induction heating micro-fluidized bed reactor [56]. (A) The system flow diagram, (B) the configuration of inductive platinum rods, and (C) the illustration of solid sample lifting process.

Fig. 7.14A shows that this MFB is heated by induction heating elements. Inside the reactor, eight platinum rods of 3 mm in diameter are placed in a circular configuration symmetrically (Fig. 7.14B). These electrically conductive rods are located in a strong magnetic field with a frequency of up to 40–500 kHz. The heating rate can be adjusted by controlling the power supplied to the rods to generate and transfer heat directly to the fluidized bed particles. As shown in Fig. 7.14C, the solid sample is placed at the bottom of the lift tube at the beginning of an experiment, where the temperature is low enough to avoid unwanted thermal degradation; when the bed temperature rises and stabilizes at the desired setpoint, the lift tube is pushed up into the heated reaction zone to initiate the desired reactions. Using this induction heating micro-fluidized bed reactor, Latifi et al. [56] investigated co-combustion of pulverized coal with a sorbent-containing in situ SO_2 capturable fuel feedstock, which demonstrated the advantages of the induction heating in quick heating and cooling of the reactor.

7.3.3 External force assistance

Gas–solid fluidization typically favors Geldart group A and B powders. Fine particles classified into the group C are difficult to fluidize because of strong interparticle forces (cohesive forces). There have been many studies on the fluidization of ultrafine particles in conventional fluidized beds, and readers can reference the research and review articles for more details about ultrafine particle fluidization, e.g., [58,59]. Similar to macro fluidized beds, micro-fluidized beds can be assisted by using external forces such as acoustic waves, vibration, stirring, pulsed gas flow, magnetic or electric fields to improve their fluidization performance, especially when cohesive Geldart group C powders are fluidized. Notably, the acoustic intensified [60,61] and magnetically assisted [62] micro-fluidized beds offer good gas–solid contact, constant pressure drop, and bubble-free fluidization. Li et al. [62] reported that a magnetic fluidized bed reactor could significantly enhance the CO methanation over a nanosized NiCo aerogel catalyst (NiCo) for synthetic natural gas production. The magnetic-enhanced micro-fluidized bed system is shown in Fig. 7.15. The MFB reactor is located in an axial uniform magnetic field generated by four parallel solenoids. The magnetic field intensity is controlled by adjusting the power supply's current. It shows that when the magnetic field is introduced to the micro-fluidized bed reactor, the fluidization quality of the Ni/Al_2O_3 catalyst is significantly improved. Compared with that in the conventional fluidized bed, the maximum bed expansion ratio in the magnetic micro-fluidized bed reactor increases from 1.6 to 2.6. Experimental results demonstrate that homogeneous fluidization achieved in the magnetic micro-fluidized bed is beneficial to improving the efficiency of the CO methanation reaction.

7.3.4 Micro-spouted bed reaction analyzer

A micro-spouted bed (or spouted fluidized bed) reactor, illustrated in Fig. 7.16A, is a particular type of micro-fluidized bed reactor [63]. The reactor has a conical bottom section designed to overcome some issues, such as defluidization, channeling, and ratholing, which have been experienced by fluidized bed reactors when wet, coarse, nonuniformly sized, and sticky particles are used. Fig. 7.16B shows the variations of gas residence time distribution in the micro-spouted bed. Experiments show that a close-to-plug flow of gas can be attained in the micro-spouted bed reactor under a high gas velocity.

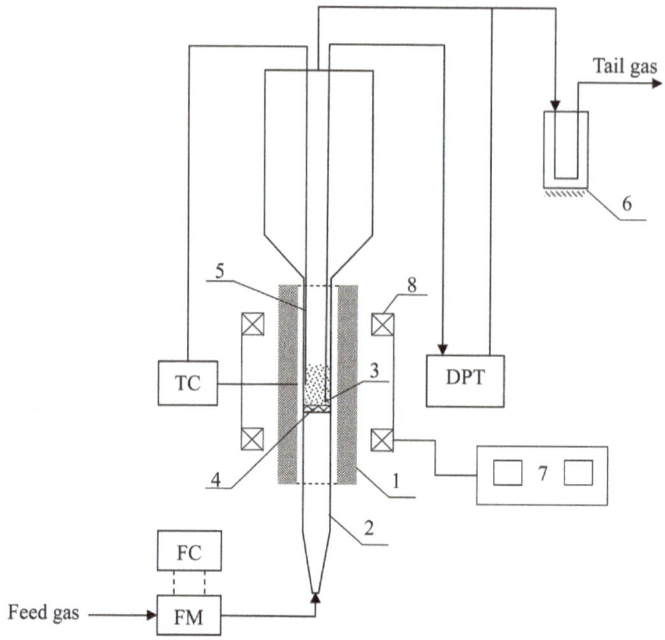

Figure 7.15 Schematic diagram of magnetic fluidized bed apparatus. *FM*, Flow meter; *FC*, flow controller; *TC*, temperature controller; *DPT*, differential pressure transducer. (1) heat furnace; (2) fluidized bed reactor; (3) pressure port; (4) gas distributor; (5) thermocouple; (6) ice trap; (7) power supply; (8) magnetic solenoids.

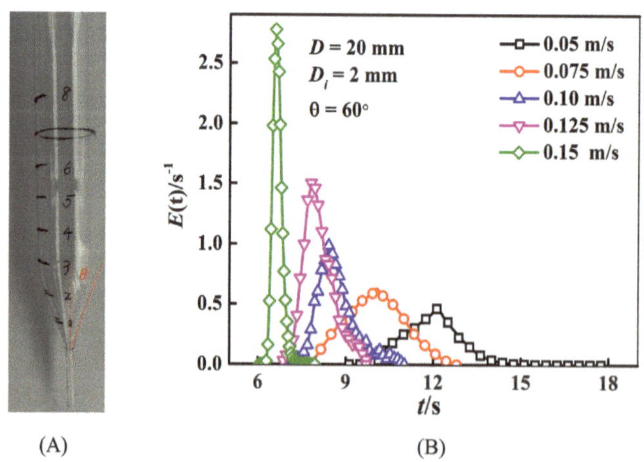

Figure 7.16 The spouted bed principle and its application in characterizing biomass pyrolysis. (A) A glass spouted bed reactor (B) the experimental results.

7.3.5 Membrane-assisted micro-fluidized beds

Membrane-assisted micro-fluidized bed reactor (MAMFBR) is a practical reactor for producing pure or ultrapure hydrogen and therefore has attracted the attention of some researchers [64–67]. The MAMFBR combines the excellent separation properties of membranes with the advantages of micro-fluidized beds, providing considerable potential for process integration and intensification. Overcoming the reaction equilibrium limitations is the most significant advantage of the membrane-assisted micro-fluidized bed reactors. Thus, MAMFBRs can obtain much higher reactant conversion and product yield by operating under virtually isothermal and close-to-plug flow conditions than fluidized beds without membrane assistances. In practical design, tubular or planar membrane tubes or compartments can be installed, as shown in Figs. 7.17A and B.

7.3.6 Other developments

As a unique microchemical reactor, the micro-fluidized bed reaction analyzer offers distinguishable characteristics such as minimized mass/heat transfer resistances, excellent gas–solid mixing, closeness-to-plug flow of gas, and ease of operation. Extensive research and development efforts have made MFBRA commercially available in several standardized designs capable of performing various reaction analyses. Recently, some progress has been made in the research and development of pressurized micro-fluidized bed reaction analyzers. In addition, a research program has been actively undertaken in Xu's group to develop micro-fluidized bed reaction analyzers that can characterize extremely fast, explosive, and corrosive reactions.

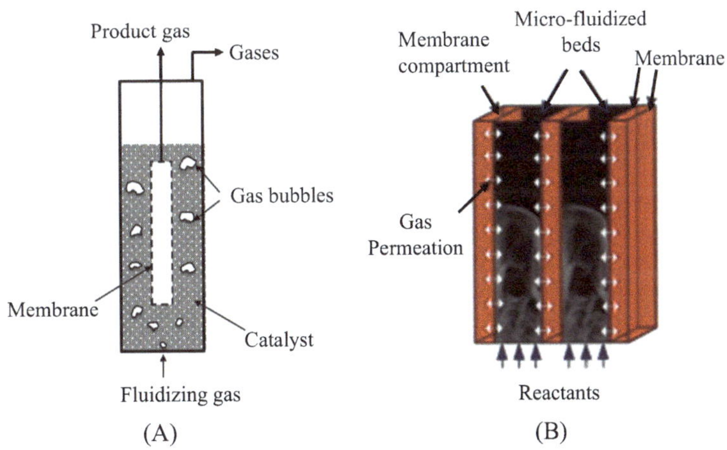

Figure 7.17 Schematics of membrane-assisted micro-fluidized beds with (A) tubular membrane tubes and (B) membrane compartments.

In conclusion, although some progress has been made in micro-fluidized bed reaction analyzers, much research and development are still required to advance the technology further. Continued efforts from academia and industry will be necessary to meet the challenges ahead.

Abbreviations

CCFBR	catalyst cell fluidized bed reactor
CI	chemical ionization
CPR	Curie point reactor
DSC	differential scanning calorimetry
DTA	differential thermal analysis
DTR	drop-tube reactor
EGA	evolved gas analysis
EI	electron impact ionization
ESI	electrospray ionization
FBR	fixed-bed reactor
FTIR	Fourier-transform infrared spectroscopy
GC	gas chromatography
GFSS	gas-flow-switching system
GPR	gas pulsed reactor
GSS	gas supply system
HGR	heated grid reactor
IR	infrared radiation
LAR	laser ablation reactor
MALDI	matrix-assisted laser desorption/ionization
MFB	micro-fluidized bed
MFB-IH	inducting heating micro-fluidized bed
MFB-TDA	micro-fluidized bed thermogravimetric analyzer
MFBR	micro-fluidized bed reactor
MFBRA	micro-fluidized bed reaction analyzer
Ms	mass spectrometry
MSBRA	micro-spouted bed reaction analyzer
MW-TGA	microwave thermogravimetric analyze
OGAS	online gas analysis system
PHASR	pulse-heated analysis of solid reaction reactor
PI	photon ionization
PP	pyroprobe
RS	reaction system
SBR	spouted bed reactor
SHR	screen heater reactor
SPI-Ms	photoionization mass spectrometer
TGA	thermogravimetric analyzer
TPR	thermal plasma reactor
TR	tandem microtubular reactor
WMR	wire mesh reactor

Nomenclature

A pre-exponential factor
C_i volumetric concentration of the gas component i
E_a activation energy
F_0 total volumetric flow rate under the standard conditions
$f(x)$ reaction model function
I_i MS signal intensity
$k(T)$ reaction rate constant
M_i molecular weight of the gas component i
m_i cumulative production amounts for the gas component i
$m_{i,f}$ cumulative production amounts for the gas component i the end of the reaction t_f
R gas constant (8.314 J/(mole · K))
r_i reaction rate of the gas component i
r_o total reaction rate
T temperature
t time
t_f end time of the reaction
V_0 moles per standard volume of gas (22.4 L/mole)
x relative conversion
x_i relative conversion of the gas component i
x_o relative total conversion

References

[1] Xu G. Micro-fluidized bed reaction kinetics analyzer. Project #2005014, China: Chinese Academy of Sciences; 2005. Available from: http://www.rcmlab.net/enclosure-mfbra[R].
[2] Xu G, Yu J, Yao M, et al. Progress in the research, development, and application demonstration of the isothermal micro-fluidized bed (gas) solid reaction analyzer. Management and Research of Scientific and Technological Achievements 2016;10:63–6.
[3] Yu J, Yue J, Liu W, et al. Thermal analysis approach and instrument for non-catalytic gas-solid reactions. Chinese Journal of Analytical Chemistry 2011;39(10):1549–54.
[4] Yu J, Zeng X, Yue J, et al. Micro-fluidized bed reaction analysis and its applications. In: Kuipers JAM, Mudde RF, Van Ommen JR, et al., editors. The 14th International conference on fluidization – from fundamentals to products. ECI symposium series; 2013. http://dc.engconfintl.org/fluidization_xiv/5.
[5] Yu J, Zhu J, Yue J, et al. Development and application of micro kinetic analyzer for fluidized bed gas-solid reactions. CIESC Journal 2009;60(10):2669–74.
[6] Yu J, Zeng X, Zhang J, et al. Isothermal differential characteristics of gas-solid reaction in micro-fluidized bed reactor. Fuel 2013;103:29–36.
[7] Guo L, Cheng W, Guo Z. Application of micro-fluidized bed in kinetics analysis of coal oxygen combustion at high temperature. Jiangxi Metallurgy 2019;39(2):14–19.
[8] Guo Y, Zhao Y, Meng S, et al. Development of a multistage in situ reaction analyzer based on a micro-fluidized bed and its suitability for rapid gas – solid reactions. Energy and Fuels 2016;30:6021–33.
[9] Han ZNZ, Yue JRJ, Zeng X, et al. Characteristics of gas-solid micro-fluidized beds for thermochemical reaction analysis. Carbon Resources Conversion 2020;3:203–18.
[10] Sun S, Guo Y, Gao D, et al. An in-situ decoupling based analytic device and method for gas-solid reactions: Chinese Patent: CN104749206B; 2017.
[11] Fang Y, Luo G, Chen C, et al. Combustion kinetics of In-situ char and cold char in micro-fluidized bed. Journal of Combustion Science and Technology 2016;22(2):148–54.

[12] Yang K, Wang J, Huang J, et al. Understanding the homogeneous reactions of primary tar from biomass pyrolysis by means of photoionization mass spectrometry. Energy and Fuels 2020;34(10): 12678–87.
[13] Yang X, Liu Y, Yu J, et al. Numerical simulation of mixing characteristics of trace sample and bed material in micro-fluidized bed reaction analyzer. CIESC Journal 2014;65(9):3323–30.
[14] Gai C, Dong Y, Fan P, et al. Kinetic study on thermal decomposition of toluene in a micro-fluidized bed reactor. Energy Conversion and Management 2015;106:721–7.
[15] Gai C, Dong Y, Lv Z, et al. Pyrolysis behavior and kinetic study of phenol as tar model compound in micro-fluidized bed reactor. International Journal of Hydrogen Energy 2015;40(25):7956–64.
[16] Gao W, Farahani MR, Jamil MK, et al. Kinetic modeling of pyrolysis of three Iranian waste oils in a micro-fluidized bed. Petroleum Science and Technology 2017;35(2):183–9.
[17] Guo Y, Zhao Y, Liu P, et al. Use of a process mass spectrometer to measure rapid change of gas concentration. Chinese Journal of Analytical Chemistry 2016;44(9):1335–41.
[18] Jia L, Le-Brech Y, Shrestha B, et al. Fast pyrolysis in a microfluidized bed reactor: effect of biomass properties and operating conditions on volatiles composition as analyzed by online single photoionization mass spectrometry. Energy and Fuels 2015;29(11):7364–74.
[19] Jia L, Dufour A, Le Brech Y, et al. On-line analysis of primary tars from biomass pyrolysis by single photoionization mass spectrometry: experiments and detailed modelling. Chemical Engineering Journal 2017;313:270–82.
[20] Pang F, Song F, Zhang Q, et al. Study on the influence of oxygen-containing groups on the performance of Ni/AC catalysts in methanol vapor-phase carbonylation. Chemical Engineering Journal 2016;293:129–38.
[21] Li Z, Zou R, Xu Y, et al. Kinetic study on continuous sampling of coal char from a micro-fluidized bed. ACS Omega 2021;6(13):9086–94.
[22] Zeng X, Wang F, Wang Y, et al. Characterization of char gasification in a micro-fluidized bed reaction analyzer. Energy and Fuels 2014;28(3):1838–45.
[23] Wang F, Zeng X, Wang Y, et al. Characterization of coal char gasification with steam in a micro-fluidized bed reaction analyzer. Fuel Processing Technology 2016;141:2–8.
[24] Zhang Y, Sun G, Gao S, et al. Regeneration kinetics of spent FCC catalyst via coke gasification in a micro-fluidized bed. Procedia Engineering 2015;102:1758–65.
[25] Zhang Y, Yao M, Gao S, et al. Reactivity and kinetics for steam gasification of petroleum coke blended with black liquor in a micro-fluidized bed. Applied Energy 2015;160:820–8.
[26] Zhang Y, Yao M, Sun G, et al. Characteristics and kinetics of coked catalyst regeneration via steam gasification in a micro-fluidized bed. Industrial and Engineering Chemistry Research 2014;53 (15):6316–24.
[27] Wang F, Zeng X, Yu J, et al. Char gasification kinetics in micro-fluidized bed reaction analyzer. Journal of Shenyang University of Chemical Technology 2014;28(3):213–19.
[28] Ji Y, Zeng X, Yu J, et al. Steam gasification characteristics of coal char in micro-fluidized bed reaction analyzer. CIESC Journal 2014;65(9):3447–56.
[29] Chen K, Zhao Y, Zhang W, et al. The intrinsic kinetics of methane steam reforming over a nickel-based catalyst in a micro-fluidized bed reaction system. International Journal of Hydrogen Energy 2020;45:1615–28.
[30] Yue J, Guan Y, Xu G, et al. An online device for analyzing gas-solid thermal reactions in micro-reactors: Chinese Patent: CN108614077A; 2018.
[31] Kunii D, Levenspiel O. Fluidization engineering. 2nd ed. Boston, MA: Butterworth—Heinemann; 1991.
[32] Wu C, Cheng Y. Downer reactors. In: Grace JR, Bi X, Ellis N, editors. Essentials of fluidization technology. Wiley-VCH Verlag GmbH & Co. KGaA, Boschstr. 12, 69469 Weinheim, Germany; 2020. p. 499–530.
[33] Yu J, Li Q, Duan Z, et al. Isothermal differential characteristics of the reaction in micro-fluidized bed. SCIENTIA SINICA Chimica 2011;41(1):152–60.
[34] Ozawa T. A new method of analyzing thermogravimetric data. Bulletin of the Chemical Society of Japan 1965;38(11):1881–6.

[35] Flynn J. The isoconversional method for determination of energy of activation at constant heating rates. Journal of Thermal Analysis and Calorimetry 1983;27:95–102.
[36] Starink MJ. The determination of activation energy from linear heating rate experiments: a comparison of the accuracy of isoconversion methods. Thermochimica Acta 2003;404(1–2):163–76.
[37] Vyazovkin S. Evaluation of activation energy of thermally stimulated solid-state reactions under arbitrary variation of temperature. Journal of Computational Chemistry 1997;18(3):393–402.
[38] Kissinger HE. Reaction kinetics in differential thermal analysis. Analytical Chemistry 1957;29 (11):1702–6.
[39] Flynn J, Wall L. A quick, direct method for the determination of activation energy from thermogravimetric data. Journal of Polymer Science. Part B: Polymer Letters 1966;4:323–8.
[40] Zelic J, Ugrina L, Jozic D. Application of thermal methods in the chemistry of cement: kinetic of portlandite from non-isothermal thermogravimetric data. In: The first international proficiency testing conference. p. 420–29.
[41] Vyazovkin S, Burnham AK, Criado JM, et al. ICTAC Kinetics Committee recommendations for performing kinetic computations on thermal analysis data. Thermochimica Acta 2011;520(1–2): 1–19.
[42] Vyazovkin S, Wight CACCACCA. Isothermal and non-isothermal kinetics of thermally stimulated reactions of solids. International Reviews in Physical Chemistry 1998;17(3):407–33.
[43] Langtry BN. Identity and spatio-temporal continuity. Australasian Journal of Philosophy 1972;50 (2):184–9.
[44] Friedman HL. Kinetics of thermal degradation of char-forming plastics from thermogravimetry. Application to a phenolic plastic. Journal of Polymer Science Part C: Polymer Symposia 2007;6 (1):183–95.
[45] Sbirrazzuoli N. Determination of pre-exponential factors and of the mathematical functions f(α) or G(α) that describe the reaction mechanism in a model-free way. Thermochimica Acta 2013;564: 59–69.
[46] Coats AW, Redfern J, Coats AW, Redfern JP. Kinetic Parameters from thermogravimetric data. Nature 1964;201(4914):68–9.
[47] Samih S, Chaouki J. Development of a fluidized bed thermogravimetric analyzer. AIChE Journal 2015;61:84–9.
[48] Chaouki J, Latifi M, Samih S. Development of a fluidized bed thermo-gravimetric analyzer (FB-TGA). In: Garcia-Perez M, Meier D, Ocone R, et al., editors. BioEnergy IV: innovations in biomass conversion for heat, power, fuels and chemicals. ECI symposium series; 2013.
[49] Samih S, Chaouki J. Coal pyrolysis and gasification in a fluidized bed thermogravimetric analyzer. The Canadian Journal of Chemical Engineering 2018;96:2144–54.
[50] Samih S, Chaouki J. Catalytic ash free coal gasification in a fluidized bed thermogravimetric analyzer. Powder Technology 2017;316:551–9.
[51] Tao L, Samih S, Sauriol P, et al. Synthesis of $Li_4Ti_5O_{12}$ negative electrode material in a fluidized bed thermogravimetric analyzer. Canadian Journal of Chemical Engineering 2021;1–13 September 2020.
[52] Li Y, Wang H, Li W, et al. CO_2 gasification of a lignite char in micro-fluidized bed thermogravimetric analysis for chemical looping combustion and chemical looping with oxygen uncoupling (CLC/CLOU). Energy and Fuels 2019;33:449–59.
[53] Li Y, Li Z, Wang H, et al. CaO carbonation kinetics determined using micro-fluidized bed thermogravimetric analysis. Fuel 2020;264:116823.
[54] Li Y, Li Z, Cai N. Micro-fluidized bed thermogravimetry combined with mass spectrometry (MFB-TG-MS) for redox kinetic study of oxygen carrier. Energy & Fuels 2020;34:11186–93.
[55] Yan J, Shen T, Wang P, et al. Redox performance of manganese ore in a fluidized bed thermogravimetric analyzer for chemical looping combustion. Fuel 2021;295:120564.
[56] Latifi M, Chaouki J. A novel induction heating fluidized bed reactor: its design and applications in high temperature screening tests with solid feedstocks and prediction of defluidization state. AIChE Journal 2015;61:1507–23.
[57] Saadatkhah N, Carillo A, Sarah G, et al. Experimental methods in chemical engineering: thermogravimetric analysis — TGA. Canadian Journal of Chemical Engineering 2020;98:34–43.

[58] van Ommen JR, Valverde JM, Pfeffer R. Fluidization of nanopowders: a review. Journal of Nanoparticle Research 2012;14:737(3):1–29.
[59] Dasgupta K, Joshi JB, Banerjee S. Fluidized bed synthesis of carbon nanotubes – a review. Chemical Engineering Journal 2011;171(3):841–69.
[60] Hao J, Guo Q. Fluidization characteristics in micro-scale fluidized beds with acoustic intensification. Journal of Chemical Engineering of Chinese Universities 2010;24(6):929–35.
[61] Li Z, Cao C. Numerical simulationof Group A and C particles in the micro-scale fluidized bed with acoustic assistance. Journal of Chemical Engineering of Guangdong 2013;40(24):22–4.
[62] Li J, Zhou L, Zhu Q, et al. Enhanced methanation over aerogel NiCo/Al_2O_3 catalyst in a magnetic fluidized bed. Industrial and Engineering Chemistry Research 2013;52:6647–54.
[63] Hu D, Zeng X, Wang F, et al. Comparison and analysis of gas backmixing in micro-fluidized and spouted beds. In: The 11th annual conference and cross-strait symposium on particle technology. Xiamen: Chinese Society of Particuology; 2020.
[64] Tan L, Roghair I, van Sint Annaland M. Simulation study on the effect of gas permeation on the hydrodynamic characteristics of membrane-assisted micro-fluidized beds. Applied Mathematical Modelling 2014;38(17–18):4291–307.
[65] Dang NTY, Gallucci F, Van Sint Annaland M. Micro-structured fluidized bed membrane reactors: solids circulation and densified zones distribution. Chemical Engineering Journal 2014;239:42–52.
[66] Dang TYNYN, Gallucci F, Van Sint Annaland M. Gas back-mixing study in a membrane-assisted micro-structured fluidized bed. Chemical Engineering Science 2014;108:194–202.
[67] Dang NTY, Gallucci F, van Sint Annaland M. An experimental investigation on the onset from bubbling to turbulent fluidization regime in micro-structured fluidized beds. Powder Technology 2014;256:166–74.

CHAPTER 8

Characteristics of micro-fluidized bed reaction analyzers

Micro-fluidized bed reaction analyzers are attractive and useful for thermal reaction analysis. This chapter presents a comprehensive summary of the major features of micro-fluidized bed reaction analyzers. Specific applications are discussed to illustrate these characteristics. Readers are advised to review the essential requirements for thermal analysis techniques presented in Chapter 6 and understand how MFBRAs are made to meet these requirements.

8.1 Approaching intrinsic kinetics

Modeling and designing a gas–solid reaction process require a better understanding of the reaction mechanism and intrinsic kinetics. In principle, for a gas–solid reaction, the intrinsic kinetics should only depend on the reaction process and not be affected by gas diffusions inside and outside particles. For this reason, all of the following requirements have to be met during thermal measurements using a microreactor:

1. Fast heating rate,
2. Negligible mass transfer inhibition both outside and inside of particles,
3. Close-to-plug flow of gas inside the reactor and between the reactor and gas analyzer, and
4. Homogenous reaction fields of flow, temperature, and concentration.

In the following, we illustrate how an MFB meets these requirements.

8.1.1 High heating and cooling rates

The heating or cooling rate significantly affects the product profile of gas–solid reactions and the structural transformation of solid particles during reactions [1,2]. Benefiting from high heat transfer rates between gas and solids, between gas–solid suspension and the reactor wall surfaces, and between bed materials, micro-fluidized beds offer significantly high heating and cooling rates. When sample particles are injected into a micro-fluidized bed (Fig. 8.1A), they are quickly heated or cooled by intimate contact with bed materials and fluidizing gas. Under high temperatures, heat transfer will also be contributed by radiation, which is intensified greatly in micro-fluidized beds because of short radiation paths. Based on a theoretical heat transfer modeling

Micro Fluidization
DOI: https://doi.org/10.1016/B978-0-443-18718-6.00008-5
© 2023 Chemical Industry Press Co., Ltd.
Published by Elsevier Inc., under an exclusive license with Chemical Industry Press Co., Ltd.

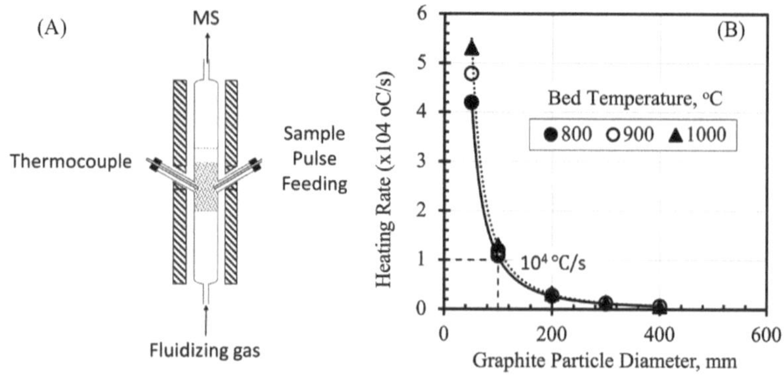

Figure 8.1 The heating rates of graphite particles estimated by the mathematic model for different sizes and bed temperatures in MFB [3]. (A) A micro-fluidized bed reactor and (B) the heating rate.

analysis, the heating rate of graphite particles in a micro-fluidized bed reactor is estimated [3]. As shown in Fig. 8.1B, the heating rate of the graphite particles increases with rising bed temperature and decreasing particle size. The heating rate can be as high as over 10^4°C/s when the particle diameter is below 100 μm. On this basis, it will take less than 0.1 seconds for sample particles in micro-fluidized beds to reach 1000°C from the ambient condition. In contrast, TGA will need approximately 10 minutes to do so because it has a heating rate typically less than 100°C/min (or <2°C/s).

The high heating rate predicted by the theoretical calculations [3] has been confirmed experimentally by Fang et al. [4]. Using a dual reactor system shown in Fig. 8.2A, Fang et al. [4] measured the heating and cooling rates of sample particles in the micro-fluidized bed. The experimental results are presented in Figs. 8.2B and C. It can be seen that heating the char particles in micro-fluidized beds from the ambient temperature to 500°C, 650°C, and 750°C, respectively, or cooling the hot char particles from 950°C to temperatures of 500°C, 650°C, and 800°C, respectively, takes less than 0.05 seconds, corresponding to a heating rate (positive) and a cooling rate (negative) of greater than 10^4°C/s. Therefore, rapid heating or cooling can be attained in micro-fluidized beds by utilizing particle bed heating or cooling.

8.1.2 Effective suppression of diffusion

Micro-fluidized beds can operate at relatively high superficial gas velocities with negligible gas backmixing due to the delayed minimum fluidization and the advanced onset of turbulence. This feature provides an opportunity to eliminate the external diffusion of particles substantially by operating the reactor at a high gas flow rate while still maintaining a good fluidization quality. Generally, effective suppression of gas diffusion can be judged by observing the effect of gas velocity (or flow rate) or

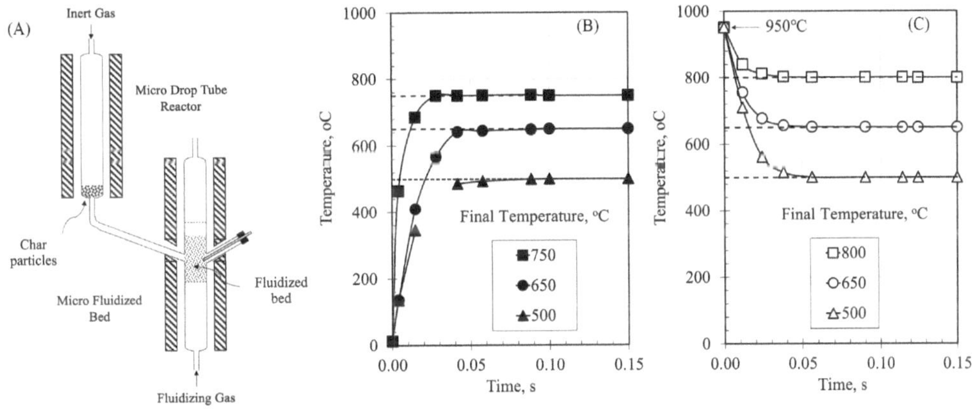

Figure 8.2 (A) A dual reactor system used to measure heating and cooling rates of sample particles in a micro-fluidized bed, (B) temperature variations of char particles during heating, and (C) temperature variations of char particles during cooling [4].

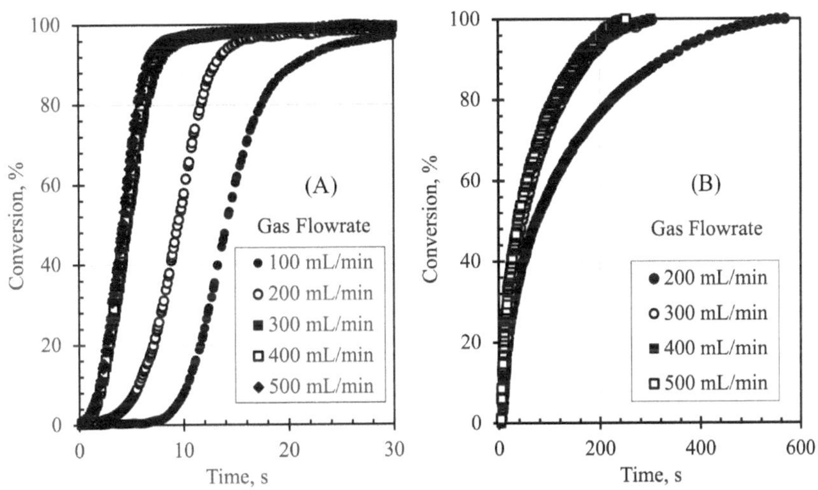

Figure 8.3 (A) Influence of gas flow rate on the conversion of biomass pyrolysis and (B) iron ore reduction.

reaction time on the reaction performance (e.g., reaction rate or conversion). In principle, if the gas flow rate or reaction time has an unnoticeable effect on the reaction performance, the external diffusion can be considered to have been eliminated substantially. As illustrated in Fig. 8.3, the conversion changes little as the gas flow rate is high enough for biomass pyrolysis [5] and iron ore reduction by CO [6]. The results clearly indicate that the surface chemical reaction is the rate-controlling step of these reactions. In other words, the effect of gas diffusion on the reaction rates becomes negligible when the gas flow rate continues to increase.

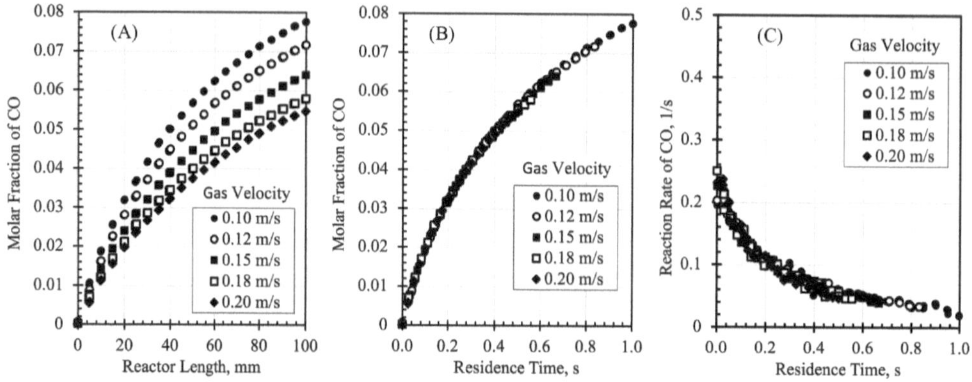

Figure 8.4 Experimental data from a methane reforming experiment in a micro-fluidized bed reactor. (A) Variation of molar fraction of CO with reactor length at different gas velocities, (B) variation of molar fraction of CO with residence time of gas at different gas velocities, and (C) reaction rate of CO with residence time of gas at different gas velocities.

The threshold flow rate, above which the chemical reaction becomes the rate-controlling step, depends on the reaction characteristics. Thus, this threshold flow rate should be determined by experiments carried out under a series of gas flow rates. In practice, the operating gas flow rate should be selected to satisfy both the above-cited threshold and the fluidization quality requirement.

Determining whether a reaction is affected by diffusion is sometimes not as straightforward as described above. Fig. 8.4A shows the axial profiles of CO mole fraction at various reactor inlet gas velocities for methane steam reforming reaction over a nickel-based catalyst in a micro-fluidized bed reactor of 20 mm in inner diameter [7]. At first glance, the axial profile of the CO mole fraction is apparently affected by the gas velocity, but without further analysis of the data, it is impossible to draw a conclusion that the gas diffusion affects the reaction from the observations only. If we convert the data shown in Fig. 8.4A–C, we can see that the CO mole fraction and the corresponding reaction rate dy_{CO}/dt (here, y_{CO} is the mole fraction of CO, and t is the reaction time calculated by dividing the reactor length by the gas velocity) are solely dependent on the reaction time, regardless of the gas velocity. Therefore, we are ensured that the intrinsic kinetics for this reaction is already obtained by using the MFBRA.

In addition to effectively suppressing the external diffusion of particles, the gas diffusion within particles (i.e., internal diffusion) must also be excluded substantially so that true or close-to-intrinsic reaction kinetics can be obtained. For catalytic gas–solid reactions, the catalyst effectiveness factor is a measure of internal diffusion, which is dependent on the Thiele modulus. When the Thiele modulus is low enough, the effectiveness factor is close to unity, corresponding to a condition that the internal diffusion can be neglected. For non-catalytic gas–solid reactions, the internal diffusion will add little resistance to the overall reaction rate when the particle size is small enough because a small particle features short pores and the resultant rapid diffusion rate within the pores (note that we assume that the particle is porous here, which is true in most practical applications). The use of fine particles at high gas velocities is known to be problematic in large-scale gas—solid fluidized beds because of increased particle elutriation rate, but it is quite possible in micro-fluidized beds due to their unique hydrodynamic characteristics. As discussed in the preceding chapters, of the many unique hydrodynamic characteristics in micro-fluidized beds, the most important characteristic is the remarkably increased minimum fluidization velocity, especially when fine particles are used. This feature means that micro-fluidized beds will operate at a reduced fluidization number (U_g/U_{mf}) for a given superficial gas velocity compared to their macro counterparts. For this reason, the particle elutriation from micro-fluidized beds is significantly reduced.

The influence of particle diameter on the conversion is shown in Fig. 8.5A for biomass pyrolysis [5] and Fig. 8.5B for an iron ore reduction [6]. It illustrates that the conversion increases with reducing particle diameter at a given reaction time. When the particle size is smaller than approximately 100 μm, the influence of the particle size on the reaction diminishes, indicating that for such small particles, the reaction rate is essentially controlled by the chemical reaction rather than the internal diffusion.

Based on the above discussions, we have verified that the intrinsic kinetics can be approximated by operating micro-fluidized beds at high gas velocities and using small particles. One may argue that true intrinsic kinetics can never be obtained in practical reactions due to the existence of molecular diffusion associated with the micropore structures of porous particles. This is true, but we have to realize that the molecular diffusion is comparatively insignificant from the reactor design and operational

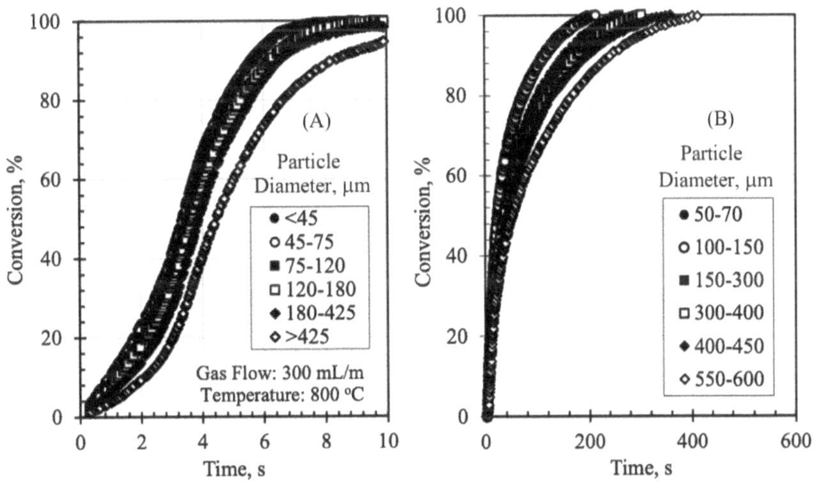

Figure 8.5 Influence of particle diameter on the reaction kinetics of (A) biomass pyrolysis and (B) iron ore reduction by CO in micro-fluidized bed reactors.

perspective. For clarity, in this book, we refer to the close-to-intrinsic kinetics that is insensitive to the gas flow rate and the particle size as "applied kinetics." We will elucidate this concept later.

8.1.3 Close-to-plug flow of gas

Gas flows in micro-fluidized beds exhibit the close-to-plug flow behavior, especially in micro-fluidized beds of Group B particles. As discussed in Chapter 3, a close-to-plug flow of gas in a micro-fluidized bed is characterized by a symmetrically shaped and narrowly spread RTD curve, with its peak height greater than 1 s^{-1} and the standard deviation less than 0.25 s^2.

For a micro-fluidized bed reaction analyzer, the gas flow from the fluidized bed reactor to the gas analyzer must also be fast enough and close-to-plug flow. The close-to-plug flow ensures that the product gas is unaltered in both the dynamic production sequence and composition when the gas flows to a gas analyzer without significant delay. Using a pipe or a capillary tube with a small enough diameter and volume can help achieve this goal because the gas flow is close-to-plug flow under this condition.

Further assisted with the quick responding process Ms or other fast-responding gas analyzers, MFBRA can accurately obtain the dynamic release sequence, composition, and yield of each gas product. Fig. 8.6 shows the typical evolution and yields of major gas species from a test of biomass pyrolysis in a micro-fluidized bed reaction analyzer [8].

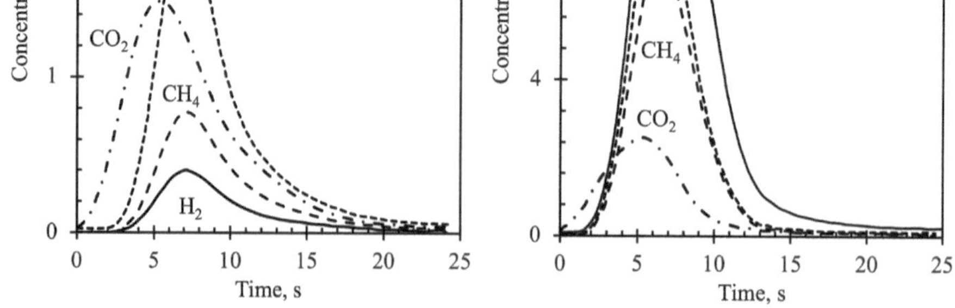

Figure 8.6 Typical evolution of major gas species from biomass (beer lees) pyrolysis in a micro-fluidized bed at temperatures of (A) 500°C and (B) 800°C [8]. Insert pie charts show the composition of gases.

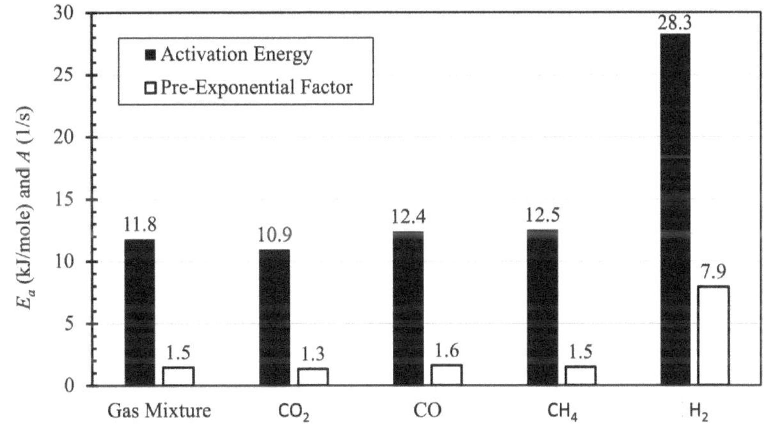

Figure 8.7 Kinetic parameters of CO_2, CO, CH_4, H_2, and their mixture produced by the pyrolysis of beer lees in a micro-fluidized bed [8].

It shows that at 500°C, CO_2 releases first, followed by CO, CH_4, and H_2; at 800°C, the release order of these gases remains the same, but the release time intervals become smaller. In producing these permanent gases, the pyrolysis reaction completes in about 10 seconds in the micro-fluidized bed, which is much shorter than that in a large-scale fluidized bed reactor. As shown in the insert charts of Fig. 8.6, the yields of H_2 and CH_4 increase as the pyrolysis temperature is elevated from 500°C to 800°C.

The reaction kinetics can be obtained by the dynamic data of product gas release based on the relative conversion concept (refer to Chapter 7). Fig. 8.7 presents the estimated kinetic parameters of producing CO_2, CO, CH_4, H_2, and their mixture. It indicates that the activation energy of gas production increases in the order of CO_2, CO, CH_4, and H_2. The in-depth discussions about the underlying mechanism are presented in the literature [5], and therefore, we will not discuss this further here.

8.1.4 Bed homogeneity

It is vital to provide a reaction field with uniform temperature and reactant concentration to obtain reaction kinetics accurately. The uniform reaction field provides identical reaction conditions for all reacting particles so that their reaction performances are the same at the same time. Thus, the reaction behavior recorded by the reaction analyzer reflects the true reaction kinetics rather than the average of reaction results of particles under different reaction conditions. At the same time, the uniformity of the reactor makes it easy to measure the reaction temperature accurately, which is of great significance in obtaining the real reaction kinetics. It is because, as the Arrhenius equation implies, the reaction rate varies exponentially with temperature. Thus, a slight deviation in the measured temperature may result in a considerable error in the derived reaction kinetics.

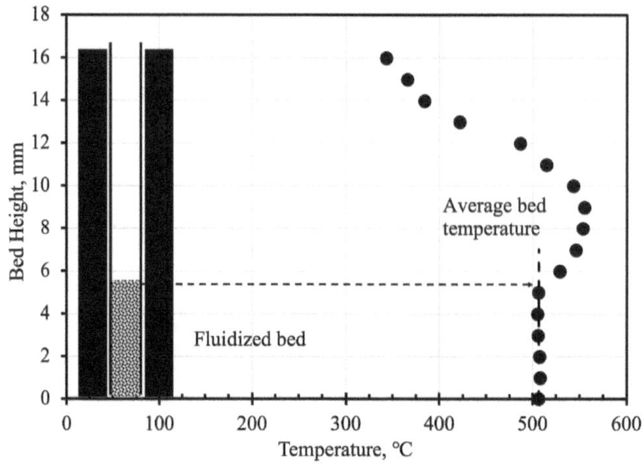

Figure 8.8 The temperature profile in a micro-fluidized bed highlighting the very good homogeneity of bed temperature [9].

As a result of good solid mixing and high heat transfer rate [9], the temperature fluctuation in a micro-fluidized bed is only a few degrees of Celsius, as shown in Fig. 8.8. In contrast, it is tens or even hundreds of degrees Celsius in other microthermal analysis reactors (e.g., fixed beds, TGAs, mesh reactors, drop tube reactors). Therefore, the micro-fluidized bed is favored for thermal analysis of gas—solid reactions, especially those sensitive to temperature variation.

8.1.5 Applied kinetics

Theoretically, the intrinsic kinetics of a gas—solid reaction may never be obtained in a practical reactor, but when the operating conditions described above are met satisfactorily, the kinetics obtained by using a micro-fluidized bed reaction analyzer will be very close to the intrinsic kinetics. To avoid confusion, we will call this close-to-intrinsic kinetics "applied kinetics." As discussed above, the applied kinetics is insignificantly affected by gas flow rate, particle size, and heating rate and is useful for designs of practical reactors.

Fig. 8.9 shows three experimental methods to illustrate the advantages of the micro-fluidized bed reaction analyzer in acquiring the applied kinetics [10]. In method I, the α-Al_2O_3 particle bed is fluidized by an Ar/CH_4 gas mixture at the preset reaction temperature. A known mass of Fe_2O_3 catalyst is then pulse-injected into the bed to undertake the catalytic methane decomposition reaction. In method II, the catalyst particles and the bed materials (α-Al_2O_3) are mixed with Ar gas at the preset temperature in the fluidized bed. The Ar gas is switched to the Ar/CH_4 gas mixture to initiate the catalytic methane decomposition when the bed is stabilized. In method III, the α-Fe_2O_3 catalyst is maintained in a fixed-bed state with the Ar gas at the preset

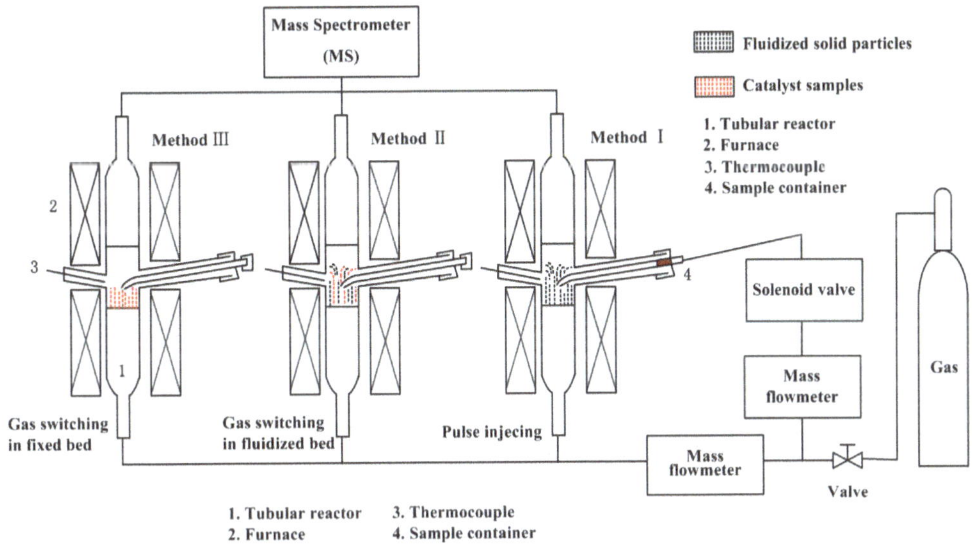

Figure 8.9 The experimental system to investigate the effect operation method on catalytic methane decomposition. (1) Tubular reactor; (2) furnace; (3) thermocouple; (4) sample holder.

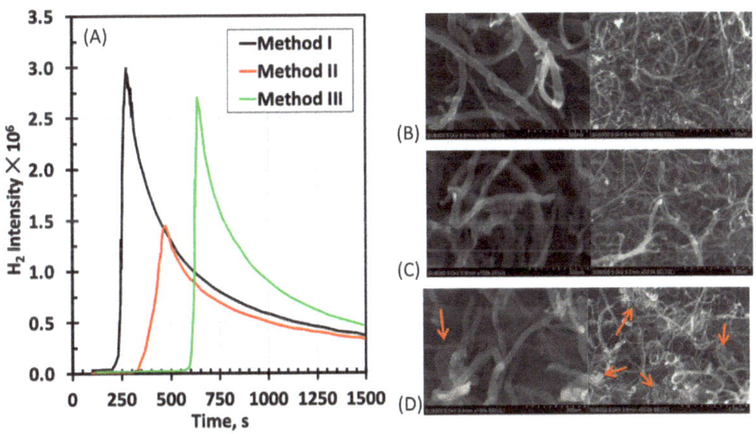

Figure 8.10 H_2 gas and carbon produced by the catalytic methane decomposition at 750°C and 0.75 atm. (A) H2 gas production, (B) carbon produced by Method I, (C) carbon produced by Method II, and (D) carbon produced by Method III.

temperature. The gas is then switched to the Ar/CH_4 gas mixture, and the catalytic methane decomposition starts.

Fig. 8.10 compares the hydrogen and carbon production characteristics of the three experimental methods at the condition of 750°C and 0.75 atm. As illustrated in Fig. 8.10A, method I produces a greater concentration of H_2 than the other two methods. Figs. 8.10B–D indicate that the carbon nanotubes (CNTs) produced in

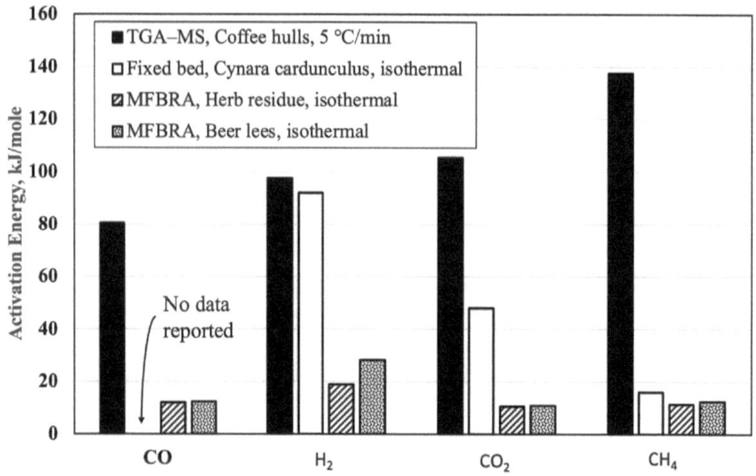

Figure 8.11 Comparison of activation energy of biomass pyrolysis reaction from TGA, fixed bed, and micro-fluidized bed.

micro-fluidized beds (methods I and II) are well dispersed, but highly agglomerated in method III. The results show that the micro-fluidized reaction analyzer with sample particle pulse injection method is more adequate than other methods to obtain the close-to-intrinsic reaction kinetics. For this reason, Geng et al. [10] conducted a series of experiments on methane decomposition over the Fe_2O_3 catalyst in MFBRA. The results show that the reaction order and activation energy of CMD are 2.27 (with respect to the partial pressure of methane) and 50 kJ/mol, respectively.

Fig. 8.11 compares the activation energy of biomass pyrolysis gas products CO, H_2, CO_2, CH_4, and their mixture obtained from a TGA [11], a fixed-bed reactor [12], and a micro-fluidized bed [8,13]. It is worth pointing out that although the biomass materials used in these tests are different, they have very similar chemical compositions, and therefore, if the same experimental analysis method is used, the pyrolysis behavior of these biomass materials should be similar and comparable. As shown in Fig. 8.11, the activation energy of gas production obtained by TGA is the highest, since TGA is seriously affected by gas diffusion. In the fixed-bed reactor, the activation energy of gas production is reduced due to the decrease in temperature distribution inhomogeneity and gas diffusion resistance compared with TGA. The activation energy of gas production obtained by MFBRA is the lowest among the three test methods because MFBRA can substantially eliminate the gas diffusion resistance and temperature distribution inhomogeneity, thus achieving the close-to-intrinsic kinetics to a large extent.

The micro-fluidized bed reaction analyzer is also valuable for studying the kinetics of slow reactions. Fig. 8.12A shows that the reaction rate of the CuO reduction by CO at 600°C in MFBRA is 8.8 times that in TGA, from 0.005 to 0.044 1/s [14], and Fig. 8.12B shows that for the gasification reaction of coal char by CO_2 [15], the reaction rate in the

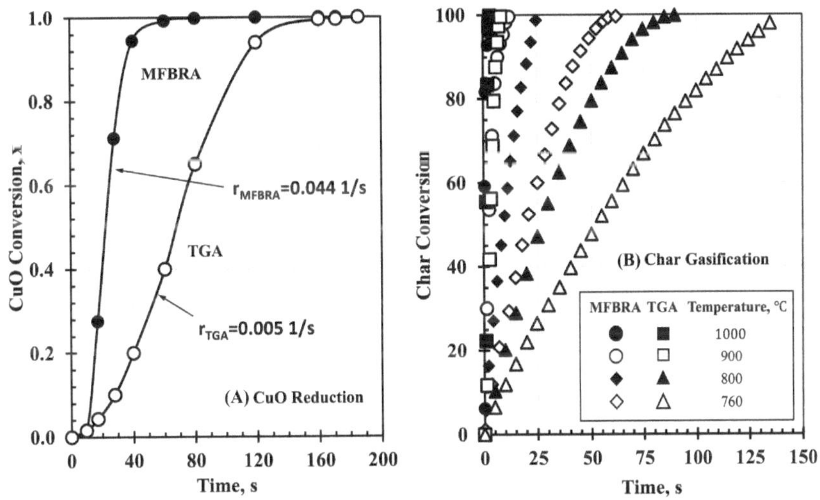

Figure 8.12 Illustration of effect of diffusion inhibition on (A) CuO reduction and (B) char gasification by comparison of experimental results in TGA and MFBRA.

micro-fluidized bed is significantly higher than that in TGA at the operating temperature below 900°C. These results indicate that the rate-controlling step of gasification reaction in TGA is mass transfer at lower temperatures, and thus the reaction rate is lower. Comparatively, the micro-fluidized bed can approach the intrinsic reaction rate at the corresponding temperature because of its effective suppression of the mass transfer limitation, yielding a higher reaction rate. When the temperature is above 900°C, the char gasification undergoes fast, so surface chemical reaction becomes the rate-controlling step; thus, no significant difference between the micro-fluidized bed and TGA is observable.

A micro-fluidized bed reaction analyzer equipped with an SPI-Ms can detect and analyze the real-time evolution of tar generated by rapid pyrolysis of biomass [9,16]. Fig. 8.13 shows typical results of such experiments, in which the pyrolysis tar products detected experimentally are classified into cellulose, hemicellulose, lignin-1, and lignin-2 groups. As we all know, in traditional experiments, to analyze the distribution and yield of tar products, it is usually necessary to obtain the liquid tar first by sampling the tar gas at a high temperature and condensing the gas sample. The liquid tar sample is then characterized by GC/Ms. This tedious process can be avoided by using the MFBRA/SPI-Ms method, a powerful tool for investigating pyrolysis and other reactions that produce high-molecular liquid products.

8.2 Understanding reaction mechanism

By integrating a rapid responding Ms, the micro-fluidized bed reaction analyzer is capable of characterizing fast reactions like pyrolysis and combustion thanks to its high

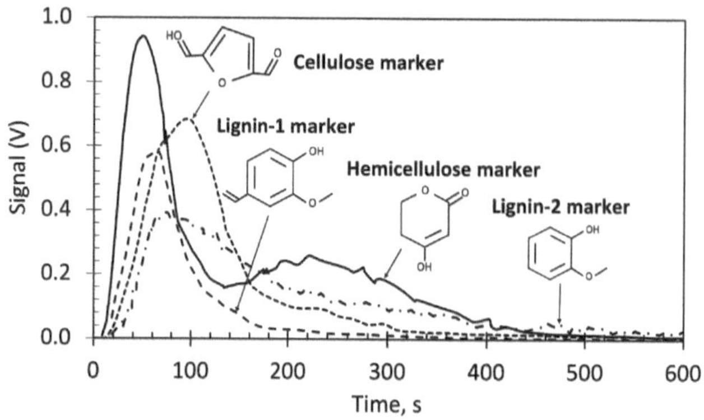

Figure 8.13 Real-time profiles of some typical markers from hemicelluloses, cellulose, and lignin during Douglas fast pyrolysis in a micro-fluidized bed equipped with SPI-Ms [9].

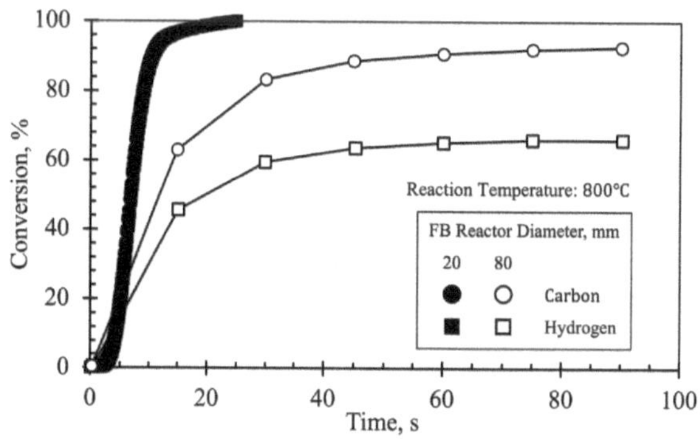

Figure 8.14 Time-dependent conversions of carbon and hydrogen elements obtained from experiments using fluidized bed reactors of two different diameters.

heating rate, closeness to plug flow of gas, and negligible gas diffusion inhibition. Most importantly, the MFBRA provides opportunities to discover and correct the misconception about a reaction caused by the limitations of traditional experimental methods. In the following, we will show a few examples to illustrate this unique characteristic of the micro-fluidized bed reaction analyzer.

8.2.1 Revealing the true character of fast reactions

Today, biomass pyrolysis is a well-known fast reaction. However, this was not certain only several years ago. Fig. 8.14 compares the conversions of carbon and hydrogen elements in biomass materials based on the experimental data reported by Murakami et al. [17] using an 80-mm-diameter laboratory fluidized bed reactor and Yu et al. [8] using a 20-mm-diameter micro-fluidized bed reactor. In the larger diameter fluidized bed reactor, the carbon and hydrogen conversions begin to increase slowly after

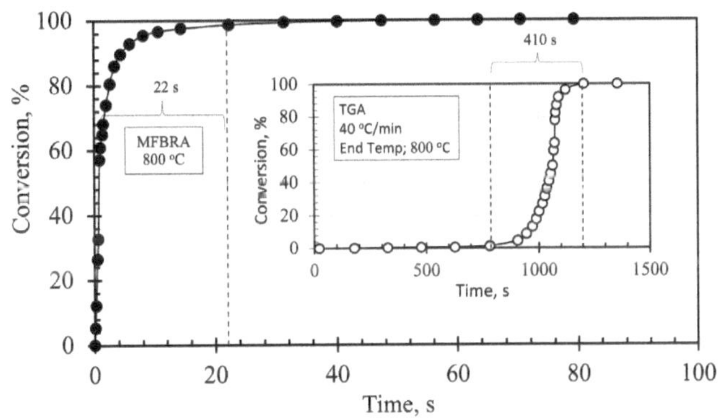

Figure 8.15 Revealing the light calcination of magnesite being a fast reaction [18].

30 seconds and do not complete when the reaction time exceeds 90 seconds. Should this be the case, biomass pyrolysis would not be considered a fast reaction. In the micro-fluidized bed reactor, however, the conversion of carbon and hydrogen is complete within 20 seconds, suggesting that biomass pyrolysis is indeed a fast reaction. Today, it has become well known that a large extent of the gas backmixing that occurs in large-scale fluidized beds can lead to misrepresentations about the kinetics of reactions, just as in the example mentioned above. Therefore, we recommend using a micro-fluidized bed reactor to reveal the true character of reactions.

1. Another example of MFBRA's capability to reveal the nature of a fast reaction is the calcination of pulverized magnesite powders, as shown in Fig. 8.15. The figure shows that the calcination of pulverized magnesite powders takes more than 410 seconds to complete under 800°C in the traditional TGA [18]. In contrast, the calcination is complete in only 22 seconds under the same temperature in the micro-fluidized bed reaction analyzer, indicating that the calcination of pulverized magnesite powders is actually a fast reaction. This example again demonstrates that gas diffusion in traditional reactors (in this case, TGA) can lead to erroneous reaction kinetics. Naturally, if misrepresented kinetics is used, different reactor designs can result, with significant consequences to product qualities and economics. It is thus advisable, when possible, to check the reaction performance in an MFBRA, even though sometimes the reactions of interest do not seem to be fast (such as the case illustrated here).

On the basis that the calcination of pulverized magnesite powders is a fast reaction, a transport calciner has been designed and commercially commissioned successfully in Haicheng, Liaoning Province of China, which will be described in Chapter 10.

8.2.2 Detecting intermediary reactions

Some reactions take several intermediate steps to complete. For example, gasification and combustion involve pyrolysis as the first step in the overall reaction process.

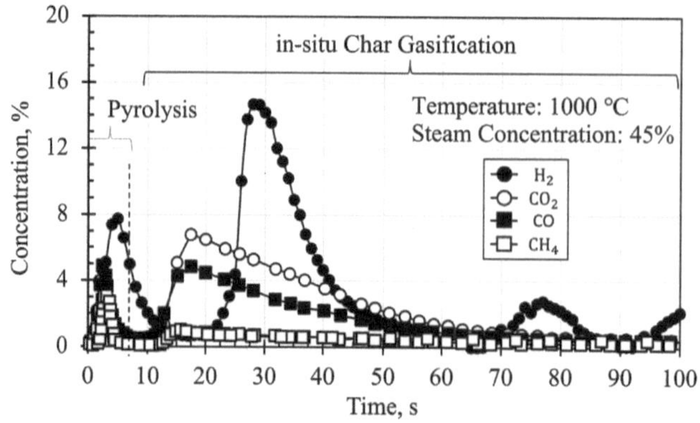

Figure 8.16 Experimental results of gas release from steam gasification of char particles in a micro-fluidized bed reaction analyzer [19].

Traditional experimental techniques often cannot detect the intermediary reaction steps, so a better understanding of the reaction mechanism and kinetics is likely difficult to be obtained. In this respect, the micro-fluidized bed reaction analyzer has shown unique advantages. Fig. 8.16 shows the release behavior of major gases during a coal gasification experiment in a micro-fluidized bed reaction analyzer [19]. The figure displays two easily distinguishable reaction steps: coal pyrolysis and steam gasification. The coal pyrolysis starts quickly after the coal powders are pulse-injected into the high-temperature bed. The pyrolysis produces a cloud cluster of pyrolysis gases and the in situ char particles, followed by in situ gasification of these pyrolysis products in the steam environment. This experiment reveals that the pyrolysis and gasification reactions, which occur in sequence, can be detected and observed in a single test by a micro-fluidized bed reaction analyzer. So far, although numerous tests on gasification and combustion have been performed and reported in the literature, none of them have experimentally observed this dynamic behavior.

8.2.3 Decoding the reaction mechanism

Online monitoring of the real-time variations of products can provide valuable insights into the mechanism of a reaction, especially when the reaction involves intermediary steps or products which may be unknown to the researchers before experiments or may not be readily detected during experiments. For instance, the capture of CO_2 by lime-based sorbent $Ca(OH)_2$, which has been regarded as an attractive technology with the advantage of simultaneously reducing the emissions of CO_2 and SO_2 from fuel combustion processes, can be expressed stoichiometrically as $Ca(OH)_2 + CO_2 \rightarrow CaCO_3 + H_2O$. According to this reaction scheme, one would observe the production of H_2O to be accompanied by the CO_2 consumption at the same. To investigate the reaction mechanism and kinetics of $Ca(OH)_2$

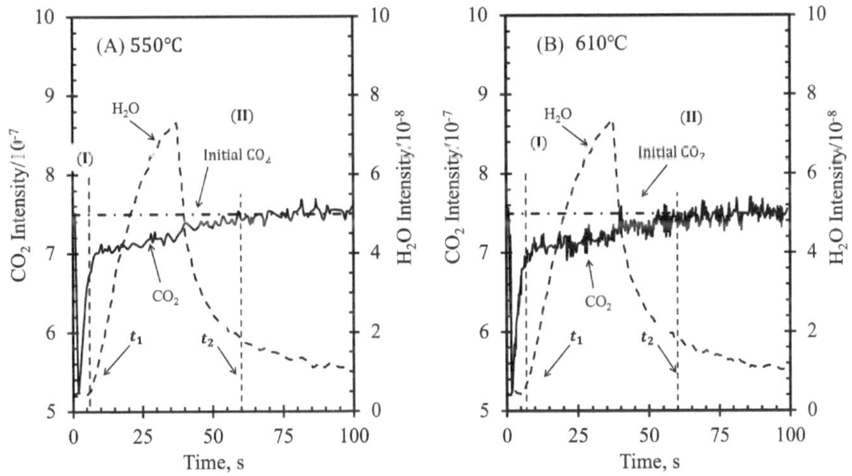

Figure 8.17 The real-time variations of CO_2 and H_2O concentrations during the reaction of $Ca(OH)_2$ and CO_2 [20] at (A) 550°C and (B) 610°C.

with CO_2, Yu et al. [20] conducted experiments in a micro-fluidized bed reactor of 20 mm in diameter. The gas products from the reaction process were measured by an online process mass spectrometer (Ms). The reactor temperature was controlled in the range of 500°C–750°C. A gas stream of 10% of CO_2 balanced with N_2 was used as the fluidizing gas. The real-time variations of CO_2 and H_2O gas concentrations recorded at the reactor exit are shown in Fig. 8.17. It shows that after several tens of milligrams $Ca(OH)_2$ are pulse-injected into the fluidized bed at the operating temperature, the CO_2 gas drops to a minimum and then rises quickly to a point after which the CO_2 increases gradually with reaction time. Interestingly, the reaction product H_2O is produced only after time t_1, 6–8 seconds later than the rapid decline of CO_2 concentration. To ensure the reliability of the result, the Ms response to a direct pulse input of air containing CO_2 and H_2O was examined, and the response speed to both gases was confirmed to be the same. Therefore, the delay in H_2O production is due to a different reaction mechanism.

Now, we divide the $Ca(OH)_2$ carbonization reaction into two reaction regimes (I and II) with the time t_1 as the transition boundary, as shown in Fig. 8.17. In regime I, the direct reaction of CO_2 and $Ca(OH)_2$ does not occur because no H_2O vapor is produced. In regime II, the CO_2 is not consumed but produced because its concentration increases as the reaction progress, especially after time t_2 when its concentration becomes equal to its initial value. A further in-depth investigation was carried out to decode the mechanism of this seemingly simple but actually complex reaction [20]. It is confirmed that before time t_1 reactions of $CO_2(g) \xrightarrow{Ca(OH)_2} CO_2(ad)$ and $2CO_2(ad) + Ca(OH)_2 \rightarrow Ca(HCO_3)_2$ take place, which produces an intermediate product $Ca(HCO_3)_2$ without H_2O. After the time t_1, the nascent intermediate $Ca(HCO_3)_2$ decomposes in situ to produce CO_2 and H_2O via the reaction $Ca(HCO_3)_2 \rightarrow CaCO_3 + H_2O + CO_2$, resulting in a gradual increase in CO_2 with reaction time.

8.2.4 Reactions with in/ex situ solid particles

Char gasification usually is the rate-determining step because of its low reaction rate. A clear and precise understanding of char gasification by CO_2, steam, oxygen, and a mixture of these gases and other oxidants is essential for the optimal design and operation of industrial gasifiers. As described previously in Chapter 6, several types of char can be prepared, namely, the slowly or rapidly cooled char, ex/in situ hot char, and in situ nascent char. Since the pyrolysis conditions (temperature and atmosphere) strongly affect the physicochemical properties of the prepared char, the char reaction (e.g., gasification, combustion) performance varies with its preparation method. As illustrated in Fig. 8.18A, the in situ nascent char has the highest gasification rate, followed by in/ex situ hot char, rapidly cooled char, and naturally cooled char. The combustion rates follow a similar order (Fig. 8.18B), although the data for in situ char are absent. The tables attached show that the char preparation methods have significant impacts on the pore structural and chemical characteristics of char particles.

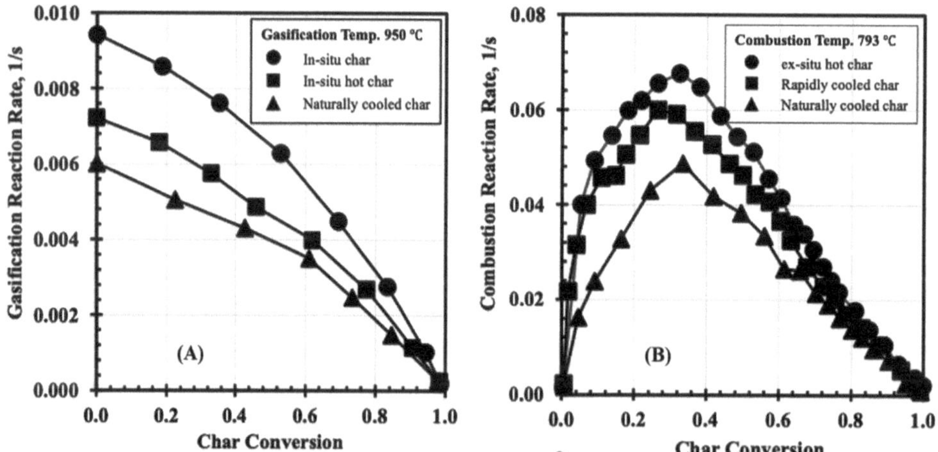

Pore Structural Parameters

	S_{BET}, m^2/g	S_{micro}, m^2/g	V_{pore}, mL/g	D_{pore}, nm
In-situ char	40.48	28.83	0.110	1.26
Hot char	34.63	24.42	0.086	1.28
Cold char	32.28	21.26	0.081	1.31

C-C and C-O Functional Group Contents

	Rapidly Cooled Char	Naturally Cooled Char
C-C	71.3	76.7
C-O	28.7	23.2

Figure 8.18 Reaction rates and structural properties of char particles prepared by different methods. (A) Gasification and (B) combustion reaction rates and structural properties of char particles prepared by different methods.

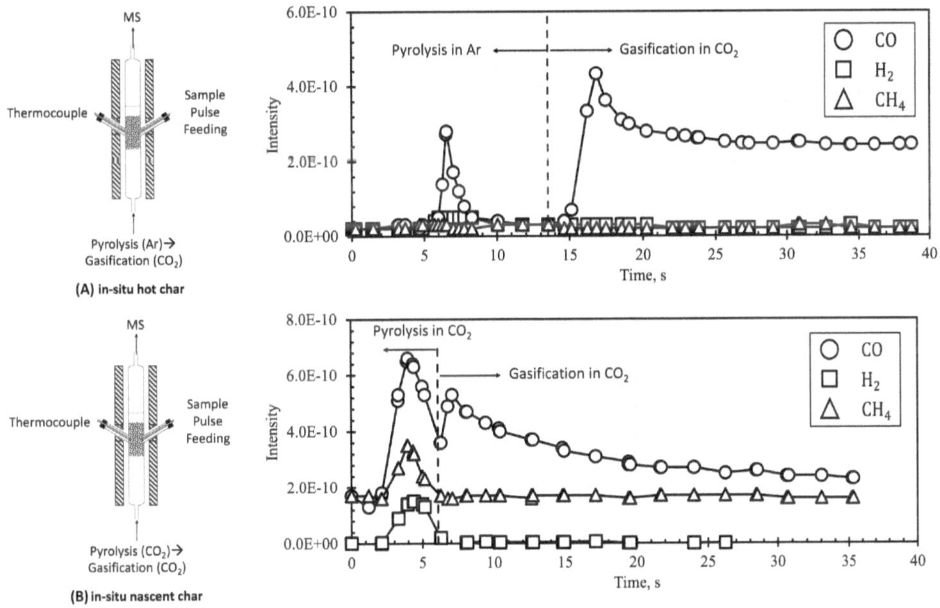

Figure 8.19 Production of (A) in situ hot char and (B) the in situ nascent char and their gasification performance at 900°C in an MFBRA [21].

Fig. 8.19 compares the gas release behavior of gasification experiments with two types of char particles in a micro-fluidized bed reactor. As shown in Fig. 8.19A, the in situ hot char particles are produced first in the reactor with Ar as the fluidizing gas, and then, the fluidizing gas is switched to CO_2 for gasification. Comparatively, the in situ nascent char particles are produced and gasified at the same CO_2 atmosphere in the micro-fluidized bed reactor [15,21], as illustrated in Fig. 8.19B. The gas release behavior is distinguishably different between these two types of char. From the perspective of designing a practical reactor, the information acquired from in situ nascent char particles bears a close resemblance to that of actual industrial gasifiers.

8.2.5 Non-isothermal differential applications

While micro-fluidized bed reaction analyzers have been used mainly as isothermal differential reactors [3,21,22], they can also be operated non-isothermally as necessary [23–25]. This feature makes MFB reactors capable of characterizing reactions with a wide heating rate range. The non-isothermal analysis exposes a sample to a reaction process in which the reaction temperature is varied by a preset control program to reveal the reaction behavior of the sample particle under different temperatures. Compared with tests using the isothermal analysis method, non-isothermal methods are less time-consuming because the dependence of reaction performance on temperature can be obtained in a single test. However, under non-isothermal conditions,

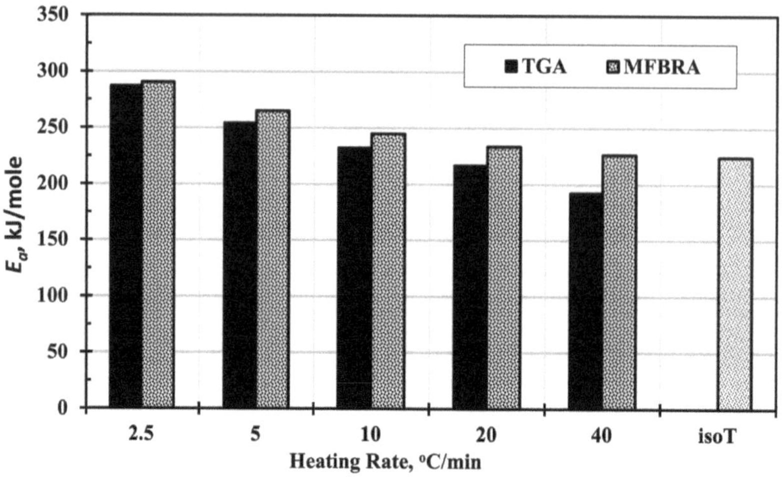

Figure 8.20 Comparison of the activation energy obtained by TGA and MFBRA under non-isothermal operations [21,23,24,26].

micro-fluidized beds differentiate from TGA in that they do not suffer from severe gas mixing and diffusion regardless of isotherm or non-isothermal operations. In this sense, the micro-fluidized bed reaction analyzers are a useful complementary tool to existing non-isothermal analyzers, such as thermogravimetric analyzers.

Fig. 8.20 compares the activation energy of CO_2 gasification of cold char particles obtained from non-isothermal experiments using a TGA and an MFBRA [21,23,24,26]. The activation energy derived from the MFBRA experiment under the isothermal condition is also indicated (denoted as isoT). It shows that the activation energy values obtained by the two experimental methods are different, and the difference enlarges as the heating rate increases. Since the gas diffusion is essentially negligible in the micro-fluidized bed reactor, the activation energy obtained by the non-isothermal experiments at heating rates above 20°C/min is close to that acquired under the isothermal condition.

8.3 Reactions under water vapor atmosphere

Some reactions need to use feedstocks with high-moisture content, add steam as a reactant, take place in a steam-containing environment, or adjust the steam concentration or flow rate in the reaction processes to get better performance. In these circumstances, micro-fluidized bed reactors offer the advantage of operability at varying steam atmospheres, which is rarely practical for other traditional thermal analysis reactors. Several techniques can be used to facilitate steam generation and to change the steam supply rate smoothly, which have been presented in Chapter 7.

8.3.1 High-moisture content feedstocks

Haruna Adamu et al. [27] carried out a series of experiments in a micro-fluidized bed, including drying, pyrolysis, gasification, and combustion, using water-soaked biomass char particles to simulate the high-temperature drying of wet biomass materials. The moisture content of wetted char particles was 25%. In a parallel study, Zeng et al. [28] investigated wet coal drying behavior using a wet coal char sample with a moisture content of 20% in a micro-fluidized bed. The use of wetted char particles, rather than wetted biomass or coal directly, was aimed at avoiding volatiles release in the drying process. The successful operations of micro-fluidized bed reactors with sample materials of such high-moisture contents greatly extend the reactors' applicability. From a practical application perspective, the ability to process high-moisture feedstocks is highly beneficial to system performance improvements [29].

8.3.2 Reactions with steam as reactants

Using steam as a gasifying agent can enhance the gasification of carbonaceous materials such as coal, char, coke, and biomass. The enhancements are characterized by significantly improved carbon conversion, lowered gasification temperature, and increased reaction rate. In addition, the produced synthesis gas has a total content of H_2 and CO up to 70%–80%, which is directly usable for the syntheses of various high-value-added chemicals. As shown in Fig. 8.21, the coke conversion in the micro-fluidized bed reactor increases after the steam replaces the CO_2 as the gasification agent. The addition of black liquor further improves the coke conversion due to the catalytic effects provided by the sodium contained in the black liquor [30–33].

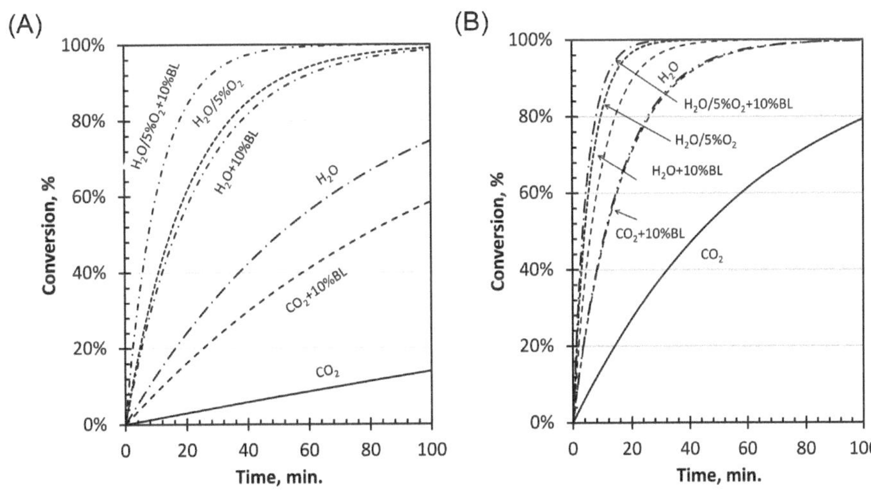

Figure 8.21 Coke conversion versus time with different gasification agents [30–33]. (A) 850°C; (B) 1000°C.

Chen et al. [7] studied the methane steam reforming (MSR) using a micro-fluidized bed of 20 mm in inner diameter with a catalyst containing more than 50 wt.% $NiO/\alpha\text{-}Al_2O_3$. The experiments were carried out at 500°C–800°C with inlet steam-to-carbon ratios (S/C) from 1.5 to 4.5. The results indicate that S/C has a pronounced effect on the reaction performance of MSR. The higher the S/C ratio, the higher the methane conversion and hydrogen yield. However, as the reaction temperature increases, the effect of S/C on the methane conversion and the yield of hydrogen declines.

8.4 Sampling and characterization of solid particles during a reaction process

The online particle sampling capability of MFBRA provides an effective means to monitor the change of catalyst performance continuously during the reaction process. Combined with the online gas measurement and analysis, MFBRA can directly relate the change of gas production performance to the performance of the catalyst in the bed. It is a powerful research tool, unmatched by other thermal analysis reaction techniques in developing and optimizing gas–solid reaction processes.

Pang et al. [34] reported the first successful application of this technique. A micro-fluidized bed reaction analyzer, capable of sampling catalyst particles during the reaction, is employed to investigate the methanol vapor-phase carbonylation with carbon monoxide over a Ni/AC catalyst. The catalyst particles in the bed are sucked into a sample bottle where they are cooled down to room temperature rapidly and are isolated from air by pulsing a large amount of inert gas (N_2) into the bottle at the end of sampling. During the reaction process, several catalyst samples are taken at specific reaction times, and each sample weighs 0.05–0.15 g. Fig. 8.22 shows variations in the yield of the goal product (i.e., the catalyst activity) and the corresponding catalysts characteristic parameters (namely, the contents of atoms O, Ni, and I, and TEM images) with the reaction time. It shows that the catalyst deactivation is caused neither by nickel particle aggregation nor by metal leaching and carbonaceous deposits, as often speculated about in the literature. The experimental results confirm that the differences in catalytic performance are exclusively related to the continuous accumulation of oxygen-containing carbonaceous deposits, a finding that has never been able to be recognized using traditional thermal analytical techniques in the past.

8.5 Multistage gas–solid reaction processes

Micro-fluidized bed reaction analyzers are applicable to characterizations of multistage reactions, benefiting from their capabilities of prompt responses to gas switching and quick heating/cooling rates [19,35]. Fig. 8.23 shows an example that MFBRA is

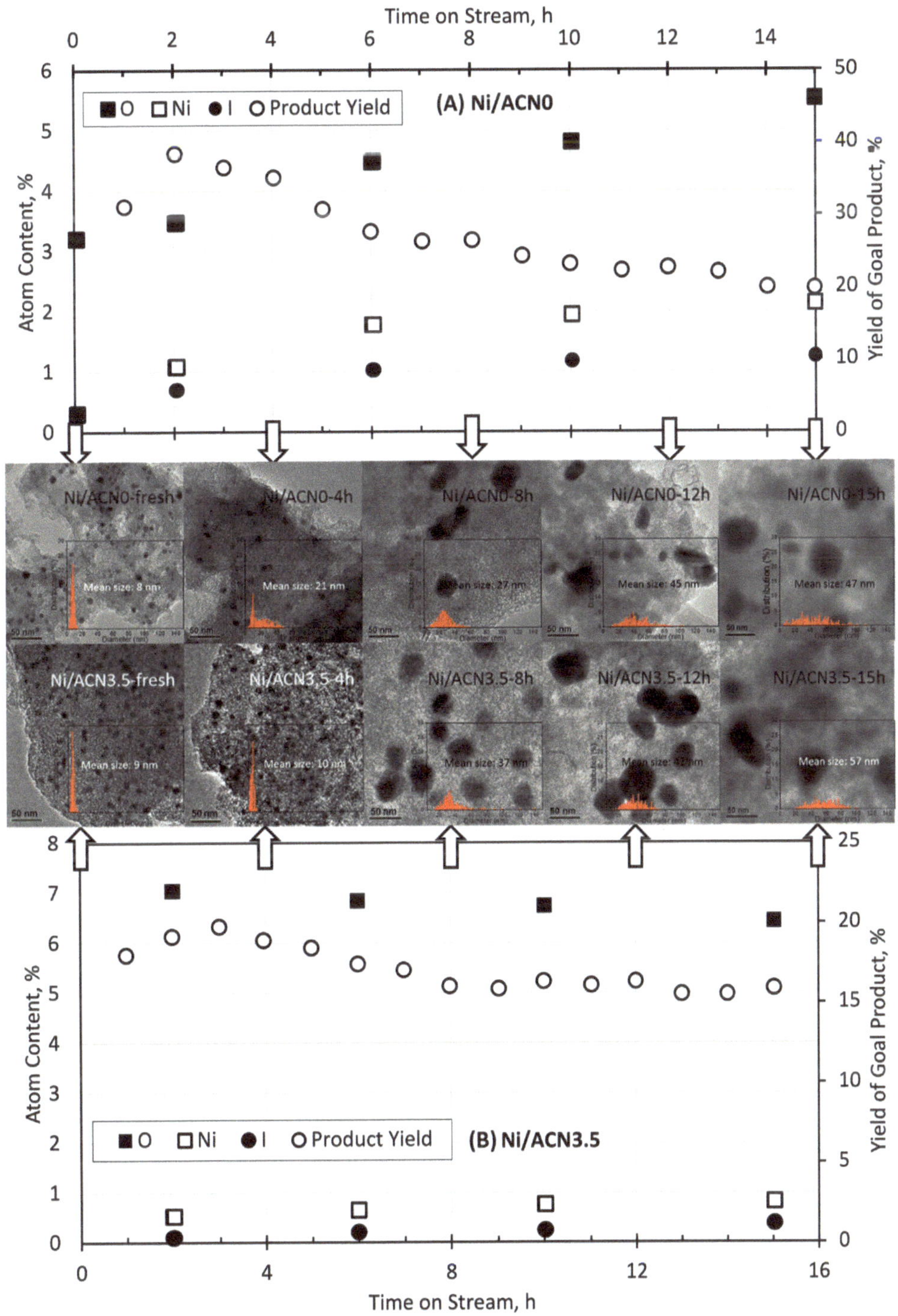

Figure 8.22 The time of stream catalytic activity and atom content with the corresponding TEM images of Ni/ACN0 and Ni/ACN3.5 [34]. (A) Ni/ACN0, (B) Ni/ACN3.5.

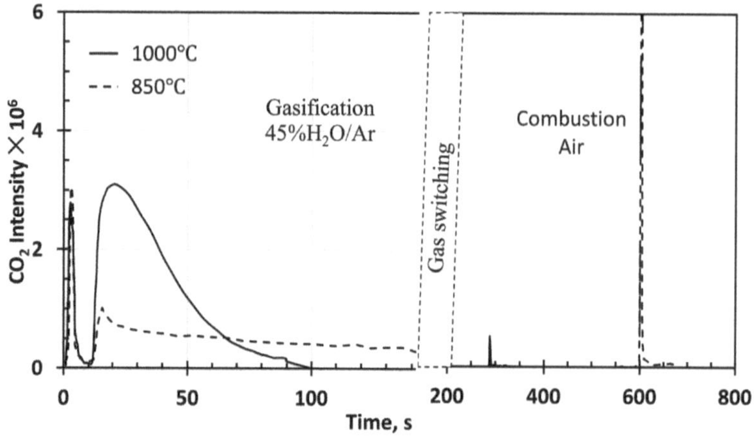

Figure 8.23 CO_2 released in consecutive coal pyrolysis, in situ char gasification, and char residue combustion processes.

employed to investigate the gasification efficiency [19]. As shown in the figure, after the coal pyrolysis (corresponding to the first peak of CO_2 intensity) and its resultant in situ char gasification (corresponding to the second peak of CO_2 intensity) are complete at the given temperature, the fluidizing gas is switched from 45% H_2O/Ar to air to combust the residue char particles. It shows, as expected, that the gasification efficiency is higher at 1000°C than at 850°C, as evidenced by the CO_2 released from char combustion. This experiment demonstrates that multistage reactions can be carried out in a micro-fluidized bed reactor by seamlessly switching the reacting gas.

Some reactions take place through several elementary reactions simultaneously in series or parallel. For example, reducing iron ore into iron with carbon monoxide is such a reaction. Generally, the iron ore reduction is expressed simply as $Fe_2O_3 \rightarrow Fe$, but the actual reduction process goes through multiple steps: $Fe_2O_3 \rightarrow Fe_3O_4 \rightarrow FeO \rightarrow Fe$. This multistep iron ore reduction process was successfully investigated by a micro-fluidized bed reactor integrated with an in situ mass spectrometer [36–38]. The results of Chen et al. [36] shown in Fig. 8.24 indicate that the overall iron ore reduction process completes in two consecutive reactions in series. The first is a single-step reaction that converts Fe_2O_3 to Fe_3O_4, corresponding to the conversion from 0 to 1/9. The second reaction converts the in situ nascent Fe_3O_4 to Fe, corresponding to conversions from 1/9 to 1.0, via two simultaneously occurring parallel reactions: $Fe_3O_4 \rightarrow FeO$ and $FeO \rightarrow Fe$. Reaction $Fe_2O_3 \rightarrow Fe_3O_4$ is fast and surface chemical reaction controlled, while the reduction from Fe_3O_4 to Fe is slow and affected by diffusion. The two reaction processes are easily recognized and distinguished by the experimental data obtained by the MFBRA, as shown in Fig. 8.24. On this basis, the kinetics of the multiple-step reactions involved in iron ore reduction is determined based on the model-fit method using the experimental data corresponding to the respective conversion range.

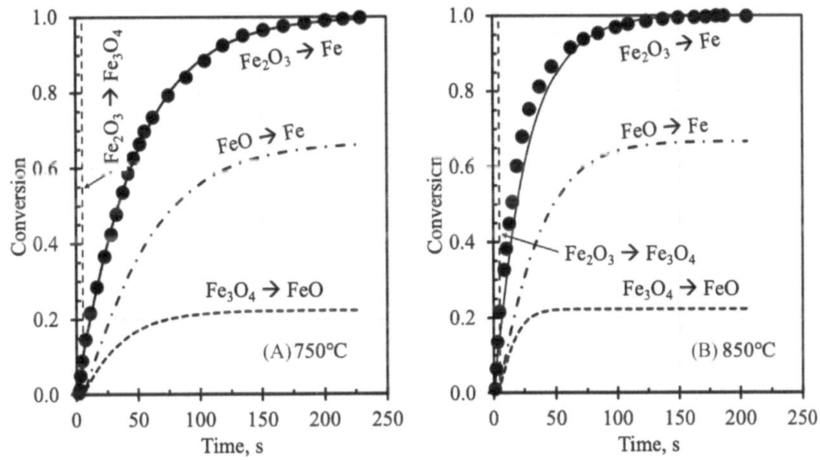

Figure 8.24 Conversion as a function of time with 100% CO at (A) 750°C and (B) 850°C.

8.6 Reaction kinetics under product gas inhibitory atmospheres

Many gas–solid reactions take place in inhibitory atmospheres containing product gases. For example, the decomposition of calcium carbonate ($CaCO_3$), which is encountered in processes such as cement manufacturing, flue gas desulfurization, and chemical looping combustion using CaO as an oxygen carrier, is conducted in CO_2-containing atmospheres. In principle, the kinetics of the reaction is affected by the presence of the reaction product gas (i.e., CO_2) in the atmosphere, but it is not easy to determine the effect of product gas inhibitory atmosphere accurately. For instance, the activation energy of over 2000 kJ/mol and a completion time of several minutes were obtained by using a TGA for $CaCO_3$ decomposition [39] in a 100% CO_2 environment. Such high activation energy is hardly explicable because the practical operation experience of fluidized bed chemical looping experiments shows that the $CaCO_3$ calcination accomplishes readily. Thus, a new approach is doubtlessly required to acquire real kinetic data of reactions inhibited by the atmosphere containing the reaction product gas.

8.6.1 Isotope-tagging method

Liu et al. [40] proposed a novel isotope-tagging method to solve the aforementioned problems. In this method, a raw reactant was intentionally selected so that the isotope of the gas product was different from that of the same gas component in the fluidizing or carrier gas. To illustrate this method, Liu et al. [40] conducted a series of experiments to calcinating calcium carbonate $Ca^{13}CO_3$ in atmospheres of varying $^{12}CO_2$ (simply CO_2) in an MFBRA. Fig. 8.25A shows the recorded production of CO_2 and $^{13}CO_2$ from the decomposition of a mixed $CaCO_3/Ca^{13}CO_3$ in the MFBRA at 700°C in the atmosphere of N_2. It shows that the CO_2 and $^{13}CO_2$ can be detected simultaneously using

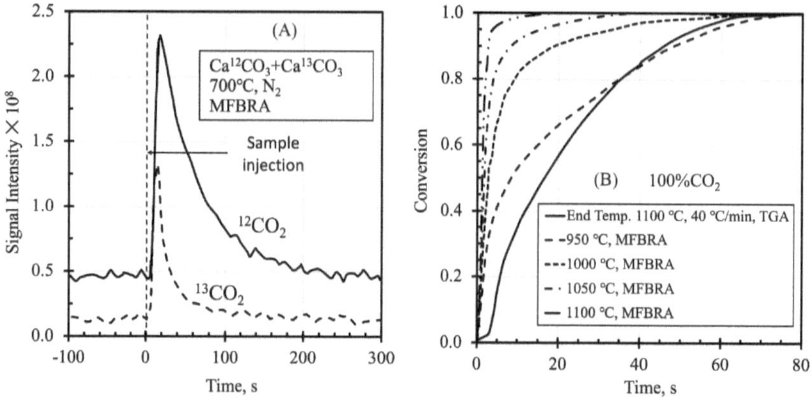

Figure 8.25 (A) The experimental-recorded CO_2 and $^{13}CO_2$ curves from the decomposition of $CaCO_3$ and $Ca^{13}CO_3$ mixture in MFBRA at 700°C in the atmosphere of N_2 and (B) variation of conversion of $Ca^{13}CO^3$ at different temperatures in a 100% CO_2 atmosphere.

the process mass spectrometry, indicating that the isotope-tagging method can be used to investigate reaction kinetics under product gas inhibitory atmospheres.

Fig. 8.25B shows the typical experimental results based on the isotope-tagging method. It shows that the conversion increases fast with time in the initial stage of decomposition and then gradually increases until decomposition completes in MFBRA. In MFBRA, the conversion increases with reaction temperature.

8.6.2 Comparisons between the micro-fluidized bed and thermogravimeter

By effectively eliminating gas diffusion, MFBRA exhibits a faster conversion rate than TGA. Fig. 8.26A compares the activation energy obtained from the TGA and the MFBRA at different CO_2 concentrations. The results indicate that the activation energy obtained by either the MFBRA or the TGA increases with increasing CO_2 concentration. The effect of CO_2 concentration in the atmosphere on the activation energy is more pronounced in the TGA than in the MFBRA. Fig. 8.26A shows that the ratio of activation energy obtained from TGA to that obtained from MFBRA, $E_{a,TGA}/E_{a,MFBRA}$, changes from 1.9 to 3.7 when CO_2 in the atmosphere is increased from 0% to 1%. With the further increase of CO_2 in the atmosphere, this ratio rises to as high as 8, indicating that the higher the CO_2 concentration is, the more seriously the TGA suffers from gas diffusion. At 100% CO_2, the activation energy obtained by the TGA is 2048.19 kJ/mol, but it is only 271.55 kJ/mol obtained by the MFBRA. For the MFBRA, however, the moderate increase in the activation energy with increasing CO_2 concentration is mainly due to the thermal equilibrium limitation, as shown in Fig. 8.26B. It shows that the initial temperatures of $CaCO_3$ decomposition measured by the TGA and the MFBRA are significantly different. The

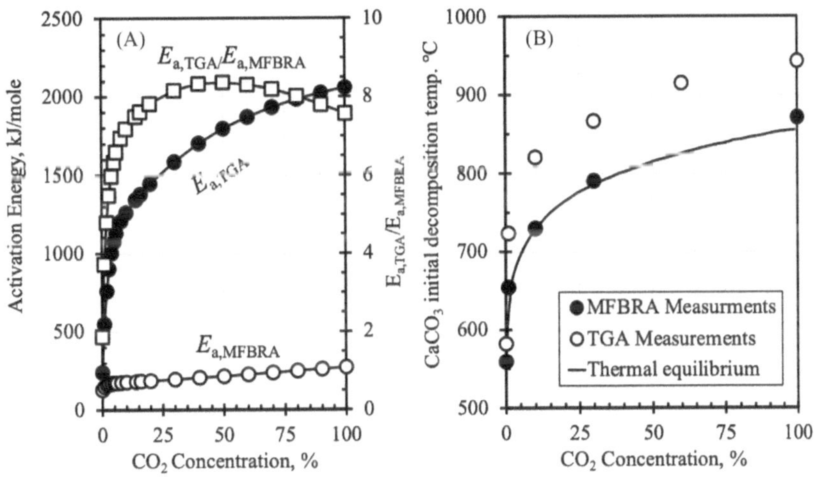

Figure 8.26 (A) Variations of the activation energy and (B) the initial temperature of $CaCO_3$ decomposition with CO_2 concentration.

difference increase with increasing CO_2 in the atmosphere. However, the initial decomposition temperatures of $CaCO_3$ measured by the MFBRA are close to those of thermal equilibrium calculations, suggesting that the increase in the initial temperature and thus the activation energy of $CaCO_3$ decomposition in the MFBRA is attributable to the thermal equilibrium limitation. Therefore, the MFBRA is superior to TGA-like conventional thermal analysis techniques for characterizing gas—solid reactions in the product gas inhibitory atmosphere.

Abbreviations

CMD	catalytic methane decomposition
CNT	carbon nanotubes
GC	gas chromatography
MFB	micro-fluidized bed
MFBRA	micro-fluidized bed reaction analyzer
Ms	mass spectrometry
RTD	residence time distribution
SPI-Ms	photoionization mass spectrometer
TGA	thermogravimetric analyzer

Nomenclature

U_g	superficial gas velocity (m/s)
U_{mf}	minimum fluidization velocity (m/s)
y_{CO}	molar fraction of CO
t	time (S)

References

[1] Zhang Y, Niu Y, Zou H, et al. Characteristics of biomass fast pyrolysis in a wire-mesh reactor. Fuel 2017;200:225–35.

[2] Trubetskaya A, Jensen PA, Jensen AD, et al. Influence of fast pyrolysis conditions on yield and structural transformation of biomass chars. Fuel Processing Technology 2015;140:205–14.

[3] Yu J, Zeng X, Zhang J, et al. Isothermal differential characteristics of gas-solid reaction in micro-fluidized bed reactor. Fuel 2013;103:29–36.

[4] Fang Y, Luo G, Chen C, et al. Combustion kinetics of in-situ char and cold char in micro-fluidized bed. Journal of Combustion Science and Technology 2016;22(2):148–54.

[5] Yu J, Zhu J, Guo F, et al. Reaction kinetics and mechanism of biomass pyrolysis in a micro-fluidized bed reactor. Journal of Fuel Chemistry and Technology 2010;38(6):666–72.

[6] Chen H, Zheng Z, Shi W. Investigation on the kinetics of iron ore fines reduction by CO in a micro-fluidized bed. Procedia Engineering 2015;102:1726–35.

[7] Chen K, Zhao Y, Zhang W, et al. The intrinsic kinetics of methane steam reforming over a nickel-based catalyst in a micro fluidized bed reaction system. International Journal of Hydrogen Energy 2020;45:1615–28.

[8] Yu J, Yao C, Zeng X, et al. Biomass pyrolysis in a micro-fluidized bed reactor: characterization and kinetics. Chemical Engineering Journal 2011;168(2):839–47.

[9] Jia L, Le-Brech Y, Shrestha B, et al. Fast pyrolysis in a microfluidized bed reactor: effect of biomass properties and operating conditions on volatiles composition as analyzed by online single photoionization mass spectrometry. Energy and Fuels 2015;29(11):7364–74.

[10] Geng S, Han Z, Hu Y, et al. Methane decomposition kinetics over Fe_2O_3 catalyst in micro fluidized bed reaction analyzer. Industrial and Engineering Chemistry Research 2018;57(25):8413–23.

[11] Huang Y, Chiueh P, Kuan W, et al. Pyrolysis kinetics of biomass from product information. Applied Energy 2013;110:1–8.

[12] Encinar J, González J, González J. Fixed-bed pyrolysis of *Cynara cardunculus* L. product yields and compositions. Fuel Process Technology 2000;68:209–22.

[13] Guo F, Dong Y, Lv Z, et al. Pyrolysis kinetics of biomass (herb residue) under isothermal condition in a micro fluidized bed. Energy Conversion and Management 2015;93:367–76.

[14] Yu J, Yue J, Liu Z, et al. Kinetics and mechanism of solid reactions in a micro fluidized bed reactor. AIChE Journal 2010;56(11):2905–12.

[15] Zeng X, Wang F, Wang Y, et al. Characterization of char gasification in a micro fluidized bed reaction analyzer. Energy and Fuels 2014;28(3):1838–45.

[16] Jia L, Dufour A, Le Brech Y, et al. On-line analysis of primary tars from biomass pyrolysis by single photoionization mass spectrometry: experiments and detailed modelling. Chemical Engineering Journal 2017;313:270–82.

[17] Murakami T, Xu G, Suda T, et al. Some process fundamentals of biomass gasification in dual fluidized bed. Fuel 2007;86:244–55.

[18] Jiang W, Hao W, Liu X, et al. Characteristic and kinetics of light calcination of magnesite in micro fluidized bed reaction analyzer. CIESC Journal 2019;70(8):2928–37.

[19] Guo Y, Zhao Y, Gao D, et al. Kinetics of steam gasification of in-situ chars in a micro fluidized bed. International Journal of Hydrogen Energy 2016;41(34):15187–98.

[20] Yu J, Zeng X, Zhang G, et al. Kinetics and mechanism of direct reaction between CO_2 and $Ca(OH)_2$ in micro fluidized bed. Environmental Science and Technology 2013;47(13):7514–20.

[21] Wang F, Zeng X, Shao R, et al. Isothermal gasification of in situ/ex situ coal char with CO_2 in a micro fluidized bed reaction analyzer. Energy and Fuels 2015;29(8):4795–802.

[22] Yu J, Li Q, Duan Z, et al. Isothermal differential characteristics of the reaction in micro fluidized bed. Scientia Sinica Chimica 2011;41(1):152–60.

[23] Wang F, Zeng X, Wang Y, et al. Non-isothermal coal char gasification with CO_2 in a micro fluidized bed reaction analyzer and a thermogravimetric analyzer. Fuel 2016;164:403–9.

[24] Wang F, Zeng X, Wang Y, et al. Comparison of non-isothermal coal char gasification in micro fluidized bed and thermogravimetric analyzer. CIESC Journal 2015;66(5):1716–22.

[25] Cai L, Liu W, Yu J, et al. Comparative study on coal pyrolysis via programmed and isothermal heating. Coal Conversion 2012;35(3):6–14.
[26] Wang F, Zen X, Wang Y, et al. Investigation on in/ex-situ coal char gasification kinetics in a micro fluidized bed reactor. Journal of Fuel Chemistry and Technology 2015;43(4):393–401.
[27] Haruna Adamu M, Zeng X, Zhang JL, et al. Property of drying, pyrolysis, gasification, and combustion tested by a micro fluidized bed reaction analyzer for adapting to the biomass two-stage gasification process. Fuel 2020;264(December 2019) Article 116827.
[28] Zeng X, Zhang J, Adamu MH, et al. Behavior and kinetics of drying, pyrolysis, gasification, and combustion tested by a micro fluidized bed reaction analyzer for the staged-gasification process. Energy and Fuels 2020;34(2):2553–65.
[29] An P, Han Z, Wang K, et al. Process analysis of a two-stage fluidized bed gasification system with and without pre-drying of high-water content coal. Canadian Journal of Chemical Engineering 2020;1–12.
[30] Zhang Y, Sun G, Gao S, et al. Regeneration kinetics of spent FCC catalyst via coke gasification in a micro fluidized bed. Procedia Engineering 2015;102:1758–65.
[31] Zhang Y, Yao M, Gao S, et al. Reactivity and kinetics for steam gasification of petroleum coke blended with black liquor in a micro fluidized bed. Applied Energy 2015;160:820–8.
[32] Zhang Y, Yao M, Sun G, et al. Characteristics and kinetics of coked catalyst regeneration via steam gasification in a micro fluidized bed. Industrial and Engineering Chemistry Research 2014;53(15): 6316–24.
[33] Yu D. Study on gasification characteristics and kinetics of petroleum koke; 2013.
[34] Pang F, Song F, Zhang Q, et al. Study on the influence of oxygen-containing groups on the performance of Ni/AC catalysts in methanol vapor-phase carbonylation. Chemical Engineering Journal 2016;293:129–38.
[35] Guo Y, Zhao Y, Meng S, et al. Development of a multistage in situ reaction analyzer based on a micro fluidized bed and its suitability for rapid gas – solid reactions. Energy and Fuels 2016;30: 6021–33.
[36] Chen H, Zheng Z, Chen Z, et al. Reduction of hematite (Fe_2O_3) to metallic iron (Fe) by CO in a micro fluidized bed reaction analyzer: a multistep kinetics study. Powder Technology 2017;316: 410–20.
[37] Chen H, Zheng Z, Chen Z, et al. Multistep reduction kinetics of fine iron ore with carbon monoxide in a micro fluidized bed reaction analyzer. Metallurgical and Materials Transactions B 2017;48B:841–52.
[38] He K, Zheng Z, Chen Z. Multistep reduction kinetics of Fe_3O_4 to Fe with CO in a micro fluidized bed reaction analyzer. Powder Technology 2020;360:1227–36.
[39] Zheng Y, Song K, Chi B, et al. Decomposition kinetics of $CaCO_3$ in CO_2 atmosphere. Huazhong University of Science and Technology (Nature Science) 2007;35(8):87–9.
[40] Liu X, Hao W, Wang K, et al. Acquiring real kinetics of reactions in an inhibitory atmosphere containing the product gases using a micro fluidized bed. AIChE Journal 2020; Article 17325.

CHAPTER 9

Applications of micro-fluidized beds

Micro-fluidized beds have great application potential in various physical and chemical processes because of their unique characteristics. This chapter presents typical applications of micro-fluidized beds, particularly the micro-fluidized bed reaction analyzers (MFBRAs) used in various thermochemical gas–solid reactions. Note that it is not the purpose of this chapter to discuss comprehensively the characteristics, mechanisms, and kinetics of the specific reactions investigated using MFBRA. Instead, we will focus on demonstrating the characteristics, advantages, and potential of MFBRA as an advanced thermochemical reaction tool by introducing the typical applications of MFBRA. Readers may refer to the original references for more details of the reaction processes.

9.1 Drying

Thermal drying is used to remove moisture from wet materials and thus can improve material handling properties. For a solid fuel, such as biomass, coal, slime, slurry, and waste mud, removing moisture from it can increase its heat content and subsequently improve the thermal efficiency of the combustion boiler where the fuel is combusted for thermal and power generation.

For controlled and gentle drying of wet solids, fluidized bed drying is one of the most optimal drying methods. The intense heat/mass transfer makes fluidized beds particularly effective for drying applications. Fluidized bed drying is also suitable for deep drying applications such as post-drying of spray granulated or extruded granular products with a low residual moisture content. Hence for the proper design of fluidized bed dryers, it is important to understand the drying kinetics. Experiences show that micro-fluidized beds can be used to acquire the real drying dynamics by monitoring the evolution of water vapor at the bed outlet [1,2]. Fig. 9.1 shows the drying results of water-soaked biomass char, which simulates the high-temperature drying of biomass in a fluidized bed dryer. It shows that increasing the drying temperature from 850°C to 950°C reduces the drying time from 600 to 313 seconds (Fig. 9.1A), indicating that rising temperature accelerates the drying process considerably. Within the initial 200 seconds, approximately 60%–90% of the moisture in the wet char is removed. A maximum drying rate is attained when the water removal is about 6%–10%. Then, the drying rate starts to decline as the moisture removal increases further. Compared with the conventional low-temperature drying process (see the inset of Fig. 9.1B), the high-temperature

Micro Fluidization
DOI: https://doi.org/10.1016/B978-0-443-18718-6.00009-7
© 2023 Chemical Industry Press Co., Ltd.
Published by Elsevier Inc., under an exclusive license with Chemical Industry Press Co., Ltd.

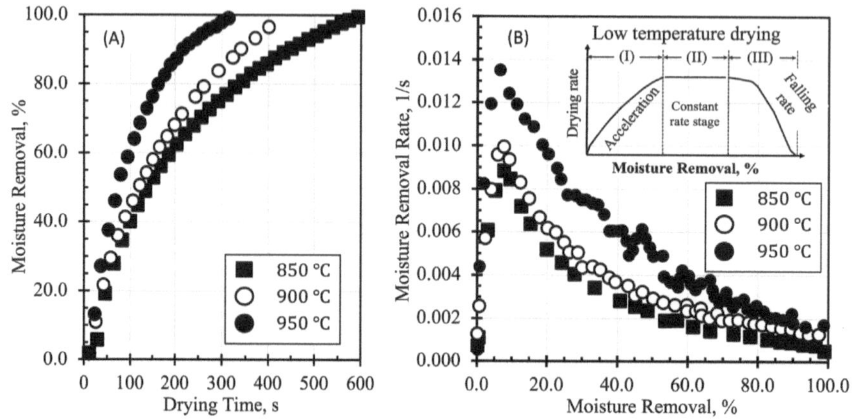

Figure 9.1 Variations in (A) the moisture removal and (B) the drying rate of wet char in a micro-fluidized bed [1].

drying in the micro-fluidized bed behaves very differently. In high-temperature micro-fluidized beds, there is no constant-rate drying stage (Fig. 9.1B). With its extremely high heat transfer rate, the micro-fluidized bed makes the water on the surface of the particles evaporate quickly so that no water saturation exists on the surface of the particles to maintain a constant-rate drying process. When all free moisture is removed, water transfer inside particles relies only on diffusion, so the drying rate decreases accordingly.

9.2 Adsorption

Global warming and climate change caused by greenhouse gas emissions have become common concerns in the past few decades. Carbon dioxide (CO_2) is considered one of the major greenhouse gases. Capturing CO_2 from power plants by using solid adsorbents may provide a viable solution for the environmental-friendly use of fossil fuels in the future. To better understand these CO_2 capture technologies, micro-fluidized beds have been used in this area in the past few years, which have generated insightful information about this increasingly important research area.

9.2.1 CO_2 capture using capsulated liquid sorbents

It is well known that amine solvents are effective for post-combustion CO_2 capture, but the estimated energy penalty reduces the efficiency of a coal power plant by 30%–40%. One solution to reduce the energy penalty is to use carbonate solutions as CO_2 sorbents. However, this solution is difficult to implement because of the slow reaction rate between the carbonate solution and CO_2. On this basis, microencapsulation of CO_2 solvents (MECS) is proposed to capture carbon dioxide from flue gas with low cost and

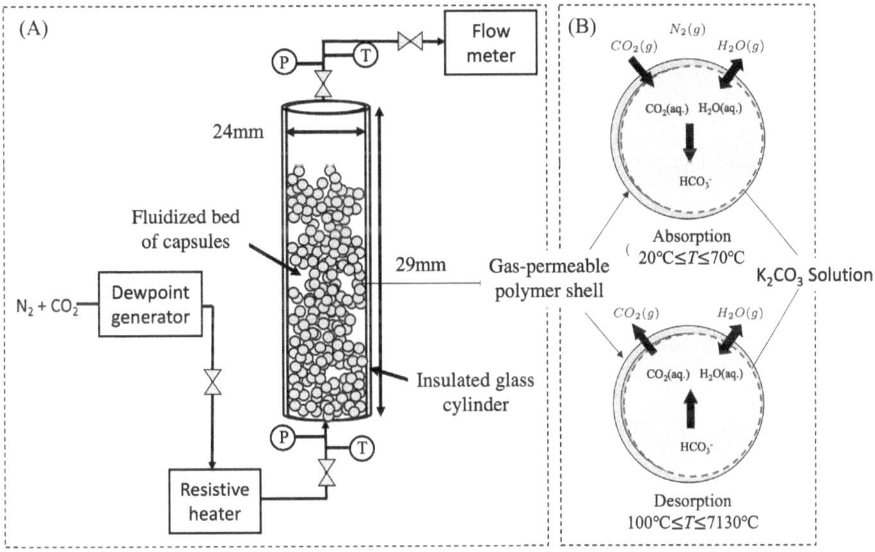

Figure 9.2 (A) The experimental apparatus and (B) the illutrations of absorption and desorption processes of capsuled Na_2CO_3 solution used for CO_2 capture.

energy consumption [3]. This approach takes advantage of the high CO_2 capture capacity of liquid adsorbents while enabling the use of fluidized beds for large-scale operations. Although the capsule shell slightly reduces the mass transfer compared to the pure liquid adsorbent, the increased specific surface area through encapsulation increases the CO_2 absorption rate by more than an order of magnitude. Finn et al. [4] conducted experimental and CFD modeling investigations on CO_2 capture by the MECS technology. Fig. 9.2A shows the schematics of the experimental apparatus, and Fig. 9.2B illustrates the principles of the MECS capsules. The results reported by Finn et al. [4] verify that the MECS technology has the potential to be deployed in fluidized beds at larger scales.

9.2.2 CO_2 capture using solid adsorbents

Fluidized beds can be used as a superior device for the adsorption of gas, such as CO_2, volatile organic compounds, and many other industrial gases. Li et al. [5,6] proposed two compact micro-fluidized bed (CMFB) reactors to capture low-concentrated CO_2 in heating, ventilation, and air conditioning (HVAC) systems using solid adsorbents. The first reactor consists of 100 micro-fluidized beds in a cuboid chamber. Each micro-fluidized bed is 10 mm in length and 10 mm in width (Fig. 9.3A). The second reactor contains multiple oblique regular hexagon-shaped micro-fluidized beds (Fig. 9.3B). The hexagonal bottom side length is 5 mm. The experimental results demonstrate that the micro-fluidized bed reactor features lower pressure drop, longer CO_2 breakthrough time, and less adsorbent attrition for CO_2 capture using solid adsorbents compared to the fixed bed, large-scale fluidized bed, and radial flow reactor.

9.2.3 CO_2 capture by gas–solid reactions

Prajapati et al. [7] carried out studies on the kinetics of CO_2 capture using K_2CO_3 supported on the activated carbon (AC) in a 25-mm-diameter micro-fluidized bed reactor. The reaction for CO_2 capture can be described by $K_2CO_3(s) + CO_2(g) + H_2O(g) \Leftrightarrow 2KHCO_3(s)$. Fig. 9.4 illustrates the effect of CO_2 concentration on the sorbent conversion (Fig. 9.4A) and the corresponding reaction rate (Fig. 9.4B). It shows that the variation of conversion with time follows a typical sigmoid relationship for a given set of parameters. The reaction rate is slow initially, increases to a maximum, and then reduces gradually. Increasing the initial CO_2 concentration increases CO_2 conversion. When the initial CO_2 concentration is higher than 70%, the reaction

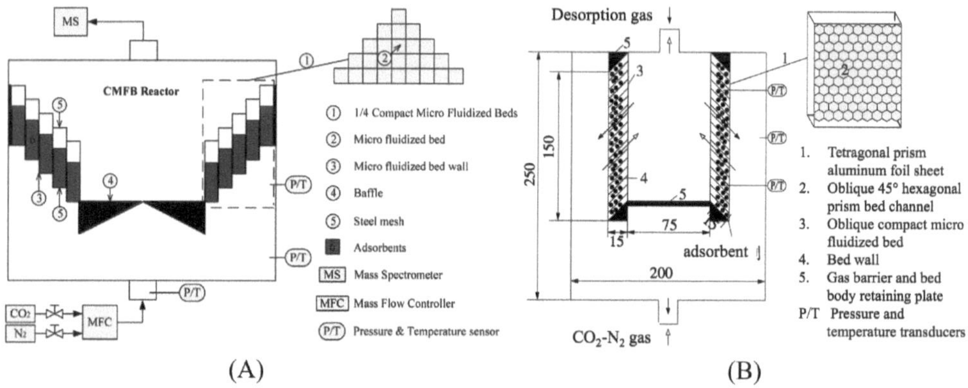

Figure 9.3 Compact micro-fluidized bed reactors for CO_2 capture. (A) A compact reactor with 100 MFBs and (B) a compact reactor with multiple oblique regular hexagon-shaped MFBs.

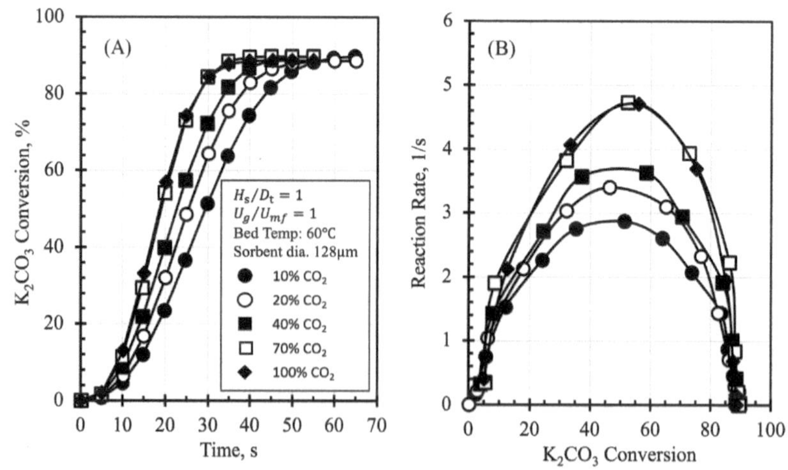

Figure 9.4 Kinetics of CO_2 capture using K_2CO_3/AC in a micro-fluidized bed reactor. (A) Conversion versus time, (B) reaction rate versus conversion.

rate changes little with CO_2 concentration, indicating that the surface chemical reaction becomes the rate-controlling step.

Amiri and Shahhosseini [8] conducted CO_2 capture experiments using the regenerable carbonate-based sorbent $K_2CO_3/\gamma\text{-}Al_2O_3$ in a micro-fluidized bed reactor of 12 mm in inner diameter. The experimental results show that the temperature and the gas flow rate are the most important variables affecting CO_2 adsorption. Increasing gas velocity intensifies gas–solid mixing and heat/mass transfer and thus enhances the reaction rate, but rising temperature influences the adsorption of CO_2 negatively because the adsorption reaction is reversible and exothermic. In addition, an appropriate steam vapor pretreatment is beneficial for the reaction, which may relate to the formation of the $K_2CO_3 \cdot 1.5H_2O$ active site, but excess vapor pretreatment may lead to a decrease in the stability of fluidization. The research demonstrates that the micro-fluidized bed reactor makes it much more manageable to study the adsorption–desorption characteristics of carbonate-based sorbents for CO_2 capture.

Due to their low cost and vast availability, lime-based sorbents, such as CaO, $CaCO_3$, and $Ca(OH)_2$, can be used to control emissions of SO_2 and CO_2 from solid fuel combustion processes [9,10]. Yu et al. [9] investigated the CO_2 capture by Ca $(OH)_2$ at temperatures from 500°C to 750°C in a 20-mm-diameter micro-fluidized bed. The experimental results reveal that the CO_2 capture is through the formation of an unstable intermediate product $Ca(HCO_3)_2$ in the reaction process between CO_2 and $Ca(OH)_2$. Fig. 9.5 shows that the conversion is affected by the temperature. At the reaction temperature below 650°C, the reaction is dominated by the chemical kinetics so that the reaction rate increases with increasing the temperature. When the reaction temperature is above 650°C, the conversion decreases with temperature because the decomposition of $CaCO_3$ becomes the rate-controlling step.

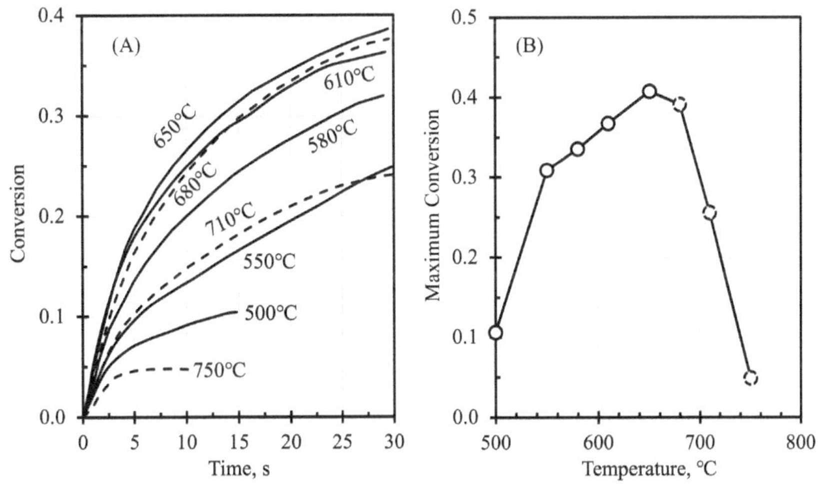

Figure 9.5 Kinetics of CO_2 capture by $Ca(OH)_2$ in a micro-fluidized bed. (A) Conversion versus time, and (B) variation of the maximum conversion with temperature.

9.3 Catalytic reaction

Strictly speaking, a catalytic gas—solid reaction refers to a reaction between reactants in the gas phase catalyzed by a solid catalyst, and a noncatalytic gas—solid reaction is a reaction between gas and solid reactants in the absence of a catalyst. Many normally noncatalytic reactions such as pyrolysis, decomposition, gasification, combustion, and reduction can also be classified as catalytic gas—solid reactions when catalysts are used to enhance their reaction performances.

9.3.1 Catalytic gas reaction

Micro-fluidized beds have been used to characterize several catalytic gas—solid reactions, including but not limited to the methanol vapor-phase carbonylation [11], methane steam reforming [12], catalytic cracking of oil [13] and methane [14], syngas methanation [15], dehydrogenation of isopentenes [16], gas-phase fructose conversion to furfural [17], methanol-to-olefin [18], steam reforming of methanol [19], and catalytic tar reforming [20]. Fig. 9.6 shows that compared to a fixed-bed reactor and a large-scale fluidized bed reactor, the micro-fluidized bed reactor yields the highest conversion (Fig. 9.6A) and selectivity (Fig. 9.6B) for the three catalytic reactions because of its high heat and mass transfer rates, uniform temperature distribution, and close-to-plug gas flow behavior. These characteristics enable close-to-intrinsic kinetics and high product selectivity to be obtained. Notably, the micro-fluidized bed reaction analyzer provides a robust tool for catalyst screening and optimization when online catalyst sampling and analysis are integrated with real-time analysis of gas products [11]. Due to the

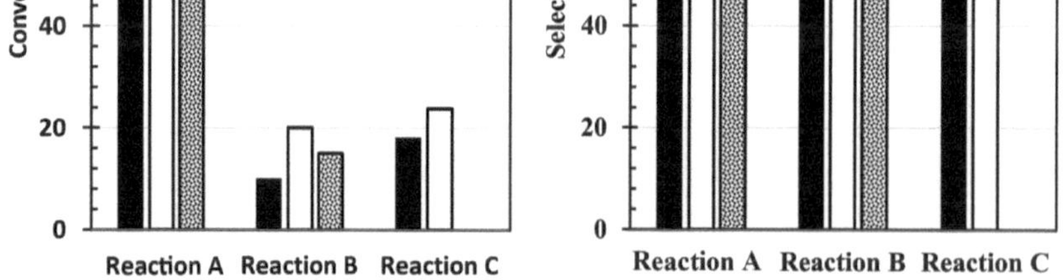

Figure 9.6 Comparison of (A) conversion and (B) selectivity of three catalytic reactions in a micro-fixed bed, a micro-fluidized bed, and a macro-fluidized bed. Reaction A: catalytic methanation [21]; Reaction B: nonoxidative aromatization of methane [21]; Reaction C: oxidative dehydrogenation of isopentenes to isoprene [16].

diversity and complexity of these reactions, interested readers are advised to read the relevant research literature for details of the reactions and the role that the micro-fluidized bed reactor played in the study.

9.3.2 Catalytic gas solid reaction

Reaction performances of thermochemical gas−solid reactions such as pyrolysis, gasification, and combustion can be improved or enhanced using catalysts. Micro-fluidized bed reaction analyzers can be used to evaluate the catalyst performance efficiently and cost-effectively.

9.3.2.1 Catalytic pyrolysis

Compared to gasification, pyrolysis of solid fuels such as coal and biomass yields synthesis gases at moderate temperatures, but the technology is not yet widely used commercially because the produced synthesis gases have low quality. To solve this problem, research on catalytic biomass pyrolysis as a direct in situ syngas upgrading solution [22−25] has gained increasing interest. Li et al. [25] investigated the pyrolysis of rice husk catalyzed by iron and nickel in a micro-fluidized bed reactor. As shown in Fig. 9.7A−D, the experimental results demonstrate that the presence of Fe and Ni reduces the production of heavy hydrocarbons due to enhanced polymerization. Fe-based catalyst generally stimulates dehydrogenation of tar to produce more H_2, while Ni-based catalyst promotes cracking and reforming reactions, leading to an increase in the production of H_2. Fe- and Ni-based catalysts reduce the production of CH_4. The increase in CO production is because both the catalysts enhance the water−gas shift ($H_2 + CO_2 \leftrightarrow CO + H_2O$) and the Boudouard reaction ($C + CO_2 \leftrightarrow 2CO$). The Fe- and Ni-based catalysts increase the CO_2 production rate. Fig. 9.7E shows that the variations in E_a are in good agreement with the above discussions.

Liu et al. [24] investigated the pyrolysis characteristics of potassium-impregnated rice husk in a two-stage micro-fluidized bed reactor under isothermal conditions. As shown in Fig. 9.8, the activation energy increases with the conversion from 0.2 to 0.8 for all the major gas components. The activation energy of the production of H_2, CO, and CO_2 decreases gradually with increasing potassium content, whereas the activation energy of CH_4 production changes with the potassium negatively. Guo et al. [23] investigated the catalytic effects of Zn, Cu, and Zn/Cu on biomass pyrolysis in micro-fluidized bed reactors. The results indicate that adding Zn and Cu/Zn reduces the activation energy for the production of H_2, CH_4, and CO_2 but increases the activation energy of CO production.

9.3.2.2 Catalytic gasification

Solid fuel gasification usually has a relatively slow reaction rate, and thus attempts have been made to use catalysts to stimulate the reaction. For example, various alkali and

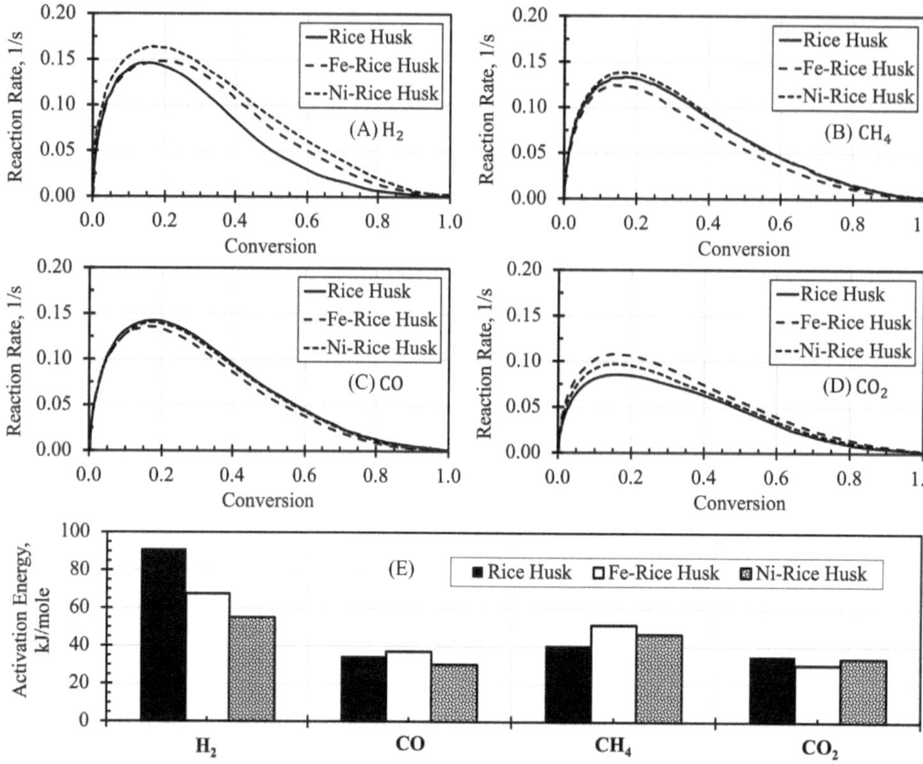

Figure 9.7 The catalytic effect of iron and nickel catalysts on the production of (A) H_2, (B) CH_4, (C) CO, and (D) CO_2 and (E) the corresponding activation energy from a rice husk (RH) pyrolysis in a micro-fluidized bed reactor.

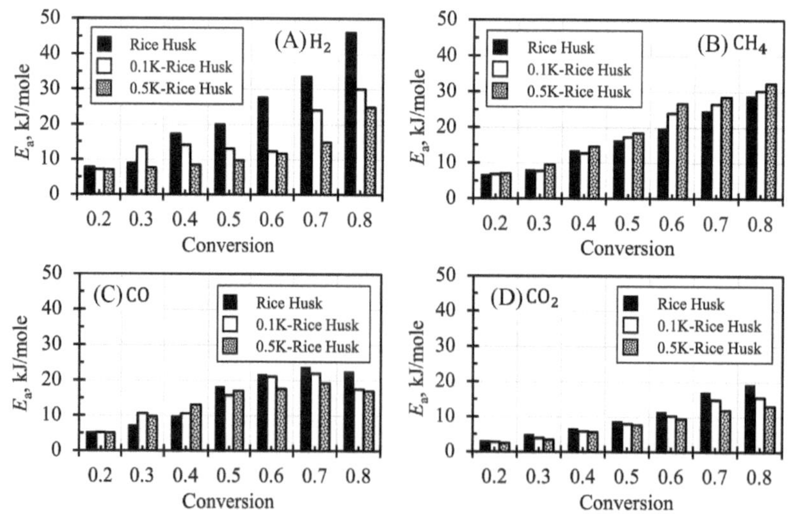

Figure 9.8 The catalytic effect of potassium on the activation energy of the production of (A) H_2, (B) CH_4, (C) CO, and (D) CO_2 from a rice husk (RH) pyrolysis in a micro-fluidized bed reactor.

alkaline earth metal compounds have been used to catalyze the gasification of carbonaceous materials, such as coal, biomass, and char [26–28]. Zhang et al. [29] used black liquor to catalyze steam gasification of petroleum coke, a byproduct of the delayed coking process in refineries, in a micro-fluidized bed reactor. This study is of great interest because it attempts to use the black liquor, a waste produced by the paper pulping industry, to convert the petroleum coke into value-added syngas products. In this gasification process, the catalytic activity is realized by alkaline matters (i.e., sodium, potassium, calcium, etc.) contained in the black liquor. Fig. 9.9A shows that the catalytic effect of the black liquor is noticeable. The completion time of gasification is reduced from over 90 minutes for pure coke (i.e., 0% black liquor) to below 30 minutes when 20 wt.% black liquor is added. Fig. 9.9B shows that when the black liquor content is below 10%, the hydrogen content in the synthesis gas increases with increasing black liquor. When the black liquor content exceeds 10%, the hydrogen

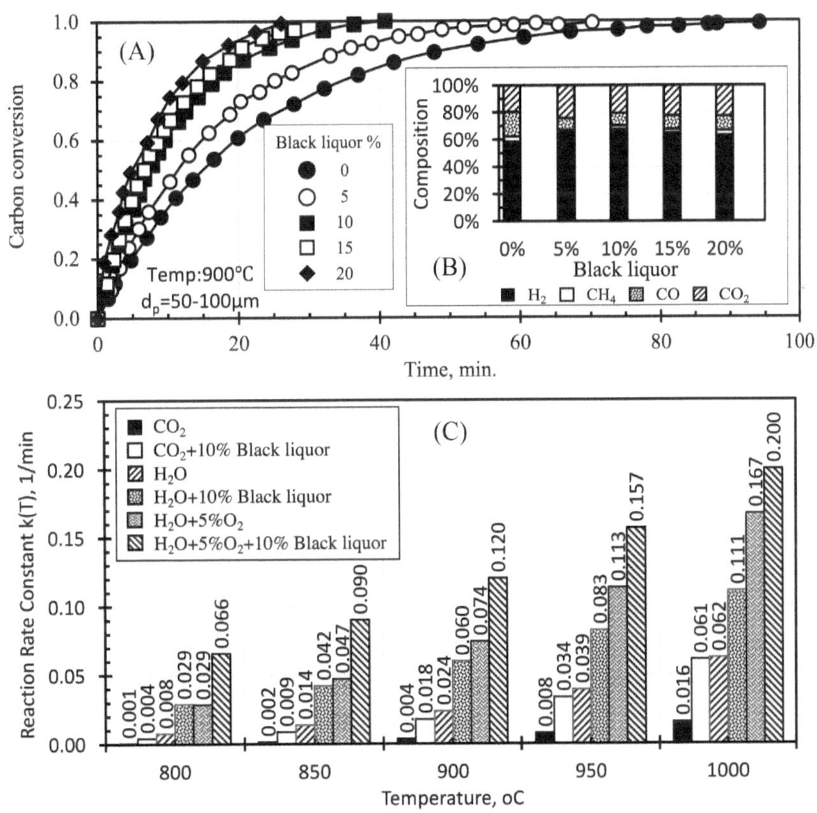

Figure 9.9 (A) Effect of black liquor loading amount on the conversion, (B) the composition of the product gas, and (C) the reaction rate constant for steam gasification of petroleum coke in a micro-fluidized bed reactor.

content decreases because the increased black liquor over-catalyzes the gasification, producing more CO_2. Thus, the sum of the valuable synthesis gases (i.e., H_2, CH_4, and CO) decreases slightly as the black liquor load increases (Fig. 9.9A).

Fig. 9.9C illustrates the catalytic effect of the black liquor on the coke gasification when steam, CO_2, and steam containing 5% of O_2 are used as gasification agents. For ease of comparison, the numerical values of the gasification reaction constant, $k(T)$, are presented in the figure for the corresponding test conditions. The results indicate that replacing CO_2 with steam as the gasification agent significantly increases the rate constant of coke gasification. Mixing 5% of O_2 into the steam further elevates the rate constant of coke gasification. Moreover, when 10% of black liquor is added, the reaction rate constant of the coke gasification increases because of the catalytic effect of sodium salts in the black liquor. It is worth noting that the tests presented here are possible thanks to the MFBRA's capability of changing the water vapor environment in the reactor flexibly.

9.3.2.3 Catalytic thermal decomposition

In recent years, hydrogen has been seen as a clean energy source and even the future energy resource because of the need to replace or minimize the use of fossil fuels to deal with global warming and climate change challenges. As one of the most promising hydrogen production technologies, catalytic methane decomposition (CMD, $CH_4 \rightarrow 2H_2 + C$) has attracted much attention due to its relatively low reaction temperature [30]. The process also produces a highly valuable byproduct, carbon nanotubes (CNTs) or carbon nanofibers (CNFs) [31]. Xu et al. [14] investigated the low-temperature synthesis of carbon nanotubes (CNTs) using Ni/MgO catalytic decomposition of CH_4 in a micro-fluidized bed reactor. It shows that in the temperature range of 500°C–550°C, the dynamic equilibrium is attained between the rate of CH_4 decomposition and the rate of carbon diffusion over Ni catalyst for continuous precipitation of CNTs in the micro-fluidized bed condition. The high-quality CNTs can be synthesized using the micro-fluidized bed. Geng et al. [32] investigated methane decomposition kinetics over Fe_2O_3 catalyst in a 20-mm-diameter micro-fluidized bed reaction analyzer. Fig. 9.10 depicts typical gas evolution results. It is interesting to see that the use of MFBRA makes revealing the complete dynamic process of CMD reaction possible in a single test. As shown in Fig. 9.10, after catalyst particles (i.e., Fe_2O_3) are pulse-injected into the micro-fluidized bed reactor under a reducing environment, their reduction starts almost immediately. The reduction process goes through a series of reactions: $Fe_2O_3 \rightarrow Fe_3O_4 \rightarrow FeO \rightarrow Fe$. In the initial period, iron oxide is in the reduction stage; thus, Fe for catalytic decomposition of CH_4 to H_2 has not been formed, so the oxygen produced by iron oxide reduction reacts with methane to produce CO_2 and water vapor. When enough Fe is produced, H_2 begins to evolve and rapidly increases as Fe is formed. As H_2 production increases, the

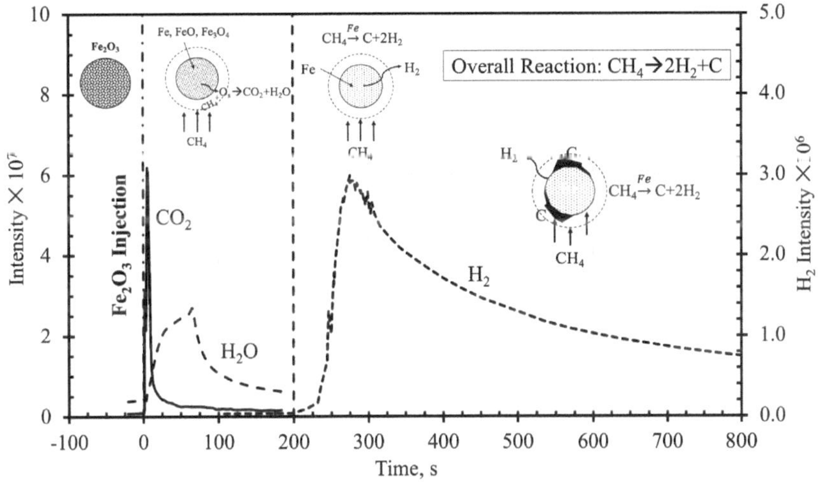

Figure 9.10 Typical gas evolution of catalytic methane decomposition at 750°C and 0.75 atm.

produced carbon accumulates on the surface of Fe, resulting in a rapid decrease in the Fe activity and a gradual reduction in the H_2 production rate. This experimental study proves that MFBRA is a powerful tool to study the reaction mechanism and kinetics.

9.4 Thermal decomposition

Thermal decomposition (sometimes referred to as thermolysis) is one of the essential thermochemical conversion processes. In thermal decomposition reactions, the heat causes one substance to break down into two or more different substances by breaking down the bonds that hold the atoms together in the original molecules, and therefore the reaction usually consumes heat energy (i.e., endothermic).

9.4.1 Liquid decomposition

Thermal decomposition can convert liquid fuels into lighter gases at high temperatures. It is an effective method to improve the quality of gaseous products during the gasification process by converting the byproduct tar into operation-friendly lighter gases. Tar is viscous when condensed into a liquid at a temperature below its dew point. Tar condensate seriously affects the stable operation of the system because of tar-induced blockages of pipelines, valves, and dust filters.

Gai et al. [33] investigated the thermal decomposition of tar in a micro-fluidized bed reactor under isothermal conditions using toluene as the tar model compound. Using the model compound overcomes the complexity of the actual tar, which contains aggregates of oxygenates, phenolic compounds and olefins, and aromatic and polyaromatic

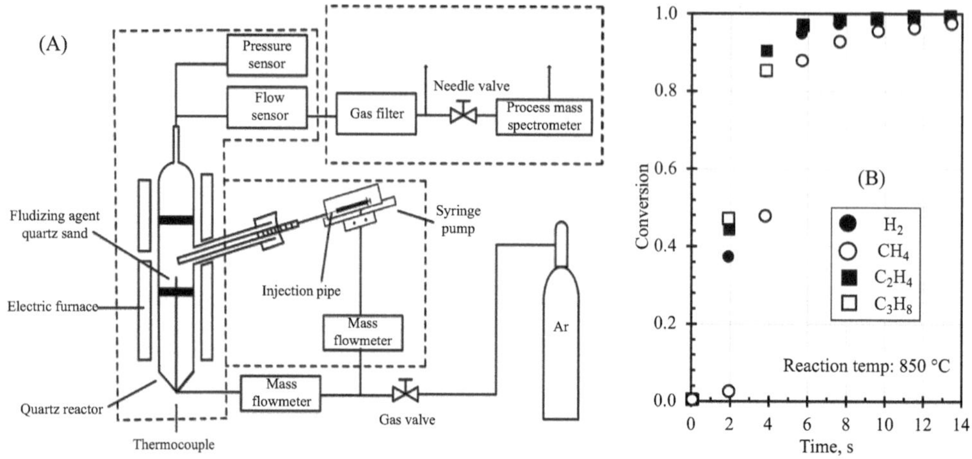

Figure 9.11 (A) A schematic diagram of the micro-fluidized bed reactor system with liquid sample feeding ability and (B) typical gas evolution profiles produced by the thermal decomposition of toluene.

hydrocarbons. In the experiments, the micro-fluidized bed of quartz sand particles was preheated to the preset temperature, and then, 10-mL toluene was pulse-injected to the bed by a syringe pump, as shown in Fig. 9.11A. The produced gas was monitored online by a mass spectrometer. Fig. 9.11B shows the typical product gas evolution profiles. The experimental results indicate that at a given reaction temperature, up to 80% of major gas products conversion (i.e., H_2, CH_4, C_2H_4, and C_3H_8) is complete in less than 4 seconds. The experimental results also show that an increase in temperature promotes toluene decomposition. This research demonstrates that MFBRA is quite capable of characterizing reactions with liquid feedstocks.

9.4.2 Solid decomposition

The thermal decomposition of minerals such as limestone and magnesite has been intensively investigated over the years [34]. Most studies are conducted using thermogravimetric analyzers, so the results are likely affected by the pronounced gas diffusion associated with the TGA technique. Fig. 9.12A and B presents the kinetic parameters of the thermal decomposition of limestone [35,36] and magnesite [37] obtained by TGA and MFB experiments. It is clear that the determined activation energy of the decomposition reaction by the micro-fluidized bed is lower than that by the thermogravimetric analysis, suggesting that MFBRA is less affected by the gas diffusion and that the results should approach the intrinsic kinetics compared to TGA.

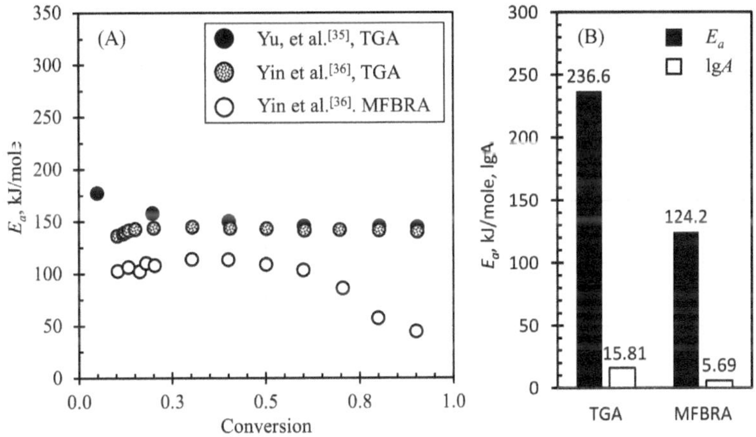

Figure 9.12 Kinetic parameters obtained by TGA and micro-fluidized bed reactor for (A) limestone and (B) magnesite thermal decomposition.

9.5 Pyrolysis

Pyrolysis is one of the thermochemical conversion methods to convert liquid and solid materials (e.g., biomass, low-rank coals, and oil shale) into gases, condensable liquids (tar), and char (solid residue) in the complete or close-to-complete absence of an oxidizing agent typically at low to moderate temperatures. In general, pyrolysis is also the first step in many thermochemical conversion processes (e.g., gasification and combustion) of solid fuels.

In thermal analysis techniques, pyrolysis often serves as a thermal decomposer to produce volatile gases, which are usually characterized by advanced gas and/or liquid analyzers. The chemical structural characterizations of the sample materials are deduced based on the characterization results. Such analytical techniques include a variety of py-GC, py-Ms, and py-GC/Ms. In this area, MFBRA, which can characterize almost all liquid or solid samples, supports a full complement of pyrolysis analytical techniques [38,39]. In addition, MFBRA adds an extra analytical ability to characterize the pyrolysis reaction process, which is often impractical by other pyrolyzers (e.g., Pyroprobe, μ-reactor, see Chapter 6). MFBRA monitors gas evolution dynamics in real time, thus obtaining the kinetics of the global reaction and the product gas. Moreover, MFBRA can acquire the product gas release sequences of fast reactions, which is particularly valuable because such information allows researchers to identify new product species, reveal new reaction paths, and decode reaction mechanisms.

9.5.1 Biomass pyrolysis

Biomass pyrolysis is an important thermochemical approach to utilizing widely available, renewable, and CO_2-neutral resources. Fig. 9.13A shows the typical product gas evolution from the herb residue pyrolysis in a micro-fluidized bed reactor [40].

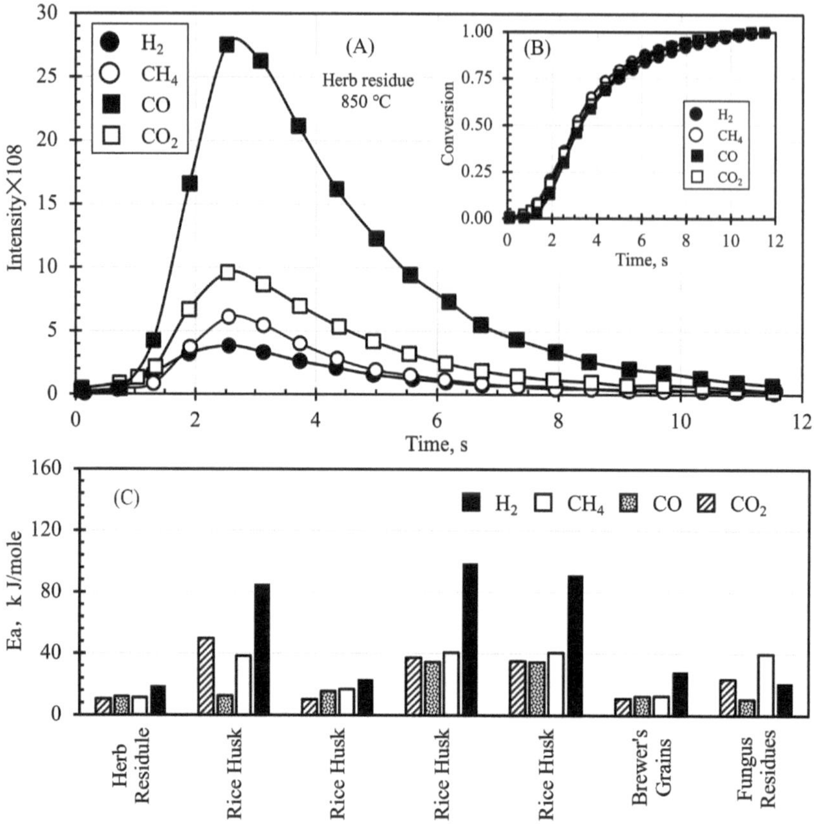

Figure 9.13 (A) Gas release behavior and (B) activation energy of biomass pyrolysis in a micro-fluidized bed reactor.

The corresponding relationship between conversion and time is displayed in Fig. 9.13B. It shows that, similar to other biomass materials, pyrolysis of herb residue is a fast reaction, since the gas production rate attains a maximum in less than 3 seconds. As further shown in Fig. 9.13C, the experimental results indicate that it is more difficult to produce H_2 than other gas components. Except for a few cases, the values of activation energy for producing H_2, CH_4, CO, and CO_2 are less than 40 kJ/mole [40–43].

9.5.2 Coal and oil shale pyrolysis

With the increased concern that petroleum resources will be depleted gradually in the future, the development of alternative energy sources has attracted more and more attention over the years [44–48]. As the first step in the thermal conversion process (i.e., gasification and combustion) of solid fuels, pyrolysis is of great significance for the utilization of abundantly reserved carbonaceous materials such as coal and oil shale. Coal and oil shale can be converted into tar, non-condensable gas, and semi-coke

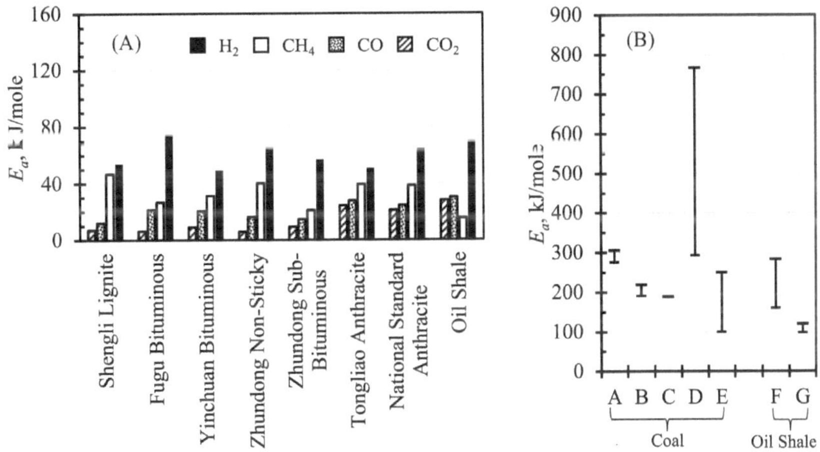

Figure 9.14 Activation energy of coal and oil shale pyrolysis in (A) micro-fluidized bed reactors and (B) TGA and fixed-bed reactors. Coal: A [51], B [52], C [53], D [54], E [55]; oil shale: F [56], G [57].

during low-temperature pyrolysis. Tar can be further upgraded to liquid fuels usable as substitutes for petroleum-derived oils.

Zhang et al. [49] investigated the characteristics and kinetics of coal and oil shale pyrolysis using a micro-fluidized bed reaction analyzer (MFBRA). Cai et al. [50] studied the pyrolysis kinetics of five differently ranked coals in a micro-fluidized bed reaction analyzer. Fig. 9.14 shows the activation energy data collected from experiments using micro-fluidized bed reactors [49,50], TGA, and fixed bed reactors [51−57]. The results show that the activation energy of primary gas production is typically lower than 80 kJ/mole, which is significantly lower than that reported in the literature using TGAs and fixed bed reactors. Again, this indicates that the MFBRA is a reliable reaction analysis tool to yield close-to-intrinsic kinetics and valid mechanisms of gas−solid reactions, especially for pyrolysis-like fast reactions.

9.5.3 Blended material pyrolysis

Co-pyrolysis of blended fuel is an attractive approach to improving pyrolysis reaction or product performance related to the pyrolysis of individual fuels. For example, co-pyrolysis of a coal mixture of different qualities may result in more efficient fuel conversion and high-quality products. It may also help solve issues related to ash slagging, fouling, or fluidity that occur when a single fuel is used, thus improving the operability and safety of the reaction process.

Fig. 9.15 shows the methane release characteristics and the corresponding activation energy of production based on the experimental data reported by Dai et al. [58]. In this experiment, two types of coal with different ranks (i.e., Xiaolongtan (XLT) lignite and Jincheng (JCH) anthracite) and their blends are pyrolyzed in a micro-fluidized

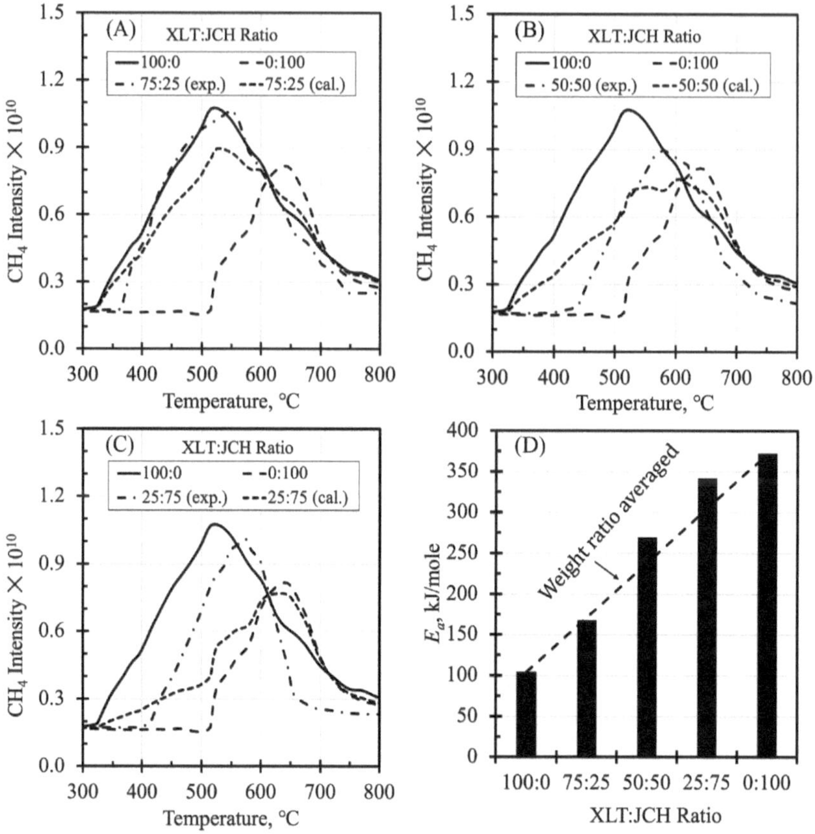

Figure 9.15 Co-pyrolysis behavior of two types of coal individually and their blends in a micro-fluidized bed reactor. (A) XLT:JCH ratio: 75:25, (B) XLT:JCH ratio: 50:50, (C) XLT:JCH ratio: 25:75, and (D) variation of activation energy with XLT:JCH ratio. *XLT*, Xiaolongtan coal; *JCH*, Jincheng coal.

bed reaction analyzer. The results indicate that compared to the low-rank highly volatile lignite coal, the high-rank low volatile anthracite coal is hard to pyrolyze judged by the following observations: (1) a higher initial temperature at which methane production starts; (2) a lower maximum releasing rate; (3) a smaller total amount of methane production; and (4) higher activation energy of methane production. When blended with lignite coal, the anthracite coal is shown to have a promoted pyrolysis performance, and the promotion effect increases with increasing weight ratio of lignite coal in the coal mixture, as shown in Fig. 9.15A–C. However, the enhancement effect of fuel blending is unequal to the weighted average of the corresponding individual fuels. This can be seen from the difference between the experimental-measured and weight-averaged values of methane release intensity and activation energy. Therefore, it is reasonable to believe that some synergistic effects would occur when fuels of different qualities are co-pyrolyzed under the same operating condition.

With high oxygen containment, the oil produced by the pyrolysis of biomass needs to be further upgraded before being used as a substitute for petroleum oil because oxygen-rich oil causes deleterious properties, such as low heating value, corrosion of equipment, and operation instability. Co-pyrolysis of biomass and other fossil fuels such as coal and waste plastics may provide a viable approach to improving oil quality [59–61]. As shown in Fig. 9.16, the gas production characteristics of the co-pyrolysis of Zhundong lignite and pine sawdust are different from those of the constituents of the fuel mixture [61]. The results demonstrate that the synergistic effects between biomass and coal during co-pyrolysis can be analyzed using micro-fluidized bed reactors. Unfortunately, there is no report on the quality of oil produced by micro-fluidized bed co-pyrolysis, so future research may be needed to address this issue.

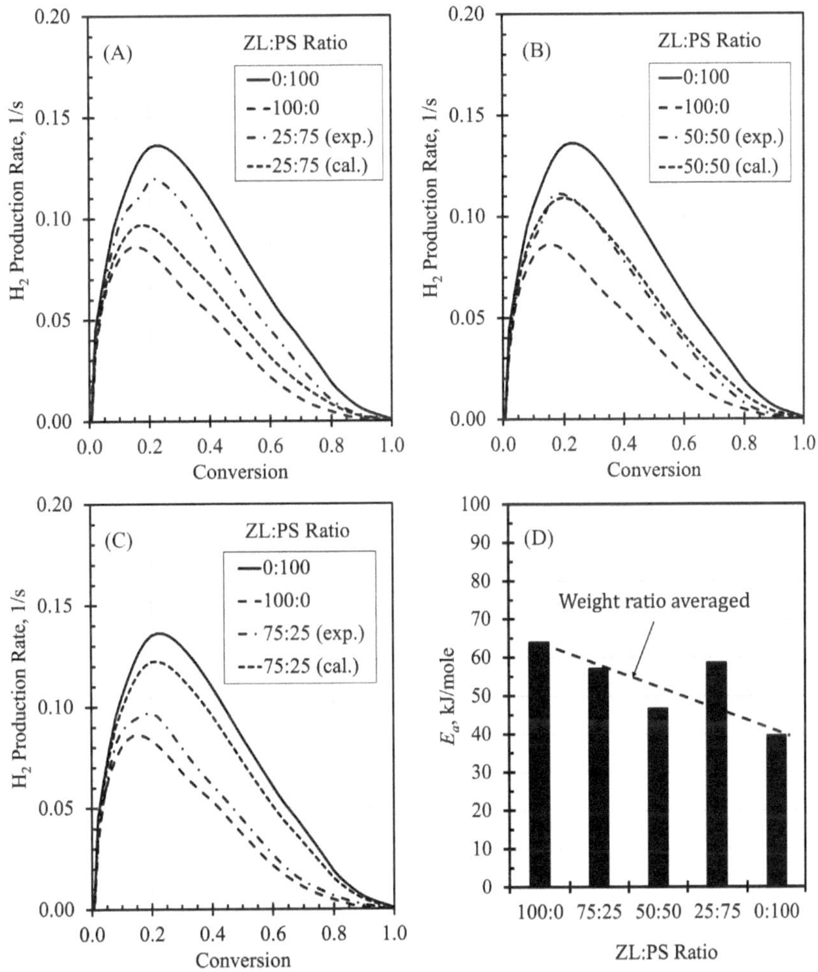

Figure 9.16 Co-pyrolysis behavior of Zhundong lignite (ZL) coal and pine sawdust (PS). (A) ZL:PS ratio: 25:75, (B) ZL:PS ratio: 50:50, (C) ZL:PS ratio: 75:25, and (D) variation of activation energy with ZL:PS ratio.

9.6 Thermal cracking

Thermal cracking is a process in which larger and heavier nonvolatile hydrocarbon molecules (e.g., crude oil, tar) are broken down into smaller and lighter molecules by the heat at high temperatures and pressures with or without the presence of catalysts. It is one of the most critical steps for oil refineries to convert crude oil into lighter petroleum products such as gasoline, diesel, and gases (e.g., C_1-C_4). Gross et al. [13] reported that for all the three oil feedstocks investigated, the conversion in a micro-fixed bed was higher than that in a 40.6 mm (i.e., 1.6 in.) fluidized bed, and the catalyst deactivation and gasoline selectivity were independent of reactor configuration. The FCC zeolite catalyst has an average particle size of 60−75 μm, so the ratio D_t/d_p is estimated to be from 541 to 677. According to the definition in Chapter 4, the fluidized bed used by Gross et al. [13] is a macro-fluidized bed based on the criterion of $D_t/d_p > 150$. Understandably, lower conversion and selectivity obtained by Gross et al. [13] are due to gas backmixing in the macro-fluidized bed.

Hu et al. [20] carried out experiments in a micro-fluidized bed reactor to thermally crack tar into lighter gases. The micro-fluidized bed reactor employed is schematically shown in Fig. 9.17A. The reactor has three stages: the bottom for gas preheating, the middle for tar

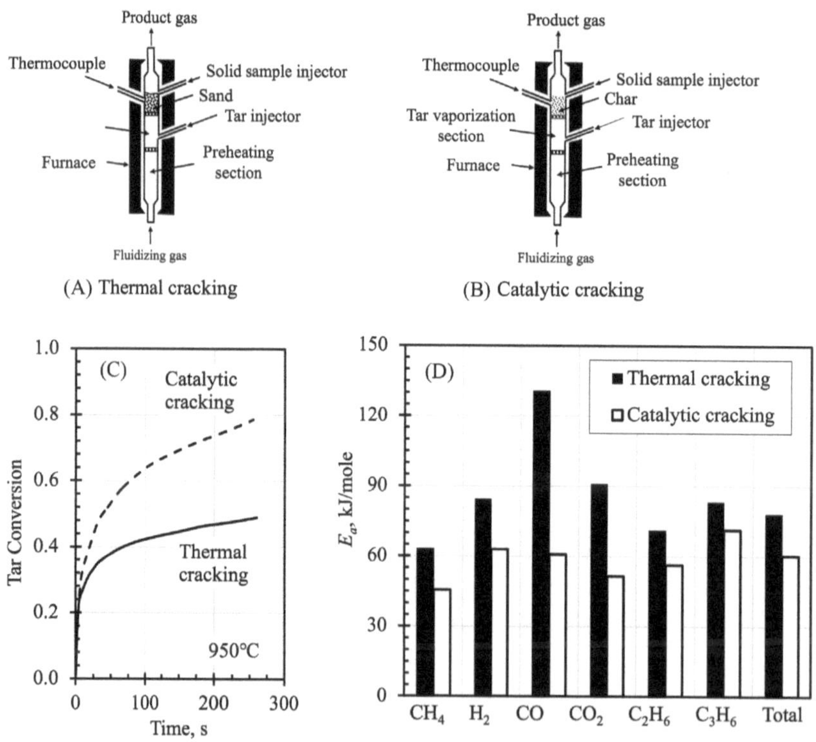

Figure 9.17 Thermal and catalytic tar cracking in a micro-fluidized bed reactor. (A) Thermal cracking reactor, (B) Catalytic cracking reactor, (C) tar conversion, and (D) activation energy.

vaporization, and the upper for tar cracking. In the thermal (Fig. 9.17A) and catalytic (Fig. 9.17B) cracking experiments, sand and char particles are used as bed materials, respectively. Fig. 9.17C shows a typical total gas evolution from the thermal and catalytic cracking experiments at 950°C. Fig. 9.17D depicts the activation energy for the gas components of CH_4, H_2, CO, CO_2, C_2H_6, C_3H_6, and their mixture. The results demonstrate that the presence of char particles increases the tar conversion due to the catalytic effect of char particles. The MFBRA makes the experiments easy to accomplish in order to uncover the differences between thermal and catalytic cracking of tar under the same operating condition.

9.7 Gasification

Gasification is a thermochemical process in which carbonaceous materials are converted into a gas mixture consisting mainly of carbon monoxide, hydrogen, methane, and carbon dioxide. The gas mixture, also called producer gas or synthesis gas, can be utilized for heat and power generation, hydrogen production, and the synthesis of various chemicals. In principle, gasification is a partial oxidation process in the deficiency of an oxidant. In recent years, there has been an increasingly renewed interest in gasification technology as it provides alternatives to the clean and efficient utilization of both renewable materials and fossil fuels.

Micro-fluidized beds have been used to characterize several gasification processes, including gasification of coal [48], coal char [62–73], and biomass [1,74]. Notably, it can initiate an in situ reaction, making it possible to investigate the differences in gasification performance between in situ and ex situ [69,72] char produced by biomass and coal materials. The excellent adaptability of the water vapor atmosphere in micro-fluidized beds also permits the successful investigations of gasification with steam [29,64,65,70,73,75,76]. Furthermore, MFBRA has been used extensively in recent years to assist in the development of advanced thermochemical conversion processes, such as dual fluidized bed gasification processes, decoupled pyrolysis gasification, and decoupled combustion. As introduced in the preceding "catalytic gas–solid reaction" section, MFBs have also been used for catalytic gasification of deposited coke on catalysts [75–77] and petroleum coke mixed with black liquor [29].

9.7.1 Biomass gasification

Biomass gasification has captured interest in the past 20–30 years, driven by the increasing demand to produce bio-oil and chemicals from abundantly renewable biomass as substitutes for depleting fossil resources. However, biomass gasification technology still has problems, such as the high tar content in the resulting synthesis gas. Therefore, a great deal of research has been attempted over the years to develop biomass gasification into a mature and competitive thermochemical conversion technology [78–85].

Gao et al. [74] studied the gasification of biomass particles with three sizes (i.e., 50, 100, and 150 μm) in a micro-fluidized bed reactor using pure oxygen as a gasification reagent. The experimental results show that particle size is an influential factor affecting syngas and char production profiles, and the small-sized particles result in increased syngas production. For the biomass gasification process, the char conversion is the rate-controlling step. The char conversion rate achieved in the micro-fluidized bed strongly depends on char characteristics produced in the pyrolysis stage.

9.7.2 Coal gasification

Coal gasification is a clean coal utilization technology for power, liquid fuels, chemicals, and hydrogen. Coal gasification is mature in theory and is already in use industrially, but the challenge has always been how to make this decades-old technology more efficient and economical. Problems associated with coal gasification include low efficiency, high tar content, and expensive capital and operating costs. To solve these problems, using catalysts may be a better solution, but catalysts will deactivate because they react with the coal ash in gasifiers. Samih et al. [47] investigated the gasification characteristics of ash-free coal in a micro-fluidized bed thermogravimetric analyzer. Fig. 9.18 illustrates the effects of ash and catalysts on carbon conversion and higher heating value (HHV) of synthesis gas. It demonstrates that removing ash from coal reduces carbon conversion but increases the HHV of the gas product. Adding the catalyst K_2TiO_3 to the ash-free coal has varying effects on carbon conversion and heating value of the gas product, depending on the gasification temperature.

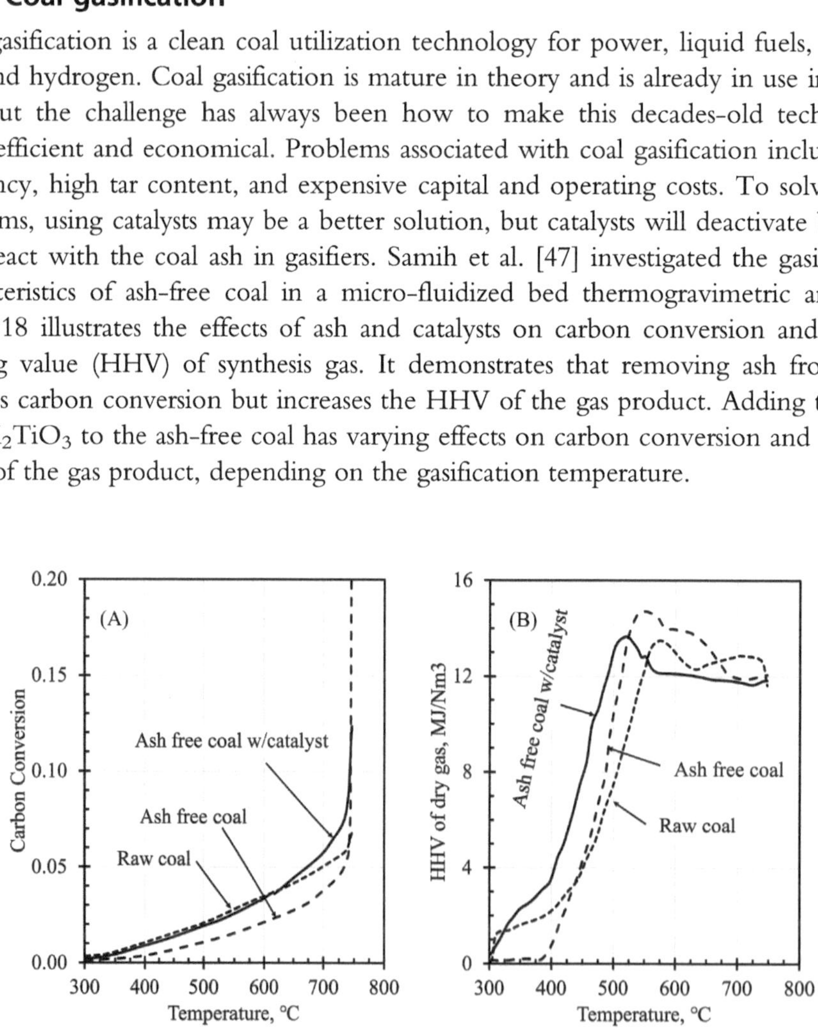

Figure 9.18 Comparison of (A) carbon conversion and (B) HHV of synthesis gas obtained by the raw coal, the ash-free coal, and the ash-free coal with catalyst.

9.7.3 In/ex situ char gasification

As discussed previously, only in situ nascent char gains the right insights into char gasification performance in reaction characterizations. To realize it, one needs to produce the char and gasify it in the same reactor at the same temperature and atmosphere. Furthermore, the reactor system needs to have a precisely controlled and highly uniform reaction temperature, a close-to-plug flow of gas, and negligible gas backmixing. Experimental results have verified that MFBRA meets these requirements satisfactorily [67,69,72]. Fig. 9.19 summarizes the activation energy data of gasification of char particles prepared by different methods. The results show that the gasification activity of the in situ nascent char is the highest, the ex situ cold char is the lowest, and the in situ hot char is in between the others. Investigations confirm that the in situ nascent char features the largest specific surface area, the smallest average pore size, the weakest graphitization degree, and the most surface-active sites. The results also show that MFBRA can generate in situ nascent char particles to facilitate the study of the char gasification properties, providing an effective tool to understand the effect of char preparation methods on the physical and chemical structures and reactivities of char particles.

Moreover, using MFBRA can study the evolution of char morphologic structure during gasification reactions. Li et al. [86] first prepared the char in a fixed bed reactor and then gasified it in a micro-fluidized bed reactor at ambient pressure and 1000°C. A vacuum-assisted online particle sampling system sampled the in-bed char particles at different time points. The reactivity of the char samples was evaluated in a thermogravimetric analyzer. Fig. 9.20 shows the variation of gasification reaction rate constant of char particles sampled at different gasification times. It is observed that char

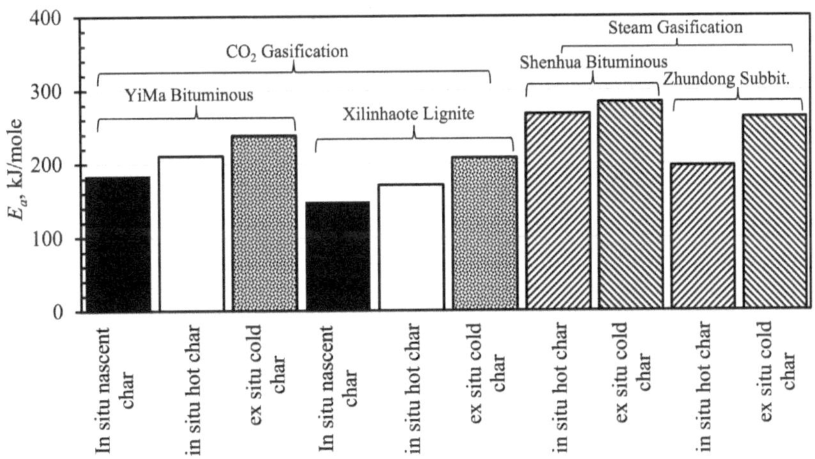

Figure 9.19 Comparison of gasification activation energy of in situ char, in situ hot char, and cold char in micro-fluidized bed reactors.

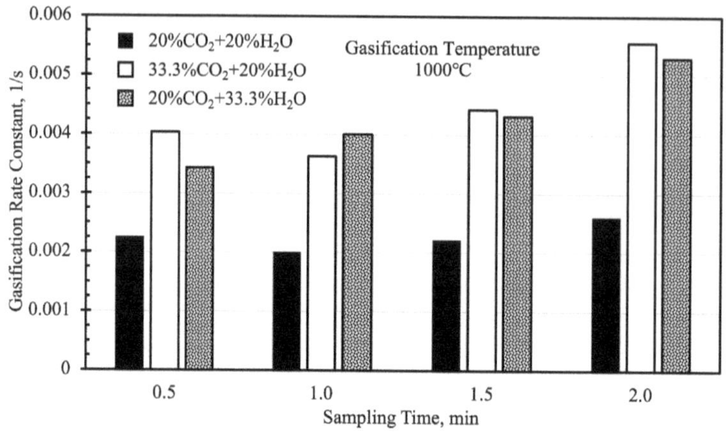

Figure 9.20 Gasification reaction rate constant of char sampled at different times.

morphology evolves with gasification time. The development of char structure increases the gasification rate, and the increase of CO_2 and H_2O concentrations in the gasifying reagent enhances the gasification reactivity.

9.8 Combustion

Combustion, or burning, is a thermochemical process in which a substance reacts rapidly with oxygen to produce a flue gas mixture while giving off heat. Micro-fluidized beds have been used to investigate the combustion performance of graphite [87], activated carbon [88], raw and ash-free coal [89], and coal char [90]. In recent years, with the growing demand for protecting the environment from pollution caused by the combustion of fossil fuels, new combustion technologies, such as oxy-fuel combustion [91,92], decoupling combustion [1], and chemical looping combustion [93–95], have been investigated by using micro-fluidized bed reactors.

9.8.1 Decoupling combustion

Decoupling combustion is an advanced low-NOx combustion technology in which fuel is first pyrolyzed to produce volatiles and char, and then, the in situ char and the volatiles are used to reduce the NOx produced during their combustion process [96]. Guo et al. [97] used the micro-fluidized bed reaction analyzer as an in situ decoupling analytical reactor to characterize decoupling combustion. Fig. 9.21 shows the gas release characteristics during graphite and coal combustions in a micro-fluidized bed reactor with an inner diameter of 20 mm. It shows that devolatilization occurs after the coal particles are injected into the reactor. The combustion of in situ produced char begins after gas switching, and no devolatilization occurs during graphite

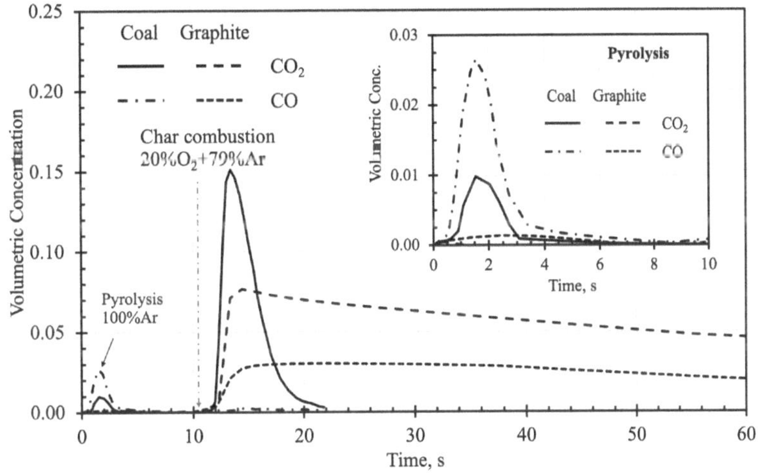

Figure 9.21 Real-time CO_2 and CO release characteristics during decoupling combustion of graphite and coal particles in a micro-fluidized bed reactor.

pyrolysis. Compared to graphite, the in situ coal char produces little CO in the combustion process because the elemental hydrogen in the char particles accelerates CO oxidation.

9.8.2 Oxy-fuel combustion

Oxy-fuel combustion, i.e., oxygen-enriched combustion, is one of the CO_2 capture technologies that are potentially helpful in reducing carbon emissions from coal-fired power plants. In oxy-fuel combustion, coal is burned under oxygen-enriched conditions in order to increase the volumetric fraction of CO_2 in combustion products to 90% or higher, thus helping to efficiently separate and capture CO_2 from flue gases. Su et al. [92] and Guo et al. [91] independently investigated oxy-coal combustion characteristics using micro-fluidized bed reactors. Fig. 9.22 shows the dynamic release results of CO and CO_2 produced in a pulverized coal combustion test at 1300°C and 60% oxygen concentration [91]. This experiment demonstrates that the MFBRA can capture the typical bimodal gas release behavior of coal combustion. We believe that the first peak is mainly generated by the rapid pyrolysis of pulverized coal, and the second peak is formed by the combustion of the in situ nascent volatiles; finally, the long tail after the second peak comes from the combustion of residual char particles.

9.8.3 Chemical looping combustion

Chemical looping combustion (CLC) is a novel combustion technology in which a carbonaceous fuel is burned indirectly with air into a sequestration-ready CO_2 stream at a low cost. The CLC process occurs in two interconnected reactors: a fuel reactor

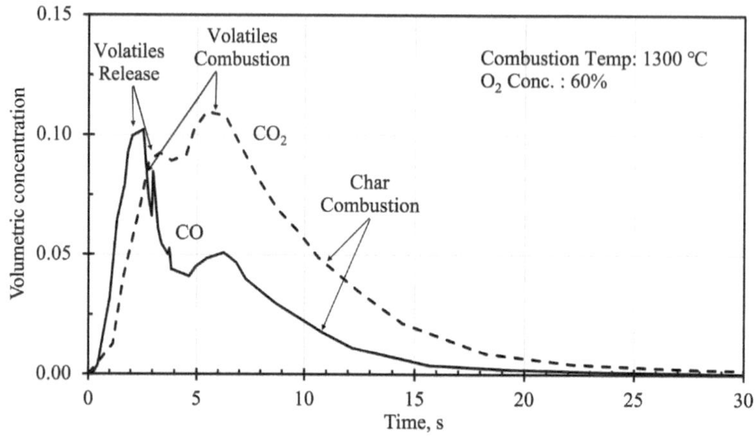

Figure 9.22 Dynamic release of CO and CO_2 from pulverized coal combustion in a micro-fluidized bed reactor.

and an air reactor. Oxygen carriers (typically metal oxides) are reduced in the fuel reactor to provide the oxygen for combustion; then, the reduced metal is transferred to the air reactor to be oxidized by air; the reoxidized oxygen carrier is recycled back to the fuel reactor to continue the reduction–oxidation cycle.

The selection of the oxygen carrier is an important aspect of determining the overall performance of the chemical looping combustion, which requires a better understanding of the kinetics of heterogeneous reduction–oxidation (redox) reactions and the time-dependent variations of physicochemical properties of the oxygen carrier particles. Having recognized the limitations of other microthermal analyzers such as TGA and fixed-bed reactor, Li et al. [98] and Yan et al. [94] used the MFB-TGA as an effective tool to investigate redox kinetics as well as attrition, sintering, and agglomeration mechanism of oxygen carrier in the condition close to a real CLC's operating environment. Fig. 9.23A schematically illustrates that the reduction–oxidation cycle performance can be readily experimented with by quickly changing the MFB reactor environment from reducing to neutral to oxidizing and back to reducing. Fig. 9.23B shows the experimental results of 444 redox cycles using manganese as the oxygen carrier at 900°C in the FB-TGA. The results identify that the redox process experiences the following stages: (1) the oxygen release of Mn_2O_3, (2) the mutual conversion between Mn_3O_4 and MnO, (3) aggravated attrition due to the cyclic thermal stress on oxygen carrier particles, and (4) the deep reduction. At the end of the 444th cycle, the sintered and agglomerated particles accumulated during the cyclic operations cause the bed to completely defluidize. The typical cycle performance for each redox reaction stage is shown in Fig. 9.23C. The research confirms that the MFB-TGA is a convenient and efficient method to investigate oxygen carriers close to the real operating environment. Therefore, the MFB-TGA is a powerful and efficient tool for screening

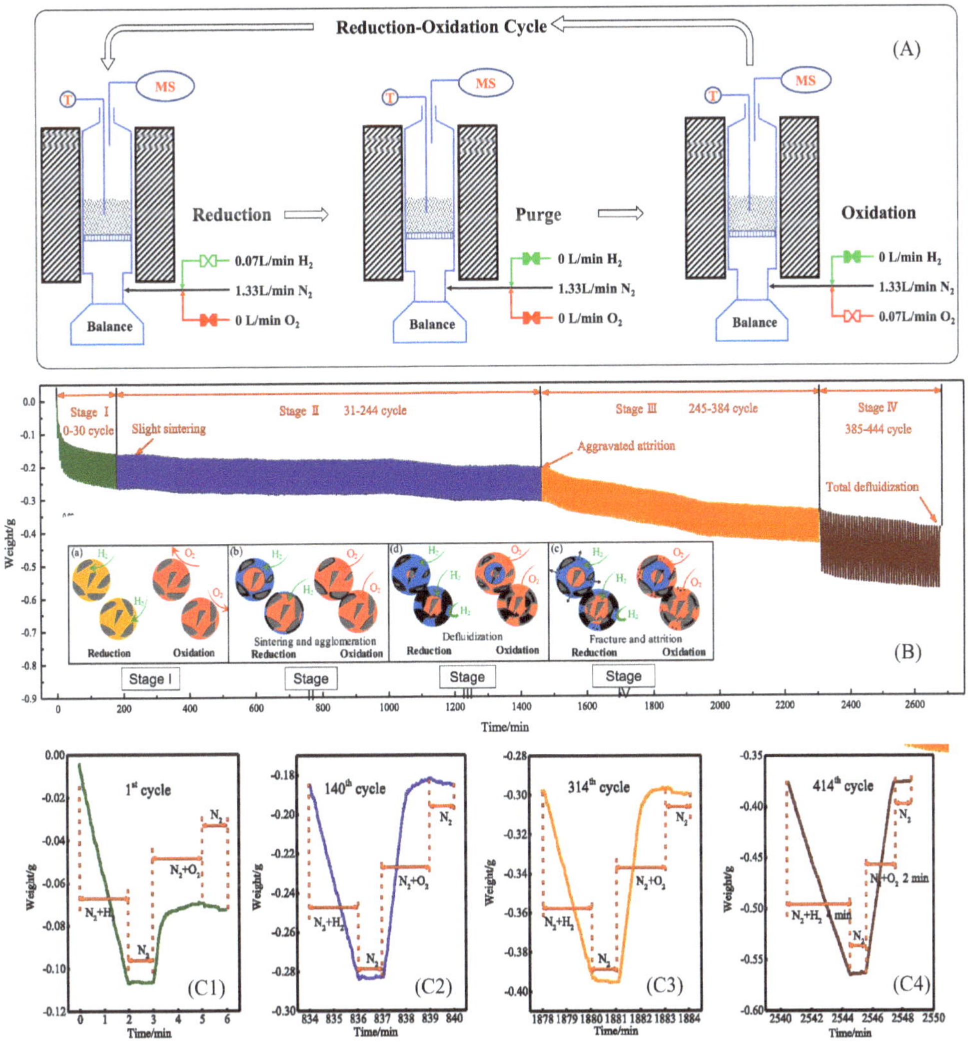

Figure 9.23 Results of redox reaction cycles in a micro-fluidized bed thermogravimetric analyzer (oxidation time: 2 min; reduction time: 2 min for 1–384 cycle and 4 min for 385–444 cycle; purging time: 1 min). (A) Illustration of cycle operation, (B) results of redox reaction, and (C1–C4) cycles of different stages.

oxygen carriers, understanding the redox cycle performance, and generating essential data to assist in operating industrial CLC processes.

9.8.4 In/ex situ chart combustion

The combustion process of solid fuels (e.g., coal and biomass) usually goes through two steps: the rapid pyrolysis and the subsequent combustion of the char and volatiles produced

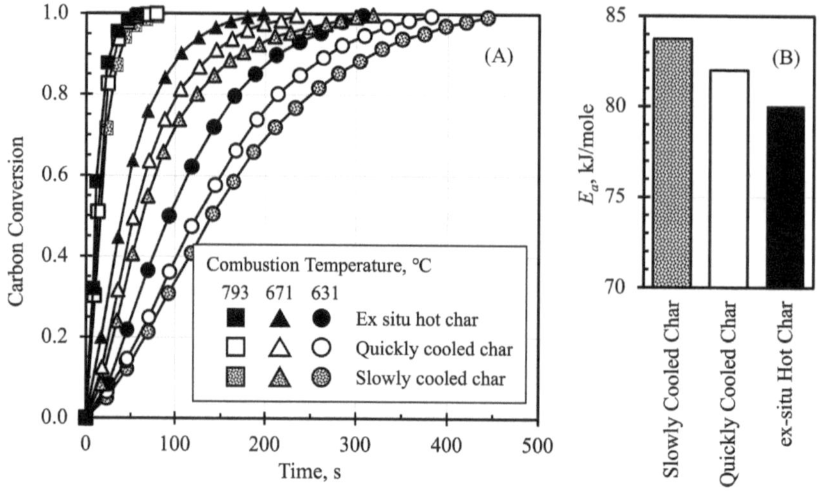

Figure 9.24 (A) Conversion and (B) activation energy of char combustion determined by micro-fluidized bed reactors.

by pyrolysis. The burnout time of char accounts for 70%–90% of the entire combustion time [99]. Therefore, the study of char combustion kinetics is of great significance to the design and operation of the actual combustion process. Fig. 9.24 shows that the combustion kinetics of ex situ hot char prepared by a dual-bed micro-fluidized bed reaction analyzer is quite different from those of slowly and quickly cooled char particles. It suggests that the char preparation method dramatically affects the char combustion performance [90].

Li et al. [86] employed a vacuum sampling system to collect char particles to understand the evolution characteristics of char in the combustion process. Fig. 9.25A–E shows the surface morphologies of the sampled char at different combustion times. It can be seen that the macro-structured pores are developing during the combustion process until the combustion reaches a moment, at which the macropores start to break down and collapse on the surface. Fig. 9.25F shows that the char combustion rate increases with temperature for all char particles sampled at different times. The activation energy of char combustion increases with the combustion time before 5.5 minutes and then tends to be stable.

9.9 Reduction

A reduction reaction is a chemical reaction in which a reactant gains one or more electrons. A reduction reaction always occurs in conjunction with an oxidation reaction to form the so-called redox reaction. To date, micro-fluidized bed reactors have been successfully used for several reduction reactions, and we will discuss some of these reactions in the next section.

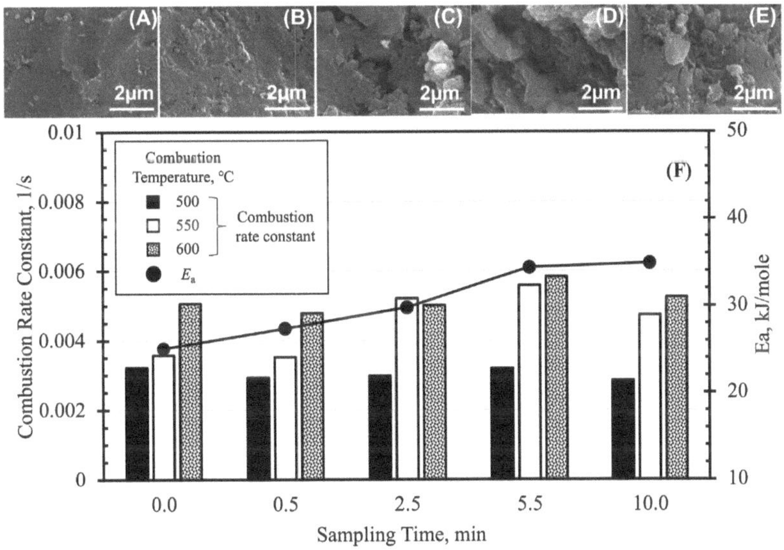

Figure 9.25 Evolution of surface morphologies of char at (A) 0 min, (B) 0.5 min, (C) 2.5 min, (D) 5.5 min, and (E) 10 min and (F) the combustion reaction rate constant during combustion process in a micro-fluidized bed.

9.9.1 Iron ore reduction

As one of the most competitive ironmaking processes, direct reduction of iron (DRI) ore by reducing gases (e.g., CO, H_2, and CH_4) in fluidized beds can produce sponge iron. The newly emerging DRI process significantly simplifies the ironmaking process because it eliminates the agglomeration and coke production steps involved in the traditional ironmaking process. In recent years, a great deal of effort has been devoted to understanding the reduction kinetics of iron oxides using micro-fluidized beds [100–108]. The results indicate that the overall iron reduction completes via multiple reactions. Table 9.1 summarizes the results of the estimated multistep kinetics. According to the reported results, the activation energy of reducing iron ore to metallic iron varies from 14 to 110 kJ/mole, which is generally lower than the data obtained by TGA and fixed-bed reactors [105]. This once again confirms that MFBRA is an effective thermal reaction analyzer to investigate the mechanism and kinetics of reactions.

Micro-fluidized beds have also been used to investigate reductions of other metal oxides, including NiO [109] and CuO [35]. Li et al. [109] employed a micro-fluidized bed reactor to evaluate a novel concept that a two-stage fluidized bed reactor is used to produce high-purity Ni powder. In the experiments, N_2 was used as the fluidizing gas to purge the reactor for 15 minutes, and then, the fluidizing gas was switched to a reactant gas (50 vol% H_2 in N_2) to initiate the prereduction of NiO at a lower temperature between 360°C and 400°C for approximately 15 minutes. Following the prereduction process, the reactor was quickly heated to 500°C or 600°C, under which the NiO was

Table 9.1 Summary of direct reduction of iron ore in micro-fluidized bed reactors.

Sample material	Temperature (°C)	Reducing gas	E_a (kJ/mole)
Brazilian hematite [100]	700–850	50%CO-N_2	$Fe_2O_3 \rightarrow Fe_3O_4$: 28.7 $Fe_3O_4 \rightarrow FeO$: 40.2 $FeOx \rightarrow Fe$: 47.1
Brazilian hematite [101]	700–850	CO	Overall reaction: 29.7–109.6 $Fe_2O_3 \rightarrow FeO$: 83.6 $FeO \rightarrow Fe$: 80.4
α-Fe_2O_3 [106]	750–950	(40–100)%CO-Ar	$Fe_2O_3 \rightarrow Fe_3O_4$: 30.6–53.0 $Fe_3O_4 \rightarrow FeO$: 52.4–80.8 $FeO \rightarrow Fe$: 45.7–92.1
Fe_2O_3 [103]	700–850	CO	76.8
Natural hematite ore [105]	500–600	20%CO-CO_2	48.7
Fe_3O_4 [107]	550–800	60%CO-Ar	$Fe_4O_3 \rightarrow FeO$: 600°C–800°C: 35.0 $FeO \rightarrow Fe$: 600°C–700°C: 42.8 760°C–800°C: 14.3
Australian iron ore [108]	700–950	CO	26.3

further reduced for 2 minutes. For these experiments, the use of MFBRA enabled quick changes in the reacting gas and reactor temperature. Based on the results of this study, they believed that the newly developed method was efficient to produce ultrafine Ni and might be extended to the synthesis of other nanoparticle materials.

9.9.2 Nitrogen oxide reduction by tar

In the development of circulating fluidized bed decoupling combustion technology, MFBRA was used to investigate the NO reduction characteristics of biomass char and tar [110]. The experiments demonstrate that both char and tar can reduce NO considerably and that their reaction activity and reduction efficiency increase with increasing temperature. Under the same conditions, the NO reduction efficiency is higher by tar than by char. By using acetic acid, toluene, phenol, naphthalene, and 1-hydroxy-naphthalene as the tar model compounds, the experiments further reveal that the aromatic compounds are more effective in reducing NO, especially at high temperatures. This research has led to the successful development of low-NOx decompiling combustion technology [111].

9.9.3 WO_3 reduction–sulfurization

Recently, the use of fluidized bed reactors for the synthesis of fullerene-like and other nanoparticles has been actively explored. Li et al. [112] conducted experiments to synthesize fullerene-like WS_2 nanoparticles through a WO_3 reduction–sulfurization reaction in a micro-fluidized bed reactor. A high yield of WS_2 nanoparticles was achieved because of enhanced mass transfer in the MFB reactor.

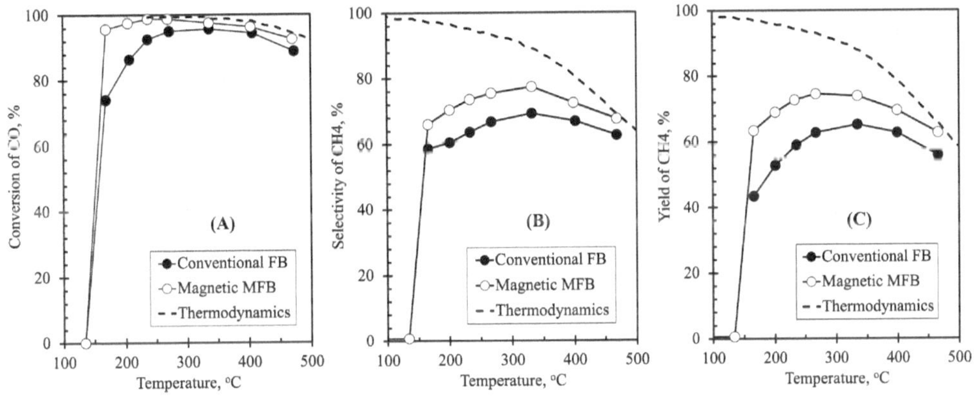

Figure 9.26 Intensification of syngas methanation by a magnetically assisted micro-fluidized bed reactor. (A) Conversion of CO, (B) selectivity of CH_4, and (C) yield of CH_4.

9.10 Other reactions

As a result of the compacted sizes, micro-fluidized bed reactors can be useful for many other applications. For example, syngas methanation was successfully conducted in a magnetic micro-fluidized bed of 20 mm in inner diameter [15]. In this work, an axial uniform magnetic field of 320 Oe in intensity was generated through four parallel solenoids around the micro-fluidized bed. The experiments show that the magnetic MFB can achieve higher conversion of syngas and selectivity to methane than the conventional fluidized bed (Fig. 9.26), suggesting that the application of magnetic field improves the fluidization quality of hard-to-fluidize catalysts.

References

[1] Haruna Adamu M, Zeng X, Zhang JL, et al. Property of drying, pyrolysis, gasification, and combustion tested by a micro fluidized bed reaction analyzer for adapting to the biomass two-stage gasification process. Fuel 2020;264(December 2019). Article 116827.

[2] Zeng X, Zhang J, Adamu MH, et al. Behavior and kinetics of drying, pyrolysis, gasification, and combustion tested by a micro fluidized bed reaction analyzer for the staged-gasification process. Energy Fuels 2020;34(2):2553–65.

[3] Vericella JJ, Baker SE, Stolaroff JK, et al. Encapsulated liquid sorbents for carbon dioxide capture. Nat Commun 2015;6:1–7.

[4] Finn JR, Galvin JE, Hornbostel K. CFD investigation of CO_2 absorption/desorption by a fluidized bed of micro-encapsulated solvents. Chem Eng Sci 2020;6. Article 100050.

[5] Li X, Wang L, Jia L, et al. Design and performance of oblique compact micro fluidized bed for low-concentration CO_2 capture. J Southeast Univ (Nat Sci Ed) 2016;46(4):770–5.

[6] Li X, Wang L, Jia L, et al. Numerical and experimental study of a novel compact micro fluidized beds reactor for CO_2 capture in HVAC. Energy Build 2017;135:128–36.

[7] Prajapati A, Renganathan T, Krishnaiah K. Kinetic studies of CO_2 capture using K_2CO_3/activated carbon in fluidized bed reactor. Energy Fuels 2016;30(12):10758–69.

[8] Amiri M, Shahhosseini S. Optimization of CO_2 capture from simulated flue gas using K_2CO_3/Al_2O_3 in a micro fluidized bed reactor. Energy Fuels 2018;32(7):7978–90.

[9] Yu J, Zeng X, Zhang G, et al. Kinetics and mechanism of direct reaction between CO_2 and Ca $(OH)_2$ in micro fluidized bed. Environ Sci Technol 2013;47(13):7514–20.
[10] Li Y, Li Z, Wang H, et al. CaO carbonation kinetics determined using micro-fluidized bed thermogravimetric analysis. Fuel 2020;264:116823.
[11] Pang F, Song F, Zhang Q, et al. Study on the influence of oxygen-containing groups on the performance of Ni/AC catalysts in methanol vapor-phase carbonylation. Chem Eng J 2016;293:129–38.
[12] Chen K, Zhao Y, Zhang W, et al. The intrinsic kinetics of methane steam reforming over a nickel-based catalyst in a micro fluidized bed reaction system. Int J Hydrog Energy 2020;45:1615–28.
[13] Gross B, Nace DM, Voltz SE. Application of a kinetic model for comparison of catalytic cracking in a fixed bed microreactor and a fluidized dense bed. Ind Eng Chem Process Des Dev 1974;13(3):199–203.
[14] Fu X, Cui X, Wei X, et al. Investigation of low and mild temperature for synthesis of high quality carbon nanotubes by chemical vapor deposition. Appl Surf Sci 2014;292:645–9.
[15] Li J, Zhou L, Zhu Q, et al. Enhanced methanation over aerogel NiCo/Al_2O_3 catalyst in a magnetic fluidized bed. Ind Eng Chem Res 2013;52:6647–54.
[16] Wang Q, Zhang C, Zhu Z, et al. Comparison study for the oxidative dehydrogenation of isopentenes to isoprene in fixed and fluidized beds. Catal Today 2016;276:78–84.
[17] Carnevali D, Guévremont O, Rigamonti MG, et al. Gas-phase fructose conversion to furfural in a microfluidized bed reactor. ACS Sustain Chem Eng 2018;6(4):5580–7.
[18] Kaarsholm M, Rafii B, Joensen F, et al. Kinetic modeling of methanol-to-olefin reaction over ZSM-5 in fluid bed. Ind Eng Chem Res 2010;49(1):29–38.
[19] Dias F, Dias J. Mathematical modelling of the solar—driven steam reforming of methanol for a solar thermochemical micro—fluidized bed reformer: thermal performance and thermochemical conversion. J Braz Soc Mech Sci Eng 2020;42(447):1–28.
[20] Hu D, Zeng X, Wang F, et al. Comparison of tar thermal cracking and catalytic reforming by char in a micro fluidized bed reaction analyzer. Fuel 2021;290(October 2020):120038.
[21] Yu J, Geng S, Liu J, et al. A multi-channel micro fluidized bed and its applications: Chinese Patent No. CN105921082B; 2016.
[22] Feiqiang G, Yuan L, Chenglong G, et al. Influence of AAEM on kinetic characteristics of rice husk pyrolysis in micro-fluidized bed reactor. CIESC J 2017;68(10):3795–804.
[23] Guo F, Peng K, Zhao X, et al. Influence of impregnated copper and zinc on the pyrolysis of rice husk in a micro-fluidized bed reactor: Characterization and kinetics. Int J Hydrog Energy 2018;3:21256–68.
[24] Liu Y, Wang Y, Guo F, et al. Characterization of the gas releasing behaviors of catalytic pyrolysis of rice husk using potassium over a micro-fluidized bed reactor. Energy Convers Manage 2017;136:395–403.
[25] Liu Y, Guo F, Li X, et al. Catalytic effect of iron and nickel on gas formation from fast biomass pyrolysis in a micro fluidized bed reactor: a kinetic study. Energy Fuels 2017;31(11):12278–87.
[26] Kitsuka T, Bayarsaikhan B, Sonoyama N, et al. Behavior of inherent metallic species as a crucial factor for kinetics of steam gasification of char from coal pyrolysis. Energy Fuels 2007;21(2):387–94.
[27] Kajita M, Kimura T, Norinaga K, et al. Catalytic and noncatalytic mechanisms in steam gasification of char from the pyrolysis of biomass. Energy Fuels 2010;24(1):108–16.
[28] Perander M, DeMartini N, Brink A, et al. Catalytic effect of Ca and K on CO_2 gasification of spruce wood char. Fuel 2015;150:464–72.
[29] Zhang Y, Yao M, Gao S, et al. Reactivity and kinetics for steam gasification of petroleum coke blended with black liquor in a micro fluidized bed. Appl Energy 2015;160:820–8.
[30] Abbas HF, Wan, Daud WMA. Hydrogen production by methane decomposition: a review. Int J Hydrog Energy 2010;35(3):1160–90.
[31] Qian JX, Chen TW, Enakonda LR, et al. Methane decomposition to produce COx-free hydrogen and nano-carbon over metal catalysts: a review. Int J Hydrog Energy 2020;45(15):7981–8001.
[32] Geng S, Han Z, Hu Y, et al. Methane decomposition kinetics over Fe_2O_3 catalyst in micro fluidized bed reaction analyzer. Ind Eng Chem Res 2018;57(25):8413–23.
[33] Gai C, Dong Y, Fan P, et al. Kinetic study on thermal decomposition of toluene in a micro fluidized bed reactor. Energy Convers Manage 2015;106:721–7.

[34] Georgieva V, Vlaev L, Gyurova K. Non-isothermal degradation kinetics of $CaCO_3$ from different origin. J Chem 2013;2013.
[35] Yu J, Yue J, Liu Z, et al. Kinetics and mechanism of solid reactions in a micro fluidized bed reactor. AIChE J 2010;56(11):2905–12.
[36] Yin J, Fang Y, Zhu X, et al. Kinetic study on $CaCO_3$ decomposition via MFB-IR thermal analyzer. J Eng Thermophys 2011;35(6):1216–20.
[37] Jiang W, Hao W, Liu X, et al. Characteristic and kinetics of light calcination of magnesite in micro fluidized bed reaction analyzer. CIESC J 2019;70(8):2928–37.
[38] Funazukuri T, Hudgins RR, Silveston PL. Product distribution in pyrolysis of cellulose in a micro-fluidized bed. J Anal Appl Pyrolysis 1986;9(2):139–58.
[39] Jia L, Le Brech Y, Mauviel G, et al. Online analysis of biomass pyrolysis tar by photoionization mass spectrometry. Energy Fuels 2016;30(3):1555–63.
[40] Guo F, Dong Y, Lv Z, et al. Pyrolysis kinetics of biomass (herb residue) under isothermal condition in a micro fluidized bed. Energy Convers Manage 2015;93:367–76.
[41] Yu J, Zhu J, Guo F, et al. Reaction kinetics and mechanism of biomass pyrolysis in a micro fluidized bed reactor. J Fuel Chem Technol 2010;38(6):666–72.
[42] Bin W, Yuping D, Yebing MAO, et al. Fast pyrolysis behavior of fungus residues in a fluidized bed reactor. Chem Ind Eng Prog 2017;36(3):1113–19.
[43] Yu J, Yao C, Zeng X, et al. Biomass pyrolysis in a micro-fluidized bed reactor: characterization and kinetics. Chem Eng J 2011;168(2):839–47.
[44] Tyler RJ. Flash pyrolysis of coals. 1. Devolatilization of Victorian brown coal in a small fluidized-bed reactor. Fuel 1979;58:680–6.
[45] Tyler RJ. Flash pyrolysis of coals. Devolatilitation of bituminous coals in a small fluidized-bed reactor. Fuel 1980;59:218–26.
[46] Samih S, Chaouki J. Coal pyrolysis and gasification in a fluidized bed thermogravimetric analyzer. Can J Chem Eng 2018;96:2144–54.
[47] Samih S, Chaouki J. Catalytic ash free coal gasification in a fluidized bed thermogravimetric analyzer. Powder Technol 2017;316:551–9.
[48] Zhang ZG, Scott DS, Silveston PL. Steady-state gasification of an alberta subbituminous coal in a microfluidized bed. Energy Fuels 1994;8(3):637–42.
[49] Zhang Y, Zhao M, Linghu R, et al. Comparative kinetics of coal and oil shale pyrolysis in a micro fluidized bed reaction analyzer. Carbon Resour Convers 2019;2(3):217–24.
[50] Cai L, Liu W, Yu J, et al. Comparative study on coal pyrolysis via programmed and isothermal heating. Coal Convers 2012;35(3):6–14.
[51] Jayaraman K, Gokalp I, Bostyn S. High ash coal pyrolysis at different heating rates to analyze its char structure, kinetics and evolved species. J Anal Appl Pyrolysis 2015;113:426–33.
[52] Wang M, Li Z, Huang W, et al. Coal pyrolysis characteristics by TG-MS and its late gas generation potential. Fuel 2015;156:243–53.
[53] Scaccia S. TG-FTIR and kinetics of devolatilization of Sulcis coal. J Anal Appl Pyrolysis 2013;104:95–102.
[54] Geng C, Li S, Yue C, et al. Pyrolysis characteristics of bituminous coal. J Energy Inst 2016;89(4):725–30.
[55] Zhang S, Zhu F, Bai C, et al. Thermal behavior and kinetics of the pyrolysis of the coal used in the COREX process. J Anal Appl Pyrolysis 2013;104:660–6.
[56] Pan L, Dai F, Li G, et al. A TGA/DTA-MS investigation to the influence of process conditions on the pyrolysis of Jimsar oil shale. Energy 2015;86:749–57.
[57] Al-Ayed OS, Matouq M, Anbar Z, et al. Oil shale pyrolysis kinetics and variable activation energy principle. Appl Energy 2010;87(4):1269–72.
[58] Dai C, Ma S, Liu X, et al. Study on the pyrolysis kinetics of blended coal in the fluidized-bed reactor. Procedia Eng 2015;102:1736–41.
[59] Xue Y, Kelkar A, Bai X. Catalytic co-pyrolysis of biomass and polyethylene in a tandem micropyrolyzer. Fuel 2015;166:227–36.
[60] Mao Y, Dong L, Dong Y, et al. Fast co-pyrolysis of biomass and lignite in a micro fluidized bed reactor analyzer. Bioresour Technol 2015;181:155–62.

[61] Guo F, Liu Y, Wang Y, et al. Characterization and kinetics for co-pyrolysis of Zhundong lignite and pine sawdust in a micro fluidized bed. Energy Fuels 2017;31(8):8235–44.
[62] Wang F, Zeng X, Wang Y, et al. Non-isothermal coal char gasification with CO_2 in a micro fluidized bed reaction analyzer and a thermogravimetric analyzer. Fuel 2016;164:403–9.
[63] Wang F, Zeng X, Wang Y, et al. Comparison of non-isothermal coal char gasification in micro fluidized bed and thermogravimetric analyzer. CIESC J 2015;66(5):1716–22.
[64] Ji Y, Zeng X, Yu J, et al. Steam gasification characteristics of coal char in micro-fluidized bed reaction analyzer. CIESC J 2014;65(9):3447–56.
[65] Zeng X, Wang F, Han J, et al. Micro fluidized bed reaction analysis and its coal char gasification kinetics. CIESC J 2013;64(1):289–96.
[66] Wang F, Zeng X, Han J, et al. Comparison of char gasification kinetics studied by micro fluidized bed and by thermogravimetric analyzer. J Fuel Chem Technol 2013;41(4):407–13.
[67] Guo Y, Zhao Y, Gao D, et al. Kinetics of steam gasification of in-situ chars in a micro fluidized bed. Int J Hydrog Energy 2016;41(34):15187–98.
[68] Li Y, Wang H, Li W, et al. CO_2 gasification of a lignite char in micro fluidized bed thermogravimetric analysis for chemical looping combustion and chemical looping with oxygen uncoupling (CLC/CLOU). Energy Fuels 2019;33:449–59.
[69] Wang F, Zeng X, Shao R, et al. Isothermal gasification of in situ/ex situ coal char with CO_2 in a micro fluidized bed reaction analyzer. Energy Fuels 2015;29(8):4795–802.
[70] Wang F, Zeng X, Wang Y, et al. Characterization of coal char gasification with steam in a micro-fluidized bed reaction analyzer. Fuel Process Technol 2016;141:2–8.
[71] Zeng X, Wang F, Wang Y, et al. Characterization of char gasification in a micro fluidized bed reaction analyzer. Energy Fuels 2014;28(3):1838–45.
[72] Wang F, Zen X, Wang Y, et al. Investigation on in/ex-situ coal char gasification kinetics in a micro fluidized bed reactor. J Fuel Chem Technol 2015;43(4):393–401.
[73] Wang F, Zeng X, Yu J, et al. Char gasification kinetics in micro fluidized bed reaction analyzer. J Shenyang Univ Chem Technol 2014;28(3):213–19.
[74] Gao W, Farahani MR, Rezaei M, et al. Kinetic modeling of biomass gasification in a micro fluidized bed. Energy Sources Part A: Recovery, Utilization Environ Eff 2017;39(7):643–8.
[75] Zhang Y, Sun G, Gao S, et al. Regeneration kinetics of spent FCC catalyst via coke gasification in a micro fluidized bed. Proc Eng 2015;102:1758–1765. Elsevier Ltd. https://doi.org/10.1016/j.proeng.2015.01.312.
[76] Zhang Y, Yao M, Sun G, et al. Characteristics and kinetics of coked catalyst regeneration via steam gasification in a micro fluidized bed. Ind Eng Chem Res 2014;53(15):6316–24.
[77] Zhang Y, Sun G, Gao S, et al. Characteristics and kinetics of steam-gasification regeneration for coked FCC catalyst in a micro-fluidized bed reactor. ACTA Petrolei Sin (Pet Process Sect) 2014;30(6):1043–51.
[78] Molino A, Chianese S, Musmarra D. Biomass gasification technology: the state of the art overview. J Energy Chem 2016;25(1):10–25.
[79] Alauddin ZABZ, Lahijani P, Mohammadi M, et al. Gasification of lignocellulosic biomass in fluidized beds for renewable energy development: a review. Renew Sustain Energy Rev 2010;14(9):2852–62.
[80] Asadullah M. Biomass gasification gas cleaning for downstream applications: a comparative critical review. Renew Sustain Energy Rev 2014;40:118–32.
[81] Watson J, Zhang Y, Si B, et al. Gasification of biowaste: a critical review and outlooks. Renew Sustain Energy Rev 2018;83(October 2017):1–17.
[82] Sansaniwal SK, Rosen MA, Tyagi SK. Global challenges in the sustainable development of biomass gasification: an overview. Renew Sustain Energy Rev 2017.
[83] Widjaya ER, Chen G, Bowtell L, et al. Gasification of non-woody biomass: a literature review. Renew Sustain Energy Rev 2018;89(September 2016):184–93.
[84] Sansaniwal SK, Pal K, Rosen MA, et al. Recent advances in the development of biomass gasification technology: a comprehensive review. Renew Sustain Energy Rev 2017.
[85] Kumar A, Jones D, Hanna M. Thermochemical biomass gasification: a review of the current status of the technology. Energies 2009;2:556–81.
[86] Li Z, Zou R, Xu Y, et al. Kinetic study on continuous sampling of coal char from a micro fluidized bed. ACS Omega 2021;6(13):9086–94.

[87] Yu J, Li Q, Duan Z, et al. Isothermal differential characteristics of the reaction in micro fluidized bed. Scientia Sinica Chimica 2011;41(1):152–60.

[88] Liu W, Yu J, Zhang J, et al. Kinetic study of reaction of porous solids. Scientia Sinica Chimica 2012;42(8):1210–16.

[89] Zhang W, Wang P, Sun S, et al. Effects of demineralization methods on structure and reactivity of Zhundong sub-bituminous coal. CIESC J 2017;68(8):3291–300.

[90] Fang Y, Luo G, Chen C, et al. Combustion kinetics of In-situ char and cold char in micro-fluidized bed. J Combust Sci Technol 2016;22(2):148–54.

[91] Guo L, Cheng W, Guo Z. Application of micro-fluidized bed in kinetics analysis of coal oxygen combustion at high temperature. Jiangxi Metall 2019;39(2):14–19.

[92] Su W, Liu Q, Zhong W, et al. Oxy-coal combustion characteristics in micro-fluidized bed reactor. J Southeast Univ (Nat Sci Ed) 2020;50(5):896–903.

[93] Zheng M, Shen L, Feng X. Study on chemical-looping combustion of coal with $CaSO_4$ oxygen carrier assisted by CaO addition. J Fuel Chem Technol 2014;42(4):399–407.

[94] Yan J, Shen T, Wang P, et al. Redox performance of manganese ore in a fluidized bed thermogravimetric analyzer for chemical looping combustion. Fuel 2021;295(February):120564.

[95] Li Y, Li Z, Cai N. Support information for micro-fluidized bed thermogravimetry combined with mass spectrometry. Energy Fuel 2020;34:11186–93.

[96] Li J, Xu G, Yang L, et al. A NOx-suppressed smokeless coal combustion technique and stove: Chinese patent ZL95102081.1. China; 1995.

[97] Guo Y, Zhao Y, Meng S, et al. Development of a multistage in situ reaction analyzer based on a micro fluidized bed and its suitability for rapid gas – solid reactions. Energy Fuels 2016;30:6021–33.

[98] Li Y, Li Z, Cai N. Micro-fluidized bed thermogravimetry combined with mass spectrometry (MFB-TG-MS) for redox kinetic study of oxygen carrier. Energy Fuel 2020;34:11186–93.

[99] Zhang Z, Li Z, Cai N. Comparison and analysis of different models of pulverized coal char combustion. J Combust Sci Technol 2014;20(5):393–400.

[100] Chen H, Zheng Z, Shi W. Investigation on the kinetics of iron ore fines reduction by CO in a micro-fluidized bed. Procedia Eng 2015;102:1726–35.

[101] Chen H, Zheng Z, Chen Z, et al. Multistep reduction kinetics of fine iron ore with carbon monoxide in a micro fluidized bed reaction analyzer. Metall Mater Trans B 2017;48B:841–52.

[102] Lin Y, Guo Z, Tang H. Reduction behavior with CO under micro-fluidized bed conditions. J Iron Steel Res Int 2013;20(2):8–13.

[103] Lin Y, Guo Z, Tang H, et al. Kinetics of reduction reaction in micro-fluidized bed. J Iron Steel Res Int 2012;19(6):6–8.

[104] Lin Y, Guo Z, Tang H. Effect of atmosphere on degree of reduction in micro-fluidized bed. J Iron Steel Res 2014;26(4):18–23.

[105] Yu J, Han Y, Li Y, et al. Mechanism and kinetics of the reduction of hematite to magnetite with $CO-CO_2$ in a micro-fluidized bed. Minerals 2017;7(11):1–12.

[106] Chen H, Zheng Z, Chen Z, et al. Reduction of hematite (Fe_2O_3) to metallic iron (Fe) by CO in a micro fluidized bed reaction analyzer: a multistep kinetics study. Powder Technol 2017;316:410–20.

[107] He K, Zheng Z, Chen Z. Multistep reduction kinetics of Fe_3O_4 to Fe with CO in a micro fluidized bed reaction analyzer. Powder Technol 2020;360:1227–36.

[108] Wang Q, Shao J, Lin Y, et al. An experimental study on the kinetics of iron ore fine reduced by CO in micro fluidized bed. J Iron Steel Res 2012;24(4):6–9.

[109] Li J, Liu X, Zhou L, et al. A two-stage reduction process for the production of high-purity ultrafine Ni particles in a micro-fluidized bed reactor. Particuology 2015;19:27–34.

[110] Song Y, Wang Y, Yang W, et al. Reduction of NO over biomass tar in micro-fluidized bed. Fuel Process Technol 2014;118:270–7.

[111] Han Z, Zeng X, Yao C, et al. Comparison of direct combustion in a circulating fluidized bed system and decoupling combustion in a dual fluidized bed system for distilled spirit lees. Energy Fuels 2016;30(3):1693–700.

[112] Li J, Ma T, Zhou L, et al. Synthesis of fullerene-like WS2 nanoparticles in a particulately fluidized bed: kinetics and reaction phase diagram. Ind Eng Chem Res 2014;53(2):592–600.

CHAPTER 10

Applications of micro-fluidized bed reactor in industrial process development

The characteristics, analytical methods, and applications of micro-fluidized bed reaction analyzers have been introduced in detail in the previous chapters. In this chapter, we will take low-NOx decoupling combustion, low tar two-stage gasification, and light calcination of magnesite as examples to demonstrate how the use of micro-fluidized beds reaction analysis can accelerate its research, development, and commercialization processes.

10.1 Advanced combustion with low-NOx emissions

10.1.1 Low-NOx combustion technology

Thermochemical fuel conversions, such as combustion, gasification, and cracking, involve many interconnected or interactive individual reactions. In each of these conversion processes, not a single but a series of reactions constituting a complex network of chemical reactions bring about the observed explicit chemical changes. For example, pyrolysis is the first step of fuel combustion, followed by the subsequent combustions of the generated char and volatiles. Apart from the above basic reactions involved in the combustion process, some interactions between/among intermediates or products co-occur, significantly impacting the composition of flue gas or the pollutants emission. Typically, the NOx generation during combustion is a complex process because, on the one hand, NOx can be produced by volatiles and char combustions and, on the other hand, NOx can also be reduced to nitrogen by these same reactants simultaneously, which means that there is a competition between fuel-N oxidation and NOx reduction. In a conventional combustion process, such as fixed bed, fluidized bed, and circulating fluidized bed combustion, the basic reactions and interactions in relation to NOx generation take place in the same reaction space; thus, it is difficult to maximize the NOx reduction only through flue gas recirculation and air staging. To overcome this issue, it is highly desirable to separate the related reactions to provide optimal conditions for each reaction independently. This idea of reaction control is called "reaction decoupling" [1–3]. Fig. 10.1 schematically illustrates the concept of combustion reaction decoupling.

Compared to traditional combustion, combustion decoupling is featured by separating and undertaking pyrolysis and combustion in different reactors or zones of a

Micro Fluidization
DOI: https://doi.org/10.1016/B978-0-443-18718-6.00010-3
© 2023 Chemical Industry Press Co., Ltd.
Published by Elsevier Inc., under an exclusive license with Chemical Industry Press Co., Ltd.

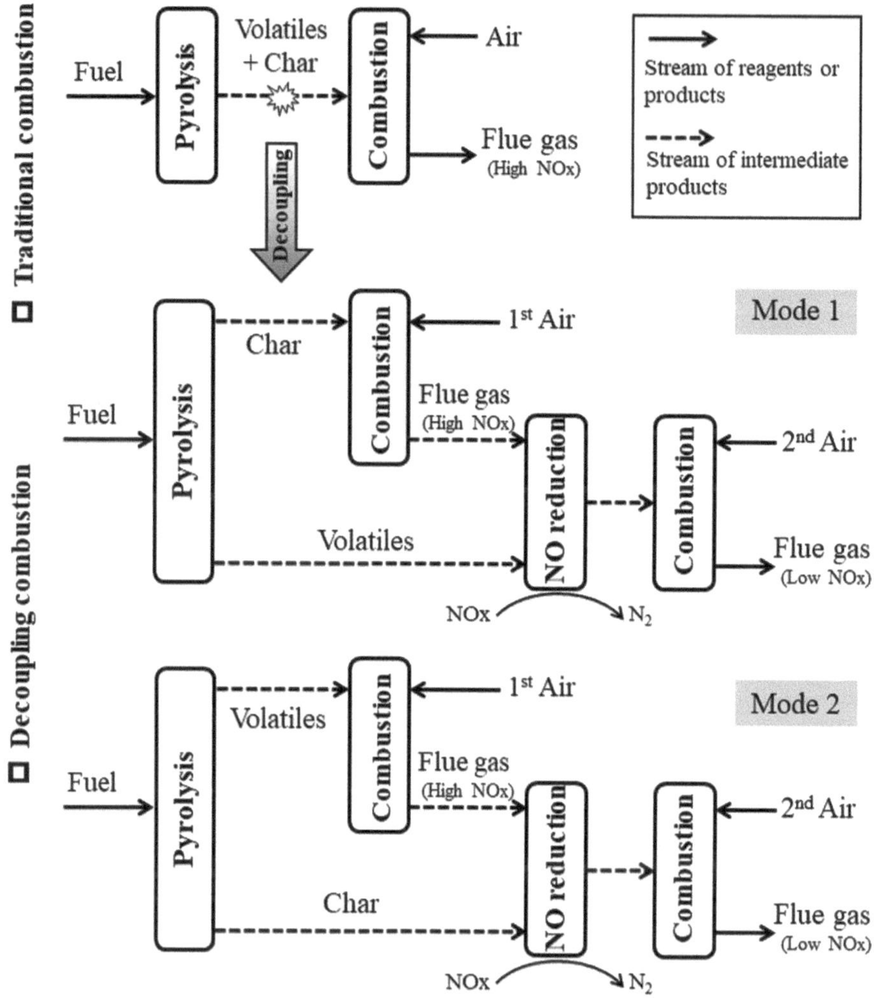

Figure 10.1 Process illustration of combustion decoupling.

reactor so that fuel devolatilization and combustions of volatiles and char can be controlled in a desired sequence and condition. In this way, the NOx generated from the combustion of one pyrolysis product (i.e., char or volatiles) can be explicitly and efficiently reduced by the other products under optimal conditions, making low-NOx combustion achievable with two types of decoupling combustion mode, as shown in Fig. 10.1. In each mode, the NOx reduction efficiency is determined by the reactivities of volatiles and char. Although both can reduce NOx emissions [4–6], it is necessary to comprehensively analyze the reduction characteristics of NOx of these two pyrolysis products to optimize NOx reduction. For this purpose, MFB provides a powerful tool to help find the optimal decoupling mode and operation conditions useful for developing and designing the decoupling combustion systems.

10.1.2 NOx reduction by pyrolysis products

In an MFBRA system, the solid or liquid samples are fed quickly to the fluidized bed reactor held at a predetermined temperature by a compressed gas pulse. Due to the extraordinarily high heat transfer rate and uniformity of bed temperature distribution, the samples are almost instantaneously heated to the bed temperature with a heating rate of as high as 10^{4}°C/s [7]. Besides, MFB features a close-to-plug flow of gas and negligible mass transfer between gas and particle external surface, allowing the close-to-intrinsic reaction characteristic to be acquired [8,9]. Benefiting from the features of online feeding, isothermal reaction, and close-to-plug flow of gas, MFBRA has been used to reveal the characteristics of NOx generation and reduction that occur in actual industrial processes.

Pyrolysis-produced char (simply called pyrolysis char in this chapter) has been widely tested as an effective heterogeneous reducing reagent for NOx reduction. It is reported that the NO−char reaction plays a vital role in suppressing the conversion of fuel-N into NO in fuel combustions [10,11]. The NO reduction activity of char has been correlated to the specific surface area and porosity of the char particles. The pyrolysis-produced tar (simply called pyrolysis tar in this chapter) is generally composed of many hydrocarbons, including phenols, BTX, and other compounds of straight chains, aromatic rings, and C=O groups, but there is almost no study on the NO reduction by pyrolysis tar. Luo et al. [12−14] selected phenol, benzene, toluene, and styrene as tar model compounds to investigate their NO reduction capabilities. Their results show that the tar components are more efficient in increasing NO reduction than low-molecular reducing gases and char. Thus, for realizing low-NOx combustion, the reaction sequence of char and tar in a combustor needs to be appropriately arranged based on their NO-reducing capacity. By utilizing micro-fluidized bed analyzers, one can determine and compare the characteristics of NO reduction by biomass-derived tar and char. The instantaneous and total NO reduction efficiencies, η and η_T, are defined as

$$\eta = \left(1 - \frac{[NO]_{out}}{[NO]_{in}}\right) \times 100 \tag{10.1}$$

$$\eta_T = \frac{\int_0^{t_{end}} \eta}{1 \times t_{end}} \times 100 \tag{10.2}$$

where $[NO]_{in}$ and $[NO]_{out}$ are the inlet and outlet NO concentrations (in ppm) at any reaction time, and t_{end} is the reaction end time.

Fig. 10.2 presents the resulting NO reduction efficiency (η) for char and tar reagents at different temperatures. The instantaneous NO reduction efficiency (η) at any reaction time is calculated by Eq. (10.1). The tar and char used in the experiments were prepared by pyrolyzing distilled spirit lees (DSL), a typical biomass with a high fuel-N content, at 1123K in an N_2 atmosphere. It shows that the char has a low reaction activity to reduce NO at 1023K, but the NO reduction increases significantly with increasing the reaction temperature above 1023K. For NO reduction, η is higher and depends more on temperature by tar than by char. Moreover, the required time to complete NO reduction reactions is much shorter, and the half-peak width of the instantaneous efficiency curve is narrower in Fig. 10.2B than in Fig. 10.2A. The total reaction time is shorter than 200 seconds for tar but about 1000 seconds for char, indicating that the performance of tar for NO reduction is better than that of char. Tar has a better NO reduction capability because it has numerous active low-carbon hydrocarbons decomposable to produce more reductive species and radicals at high temperatures [13,14]. In contrast, char is composed of stable aromatic carbon of low reactivity.

Fig. 10.3A and B compares the peak instantaneous NO reduction efficiency (η_P) and total NO reduction efficiency (η_T) estimated from data shown in Fig. 10.2. By their definitions, η_P and η_T represent the reactivity (activity) and capacity of the tested char or tar reagent for NO reduction, respectively. Fig. 10.3 shows that η_P increases almost linearly with reaction temperature for both reagents, but η_P achieved is higher by tar than by char at all the tested temperatures. It can also be seen that the variation

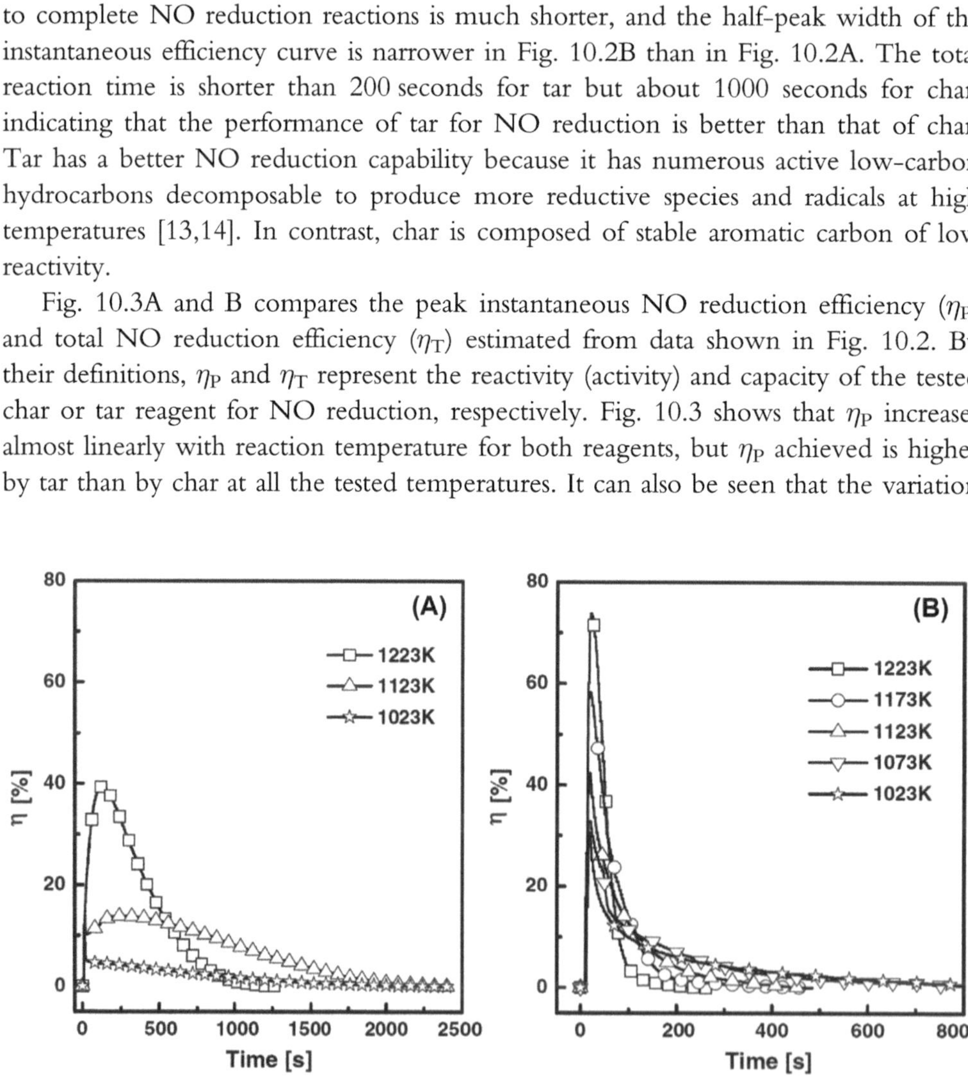

Figure 10.2 Instantaneous NO reduction curves at different temperatures for (A) char and (B) tar agents (fluidizing gas: 1000 mL/min NO-Ar with 1800 ppm NO) [4].

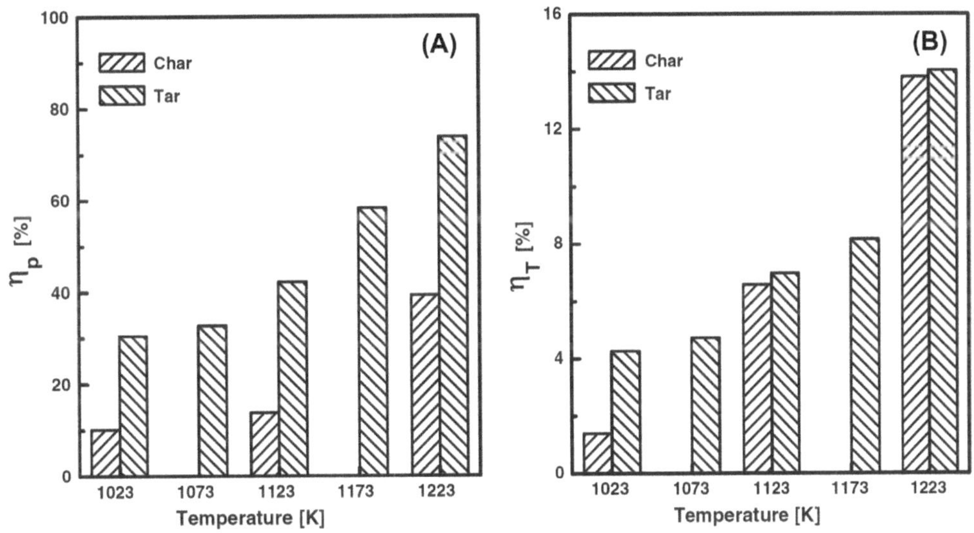

Figure 10.3 Comparison of (A) peak and (B) total NO reduction efficiencies realized by char and tar at different temperatures [4].

tendency in η_T for the char and tar reagents is the same as that in η_P. At all temperatures, η_T of tar is higher than that of char. The highest total NO reduction efficiency is 13.8% for char and 14.0% for tar. It is also important to note that the reaction time required to achieve its NO reduction efficiency is much longer for char than for tar.

The results presented above show that tar has higher reactivity and capacity for reducing NO than char, suggesting that the gas-phase homogeneous NO reduction with gaseous tar at high temperatures is more efficient than the heterogeneous gas (NO)—solid (char) reduction. The reactivity distinction is possibly related to the fact that tar releases more active gaseous species, including unsaturated hydrocarbons, fractured benzene rings, and OH^-, H^+, and CH radicals via thermal cracking. Besides, the interactive contacts between reactants are more intimate and extensive in gas—gas homogeneous reactions, contributing to tar's higher NO reduction efficiency. In contrast, gas—solid heterogeneous reactions occur only on the active sites presented on solid reactant (char) surfaces.

The research results above revealed by MFBRA demonstrate that the "decoupling mode 1" (Fig. 10.1) is more efficient in achieving low-NOx combustion. Under this guidance, a low-NOx decoupling combustion process is proposed and developed, and Fig. 10.4 shows the schematic flow diagram of the process. The process consists of two fluidized bed (FB) reactors: a FB pyrolyzer and a CFB combustor. Heat carrier particles (HCPs) are circulated between the two reactors to transfer the required heat from the combustor to the pyrolyzer for fuel drying and pyrolysis. Generally, the air is used as the pyrolyzer's fluidization gas to reduce operation costs. After pyrolysis, the

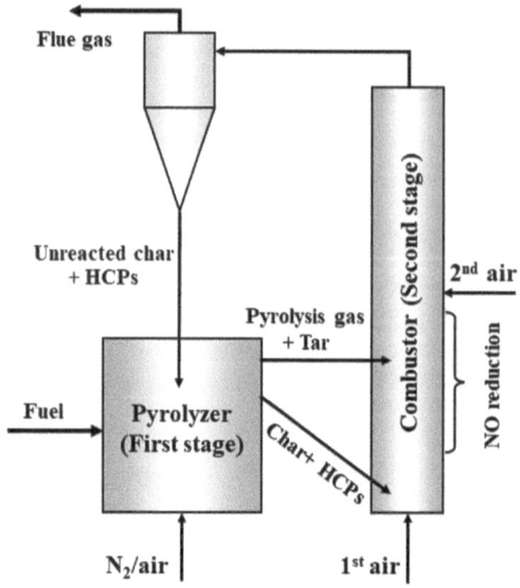

Figure 10.4 A process diagram for optimal decoupling combustion [3].

char is directed into the bottom of the CFB combustor; the tar and pyrolysis gas, as NOx-reducing reagents, are sent to the middle part of the combustor to reduce the NOx generated from char combustion. The unreacted tar, pyrolysis gas, and char are burnt out by the secondary air at the upper section of the combustor. In operations, the pyrolysis conditions are controlled preferably so that a desired amount of tar is yielded to maximize the NOx reduction.

Fig. 10.5 shows the effect of pyrolysis temperature on the combustion rate of 1.0–1.5 mm char particles under 900°C. Compared to the raw material (DSL), the combustion rates of char prepared under all pyrolysis conditions are low due to significantly reduced devolatilization; thus, char combustion is a more stable and controllable combustion process. When the conversion is below 0.5, the char combustion rates are approximately constant, and the pyrolysis temperature seems to have a negligible effect on the reaction rate. As the conversion exceeds 0.5, the combustion rates decrease sharply with conversion. In this stage, the char combustion rate, especially in the conversion range of 0.8–1.0, is much lower when char is prepared under higher temperatures (such as 800°C) than under lower temperatures. Thus, in terms of either maximizing NO reduction or promoting char combustion, a mild pyrolysis condition is more beneficial for decoupling combustion technology.

10.1.3 Pilot experiments and commercial application

For industrial processes involving fluidized bed fuel pyrolyzers, such as coal topping and pyrolysis with solid heat carriers [2,3,15–18], the pyrolyzers are generally

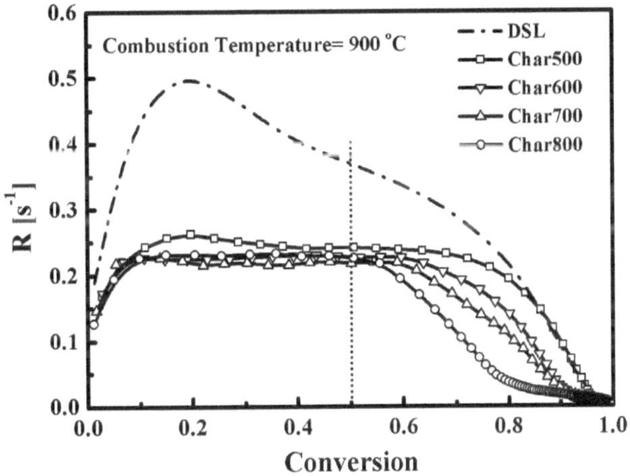

Figure 10.5 Effect of pyrolysis temperature on char combustion rate.

large-sized bubbling fluidized beds or moving beds with large sectional areas and reaction volumes. In this way, the residence time of fuel in the pyrolyzer can be extended to dozens of minutes or even several hours to ensure complete conversion. Using MFBRA, Yu et al. [19] and Zeng et al. [20] found that fuel pyrolysis with a particle size of 75–500 μm was a fast reaction under 600°C–900°C because the reaction was completed in shorter than 10 seconds. Hence, as long as the particle size distribution of fuel is adequately designed and controlled, the FB pyrolyzer does not need such a large reactor volume to provide a long residence time of fuel. In some cases, a transport bed or riser may also be a superior pyrolysis reactor.

Fig. 10.6 displays the process diagram of the decoupling combustion in a pilot plant with a bubbling fluidized bed as the pyrolyzer. The designed processing capacity is about 1000 t/a. The feedstock used is the distilled spirits lees (DSL), which has a fuel-nitrogen content of 3.4–4.0 wt.% on a dry ash-free basis. Meanwhile, DSL is also combusted in typical CFB operation mode in this plant to demonstrate the advantages of decoupling combustion. During stable operations, the riser combustor and BFB pyrolyzer temperatures are maintained at 800°C–950°C and 450°C–600°C, respectively. Fig. 10.7 shows NOx emissions in decoupling combustion and CFB combustion of DSL. For traditional CFB combustion, the NOx concentration in flue gas is 600–950 ppm, far higher than the emission standard; thus, the flue gas needs further denitration treatment before discharge. In the case of decoupling combustion, the NOx concentration in flue gas is about 100 ppm most of the time, thus proving that decoupling combustion can reduce 80%–90% NOx emission compared to conventional combustion.

In 2012, a commercial-scale demonstration plant with a capacity of 60,000 t/a was designed and built by IPE, CAS at Luzhou Laojiao Co. Ltd., for decoupling

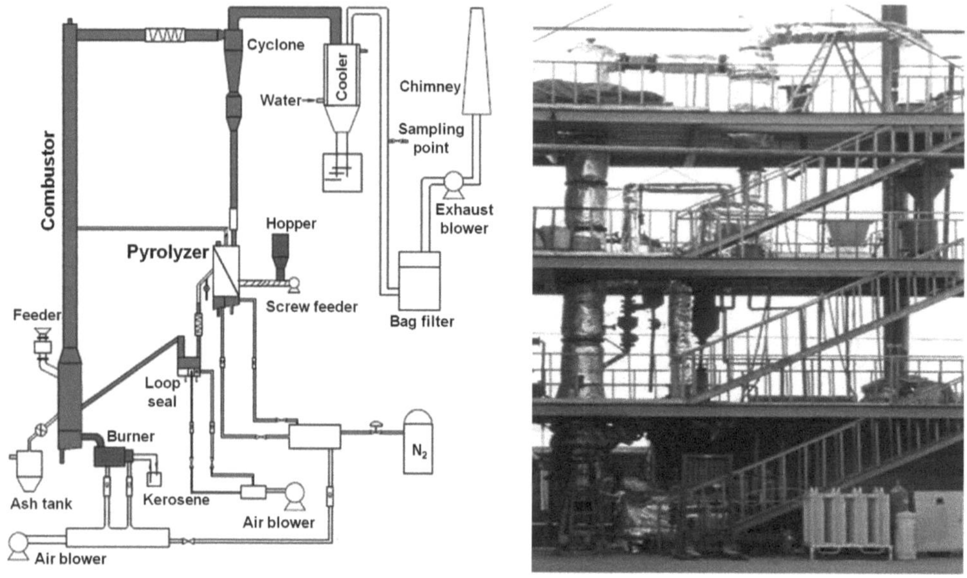

Figure 10.6 A pilot plant of decoupling combustion with a bubbling fluidized bed as the pyrolyzer [21].

combustion of distilled spirit lees (moisture content: 30–40 wt.%). Fig. 10.8 is a photograph of this demonstration plant. The combustion system consists of a riser pyrolyzer and a riser combustor. The river sand smaller than 3 mm in size was used as bed material and heat carrier particles. The temperature distribution in the combustor is in the range of 700°C–900°C. The pyrolysis temperature is kept within 500°C–600°C to maximize the tar yield by controlling the inflow of high-temperature bed material. The residence time of DSL in the riser pyrolyzer is no more than 10 seconds. The flue gas compositions of this demonstration plant are shown in Fig. 10.9. During stable running, the NOx concentration in flue gas is below 90 ppm, much less than 600–950 ppm for CFB combustion, as shown in Fig. 10.7A. In conclusion, the pilot and demonstration test results confirm the low-NOx emission feature of decoupling combustion and the rationality of the system and pyrolyzer design with reference to the findings obtained by MFBRA.

10.2 Reaction characteristics tested by MFBRA for biomass-staged gasification

10.2.1 Staged gasification process analysis using MFBRA

Gasification is a promising and competitive option for converting carbon-containing solid fuel and waste into fuel gas or syngas. Due to high efficiency, cleanness, and low

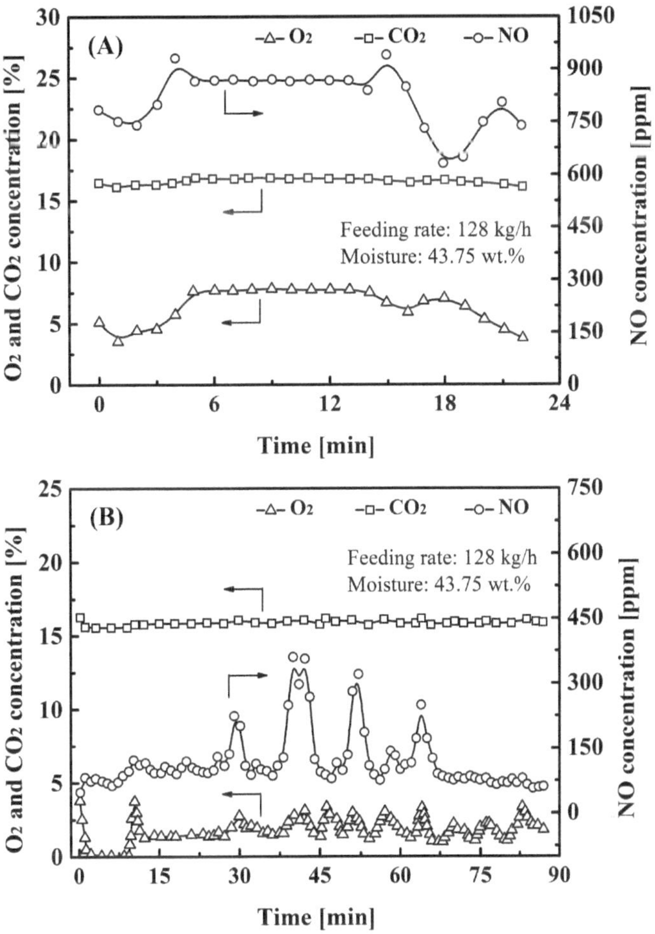

Figure 10.7 NOx emissions in (A) CFB combustion and (B) decoupling combustion of distilled spirits lees containing 43.8 wt.% water [21].

emission of CO_2, gasification has been used widely in IGCC, chemical production, and Fischer–Tropsch (FT) synthesis. Among numerous gasification technologies, staged gasification is perceived to be a good choice for producing clean fuel gas. Based on the principle of reaction decoupling, staged gasification divides the complex gasification processes into some subprocesses that take place separately in different reactors. As such, each subprocess is independently controllable so as to optimize the gasification process and achieve the desired performance, such as high gasification efficiency, low tar generation, and wet feedstock processability.

Fig. 10.10 shows a fluidized bed two-stage gasification (FBTSG) process proposed by the Institute of Process Engineering (IPE), Chinese Academy of Sciences (CAS). The FBTSG system consists of a fluidized bed pyrolyzer (the first stage) and a fluidized

Figure 10.8 A demonstration plant of decoupling combustion with a riser as the pyrolyzer [3,22].

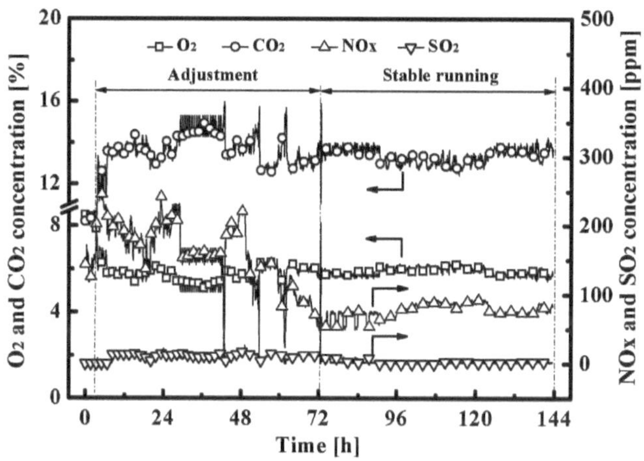

Figure 10.9 Flue gas compositions of demonstration plant of decoupling combustion.

bed gasifier (the second stage). The feedstock is added to the pyrolyzer to undergo water evaporation and fuel devolatilization by contacting high-temperature heat carrier particles (i.e., char residue and ash from the gasifier). The generated pyrolysis products, namely, tar, char, pyrolysis gas, and water vapor, are transported to the fluidized bed gasifier to undertake char gasification and tar catalytic reforming. The energy required for the gasification process is provided by partial combustion of char at the bottom of the gasifier. Fuel gas is separated from the residue char and ash particles and then discharged for downstream applications, while the residue char and ash particles are circulated to the pyrolyzer to provide the heat required for water evaporation and fuel devolatilization.

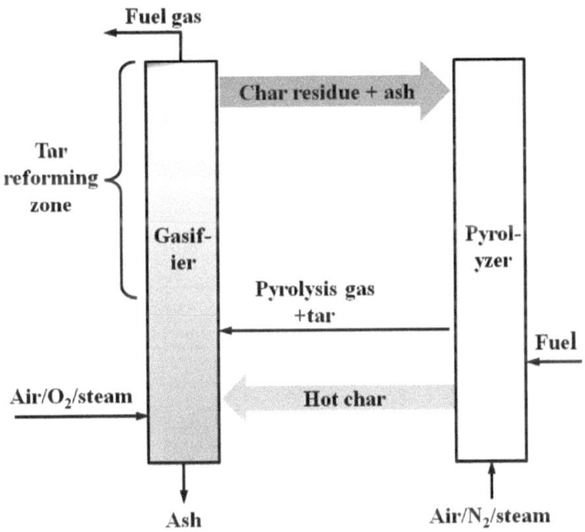

Figure 10.10 Typical staged gasification processes from IPE, CAS.

Designing a two-staged gasification system requires a better understanding of the reaction behavior and kinetics of the subprocesses operating at conditions that approximate the actual processes. However, the staged gasification behavior and kinetics have not been well understood. In addition, almost all studies on biomass drying have been conducted at temperatures below 473K, which is significantly lower than the temperature experienced by biomass in actual staged gasifiers.

In the next section, we briefly introduce how MFBRA has been used to help understand the dynamics of drying, pyrolysis, char gasification, and combustion under conditions that mimic actual gasifiers and how these experimental data have helped design an FBTSG system that processes 10,000 tons of biomass annually.

10.2.2 Characteristics of gasification subprocesses by MFBRA

Fig. 10.11A shows a three-staged MFB reactor with an ID of 20 mm used in the experiments. The experimental samples used were the distilled grains (DGs) from a traditional Chinese spirit company. Fig. 10.11B schematically illustrates the preparation procedure of the experimental samples. Some of the distilled grains were dried and used as the sample particles for pyrolysis tests, and some were pyrolyzed in a horizontal electrical heated tubular furnace at 1273K in Ar for 60 minutes to produce char sample particles. Part of the char sample particles was used for char gasification experiments, and part was humidified to 25% of moisture to be used in drying studies. The particle size of all the sample particles was

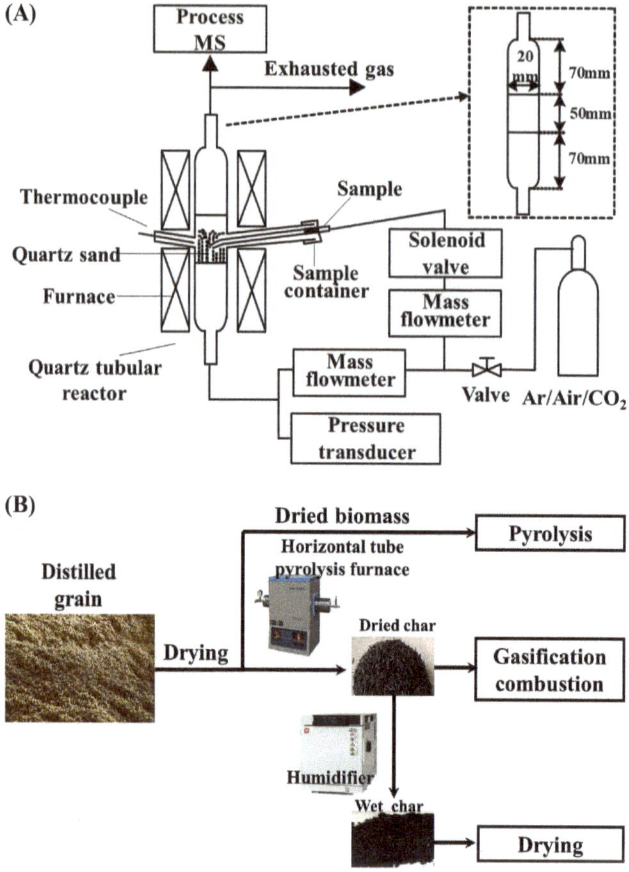

Figure 10.11 Schematic diagram of (A) MFBRA and (B) feedstock preparation [20].

in the range of 0.3–0.4 mm. In each experiment, about 30 mg of sample particles was loaded to the MFB reactor held at a preset temperature, and the released gas products were monitored by an online MS. For biomass pyrolysis, char gasification, and char residue combustion, all the gaseous products were collected in gas bags from the beginning to the end of the experiment to be analyzed by a micro-GC (Agilent 3000A).

Taking steam, CO, and CO_2 as the main products of drying, gasification, and combustion, respectively, we can determine the start time (t_0) and end time (t_e) according to their release curves and the corresponding baselines recorded by MS, as shown in Fig. 10.12A. The conversion X_i of the product (i.e., steam, CO, and CO_2) at time t_i can be calculated by the ratio of the accumulative mass at t_i to the total mass at t_e. Then, the reaction rate (R) is calculated by dX/dt.

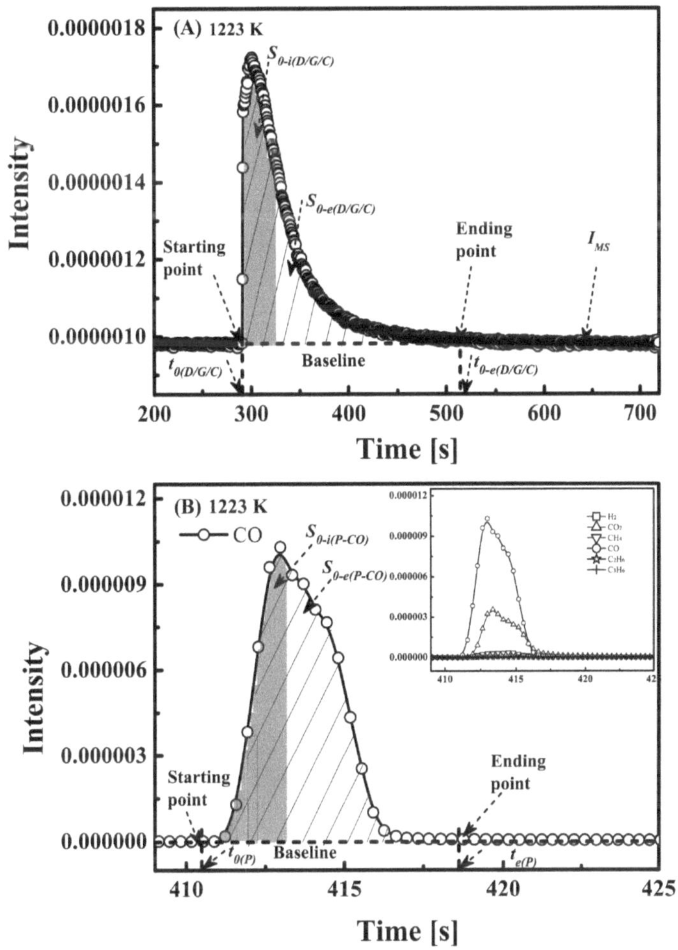

Figure 10.12 Analysis method of (A) drying/gasification/combustion and (B) pyrolysis in MFBRA [20].

$$m_{i(\mathrm{D/G/C})} = m_{e(\mathrm{D/G/C})} \times \frac{S_{i(\mathrm{D/G/C})}}{S_{e(\mathrm{D/G/C})}} \tag{10.3}$$

$$m_{e(\mathrm{D/G/C})} = \frac{F \times T_{0\rightarrow e(\mathrm{D/G/C})} \times \overline{C} \times M}{22.4} \tag{10.4}$$

$$S_i = \int_{t_0(\mathrm{D/G/C})}^{t_i(\mathrm{D/G/C})} \left(I_{MG}^{t_i(\mathrm{D/G/C})} - I_{MS}^{t_0(\mathrm{D/G/C})} \right) \mathrm{d}t \tag{10.5}$$

$$S_e = \int_{t0(D/G/C)}^{t_e(D/G/C)} (I_{MS}^{t_i(D/G/C)} - I_{MS}^{t_0(D/G/C)})dt \tag{10.6}$$

$$X_{i(D/G/C)} = \frac{m_{i(D/G/C)}}{m_{e(D/G/C)}} \tag{10.7}$$

$$R_{i(D/G/C)} = \frac{dX_{i(D/G/C)}}{dt} \tag{10.8}$$

where m_i is the mass of product in a specific process (i.e., moisture in drying [D], CO in char gasification [G], and CO_2 in combustion [C]) at time t_i, and m_e is the total mass of the product from t_0 to t_e. Here, F, $\overline{C}$, and M are the flow rate of carrier gas at the standard state, the average concentration of each gas component determined by GC, and the molecular weight of each gas component, respectively. S_i and S_e are the integrating areas between the releasing curve of the product (i.e., steam, CO, or CO_2) and the corresponding baseline from time t_0 to t_i and t_0 to t_e, respectively. I_{MS}^t denotes the signal intensity of the specific product (i.e., steam, CO, and CO_2) measured by MS at time t (i.e., t_0, t_i, and t_e).

For biomass pyrolysis (P), it is necessary to calculate the mass of each gas component in the pyrolysis gas. The mass of pyrolysis gas from the starting time to a given time is the accumulation of m_i over the period. The total mass of pyrolysis gas is the accumulation of m_i from the starting time to the ending time. Fig. 10.12B schematically shows the definitions of these parameters. Then, the conversion of pyrolysis gas (Xp) and the reaction rate (Rp) can be calculated by Eqs. (10.9) to (10.14), in which the subscripts $1-n$ denote the number of gas components in the produced pyrolysis gas.

$$m_{i(P)} = \mathrm{m}_{e(p)-1} \times \frac{S_{i(P)-1}}{S_{e(P)-1}} + m_{e(p)-2} \times \frac{S_{i(P)-2}}{S_{e(P)-2}} \ldots + m_{e(p)-n} \times \frac{S_{i(P)-n}}{S_{e(P)-n}} \tag{10.9}$$

$$m_{e(P)-1} = \frac{F \times T_e \times \overline{C_{p-1}} \times M_{p-1}}{22.4} \tag{10.10}$$

$$S_{0(p)-1} = \int_{t_0(p)}^{t_i(p)} (I_{MS}^{t_{i(p)-1}} - I_{MS}^{t_{0(p)-1}})dt \tag{10.11}$$

$$S_{e(p)-1} = \int_{t_0(p)}^{t_e(p)} (I_{MS}^{t_{e(p)-1}} - I_{MS}^{t_{0(p)-1}})dt \tag{10.12}$$

$$X_{i(p)} = \frac{m_{i(p)-1} + \ldots m_{i(p)-n}}{m_{0(p)-1} + \ldots m_{0(p)-n}} \tag{10.13}$$

$$R_{i(p)} = \frac{dX_{i(p)}}{dt} \tag{10.14}$$

Fig. 10.13 plots the conversion as a function of time for drying, pyrolysis, gasification, and combustion processes measured by MFBRA. It shows that the temperature has a much more significant effect on drying and char gasification than on pyrolysis and combustion processes. From the $X-t$ curve, one can see that pyrolysis and combustion are completed in the order of seconds, while drying and char gasification take

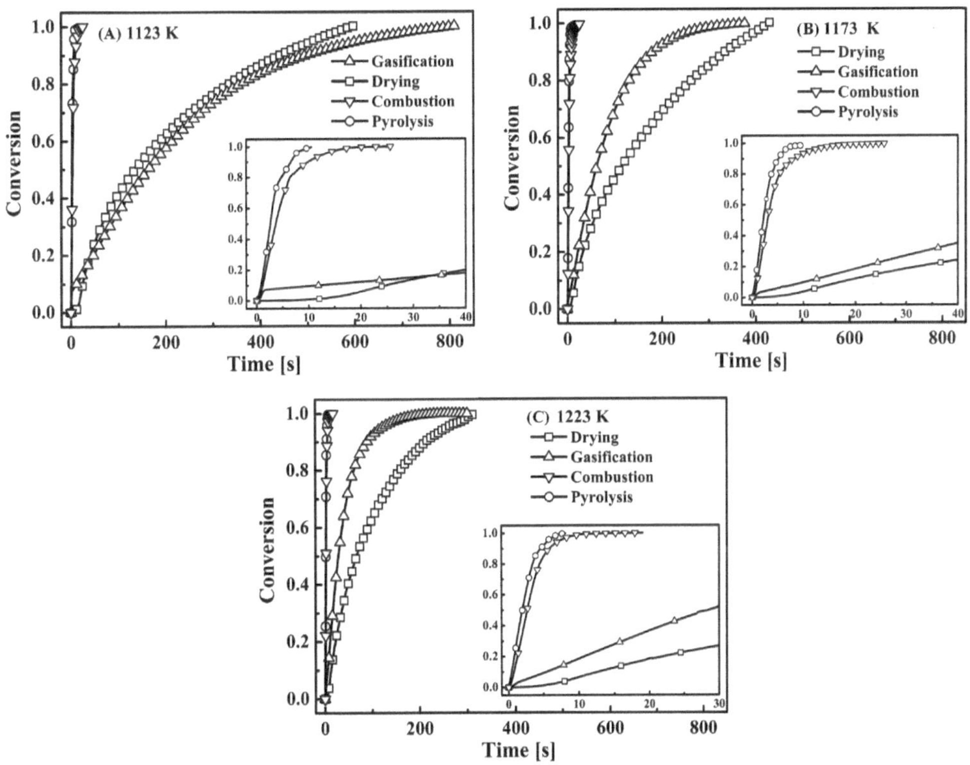

Figure 10.13 Conversions of drying, pyrolysis, gasification, and combustion in MFBRA [20]. (A) 1123 K, (B) 1173 K, and (C) 1223 K.

much longer time (in minutes) to complete conversion. For example, the time for char gasification, drying, combustion, and pyrolysis reduces from 801, 600, 26, and 11 seconds at 1123K (Fig. 10.13A) to 379, 430, 25, and 9 seconds at 1173K (Fig. 10.13B), respectively. The results demonstrate that char gasification or drying is the rate-limiting step of the biomass gasification process in gasifiers. In conventional gasification processes, the feedstock moisture is generally controlled to be below 10% in order to reduce its adverse effect on feedstock handling ability and gasification performance. In the FBTSG process, however, the feedstock with a moisture content of up to 30% can be stably processed and is often preferred, since char gasification and tar reduction can be enhanced to some extent under a high water vapor environment.

Fig. 10.14 presents the reaction rates of drying, pyrolysis, gasification, and combustion derived from MFBRA experimental data. It shows that for all these reaction processes, the reaction rate increases with conversion initially, reaches a maximum at a moderate conversion, and then declines when conversion continues to increase. For pyrolysis and combustion, the maximum reaction rates are attained in the conversion range of 0.2–0.5 and 0.1–0.4, respectively. Comparatively, the rates of drying and

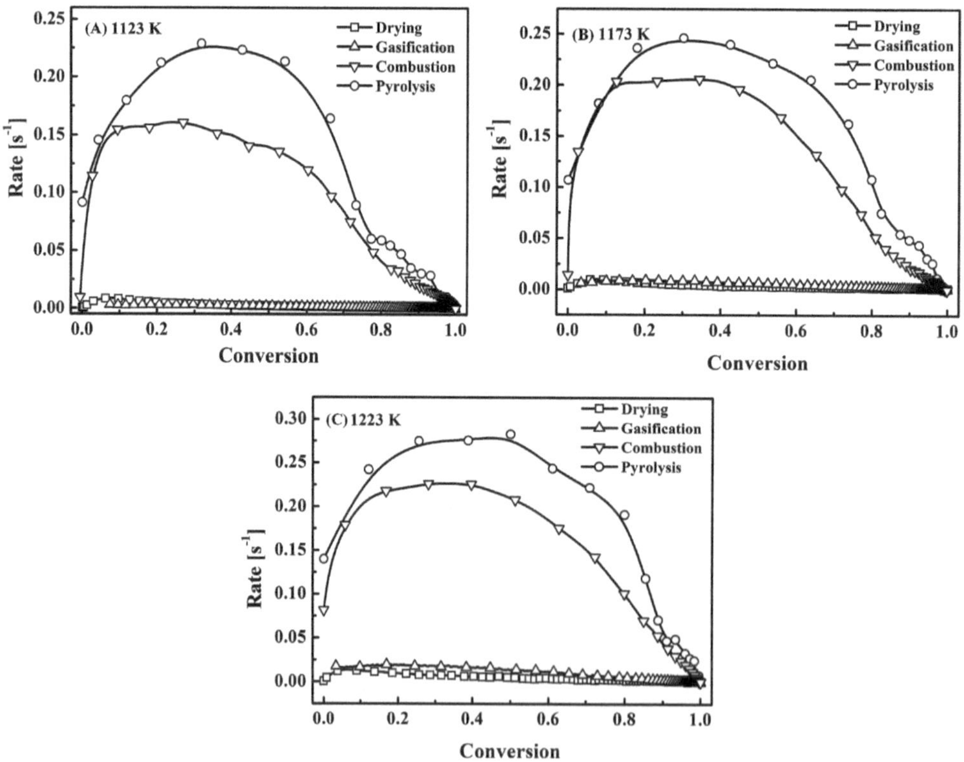

Figure 10.14 Rates of drying, pyrolysis, gasification, and combustion in MFBRA [23]. (A) 1123 K, (B) 1173 K, and (C) 1223 K.

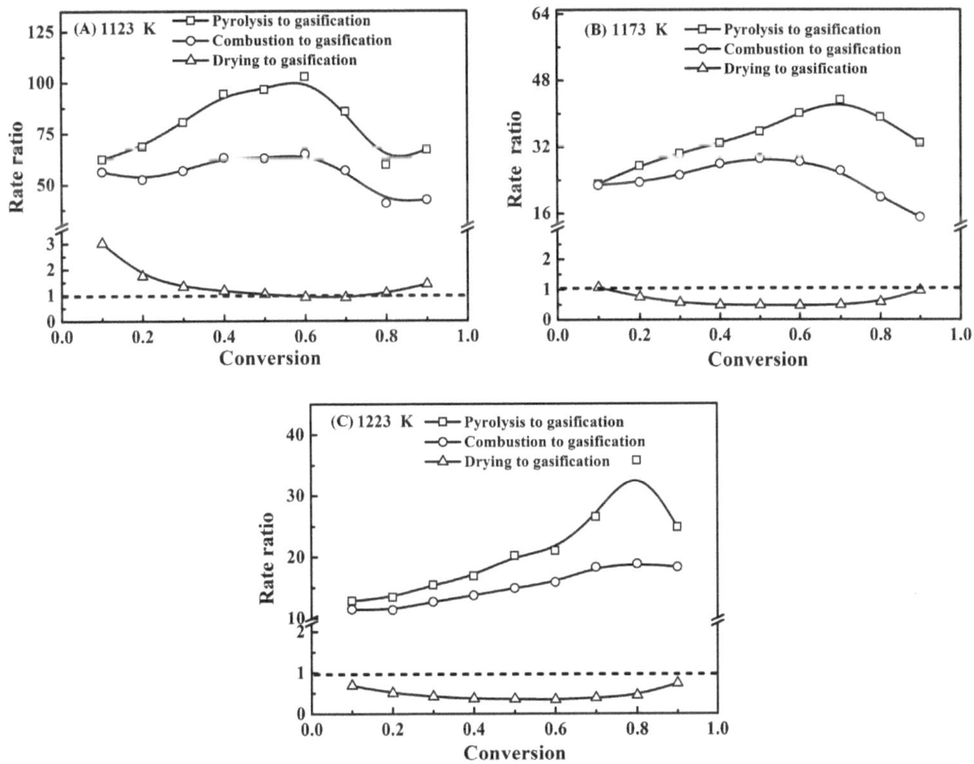

Figure 10.15 Comparison of rate ratio at different temperatures [20]. (A) 1123 K, (B) 1173 K, and (C) 1223 K.

gasification are much lower than those of combustion and pyrolysis. At 1123K, the reaction rate decreases in the order of pyrolysis, combustion, drying, and gasification, but it becomes pyrolysis, combustion, gasification, and drying at 1173K.

Fig. 10.15 shows the reaction rates of drying, pyrolysis, gasification, and combustion. Three parameters, namely, $R_{D/G}$, $R_{C/G}$, and $R_{P/G}$, are introduced, defined as the ratio of reaction rate of drying/combustion/pyrolysis to gasification. At a given temperature (e.g., 1123K), the $R_{P/G}$ and $R_{D/G}$ have the maximum and minimum values, respectively. With an increase in reaction temperature, the value of each ratio gradually decreases. For example, by raising the reaction temperature from 1123K to 1273K, the maximum $R_{P/G}$ changes from 105 to 35, while $R_{P/G}$ evolves from 60 to 18. Moreover, at 1123K, the value of $R_{D/G}$ is above 1, while above 1173K, it becomes below 1, further verifying the limitation step of biomass drying and char gasification.

For pyrolysis, gasification, and combustion in MFBRA, the E_a can be calculated by the Arrhenius equation. In theory, the relationship between Ln(k) and $1/T$ is a linear straight line with the slope corresponding to $-E_a/RT$. The effective moisture diffusivity coefficient (D_{eff}) and activation energy are two critical parameters of drying

Table 10.1 Kinetics of drying, pyrolysis, gasification, and combustion tested by MFBRA [20].

Process	E_a (kJ/mol)	A (s^{-1})	D_0 (m^2/s)
Drying	83.75	—	3.61E − 07
Pyrolysis	29.51	19.95	—
Combustion	62.39	7.03E + 02	—
Gasification	210.23	2.96E + 07	—

kinetics. The apparent diffusion coefficient (D_{eff}) can be calculated by Fick's second law as follows:

$$D_{eff} = D_0 \times e^{-\frac{E_a}{RT}} \tag{10.15}$$

Based on the calculated results, it is found that the E_a of the four main subprocesses increases in the following order: pyrolysis < combustion < drying < gasification, as shown in Table 10.1. The results further demonstrate that char gasification or biomass drying is the rate-limiting step, which is consistent with the above analysis.

The reaction kinetics of biomass drying, pyrolysis, and char gasification/combustion are of great significance for the design and optimization of the gasification process, especially in the FBTSG. Because that biomass pyrolysis is a fast reaction, and mixing between high-temperature bed materials and biomass is quick, a fluidized bed reactor with a small volume is suitable for use as a pyrolyzer. To accelerate the time-consuming biomass drying in the pyrolyzer, it is preferable to operate the pyrolyzer at relatively high temperatures. In the transport bed gasifier, since the char gasification and combustion processes take place simultaneously, a high gas velocity is needed to reduce the residence time in order to avoid excessive char oxidation. Moreover, steam and/or O_2-rich reagents can be used to stimulate the rate-limiting gasification process. To ensure operability when using a feedstock of high moisture content, char produced in the fluidized bed pyrolyzer is sent to the bottom combustion zone of the transport bed gasifier, and the pyrolysis gas, steam, and tar (vapor phase at the operating temperature) are directed to the middle section of the gasifier. Note that this reactor configuration permits the large quantity of steam produced in the pyrolyzer to be used as the gasification reagent in the transport bed gasifier, as shown in Fig. 10.16.

10.2.3 Design of an industrial FBTS gasification process

Herb residue, a by-product produced in the traditional Chinese medicine industry, amounts to about 1.5 million tons per year in China. Most of herb residue is landfilled because of lacking efficient and economical utilization approaches. However, landfilled herb residue is a potential source of pollution when it degrades, releasing undesirable

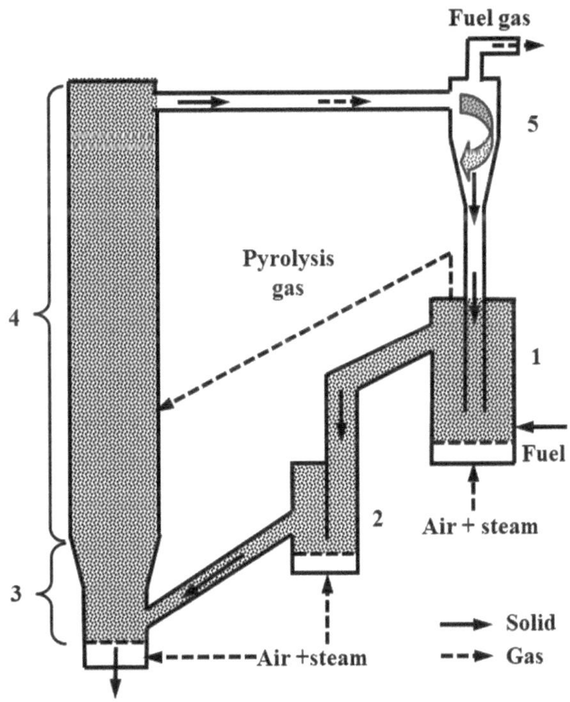

Figure 10.16 The FBTS gasification process for distilled grains [20]. 1: Fluidized bed pyrolyzer; 2: loop seal; 3: char combustion section in the transported bed gasifier; 4: char gasification section in the transported bed gasifier; 5: cyclone separator.

chemicals into soil and air. On the other hand, herb residue contains high volatiles and low ash and is especially rich in cellulose and lignin. In terms of heating value, it is almost the same as some low-quality coals and thus can be a valuable source of renewable energy.

As an attempt to beneficially utilize herb residue, an FBTSG process was designed based on the principle of decoupling gasification. Figs. 10.17 and 10.18 show the process diagram and photographs of an industrial-scale demonstration plant.

The industrial demonstration FBTSG process has a capacity of 1 t/h. The pyrolysis reactor has a cuboid structure with a cross section of 0.3 m^2 and a height of 1.2 m. The gasification reactor is a cylindrical column with a cross section of 0.25 m^2 and a height of 13 m. The herb residue from a Chinese medicine company located in Henan Province of China was processed as the feedstock. The proximate analysis, ultimate analysis, and heating value of the feedstock are listed in Table 10.2. Before experiments, the herb residue was mechanically dewatered, crushed, and dried, reducing the moisture content from 70% to 15%. The particle size of the herb residue ranged from 0 to 4 mm. During experiments, silica sand with a particle size range of 0.2–0.4 mm and a particle density of 2100 kg/Nm^3 was used as the bed material.

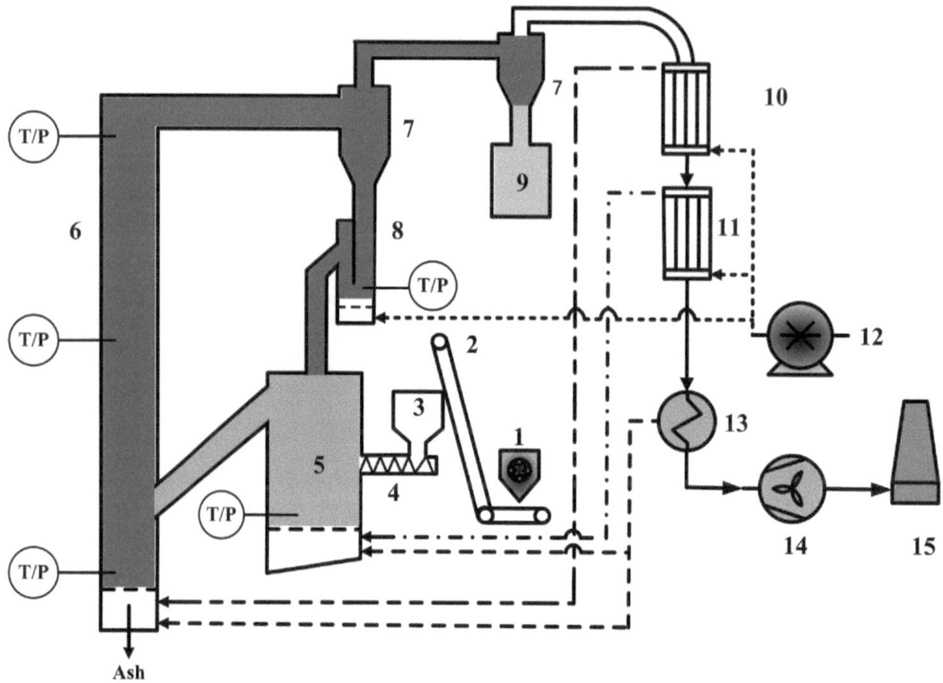

Figure 10.17 Process diagram of the pilot test apparatus [24]. 1: Herb residue; 2: conveyor; 3: hopper; 4: screw feeder; 5: pyrolyzer; 6: gasifier; 7: primary cyclone; 8: loop seal; 9: secondary cyclone; 10: primary heat exchanger; 11: secondary heat exchanger; 12: air compressor; 13: boiler; 14: draught fan; 15: chimney.

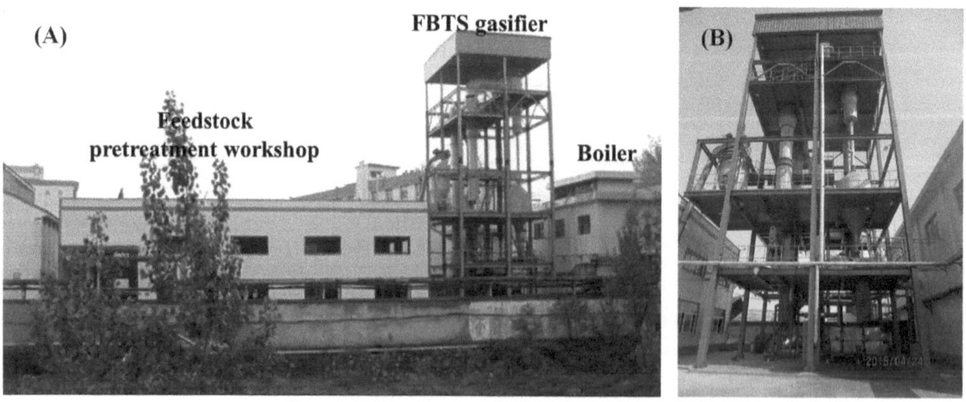

Figure 10.18 Photograph of (A) the industrial demonstration plant and (B) the gasifier [23].

Table 10.2 Property of herb residue used in experiments.

Proximate analysis (wt.%)					Ultimate analysis (wt.%)					
M_{ad}	A_{ad}	V_{ad}	FC_{ad}		C_{daf}	H_{daf}	S_{daf}	O_{daf}	N_{daf}	LHV (MJ/kg)
14.12	4.32	66.98	14.58		51.64	5.60	0.16	11.22	1.38	16.37

ad, air-dried basis; *daf*, dry and ash-free basis.

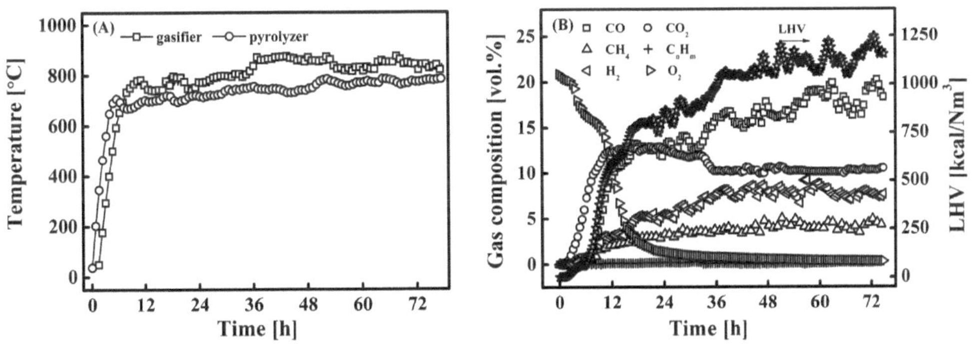

Figure 10.19 Typical running results of the industrial demonstration plant [23].

Fig. 10.19A shows that in stable operation conditions, the temperatures in the pyrolyzer and gasifier can be maintained at around 700°C and 850°C, respectively. The heating value of the produced fuel gas reaches 1200 kcal/Nm^3 (Fig. 10.19B) with a tar content of less than 0.1 g/Nm^3 [23]. The experimental results demonstrate that FBTSG is capable of processing wet herb residue with satisfactory stability and syngas quality.

10.3 Light calcination of magnesite using transported bed

MFBRA has played an essential role in developing industrial light calcination of magnesite using a transported bed reactor. Magnesite is one of the most crucial raw materials globally for refractory magnesia production [25]. In the magnesite-based industry, light calcination of magnesite is conducted at about 1000°C to produce caustic calcined magnesia (CCM, mainly composed of MgO). The chemical reaction of calcination is as follows: [26–29]

$$MgCO_3(s) \rightarrow MgO(s) + CO_2(g)\ \Delta H = +99.66\ \text{kJ/mol} \qquad (10.16)$$

CCM is a valuable feedstock for producing a variety of value-added products, such as silicon steel magnesium oxide, magnesium hydroxide, and magnesium cement [30–36]. This section describes the essential role of MFB in the industrial development of light calcination of magnesite using a transported bed.

10.3.1 Existing technology and equipment

For many decades, reverberatory furnace (RF) has been the dominant calciner of magnesite, because it is easy to operate. However, it suffers from high energy consumption, prolonged reaction time, poor feedstock adaptability, unstable product quality, and severe environmental pollution [37]. It is because, as shown in Fig. 10.20, RF is a fixed bed reactor [38], in which high-temperature flue gases produced from fuel gas combustion provide the heat required for the calcination of magnesite. In order to ensure the gas permeability through particles with acceptable pressure drop, RF can only use large-sized magnesite particles, typically 30–80 mm. As a result, the reaction time in RF is as long as 3–5 hours, limiting the production capacity per furnace to less than 30 t/d. Because of the formation of hot spots in RE, overburning on the surface and incomplete burning inside particles occur and thus reduce the product quality [39]. Besides, the inability of RF to process small-sized particles has made such materials produced in mining and processing disposed of as waste. Thus, there is a need for advanced CCM production technology.

10.3.2 Kinetic analysis of magnesite calcination

Calcination of magnesite produces CO_2, and thus, the presence of CO_2 in the calcination affects the calcination kinetics. In order to acquire the intrinsic calcination kinetics for such reactions under the product gas atmosphere, Liu et al. [40] investigated the calcination kinetics of calcium carbonate $Ca^{13}CO_3$ in $^{12}CO_2$ atmospheres by

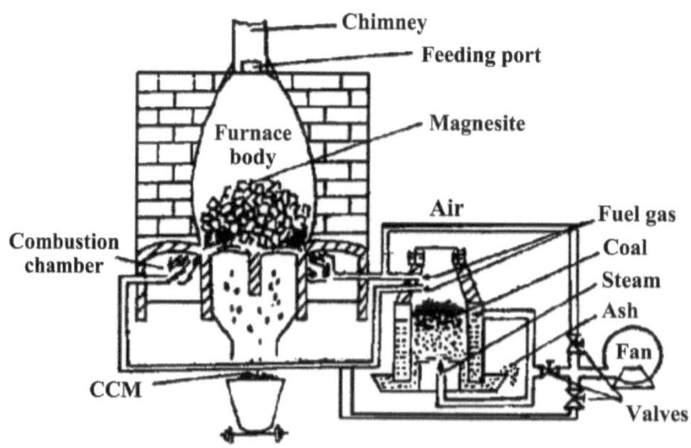

Figure 10.20 A schematic RF process [38].

monitoring $^{13}CO_2$ and $^{12}CO_2$ simultaneously in an MFBRA. The results indicate that the presence of CO_2 in the reaction atmosphere increases the apparent activation energy. Jiang et al. [41] also studied fluidization calcination of magnesite using a micro-fluidized bed. Fig. 10.21 presents the CO_2 formation curve of magnesite calcination detected using a process mass spectrometer in MFBRA. It clearly shows that the decomposition completion time decreases with the increase in calcination temperature. The magnesite powder in 100–200 μm can be completely decomposed in about 3 seconds at 900°C, clearly indicating that powdered magnesite calcination in fluidized bed conditions is a fast reaction [42].

Moreover, the reaction kinetics of magnesite calcination in N_2 and air atmospheres using TGA and MFBRA are determined, and the results are summarized in Table 10.3. The results clarify that the kinetic data in nitrogen and air are almost the same for both instruments. The activation energy of magnesite calcination measured in the nitrogen and air is approximately the same, indicating that the oxygen in the air has little effect on the calcination of magnesite. Table 10.3 also shows that both activation energy and pre-exponential factor are significantly lower measured in

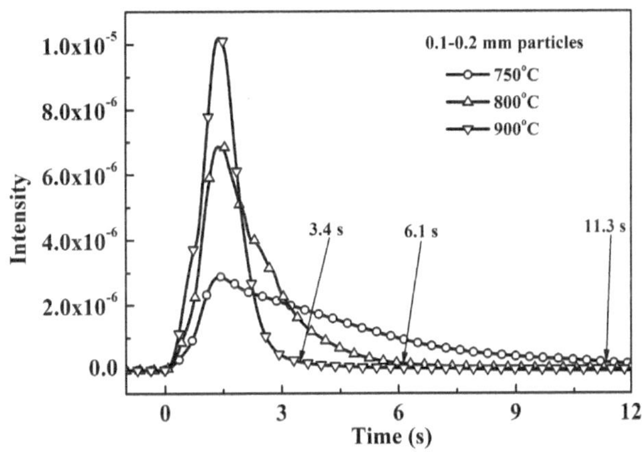

Figure 10.21 Calcination of magnesite in MFBRA: CO_2 formation curve detected using a process mass spectrometer [42].

Table 10.3 Kinetic results of magnesite calcination in N_2 and air atmospheres measured using TGA and MFBRA [41].

Analyzer	Ea (kJ/mol)	lg *A*
TGA (N_2)	205.94	15.68
TGA (air)	236.60	15.81
MFBRA (N_2)	125.74	5.61
MFBRA (air)	124.16	5.69

MFBRA (125 kJ/mol and 10^5 s^{-1}) than in TGA (200 kJ/mol and 10^{15} s^{-1}). Based on the results, the magnesite calcination kinetic equation can be expressed as follows:

$$\frac{dX}{dt} = 10^{5.69}\exp\left(-\frac{124160}{RT}\right) \times (1 - X) \tag{10.17}$$

where X is the conversion for magnesite calcination; t and T are the reaction time and temperature. This equation has been used to design an advanced transport bed calcination system, as discussed in the next section.

10.3.3 Advanced transport bed calcination process

Fig. 10.22 schematically shows the transport bed flash calcination (TBFC) process for the production of CCM developed by the Shenyang University of Chemical Technology, based on the MFBRA studies [41]. In this advanced process, a transport bed reactor is used for calcining small magnesite powders. A fuel gas combustor is located at the reactor bottom section. The high-temperature flue gas produced flows upward in the transport bed reactor to decompose fine magnesite powders quickly.

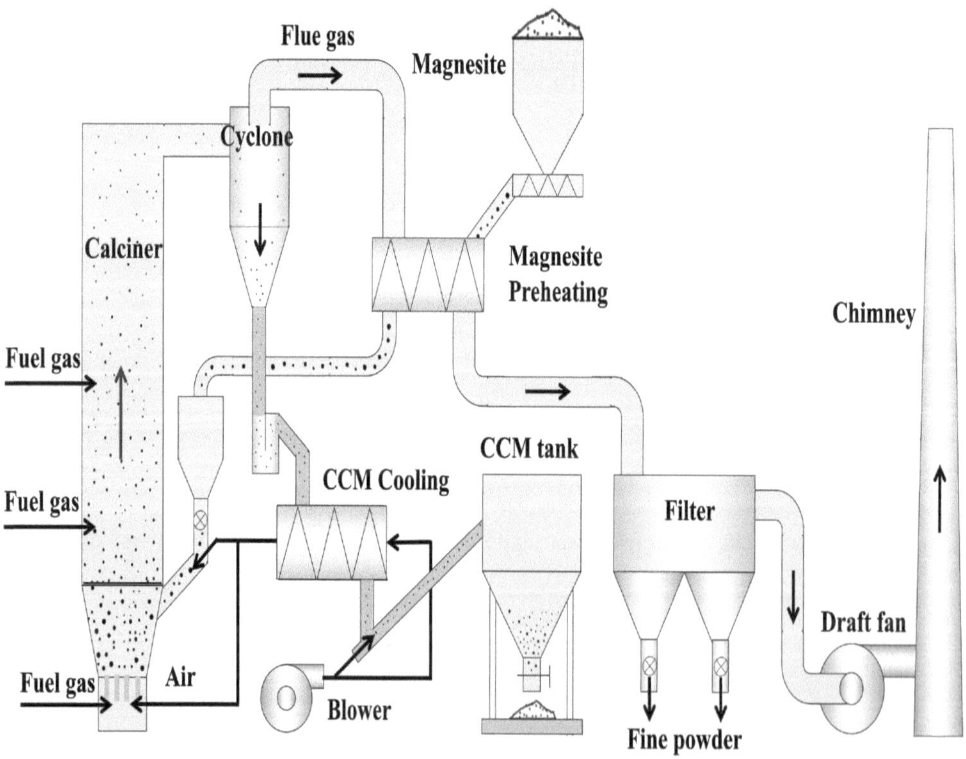

Figure 10.22 A schematic diagram of the transport bed flash calcination process.

To validate the design concept, Sun et al. [43] investigated the calcination of magnesite in a laboratory-scale transport bed. They found that 98% of the conversion of magnetite powder (<150 μm) could be reached in 1–2 seconds, and the product had a higher activity when compared to products from fixed bed calciners. The experiments confirmed the feasibility of TBFC for high-efficiency calcination of magnesite. According to kinetic studies in MFBRA and pilot investigations in the laboratory-scale transport bed, CCM products with the best product activity can be obtained at 850°C–1000°C [43]. At these temperatures, the reaction time required to achieve at least 99% conversion is estimated to be about 3 seconds based on the kinetic Eq. (10.17). To obtain the desired hydrodynamic characteristics in transport reactors, an operating gas velocity of 4–6 m/s in the transport bed reactor is chosen. Thus, the height of the transport bed reactor is determined to be 20 m. It should be noted that the reaction time is critical for the magnesite calcination in the transport bed. In practical operations, the calcination time, operating gas velocity, and temperature can be adjusted to meet production capacity and product quality needs.

In addition to the transport bed reactor, the kinetic data obtained by using MFBRA are also required to optimize TBFC process designs, particularly the heat recovery system, which is essential to improve the TBFC system's energy efficiency. Fig. 10.23 shows a TBFC process featured with the staged fuel combustion and recoveries of the sensible heat carried by the flue gas and the CCM product. By integrating a multistage heat

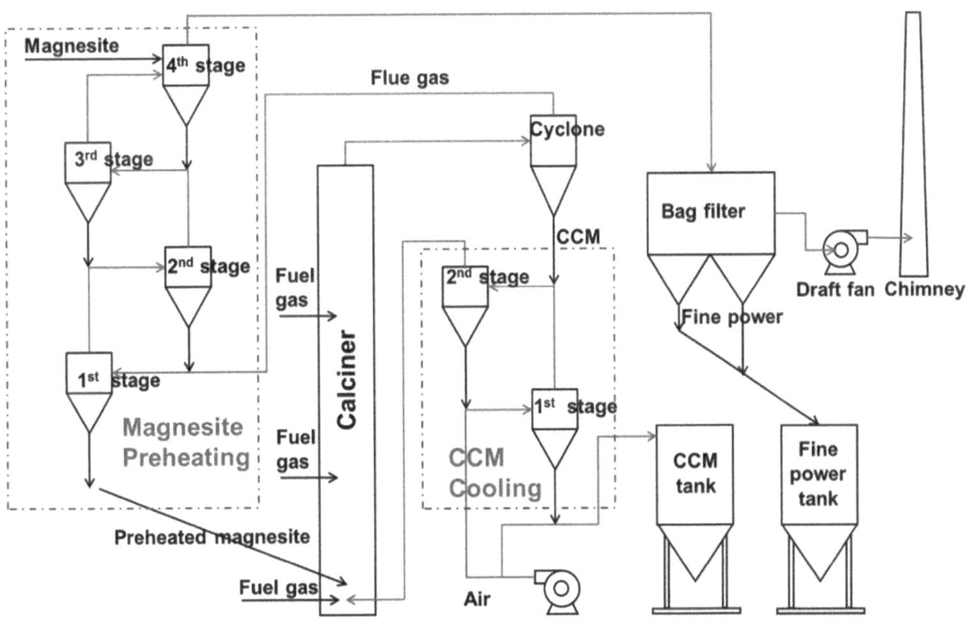

Figure 10.23 A schematic diagram of the TBFC process with two-stage cooling and four-stage preheating [42].

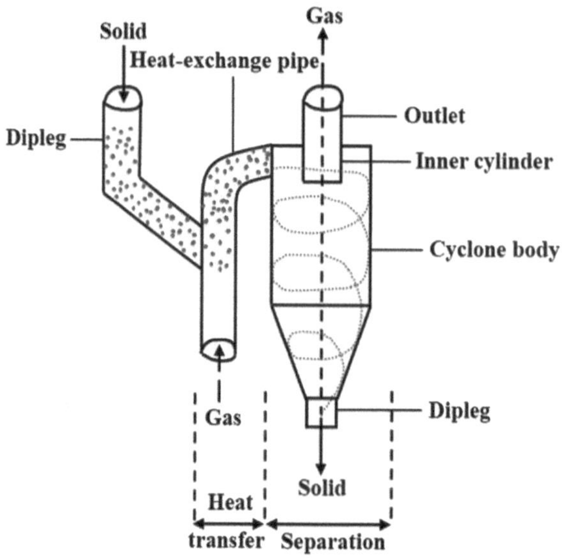

Figure 10.24 A schematic drawing of the cyclone preheater [42].

recovery system, the TFBC can increase the temperature of raw magnesite to about 500°C, at which magnesite starts to decompose [41]. With the calcination kinetics obtained using MFBRA, the decomposition of magnesite in the preheating process can be accurately analyzed [41]. In the TBFC process, heat transfer between the flue gas and raw magnesite particles is accomplished in a cyclone-type preheater, as shown in Fig. 10.24. Based on process design calculations, the temperatures of preheaters are in the range of 450°C–750°C. Under these temperatures, the residence time of particles in cyclone preheaters is estimated to be 0.2 seconds. According to the equation (10.17), the magnesite conversion in 0.2 seconds is lower than 4.3%.

An et al. [42] systematically investigated the TBFC process through process simulation by considering the pre-decomposition of magnesite during preheating. The simulation aims to establish an energy-saving strategy for the process, which is essential for designing and operating industrial-scale TFBC systems. The results reveal that the pre-decomposition occurs mainly in the first preheating stage, and the maximum conversion in the cyclone preheating process is about 13%. The occurrence of pre-decomposition slightly lowers the energy consumption and increases the energy efficiency of the process. At the same time, the magnesite residence time of about 1 seconds in the cyclone preheaters has a limited effect on the pre-decomposition rate. For the TBFC, the optimum process design has four-stage preheating for magnesite and two-stage cooling for CCM. With this design, the energy consumption is about 4100 kJ/kg-CCM, and the energy efficiency is 66.8%. This energy efficiency is almost twice that of the traditional light calcination furnaces such as RF [42].

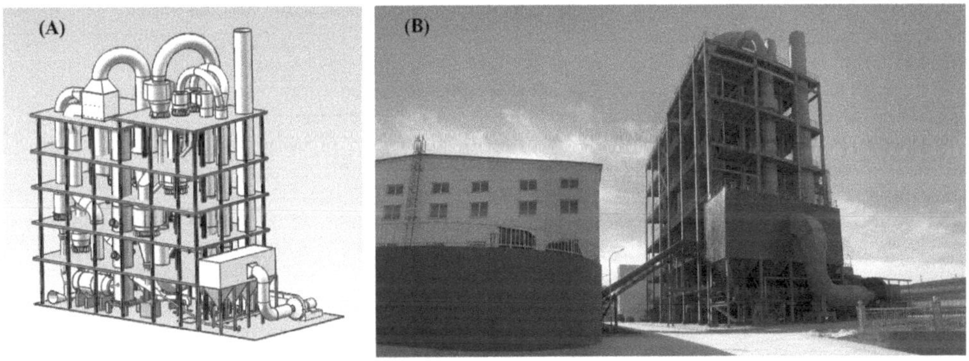

Figure 10.25 A demonstration plant for TBFC: (A) three-D design and (B) a picture of the plant site [42].

10.3.4 Engineering commissioning of a 400 kt/a industrial process

After a series of studies, including kinetic analysis of magnesite calcination, flash calcination experiments of magnesite in a laboratory-scale transport bed, and process simulations, a demonstration plant of TBFC with a capability of 400,000 t/a magnesite has been successfully designed by the Shenyang University of Chemical Technology and built in Haicheng City, Liaoning Province, China, (shown in Fig. 10.25). The powdered magnesite supplied from a flotation plant is used as raw material, which has a particle size of 0−200 μm and a moisture content of 10−15 wt.%. The fuel gas produced from a coal gasification plant or natural gas is used as fuel to provide heat for the magnesite calcination process. Before entering the transport bed calciner, the powdered magnesite undergoes four stages of cascaded preheating by leaving high-temperature flue gas from the calciner. The calcination temperature in the reactor is controlled at around 900°C, and the solid residence time is about 3 seconds. After exiting the reactor and being separated by a cyclone, the CCM powders are successfully cooled by going through two cooling stages by air. The recovered products are delivered to a silo by pneumatic conveying. The TBFC demonstration plant is under commissioning and will be in full operation by the end of 2022.

References

[1] Li C-Z, Xu GW. Decoupled thermochemical conversion − preface. Fuel 2013;112:607−8.
[2] Zhang JW, Wu RC, Zhang GY, et al. Technical review on thermochemical conversion based on decoupling for solid carbonaceous fuels. Energy & Fuels 2013;27(4):1951−66.
[3] Han ZN, Geng SL, Zeng X, et al. Reaction decoupling in thermochemical fuel conversion and technical progress based on decoupling using fluidized bed. Carbon Resources Conversion 2018;1 (2):109−25.
[4] Song Y, Wang Y, Yang W, et al. Reduction of NO over biomass tar in micro-fluidized bed. Fuel Processing Technology 2014;118:270−7.

[5] Cai LG, Shang X, Gao SQ, et al. Low-NOx coal combustion via combining decoupling combustion and gas reburning. Fuel 2013;112:695–703.
[6] Do H-S, Bunman Y, Gao SQ, et al. Reduction of NO by biomass pyrolysis products in an experimental drop-tube. Energy & Fuels 2017;31(4):4499–506.
[7] Yu J, Zeng X, Zhang JW, et al. Isothermal differential characteristics of gas-solid reaction in micro-fluidized bed reactor. Fuel 2013;103(1):29–36.
[8] Yu J, Yue JR, Liu ZE, et al. Kinetics and mechanism of solid reactions in a micro fluidized bed reactor. AIChE Journal 2010;56(11):2905–12.
[9] Wang F, Zeng X, Geng SL, et al. Distinctive hydrodynamics of a micro fluidized bed and its application to gas-solid reaction analysis. Energy & Fuels 2018;32(4):4096–106.
[10] Thomas KM. The release of nitrogen oxides during char combustion. Fuel 1997;76(6):457–73.
[11] Tullin CJ, Goel S, Morihara A, et al. Nitrogen oxide (NO and N_2O) formation for coal combustion in a fluidized bed: effect of carbon conversion and bed temperature. Energy & Fuels 1993;7 (6):1847–53.
[12] Duan J, Luo YH, Yan NQ, et al. Effect of biomass gasification tar on NO reduction by biogas reburning. Energy & Fuels 2007;21(3):1511–16.
[13] Liu CY, Luo YH, Jia D, et al. Experimental study on the effect of NO reduction by tar model compounds. Energy & Fuels 2009;23(4):4099–104.
[14] Zhang RZ, Liu CY, Yin RH, et al. Experimental and kinetic study of the NO-reduction by tar formed from biomass gasification using benzene as a tar model component. Fuel Processing Technology 2011;92(1):132–8.
[15] Lv QG, Yu KS, Zhu ZP, et al. Pilot plant research on fast pyrolysis of coal in circulating fluidized bed with hot char carrier (in Chinese). Journal of the China Coal Society 2012;37(9):1591–5.
[16] Fang MX, Cen JM, Shi ZJ, et al. Experimental study on 75 t/h circulating fluidized bed poly-generation system (in Chinese). Proceedings of the Chinese Society of Electrical Engineering 2010;30(29):9–15.
[17] Qu X, Zhang R, Bi JC. Research on poly-generation process combined CFB combustion with coal pyrolysis (in Chinese). Taiyuan Science & Technology 2008;2:57–60.
[18] Wang JG, Lu XS, Yao JZ, et al. Experimental study of coal topping process in a downer reactor. Industrial & Engineering Chemistry Research 2005;44(3):463–70.
[19] Yu J, Yao CB, Zeng X, et al. Biomass pyrolysis in a micro-fluidized bed reactor: Characterization and kinetics. Chemical Engineering Journal 2011;168(2):839–47.
[20] Adamu MH, Zeng X, Zhang JL, et al. Property of drying, pyrolysis, gasification, and combustion tested by a micro fluidized bed reaction analyzer for adapting to the biomass two-stage gasification process. Fuel 2020;264:116827.
[21] Yao CB. Dual fluidized bed decoupling combustion of distilled spirits lees (in Chinese). Beijing: Chinese Academy of Sciences 2011.
[22] Han ZN, Zeng X, Yao CB, et al. Comparison of direct combustion in a circulating fluidized bed system and decoupling combustion in a dual fluidized bed system for distilled spirit lees. Energy & Fuels 2016;30(3):1693–700.
[23] Zeng X, Ueki Y, Yoshiie R, et al. Recent progress in tar removal by char and the applications: A comprehensive analysis. Carbon Resources Conversion 2020;3:1–18.
[24] Zeng X, Shao RY, Wang F, et al. Industrial demonstration plant for the gasification of herb residue by fluidized bed two-stage process. Bioresource Technology 2016;206:93–8.
[25] Wang L, Tai PD, Jia CY, et al. Magnesium contamination in soil at a magnesite mining region of Liaoning Province, China. Bulletin of Environmental Contamination & Toxicology 2015;95(1):90–6.
[26] José N, Ahmed H, Miguel B, et al. Magnesia (MgO) production and characterization, and its influence on the performance of cementitious materials: a review. Materials 2020;13(21):4752.
[27] Pilarska AA, Klapiszewski L, Jesionowski T. Recent development in the synthesis, modification and application of Mg(OH)2 and MgO: a review. Powder Technology 2017;319:373–407.
[28] Li GY, Li ZJ, Guo YX, et al. Effect of calcination temperatures on microstructure and activity of light burned magnesite powder (in Chinese). Refractories 2016;50(5):367–9.

[29] Ren WX, Xue B, Lu CP, et al. Evaluation of GHG emissions from the production of magnesia refractory raw materials in Dashiqiao, China. Journal of Cleaner Production 2016;135:214–22.
[30] Mo LW, Deng M, Tang MS, et al. MgO expansive cement and concrete in China: past, present and future. Cement and Concrete Research 2014;57(3):1–12.
[31] Zhu J, Ye N, Yang JK. Evaluation methods on corrected hydration activity in preparation of active MgO with calcined magnesite (in Chinese). Metal Mine 2013;11:88–91.
[32] Gao QJ, Guo W, Jiang X, et al. Characteristics of calcined magnesite and its application in oxidized pellet production. Journal of Iron and Steel Research (International) 2014;21(4):408–12.
[33] Tsiplakou E, Pappas AC, Mitsiopoulou C, et al. Evaluation of different types of calcined magnesites as feed supplement in small ruminant. Small Ruminant Research 2017;149:188–95.
[34] Yang N, Ning P, Li K, et al. MgO-based adsorbent achieved from magnesite for CO_2 capture in simulate wet flue gas. Journal of the Taiwan Institute of Chemical Engineers 2018;86:73–80.
[35] Stefanidis SD, Karakoulia SA, Kalogiannis KG, et al. Natural magnesium oxide (MgO) catalysts: a cost-effective sustainable alternative to acid zeolites for the in situ upgrading of biomass fast pyrolysis oil. Applied Catalysis B Environmental 2016;196:155–73.
[36] Mo LW, Panesar DK. Effects of accelerated carbonation on the microstructure of Portland cement pastes containing reactive MgO. Cement and Concrete Research 2012;42(6):769–77.
[37] Ba HJ, Bai LM, Zhao WQ, et al. Review on preparation and processing of caustic calcined magnesite (in Chinese). Conservation and Utilization of Mineral Resources 2017;1:84–9.
[38] Li HM, Li ML. The choice and application of pollution control technique on flue dust from light sintering magnesium kiln (in Chinese). Environmental Protection Science 1997;23(2):8–10.
[39] Li ZJ. Thoughts on magnesia refractory raw materials of Liaoning Province (in Chinese). Refractories 2011;45(5):382–5 389.
[40] Liu XJ, Hao WQ, Wang KX, et al. Acquiring real kinetics of reactions in the inhibitory atmosphere containing product gases using micro fluidized bed. AIChE Journal 2021;67(9):e17325.
[41] Jiang WW, Hao WQ, Liu XJ, et al. Characteristic and kinetics of light calcination of magnesite in micro fluidized bed reaction analyzer (in Chinese). CIESC The Journal 2019;70(8):2928–37.
[42] An P, Han ZN, Wang KJ, et al. Energy-saving strategy for a transport bed flash calcination process applied to magnesite. Carbon Resources Conversion 2021;4(1):122–31.
[43] Sun C, Yan BW, Cai CY, et al. Characteristics of reaction and product microstructure during light calcination of magnesite in transport bed (in Chinese). CIESC the Journal 2020;71(12):5735–44.

CHAPTER 11

Characterization of liquid–solid micro-fluidized beds

Liquid fluidization of particulate solids is a very old technology for processing solids, predating the now commonly applied gas–solid fluidization. In microscales, the concepts and the subsequent research of liquid–solid and gas–solid micro-fluidized beds started almost simultaneously. All preceding chapters of this book dealt with gas–solid micro-fluidization. This chapter concentrates on the fundamentals and applications of liquid–solid micro-fluidized beds.

11.1 Introduction

Since the liquid–solid micro fluidized bed (MFB) was proposed by Potic et al. [1] in 2005, it has received much attention for its capabilities of fast screening of solid particle processes and bioprocesses with low experimental costs. The miniaturization of fluidized beds increases the liquid–solid contact area per unit volume in the fluidized bed, thus improving mixing, mass, and heat transfer efficiency. However, it is debatable whether micro-fluidized beds should be defined by a column diameter of less than 1 millimeter, which is in line with the general microfluidics, or by a small lab-scale fluidized bed that exhibits significant wall effects compared with conventional macro-fluidized beds.

As mentioned above, micro-fluidized beds have been defined according to their hydraulic diameters. The hydraulic diameter that defines MFBs varies from less than 500 μm [2] to several centimeters [1,3], although its exact value for liquid–solid fluidization systems is different from that for gas–solid fluidization systems [2,4]. Xu et al. [5] studied gas backmixing characteristics in gas–solid beds and proposed an MFB definition based on the bed-to-particle ratio. Specifically, they suggested that micro-fluidization behavior appears when the bed-to-particle ratio (d_t/d_P) is less than 150 [6], under which the wall effects become strong. The strong wall effects were also observed in liquid–solid micro-fluidized systems [1,2], characterized by general hydrodynamic behavior. Therefore, fluidized beds of 5 cm inner diameter fluidizing particles of 2–3 mm in size [7,8] should be considered to be micro-fluidized beds because of the corresponding low bed-to-particle ratio. Another factor distinguishing micro-fluidization from macro-fluidization is the increased importance of surface forces relative to volumetric forces, such as gravity. The surface forces can cause defluidization

Micro Fluidization
DOI: https://doi.org/10.1016/B978-0-443-18718-6.00011-5
© 2023 Chemical Industry Press Co., Ltd.
Published by Elsevier Inc., under an exclusive license with Chemical Industry Press Co., Ltd.

(i.e., adhesion of particles to the walls) when they are much bigger than the gravity [9,10]. The surface forces become important for micron-sized particles in milli- and microscale fluidized beds. In this chapter, we define the beds of <50 mm ID with d_b/d_P <150 as liquid–solid micro-fluidized beds.

11.2 Hydrodynamics properties

11.2.1 Manufacturing methods

To date, different types of micro-fluidized beds have been reported with various fabrication approaches, such as soft lithography techniques [2,11], additive manufacturing [10], micromachining [12], and capillary system [13], as illustrated in Fig. 11.1. The rectangular or cylindrical bed column with different hydraulic diameters strongly affects the fluidization performance, so it is essential to investigate the hydrodynamic characteristics of MFB manufactured by various bed design and optimization methods. As shown in Fig. 11.1A, Zivkovic et al. [16] developed a microfluidic bed with microchannels formed from a PDMS chip by a simple, cheap, and fast method known as

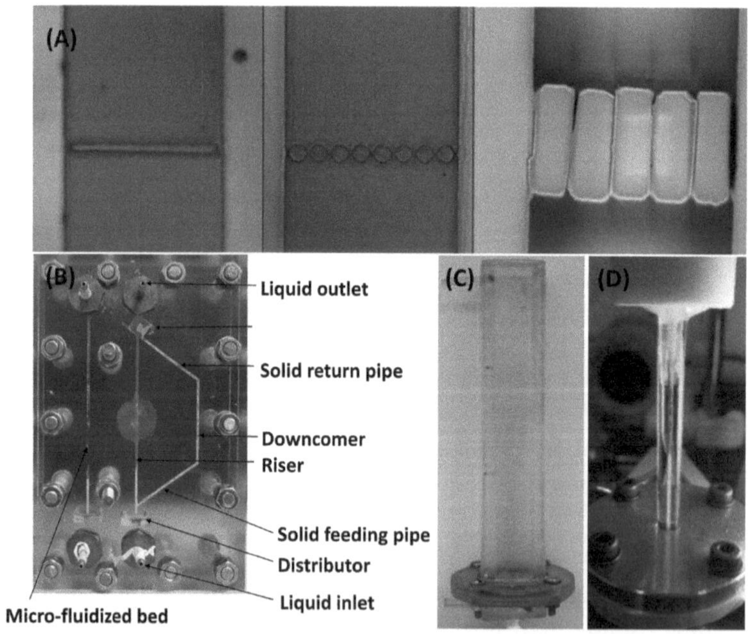

Figure 11.1 Different micro-fluidized bed systems using various fabrication methods. (A) The micro-fluidized beds with 400 × 175 μm^2 cross-section fabricated in a polydimethylsiloxane (PDMS) chip by standard soft lithography techniques [2]; (B) micro circulating fluidized bed made by milling 1 × 1 mm cross-sectional channels into Perspex [12]; (C) a 3D-printed micro-fluidized bed with a cross-sectional area of 15 × 15 mm [14,15]; (D) micro-fluidized bed made of a quartz capillary of 0.8 mm inner diameter, 6 mm outer diameter, and 60 mm length [13].

lithography [17]. do Nascimento et al. [9,18] designed and fabricated a robust MFB by milling square millimeter-sized cross-sectional channels into a Perspex block fitted with a distributor at the bottom (Fig. 11.1B). In recent years, 3D printing techniques have enabled the fabrication of complicated MFB (Fig. 11.1C), which could be hard to design and manufacture with conventional mechanical tools. The 3D printing methods make it possible to investigate the effects of the surface roughness of the printed parts on fluidization performance [18,19]. More commonly, the quartz capillaries (Fig. 11.1D) with various internal diameters are widely used as the bed column, since transparent bed walls allow visualization analysis [13]. For instance, the study of Potic et al. [1] used a quartz capillary of 1 mm ID, 6 mm OD, and a length of 1 m as a fluidized bed. Other groups also applied the same approach to set up MFB systems [6,20]. Boffito et al. [21] called this a "capillary fluidized bed." Since these MFBs cannot endure chemical reactions at high temperatures and pressures, stainless steel has been used to build MFBs by other researchers [22].

11.2.2 Minimum fluidization velocity

Minimum fluidization velocity U_{mf} is generally defined as the liquid velocity at which solid particles change from a fixed bed state to fluidization state. In a gas–solid micro-fluidized bed system, the gas flow completely supports the weight of the particles at U_{mf}, above which the bed pressure drop (ΔP) remains constant (Fig. 11.2A). Therefore, measuring the pressure drop across the bed at different gas velocities is an effective method to determine U_{mf} [6,24]. This standard method applies to liquid–solid fluidized beds, but it is challenging to use it in micro-fluidized beds because the pressure drop (in the order of several Pa) is too small to be measured precisely. Therefore, the U_{mf} in liquid–solid MFBs has been normally

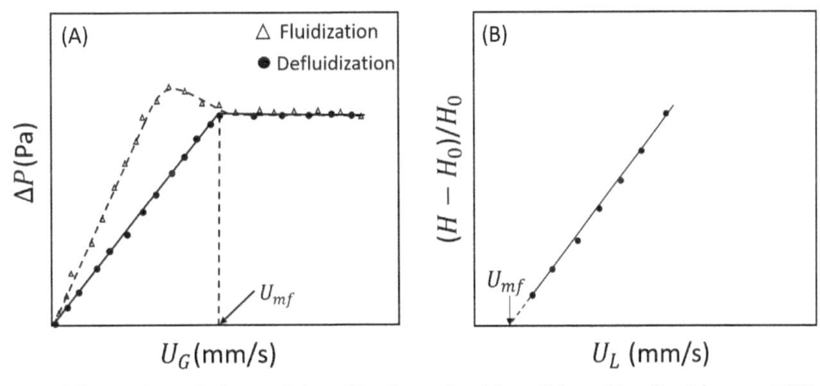

Figure 11.2 Schematics of determining U_{mf} in a liquid–solid and/or liquid–gas MFB [23]. (A) A pressure drop profile (fluidization and defluidization) generated by comparing the pressure drops at various superficial gas velocities; (B) bed expansion curve extrapolated to U_{mf}.

obtained through visual observations of bed expansion using ordinary or high-speed cameras [9,13,25,26], as shown in Fig. 11.2B.

It is worth noting that the bed expansion extrapolating method is suitable only for solids with rigid structures and smooth surfaces (such as glass beads and polymer microspheres), under which a linear relationship between the bed expansion and superficial liquid velocity close to the minimum fluidization velocity can be obtained. For solids with soft and rough surfaces (i.e., immobilized cells, fungal pellets), the determination of U_{mf} becomes complicated because the bed expansion curve may not be linear. For instance, Zhang et al. [14] reported that in a 3D-printed micro-fluidized bed of fungal pellets, the flow regime transferred from the static bed (Fig. 11.3A) to the extended bed (Fig. 11.3B and C), the partially fluidized bed (Fig. 11.3D–F), and eventually the completely fluidized bed (Fig. 11.3G) subsequently with the increase of superficial liquid velocity. These flow regimes of the extended bed and partially and fully fluidized beds are caused by the pellet agglomeration induced by the hairy external region of fungal pellets. The filamentous-linked pellets can be fluidized only when they are disengaged by the liquid flow and become dispersed. Therefore, Zhang et al. [14] defined the U_{mf} as the liquid velocity at which the full expansion and detachment of the pellets from the base occur. Although this method is dependent on visual observation and may be imprecise, it can be applied to determine the U_{mf} in liquid–solid fluidized beds when biosolid particles are fluidized.

To predict U_{mf}, several semiempirical correlations developed for macroscale fluidized beds may be used [27–31]. The most commonly utilized correlation is based on the Ergun equation [32–35].

$$\frac{\Delta P}{H} = \frac{150\mu}{\varnothing^2 d_P^2}\frac{(1-\varepsilon)^2}{\varepsilon^3} U_F + \frac{1.75\rho_F}{\varnothing d_P}\frac{(1-\varepsilon)}{\varepsilon^3} U_F^2 \tag{11.1}$$

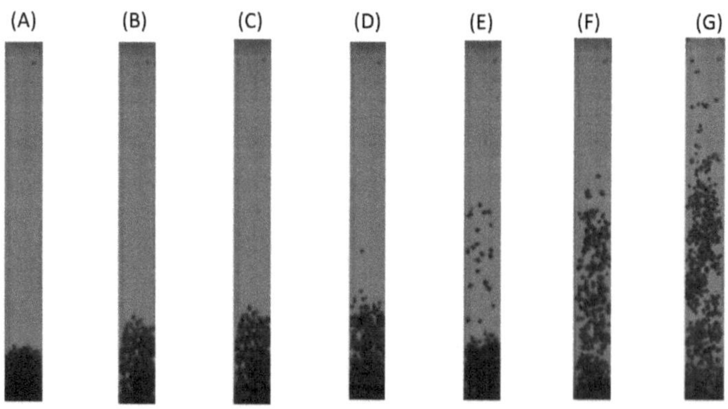

Figure 11.3 Bed expansion with the increase of liquid velocity in medium culture using 1 g of pellets [14]. (A) $U_1 = 0$ mm/s; (B) $U_1 = 1.242$ mm/s; (C) $U_1 = 2.554$ mm/s; (D) $U_1 = 3.866$ mm/s; (E) $U_1 = 4.522$ mm/s; (F) $U_1 = 5.718$ mm/s; (G) $U_1 = 6.982$ mm/s.

where H, U_F, φ_P, ρ_F, d_P, ρ_P, g, and μ refer to the height of the particle bed, superficial velocity of the fluid, particle sphericity, density of the liquid, particle diameter, particle density, acceleration of gravity, and dynamic fluid viscosity, respectively. When applied to micro-fluidization systems, the Ergun equation has to be modified to take parameters characterizing surface forces and wall effects into consideration, since the surface force and wall effect can overweight the volumetric forces (e.g., gravity) and thus affect the particle fluidization. In recent years, a few researchers investigated the influence of surface force and wall effect on U_{mf} in micro-fluidized beds by changing parameters such as the ratios of bed-to-particle diameter d_t/d_P [3,9,13,20,36], initial bed height H_0 [6,24,37], particle properties [2,38], and wall roughness [10]. Studies of surface force and wall effects in micro-fluidized beds, especially on the minimum fluidization velocity, are presented in Table 11.1. Many works indicate that U_{mf} increases with increasing H_0 or decreasing d_t/d_P. Obviously, the decrease of d_t/d_P results in an intensified wall effect due to the increased contact area between the bed wall and particles per bed volume. The reduction in d_t can cause the magnitude of bed–wall friction to exceed the magnitude of drag force between particles and fluid [10]. Similarly, the larger static bed height H_0 also means a higher bed–wall contact area, hence higher wall friction. By comparing the deviations between the experimental minimum fluidization velocity $U_{mf,e}$ and the theoretical minimum fluidization velocity $U_{mf,t}$ calculated by correlations such as the Ergun equation, we can identify and analyze the wall effect and surface force in micro-fluidized beds. For liquid–solid micro-fluidized beds, Li et al. [39] modified the Ergun equation (Eq. 11.2) to include the effects of the liquid viscous drag and wall friction, thus effectively reducing the relative error between experimental and theoretical U_{mf} to less than 15%.

$$\frac{\Delta P}{H} = \frac{2400\mu}{\varnothing^2 d_P^2}\frac{(1-\varepsilon)^2}{\varepsilon^3} U_F \left(\frac{1}{4} - \frac{d_W}{D_B} + \frac{d_W^2}{D_B^2}\right)^2 \tag{11.2}$$

where d_W is the radial thickness of the wall region, given by

$$d_W = 1.5 \times 10^{-4} - \frac{d_P}{(1-\varepsilon_0)} \tag{11.3}$$

Zhang et al. [42] studied the effect of particle size distributions (PSDs) on U_{mf} in a liquid–solid fluidized bed reactor (7 mm in ID) based on the coupled computational fluid dynamics–discrete element method (CFD-DEM) simulations. They concluded that the smallest U_{mf} resulted from the broader flat-type PSD and found that the proportion of small particles was responsible for reducing the voidage and was the key factor determining U_{mf}. Besides the d_t/d_P and H_0, the physical properties of solid particles and liquids also influenced the fluidization performance. Zivkovic et al. [43] experimentally studied micro-fluidization in a rectangular microchannel fluidized bed of $400 \times 174\ \mu m^2$. The

Table 11.1 Summary of the influences of different parameters on U_{mf} in micro-fluidized beds.

References	Properties of liquid–solid fluidized bed	Parameters	Results and conclusions
Doroodchi et al. [20]	$d_t = 0.8$, 1.2 17.1 mm; $d_p = 225$ μm.	d_t/d_p	U_{mf} increased with the decrease of d_t/d_p.
Zivkovic and Biggs [2]	$d_t = 400 \times 175$ μm; $d_p = 26.5$, 30.5, 34.5, and 38.5 μm.	Liquid/particle properties	Glass microparticles can be fluidized by ethanol instead of water in the PMMA microchannels.
do Nascimento et al. [9]	$d_t = 1$, 2 mm; $d_p = 21$ μm.	d_t/d_p	The increase in U_{mf} scales linearly with the decrease of d_t/d_p.
Li et al. [26]	$d_t = 0.8$, 1.45, and 2.3 mm; $d_p = 22–58$ μm.	d_t/d_p	U_{mf} increased with the decrease of d_t/d_p.
Tang et al. [36]	$d_t = 3.15$ and 11.6 mm; $d_p = 89–352$ μm.	d_t/d_p	U_{mf} was 1.67–5.25 times higher than the theoretical value when dp/d_t varied from 0.017 to 0.091.
Li et al. [40]	$d_t = 3$ mm; $d_p = 50–300$ μm.	d_t/d_p	U_{mf} increased with the decrease of d_t/dp.
Vanni et al. [41]	$d_t = 2$, 3, 3.2, and 5 cm; $d_p = 50$, 70, and 105 μm.	d_t/d_p	U_{mf} increased with the decrease of d_t/d_p.
Chen et al. [38]	$d_t = 46.25$ mm; $d_p = 10.31$ μm.	Particle surface hydrophobicity	U_{mf} decreased after the ultrafine particles (Al_2O_3) were treated with a surfactant.
Zhang et al. [14]	$d_t = 15$ mm; $d_p = 1–3$ mm.	Liquid/particle properties	Pellets aggregated in the packed bed but gradually became dispersed and fluidized by liquid flow.

results indicated that the glass microspheres of 30.5 μm diameter were not fluidizable by deionized water due to the strong adhesive force between glass particles and the PMMA bed wall, even when the nonionic surfactant (Tween 80) was used. The particles were fluidizable when changing the deionized water to ethanol. However, the surface force and wall effect still had impacts on U_{mf}. A subsequent study by do Nascimento et al. [9] found that U_{mf} was affected by both the surface-force-to-gravity ratio and the wall effect, but the surface forces tended to have a minimal effect in beds of about 1 mm cross section. This result was then used to set the size of micro-fluidized beds as 1 mm, which was in line with general microfluidics but larger than around 25 mm in ID defined for gas–solid micro-fluidized beds. Cunez et al. [44] noticed the jamming of solids in a liquid–solid MFB of 25.4 mm in ID when fluidizing the bonded particles, confirming the significant effect of wall friction. Besides the liquid properties, solid surface hydrophobicity also impacts fluidization. Chen [38] investigated liquid–solid fluidization using the surfactant-treated ultrafine particles (Al_2O_3) and found an improvement in fluidization quality with an increase in the liquid–solid contact angle of ultrafine particles or surface forces between particles and bed walls. Actually, it is not uncommon to use surfactants to decrease the friction or attachment between solids and bed walls. For instance, Pluronic F-68 (PF-68) is a widely used surfactant to decrease cell–bubbles interaction and attachment in cell cultivation [45]. As mentioned above, depending on the combination of wall surface, solid surface, fluid characteristics, and bed size, the adhesion forces can be an insurmountable obstacle to fluidization [20,46].

The influence of the inclination (from 0 to 10 degrees) on the micro-fluidized bed (ID $= 1 \times 1$ mm and 4×4 mm) behavior was investigated with the particle-to-bed ratio ranging from 0.025 to 0.165 [47]. Investigation of bed expansion behavior showed that the bed contracted with an increase in bed inclination due to the increased bed wall effects in inclined beds. It was observed that particles follow a circulatory motion within the bed, which led to an increased backmixing. Interestingly, at a high inclination (10 degrees) and high liquid velocity ($>$4.17 mm/s), two distinct regions were identified: a low bed voidage region in the bottom area and a high bed voidage region in the top area of the bed (Fig. 11.4E and F). This phenomenon is probably attributable to the change in particle trajectories and increased wall effects in the inclined fluidized bed [48].

Currently, the studies of wall effects and surface forces in micro-fluidized beds are mainly for academic interests. From the engineering point of view, no research has been carried out to minimize these effects. Note that wall effects vary with various parameters, including particle properties (size, shape, density, material, etc.), fluidizing medium properties (density, viscosity, wetting, etc.), bed geometry, and material types. To better understand wall effects, Zivkovic et al. [2,9] proved that the theoretical acid–base framework of van Oss, Chaudhury and Good combined with the Derjaguin approximation can be used to estimate the variation of adhesion force with the change of fluidizing medium (including use of surfactant) and the bed materials. In addition, developing some empirical correlations can be helpful for better understanding liquid–solid fluidization, as for gas–solid micro-fluidization [49].

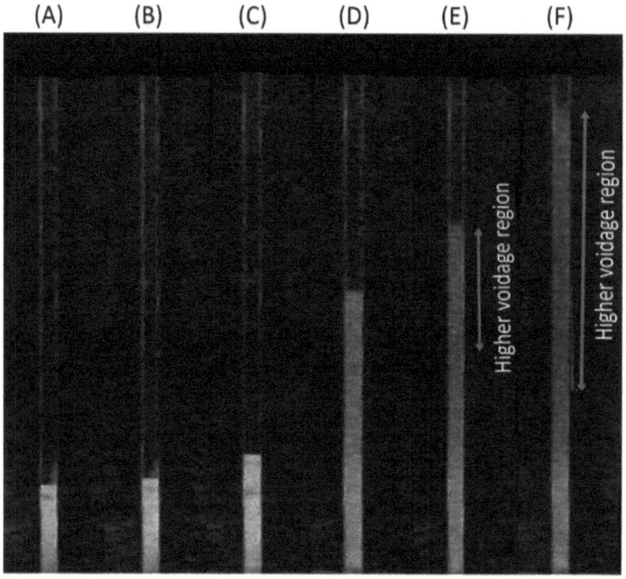

Figure 11.4 Higher voidage region with the glass microspheres size of 75–90 μm glass in an aligned 1 × 1 mm microchannel as a function of liquid velocity, U_L: (A) 0; (B) 0.17 mm/s; (C) 0.33 mm/s; (D) 2.00 mm/s; (E) 2.33 mm/s; (F) 3.00 mm/s.

11.2.3 Mixing

In chemical engineering processes, efficient mixing between different phases is required to achieve a better reaction performance [50]. The mixing process can be divided into micro-mixing (based on molecular diffusion), meso-mixing (e.g., formation and disintegration of flow eddies), and macro-mixing (e.g., flow folding and turbulent structures), according to mixing mechanisms and spatial scales [51,52]. The vigorous meso-mixing and macro-mixing of particles in fluidized bed reactors enhance the diffusive mixing at the microscale [53]. From this point of view, a micro-fluidized bed reactor is an excellent mixer because of (1) the increased surface-to-volume ratio and thus the contact area, and (2) the excellent contact time between multiple phases.

Doroodchi et al. [54] investigated the mixing performance of the micro-fluidized bed using a dye dilution technique with two miscible fluids. In general, a fluidized bed of 1.2 mm tube with borosilicate particles of ~98 μm and sodium iodide solution as the fluidizing medium has a dramatically decreased mixing time compared with the same tube without particles. The studies confirm that MFB enhances mixing performance, but fail to explain the relationship between bed voidage and mixing efficiency. Therefore, Zivkovic et al. [50] experimentally studied the mixing efficiency and confirmed that the mixing was a function of superficial velocity and bed voidage in micro-fluidized beds. In an MFB, a minimal increase in mixing efficiency was found at a low superficial liquid velocity as the bed voidage was close to that of a packed bed, and the mixing was significantly enhanced when the bed expands to a voidage of around 0.76, which was higher than that in large-scale beds (usually about 0.6–0.65). Additionally, the results demonstrate that the mixing efficiency in micro-fluidized beds is up to three times greater than that in particle-free channels, which is in agreement with the results reported by Doroodchi et al. [54]. In addition to the experimental studies, Derksen [55,56] characterized the mixing performance of micro-fluidized systems using direct numerical simulations and confirmed that scalar mixing performance in fluidized beds was more uniformly distributed than that in fixed beds.

One of the issues in liquid–solid fluidization is the limited radial dispersion of solids due to wall effects, which result in a high voidage region close to the wall and a low voidage central region, as well as adhesion problems [39,43]. To reduce the non-uniformity of solid radial distribution, Odeleye et al. [57] changed the vertical flow channels into the 45 degrees inclined ones, which generated swirling fluidization for the microcarriers and increased the particle radial velocities by up to 5.2 times. Besides the inclination of a certain part of the fluidized bed (i.e., flow channels or liquid distributor), the inclination of the whole bed to a certain angle has been widely studied [58,59]. For instance, Pereiro et al. [11] investigated an external magnetic field-assisted fluidization in a V-shaped chamber with the angle of the wall varying from 0 to 33 degrees. The results indicate that the fluidized bed with an inclined wall from 13 to 33 degrees exhibits better fluidization behavior than the vertical bed (0 degree inclination). Notably, the chamber angle of 13 degrees produces the best particle distribution and most expansion compared to chambers with smaller or larger internal wall angles. The use of external forces (e.g., magnetism, ultrasound, thermal, etc.) has also been reported to enhance the mixing efficiency [60–63]. Note that a high flow rate may result in particle segregation for solid particles with unevenly distributed density and size, thus minimizing the mixing in fluidized bed reactors [64,65]. Therefore, the non-uniformity of the solid phase needs to be considered in the actual applications.

11.2.4 Mass transfer

Mass transfer is essential for applications of fluidized bed reactors. Benzoic acid particles have been used in most experimental mass transfer studies due to easily observable mass exchange between the particles and the surrounding liquid phase. The mass transfer rate is generally determined by measuring the concentration of benzoic acid entering and leaving the fluidization column or by comparing the mass of the particles before and after dissolution experiments. The interfacial area is calculated by multiplying the area of one particle by the overall number of particles [66]. According to the literature, mass transfer experiments can be performed in a fluidized bed consisting only of active particles, as shown in Fig. 11.5A [67–69]. The other option, as illustrated in Fig. 11.5B, is that the bed can

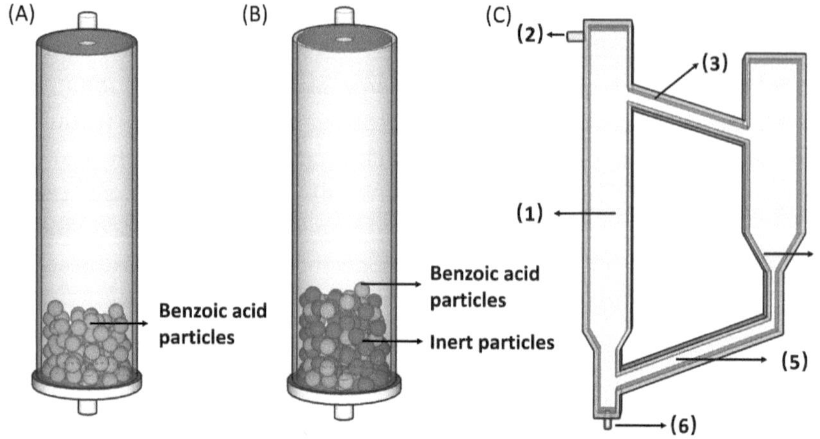

Figure 11.5 (A) Fluidized bed consists of purely benzoic acid particles; (B) fluidized bed is composed of benzoic acid pellets and inert particles; (C) simplified model of circulating fluidized bed: (1) riser column, (2) outlet, (3) top particles return pipe, (4) settler and calming column, (5) bottom particles return pipe, (6) inlet.

consist of both inert particles (e.g., glass or sand particles) and active benzoic acid granules [70,71]. Besides, mass transfer in fluidized beds can also be measured based on ion exchange methods [72–74]. In these methods, cationic resin beads are initially present in the H+ form before reacting with a dilute solution of sodium hydroxide solution, according to the following reaction:

$$R^-H^+ + NaOH \rightarrow R^-Na^+ + H_2O \tag{11.4}$$

where R^- represents the solid ion exchange matrix.

Fig. 11.5C shows a micro-circulating fluidized bed [75]. do Nascimento et al. [12,18] reported a fundamental study on the micro-circulating fluidized bed and confirmed the effects of wall confinement and surface forces on the transition velocities in micro-circulating fluidized beds (Fig. 11.6). Again, the values of U_{mf} for both PMMA and glass particles are almost the same, even though glass particles have a higher density than PMMA particles (2500 and 1200 kg/m^3, respectively). Compared to theoretical predictions, U_{mf} deviates more for PMMA microparticles than for glass beads, indicating that surface forces influence U_{mf} strongly. However, the surface forces do not significantly influence the critical transition velocity (U_{cr}), which demarcates the transition from conventional to circulating fluidized bed regime, as shown in Fig. 11.6A and B. For both types of particles, the normalized transition velocity, U_{cr}/U_t, is close to 1. The transition velocity U_a from circulating fluidized bed to transport regime is very similar in magnitude for PMMA and glass particles of the same quantity, but the relative transition velocity for PMMA particles is around 20 times particle terminal velocity while it is only 5 times for the glass beads.

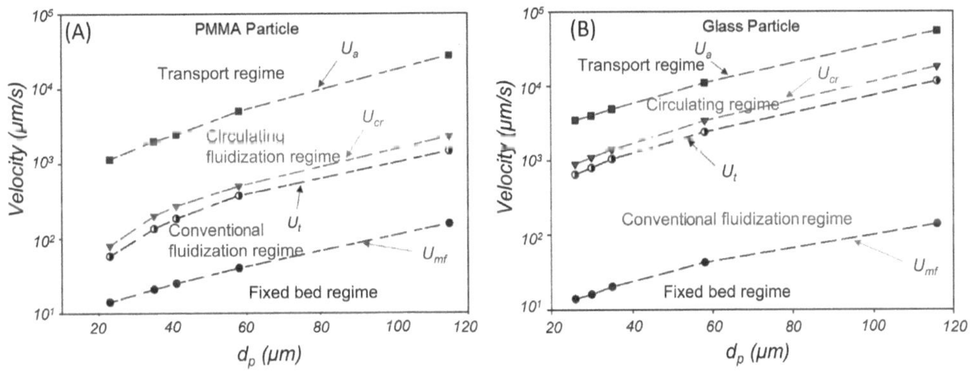

Figure 11.6 Flow regime map of solid–liquid fluidization in a micro circulating fluidized bed as a function of particle diameter and superficial velocity for (A) PMMA and (B) glass microparticles. *Adapted from Do Nascimento OL, Reay D, Zivkovic V. Study of transitional velocities of solid–liquid micro-circulating fluidized beds by visual observation. Journal of Chemical Engineering of Japan. 2018;51(4):349–355.*

To predict the mass transfer coefficient in micro-fluidized beds, some empirical/semiempirical correlations can be used. Typically, these correlations are expressed mainly by the dimensionless number of Sherwood number (Sh) as functions of Reynolds number (Re) and Schmidt number (Sc), as follows [76]:

$$Sh = k + b \times Re^{c} + Sc^{d} \tag{11.5}$$

where k, b, c, and d are experimentally determined constants. The first term on the right-hand side (k) represents mass transfer in diffusive conditions, and the second one refers to the enhancement of mass transfer caused by the convective fluid flow around the solids [77].

Ballesteros et al. [66] experimentally proved that the mass transfer coefficient K_C was independent of the diameter of solid particles. Moreover, it was also found that the column diameter and initial bed height did not influence the mass transfer rate [74,78]. Dwivedi and Upadhyay [79] reported that the bed voidage affected K_C. They compared the liquid–solid mass transfer coefficients in fixed and fluidized beds and concluded that K_C was inversely proportional to bed voidage. Yang et al. [80] found the mass transfer reached a maximum in liquid–solid micro-fluidized bed reactors when the bed voidage was around 0.7.

Note that discrepancies may appear when using empirical correlations to predict mass transfer coefficients. For example, it was reported that the diameter of the column did not affect the mass transfer rate [74], but Arters and Fan [81] specifically found a decreased mass transfer near the wall region in a liquid–solid fluidized bed of 40 mm ID compared to the central region. Arters and Fan also found that the mass transfer coefficient was independent of liquid flow rate [68], but Boskovic-Vragolovic et al. [70]

observed a slight decrease in mass transfer rate with the increase of liquid velocity in a liquid–solid fluidized bed of 40 mm ID fluidizing particles of 0.5–2.98 mm. The conflicting conclusions resulted from the limited range of parameters and different operating conditions, implying that a single correlation cannot describe the behavior over a wide range of Reynolds number or Schmidt number. Thus, more experimental work is required to elucidate further the influence of operating conditions on mass transfer in micro-fluidized beds.

11.3 Applications

11.3.1 Chemical conversions

The conversion of biomass to energy and fuels can be accomplished using various thermochemical or biochemical utilization processes, such as combustion, gasification, pyrolysis, and fermentation [82–84]. Boffito et al. [85] studied biodiesel production using novel one-step cracking/transesterification of vegetable oil in a 7 mm ID quartz micro-fluidized bed. They reported a biodiesel selectivity of 44% under optimal conditions. However, the process was only operated uninterrupted for 20 minutes, making evaluating the durability of this novel one-step process impracticable. Specifically, it was found that the catalysts of CaO/Al_2O_3 was deactivated with coke deposits on its surfaces, and the oil phase slumped the fluidized bed. Utilizing the same size fluidized bed system, Edake et al. [86] reported the production of 1,3-propanediol from glycerol hydrogenolysis using catalysts of $Pt/WO_3/Al_2O_3$ at ambient pressure. The yield of 1,3-PDO reached 14% after 2 hours at 260°C, starting from the glycerol dehydration to acrolein, followed by rehydration to 3-hydroxypropanal and then hydrogenation to 1,3-PDO. The by-products were 2-PrOH, propanal, methanol, ethylene glycol, acetone, CO, and CO_2.

11.3.2 Bioprocessing and bioproduction

The bioproduction demands high throughput with controllable and quantitative operating conditions, which led to the development of various bioreactor systems, such as flask culture, stirred tank reactor (STR), photobioreactor (PBR), and fluidized bed bioreactor. Fluidized bed bioreactors can overcome some of the limitations encountered by other reactors, such as the low mass transfer in a flask, the high shear force from the agitated impeller in an STR, and the relatively high capital cost of PBR. The fluidized bed bioreactor exploits fluidization of solids to achieve an increased mass/heat transfer, low mechanical stress on the cells, and low-pressure drops, which are all outstanding characteristics that are beneficial for bioengineering and bioproduction. To fluidize tiny microorganisms without being washed out, the cell immobilization technologies (i.e., attachment, entrapment, self-aggregation, containment) were developed to enhance cell stability in continuous fluidization conditions [87,88]. The application of micro-fluidized beds for bioengineering processes, such as wastewater treatment, biogas production, and other bio-application, is reviewed in the following sections.

11.3.2.1 Wastewater treatment

Currently, how to remove pollutants from domestic and industrial wastewater is still a tough challenge for the ecosystem and human life [89]. Micro-fluidized beds combined with cell immobilization technology have been widely applied to wastewater treatment, as summarized in Table 11.2. For example, Chowdhury et al. [90] compared the removal ability of biological nutrients using a novel liquid–solid circulating fluidized bed (LSCFB) bioreactor with and without the recirculation of particles (i.e., essentially biomass attached to lava rock particles). It was reported that the system without the particles recirculation removed 94%, 80%, and 65% of organic (chemical oxygen demand, COD), nitrogen (N), and phosphorous (P), respectively, and the system with the particles recirculation removed excess phosphorus with total removal efficiencies of 91%, 78%, and 85% for C, N, and P, respectively. Kuyukina et al. [91] tested different immobilized biocatalysts on hydrophobized carriers, such as sawdust, poly(vinyl alcohol) cryogel (cryoPVA), and poly(acrylamide) cryogel (cryoPAAG), for petroleum-contaminated water treatment in a fluidized bed reactor (14 mm ID, 120 mm length). The 70%–100% removal of n-alkanes (C10–C19) and 66%–70% removal of 2–3-ring PAHs indicated that micro-fluidized beds were efficient for biotreatment of petroleum-contaminated water, although the biotreating time lasted 2–3 weeks. Furthermore, using co-immobilized *Rhodococcus ruber* IEGM 615 and *Rhodococcus opacus* IEGM 249 strains on sawdust, Kuyukina et al. [92] achieved 70% biodegradation efficiencies for alkanes and PAHs within 2 weeks compared to corresponding 24% and 37%, respectively, without inoculations in the bed. Likewise, Banerjee et al. [7] reported petroleum wastewater biotreatment using the immobilized super-tolerant phenol strains of Bacillus cereus (AKG1 MTCC9817 and AKG2 MTCC9818) in a fluidized bed bioreactor (50 mm ID), in which a 95% reduction of the initial phenolic compounds under the continuous mode of operation was achieved. Qiu et al. [93] compared the formaldehyde biodegradation with immobilized *Methylobacterium* sp. XJLW cells in both the three-phase micro-fluidized bed (36 mm ID, 460 mm height) and the shake flask showed that the micro-fluidized bed was more efficient in wastewater treatment than other systems. The immobilized cells from the fluidized bed reactor could degrade 5 g/L formaldehyde with a maximum degradation rate of 464.5 mg/L/h after 20 batches of recycling.

For anaerobic posttreatment of wastewater, Kwak et al. [99] reported the use of a micro-fluidized bed membrane reactor (35 mm ID and 500 mm height) for low-strength ammonia wastewater treatment. Oztemur et al. [101] also reported the high

Table 11.2 Summary of wastewater treatment using different micro-fluidized bed reactor systems.

References	Column ID (mm)	Wastewater	Particle properties	Microorganisms	Results
Chowdhury et al. [90]	20, 76	Degritted municipal wastewater	Lava rock particles; $d_p = 0.3-1.0$ mm	Return activated sludge of 3511 mg TSS/L and 2810 mg VSS/L	Removal of 91%, 78%, and 85% for organic (chemical oxygen demand, COD), nitrogen (N), and phosphorous (P), respectively, with particle recirculation.
Kuyukina et al. [91]	14	Petroleum-contaminated water	Hydrophobized carriers such as sawdust, poly(vinyl alcohol) cryogel (cryoPVA), and poly(acrylamide) cryogel (cryoPAAG) $d_P = 1-3$ mm	Strains of *Rhodococcus rubber* IEGM 615 and *Rhodococcus opacus* IEGM 249	70%–100% removal of n-alkanes (C10–C19) and 46%–70% removal of 2–3-ring PAHs.
Kuyukina et al. [92]	14	Oilfield wastewater	Sawdust samples; $d_p = 1-3$ mm	Strains of *Rhodococcus ruber* IEGM 615 and *Rhodococcus opacus* IEGM 249	70% biodegradation efficiencies for alkanes and PAHs within 2 weeks; 75%–96% removal of heavy metals (Al, Cr, Cu, Fe, Hg, Zn, Mn).
Banerjee et al. [7]	50	Petroleum wastewater	Ca-alginate beads; $d_P = 3$ mm	Strains of *Bacillus cereus* (AKG1 MTCC9817 and AKG2 MTCC9818)	95% or more reduction of COD and phenolic compounds.
Qiu et al. [93]	36	Formaldehyde (H–CHO)	Gel-entrapped beads; $d_P = 4-5$ mm	Strain of *Methylobacterium* sp. XJLW	Degradation of 5 g/L formaldehyde (with a maximum degradation rate of 464.5 mg/L/h under the optimum conditions).

Ballesteros et al. [94]	20, 40	Nickel-containing synthetic wastewater	Nickel carbonate granule; $d_P = 0.50-0.15$ mm	/	98.8% of nickel removal.
Chen et al. [95]	34.4, 50	Campus domestic wastewater	Natural zeolites; $d_P = 0.2-1$ nm	Anaerobic digested sludge	Removal of COD, BOD, and SS reached 84%, 87%, and 96%, respectively, with an HRT from 3 to 4 h.
Geng et al. [96]	31	Isopropanol	Calcium alginate, polyvinyl alcohol, activated carbon, SiO_2; $d_p = 3$ mm	Strain of *Paracoccus denitrificans*	Reduction of COD from 5000 to 109 mg/L under optimum conditions; the degradation of IPA reached 98.3%.
Ismail et al. [97]	50	Furfural-laden wastewater	Natural polymers of guar gum (GG), agar-agar (AA), and sodium alginate cross-linked with polyvinyl alcohol (PVA)	*Bacillus* cells	The removal efficiency of furfural was 100%, 100%, and 95% for the first, second, and third cycles, respectively.
Kim et al. [98]	50	Synthetic wastewater with chemical oxygen demand (COD) averaging 513 mg/L	Granular activated carbon (GAC)	/	COD removals reached 99%.
Kwak et al. [99]	35	Low-strength (60 mg/L TN) synthetic wastewater	Granular activated carbon; $d_P = 0.8-1$ mm	/	Total nitrogen (TN) removal efficiency was over 82%.

(Continued)

Table 11.2 (Continued)

References	Column ID (mm)	Wastewater	Particle properties	Microorganisms	Results
Kwon et al. [100]	50	Synthetic wastewater with a COD of 300 mg/L	GAC particles; $d_p = 0.84$ mm	/	The overall removal efficiency of 82% of NH4-N.
Oztemur et al. [101]	50	Synthetic wastewater	Silica; $d_p = 1-2$ mm	/	COD oxidation and sulfate reduction efficiencies up to 98%
Patroescu et al. [8]	50	Low-pitched groundwater containing nitrates	Expanded clay granular; $d_p = 2-5$ mm	/	The denitrification rate was between 3390 and 3867 g NO_3-N/m^3/day in a fluidized bed reactor.
Wang et al. [102]	38	Synthetic municipal wastewater	Activated carbon-coated polypropylene beads; $d_p = 3.0-3.5$ mm	The dominant phyla were *Proteobacteria, Bacteroidetes*, and *Epsilonbacteraeota*	Total COD removal efficiencies of N84% were achieved concomitantly with complete nitrification. The overall nitrogen removal efficiencies were N75%.

sulfate reduction in a micro-fluidized bed membrane bioreactor, which consisted of a plexiglass column of 50 mm ID and 1250 mm height. Such membrane reactors needed no aeration requirement with less sludge production, but membrane fouling remained a significant issue [98]. As summarized in Table 11.2, the wastewater treatment is dependent on the specific microorganisms to degrade or convert certain chemical from effluent.

Micro-fluidized bed bioreactors have proven to be efficient in wastewater treatments, but the reactor design procedure and operating conditions are too complex to standardize. During wastewater processing, the cultivation conditions (i.e., pH, temperature, growth medium, concentration, etc.) for different cells or bacteria differ significantly and so do the performances of wastewater treatments. Besides, the cultivation period can be as long as several days, and the degradation time can be more than 300 hours for a wastewater treatment cycle [103,104]. In addition, the biological technology can be limited by the bioactivity of microbials, which may lose their biodegradability by toxic substances. To overcome these issues, researchers have developed electrolysis-based MFB systems for the pretreatment of poor biodegradable and even toxic wastewater, which are reported to have the advantages of low cost, ease of operation, and high efficiency without the necessity for an energy input [105,106]. However, its stability has not been thoroughly tested.

11.3.2.2 Bioproduction

Micro-fluidized bed bioreactors used for small-scale cell cultivation and fungal fermentation are controllable to obtain the desired biological products, thus benefitting parallelization, automation, and cost reduction [107]. The production of biofuels (e.g., hydrogen, ethanol) is one of the typical applications of micro-fluidized bed bioreactors. Wu et al. [108] produced biohydrogen and bioethanol in a micro-fluidized bed bioreactor (27 mm ID and a height of 120 cm) containing immobilized anaerobic sludge. The highest H_2 production rate (59 mmol/h/L) was attained with the sugar substrate of sucrose at a liquid velocity of 0.91 cm/s, and the highest H_2 yield (1.04 mol/mol-hexose) was obtained with glucose at a liquid velocity of 0.55 cm/s. On the other hand, at a liquid velocity of 0.91 cm/s, the sugar substrate of fructose was demonstrated to have the best ethanol production rate and yield of 378 mmol/h/L and 0.65 mol/mol-hexose, respectively. Cavalcante de Amorim et al. [109] found that the hydrogen production rate was linearly related to the hydraulic retention time (HRT) in an anaerobic fluidized bed reactor using expanded clay as a support carrier and glucose substrate. Furthermore, they determined that when polystyrene and expanded clay were used as carriers, the maximum hydrogen yields were 1.90 and 2.59 mol, and the highest hydrogen production rates were 0.95 and 1.21 L/h/L, respectively [110]. Other parameters (such as liquid velocity, temperatures, and HRT) affecting hydrogen production were also studied [111,112]. It was noteworthy that a high minimum fluidization velocity led to a high hydrogen output because an increase in minimum fluidization velocity probably resulted in a minor shear force on cells, thus minimizing the potential cell damage and death.

Cell encapsulation is another cell immobilization method to produce natural pigment via the cultivation of encapsulated *Monascus purpureus*. Liu et al. [113] confirmed a higher cell density in encapsulated cells than in free cell cultivation in a flask. Different cultivation modes of gel-entrapped *Sphingomonas* sp. ZUTE03 were tested in a three-phase micro-fluidized bed bioreactor to produce CoQ_{10} [114]. However, the long duration of cell cultivation through immobilized cell fluidization may result in cell detachment from carriers or cell leakage from encapsulation; thus, research on improving the stability of the immobilization is needed in the future.

Regardless of attachment or encapsulation for cell immobilization, the functionalized solids with a relatively stable size can be well fluidized if operated properly. In a study of fungal fermentation in a fluidized bed reactor (50 mm in ID) for the production of raspberry ketone, it was observed that the fungus of *Nidula niveo-tomentosa* experienced different fungal morphologies during 4 weeks of fermentation. In the experiments, the gas sparger size was kept at 0.15 mm, and the liquid flow rate was fixed at 0.25 mL/s to maintain the culture medium circulation. Meanwhile, a fixed gas flow rate of 0.15 v.v.m was used to suspend the fungal pellets and provide oxygen. Fig. 11.7 shows the snapshots of fluidization regime changes observed in the 4-week fungal fermentation test. Three different fluidization regimes were identified:

- Full fluidization regime: Pellets were fully fluidized by bubble flows in the first 2 weeks (Fig. 11.7A and B).
- Partial fluidization regime: Part of the pellets were fluidized while the other part agglomerated and settled onto the bottom of the bed in weeks 2–3 (Fig. 11.7C).

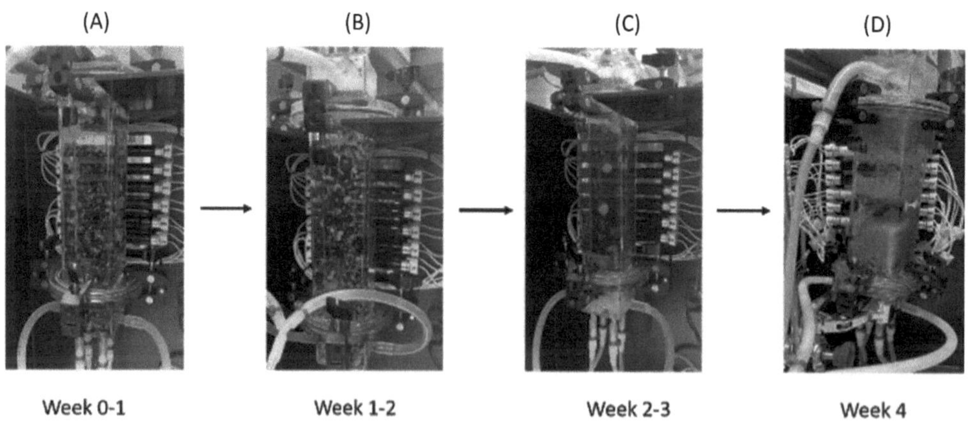

Figure 11.7 Fluidization regimes showing the change of pellet fluidization during 4 weeks of fermentation. $Q_G = 0.15$ v.v.m, $d_G = 0.15$ mm. (A–B): full fluidization regime; (C) Partial fluidization regime; and (D) Packed bed regime.

- Packed bed regime: All pellets were agglomerated and accumulated on the bottom, and channels were formed when bubbles flow through the pellets in the 4th week (Fig. 11.7D).

Change of flow regime from fully fluidized to partially fluidized and eventually to packed bed attributed to the pellet growth and biomass accumulation at low fluid flow rates. Namely, the initial fungi inoculum was small with a small amount of biomass weight, easily fluidized by gas. The continued fermentation enabled the free mycelia fungi (in week 0) to grow to mature fungal pellets (in weeks 1–2). However, the long-term gas sparging seemed to make the pellets fluffy and hairy, resulting in pellet agglomeration. Low gas flow rates (0.15–0.45 v.v.m) failed to fluidize the big clumps, leading to the aggregated pellets settling onto the bottom, forming the partially fluidized bed in week 3, and finally forming the packed bed in week 4. Therefore, compared to the immobilized cells/enzymes, the fungal pellets could be directly fluidized by the liquid–gas flow without immobilization procedures in a biocarrier, but they gradually increased size and the morphology largely impacts pellet fluidization. The change in pellet size makes it unable to maintain good fluidization at a fixed flow rate, ultimately resulting in pellet sedimentation or elutriation when the flow rate is too low or too high. Besides, the floccose pellets with filamentous structures lead to pellet agglomeration and the formation of fungal clumps, which will settle down onto the bottom and cause low mass transfer. To improve the pellet fluidization, it is suggested that the dynamic flow rate is used based on the pellet size distribution in different fermentation stages to maintain homogenous fluidization and good mass transfer.

In summary, a micro-fluidized bed bioreactor has the advantages of sufficient mass transfer and mixing, thus ensuring the medium and dissolved oxygen supply. Besides, the laminar flow in the fluidized chamber leads to less hydrodynamic shear stress for cells, which mitigates the cell damage and/or high death rate experienced in stirred tank bioreactors. However, long-term cell cultivation through immobilized cell fluidization requires that robust cell immobilization techniques are used to avoid cell detachment from carriers or cell leakage from encapsulation. In aerobic cultivation/fermentation, the air is supplied to provide oxygen in gas bubbles, which can burst to influence cell morphology and even cause cell death [115,116]. Therefore, it is worthwhile considering the effect of bubbles (velocity, size) on cell cultivation in the microfluidic system. Last but not least, the size and number of cells are growing up in fluidized beds so that the liquid and/or gas flow rate must be adjusted to maintain the ideal fluidization state, which in turn may change the operating system to affect the final product yield and rate. Such details are easily neglected but indeed play an essential role in fluidized bed bioreactors, which need to be further explored in the future.

11.3.3 Other applications

In addition to the systematic applications in chemical conversions and bioproduction, utilization of liquid–solid micro-fluidized beds in some other areas has also been reported. One of the typical cases is the magnetic-assisted micro-fluidized bed system developed by Pereiro et al. [11]. In a microfluidic chamber, the combined drag force between liquid and particles and wall effects can overweight gravitational forces; thus, the external forces such as magnetic assist in attaining stable fluidization. Using the antibody-functionalized superparamagnetic beads, Pereiro et al. [117,118] reported the application of micro-fluidized beds in capturing and detecting *Salmonella Typhimurium* in undiluted unskimmed milk without pretreatment. Besides, the same magnetic micro-fluidized bed system was also used for DNA analysis, and results showed that 93% of DNA target could be captured by hybridization, thus broadening the application potential of MFB in the medical diagnostics field [119]. In addition, Lv et al. [120] proposed an inflatable-inclined liquid–solid fluidized bed to separate fine coal, another possibility of MFB for solid processing. With the development of novel techniques and advanced methods, more research on liquid–solid MFBs for solving practical issues is expected.

11.3.4 Challenges and prospects for MFB scaling-up

Despite its demonstrated advantages, the MFB for industrial-scale applications remains in the initial stage. One of the mature commercial products is the gas–solid MFB analytical device developed by Xu's group, used for varieties of thermochemical gas–solid reactions [89,121]. Samih et al. [122] proposed an MFB thermogravimetric analyzer, which can be applied to the development and design of coal and other gasification processes. Admittedly, the direct adoption of results obtained from MFB into commercial applications is limited by the inherent scale dependence of operation parameters [123]. For instance, changing one parameter (e.g., bed size, particle size, or density) with other parameters unchanged may lead to significantly different hydrodynamics. The widely applied method for fluidized bed scaling is to use the dimensionless numbers (e.g., Reynolds number, Froude number, gas–particle density ratio, sphericity and size distribution of the particle, etc.) for keeping the hydrodynamic similarity in different sizes of reactors [124], although the wall effects and particle interactions (which are significant in MFB but negligible in large-scale fluidized bed) make the scaling-up results possibly inconsistent with those in MFBs. In addition, advanced validating tools, such as CFD simulation, can improve the scaling-up process. Finally, using MFB for commercial-scale production can be achieved by combining numerous MFB in parallel, which enlarges the production rate while maintaining the advantages of MFB. This was exampled by a novel compact micro-fluidized beds (CMFB, which contains 100 micro-fluidized beds) for CO_2 capture [125] and the micro-structured fluidized bed membrane reactors [126].

11.4 Conclusion

In this chapter, the liquid–solid micro-fluidized bed technology has been discussed, covering hydrodynamic properties and applications for chemical conversion, wastewater treatment, and microbial processing. The miniaturized reactor size enables fast screening, controllable safety, and low capital cost. Besides, micro-fluidized beds are featured with intense mixing, excellent mass, and heat transfer. Some issues related to liquid–solid micro-fluidization processes need to be further investigated, such as solids agglomeration and sintering, as well as size enlargement of biomass during the cultivation of bioprocesses.

References

[1] Potic B, Kersten SRA, Ye M, van der Hoef MA, Kuipers JAM, van Swaaij WPM. Fluidization with hot compressed water in micro-reactors. Chemical Engineering Science 2005;60(22):5982–90.
[2] Zivkovic V, Biggs MJ. On importance of surface forces in a microfluidic fluidized bed. Chemical Engineering Science 2015;126:143–9.
[3] Wang F, Fan L-S. Gas – solid fluidization in mini- and micro-channels. Industrial & Engineering Chemistry Research 2011;50(8):4741–51.
[4] Gunther A, Jensen KF. Multiphase microfluidics: from flow characteristics to chemical and materials synthesis. Lab on a Chip 2006;6(12):1487–503.
[5] Geng S, Han Z, Yue J, Li Y, Zeng X, Lai D, et al. Conditioning micro-fluidized bed for maximal approach of gas plug flow. Chemical Engineering Journal 2018;351:110–18.
[6] Liu X, Xu G, Gao S. Micro-fluidized beds: wall effect and operability. Chemical Engineering Journal 2008;137(2):302–7.
[7] Banerjee A, Ghoshal AK. Biodegradation of real petroleum wastewater by immobilized hyper phenol-tolerant strains of *Bacillus cereus* in a fluidized bed bioreactor. 3 Biotech 2016;6(2):137.
[8] Patroescu VI, Ionescu I, Constantin LA, Dinu LR, Badescu V. Nitrate removal from groundwater by denitrification in fixed and fluidized bed biofilm reactors: a comparative study. Engineering Biology 2019.
[9] do Nascimento OL, Reay DA, Zivkovic V. Influence of surface forces and wall effects on the minimum fluidization velocity of liquid-solid micro-fluidized beds. Powder Technology 2016;304:55–62.
[10] McDonough JR, Law R, Reay DA, Zivkovic V. Fluidization in small-scale gas-solid 3D-printed fluidized beds. Chemical Engineering Science 2019;200:294–309.
[11] Pereiro I, Tabnaoui S, Fermigier M, du Roure O, Descroix S, Viovy JL, et al. Magnetic fluidized bed for solid phase extraction in microfluidic systems. Lab on a Chip 2017;17(9):1603–15.
[12] Do Nascimento OL, Reay D, Zivkovic V. Study of transitional velocities of solid–liquid micro-circulating fluidized beds by visual observation. Journal of Chemical Engineering of Japan 2018;51(4):349–55.
[13] Li X, Liu M, Li Y. Bed expansion and multi-bubble behavior of gas-liquid-solid micro-fluidized beds in sub-millimeter capillary. Chemical Engineering Journal 2017;328:1122–38.
[14] Zhang Y, Ng YL, Goh K-L, Chow Y, Wang S, Zivkovic V. Fluidization of fungal pellets in a 3D-printed micro-fluidized bed. Chemical Engineering Science 2021;236.
[15] Zhang Y, Goh K-L, Ng Y-L, Chow Y, Zivkovic V. Design and investigation of a 3D-printed micro-fluidized bed. ChemEngineering 2021;5(3):62.
[16] Zivkovic VN, Kashani M, Biggs M. Experimental and theoretical study of a micro-fluidized bed. AIP Conference Proceedings 2013;1542:93–6.
[17] Becker H, Gartner C. Polymer microfabrication technologies for microfluidic systems. Analytical and Bioanalytical Chemistry 2008;390(1):89–111.
[18] do Nascimento OL, Reay DA, Zivkovic V. Solid circulating velocity measurement in a liquid-solid micro-circulating fluidised bed. Processes 2020;8(9):1159.

[19] Macdonald NP, Cabot JM, Smejkal P, Guijt RM, Paull B, Breadmore MC. Comparing microfluidic performance of three-dimensional (3D) printing platforms. Analytical Chemistry 2017;89(7):3858−66.

[20] Doroodchi E, Peng Z, Sathe M, Abbasi-Shavazi E, Evans GM. Fluidisation and packed bed behaviour in capillary tubes. Powder Technology 2012;223:131−6.

[21] Boffito DC, Neagoe C, Edake M, Pastor-Ramirez B, Patience GS. Biofuel synthesis in a capillary fluidized bed. Catalysis Today 2014;237:13−17.

[22] Lu Y, Zhao L, Han Q, Wei L, Zhang X, Guo L, et al. Minimum fluidization velocities for supercritical water fluidized bed within the range of 633−693K and 23−27MPa. International Journal of Multiphase Flow 2013;49:78−82.

[23] Zhang Y, Goh K-L, Ng YL, Chow Y, Wang S, Zivkovic V. Process intensification in micro-fluidized bed systems: a review. Chemical Engineering and Processing - Process Intensification 2021.

[24] Rao A, Curtis JS, Hancock BC, Wassgren C. The effect of column diameter and bed height on minimum fluidization velocity. AIChE Journal 2010; NA-NA.

[25] Zivkovic V, Kashani MN, Biggs MJ. Experimental and theoretical study of a micro-fluidized bed. Chemistry Materials Science 2013;93−6.

[26] Besagni G, Inzoli F, Ziegenhein T, Lucas D. Computational fluid-dynamic modeling of the pseudo-homogeneous flow regime in large-scale bubble columns. Chemical Engineering Science 2017;160:144−60.

[27] Richardson JF, Zaki WN. The sedimentation of a suspension of uniform spheres under conditions of viscous flow. Chemical Engineering Science 1954;3(2):65−73.

[28] Daizo K, Octave L. Fluidization engineering. 2nd ed Elsevier; 2013. 491 p.

[29] Miller Clark O, Logwinuk AK. Fluidization studies of solid particles. Industrial Engineering Chemistry 1951;43(5):1220−6.

[30] Leva M. Fluidization. New York; Toronto; London: McGraw-Hill; 1959.

[31] Wen CY, Yu YH. Mechanics of Fluidization. The Chemical Engineering Progress Symposium Series 1966;162:100−11.

[32] Ergun S. Fluid flow through packed columns. Chemical Engineering Process 1952;48:89−94.

[33] Yates JG. Effects of temperature and pressure on gas-solid fluidization. Chemical Engineering Science 1996;51(2):167−205.

[34] Choi YS, Kim SJ, Kim D. A semi-empirical correlation for pressure drop in packed beds of spherical particles. Transport in Porous Media 2008;.

[35] Li L, Ma W. Experimental study on the effective particle diameter of a packed bed with non-spherical particles. Transport in Porous Media 2011;89(1):35−48.

[36] Tang C, Liu M, Li Y. Experimental investigation of hydrodynamics of liquid-solid mini-fluidized beds. Particuology. 2016;27:102−9.

[37] Guo Qj, Xu Y, Yue X. Fluidization characteristics in micro-fluidized beds of various inner diameters. Chemical Engineering & Technology 2009;32(12):1992−9.

[38] Chen Y. The influence of surface hydrophobicity in fluidization of ultrafine Al_2O_3 particles. Chemical Engineering and Processing: Process Intensification 2016;110:21−9.

[39] Qie Z, Alhassawi H, Sun F, Gao J, Zhao G, Fan X. Characteristics and applications of micro-fluidized beds (MFBs). Chemical Engineering Journal 2022;428.

[40] Li Y, Liu M, Li X. Flow regimes in gas-liquid-solid mini-fluidized beds with single gas orifice. Powder Technology 2018;333:293−303.

[41] Vanni F, Caussat B, Ablitzer C, Brothier M. Effects of reducing the reactor diameter on the fluidization of a very dense powder. Powder Technology 2015;277:268−74.

[42] Zhang H, Huang Y, An X, Yu A, Xie J. Numerical prediction on the minimum fluidization velocity of a supercritical water fluidized bed reactor: effect of particle size distributions. Powder Technology 2021;389:119−30.

[43] Zivkovic V, Biggs MJ, Alwahabi ZT. Experimental study of a liquid fluidization in a microfluidic channel. AIChE Journal 2013;59(2):361−4.

[44] Cúñez FD, Lima NC, Franklin EM. Motion and clustering of bonded particles in narrow solid-liquid fluidized beds. Physics of Fluids 2021;33(2).

[45] Ma N, Chalmers JJ, Auninüs JG, Zhou W, Xie L. Quantitative studies of cell-bubble interactions and cell damage at different pluronic F-68 and cell concentrations. Biotechnology Progress 2004;20 (4):1182–91.
[46] Patroescu IV, Ionescu IA, Constantin LA, Dinu LR, Badescu VR. Nitrate removal from groundwater by denitrification in fixed and fluidized bed biofilm reactors a comparative study. Revista De Chimie 2019;70:297–300.
[47] Zhang Y, Ullah N, Law R, Reay D, Zivkovic V. Investigation into the hydrodynamics of liquid-solid inclined micro-fluidized beds. Resources Chemicals and Materials 2022.
[48] Aguilar-Corona A, Masbernat O, Figueroa B, Zenit R. The effect of column tilt on flow homogeneity and particle agitation in a liquid fluidized bed. International Journal of Multiphase Flow 2017;92:50–60.
[49] Han Z, Yue J, Geng S, Hu D, Liu X, Suleiman SB, et al. State-of-the-art hydrodynamics of gas-solid micro-fluidized beds. Chemical Engineering Science 2021;232.
[50] Zivkovic V, Ridge N, Biggs MJ. Experimental study of efficient mixing in a micro-fluidized bed. Applied Thermal Engineering 2017;127:1642–9.
[51] Mao Z, Yang C. Micro-mixing in chemical reactors: a perspective. Chinese Journal of Chemical Engineering 2017;25(4):381–90.
[52] McDonough JR, Oates MF, Law R, Harvey AP. Micromixing in oscillatory baffled flows. Chemical Engineering Journal 2019;361:508–18.
[53] Hartge EU, Luecke K, Werther J. The role of mixing in the performance of CFB reactors. Chemical Engineering Science 1999;54(22):5393–407.
[54] Doroodchi E, Sathe M, Evans G, Moghtaderi B. Liquid-liquid mixing using micro-fluidised beds. Chemical Engineering Research and Design 2013;91(11):2235–42.
[55] Derksen JJ. Scalar mixing with fixed and fluidized particles in micro-reactors. Chemical Engineering Research and Design 2009;87(4):550–6.
[56] Derksen JJ. Simulations of liquid-to-solid mass transfer in a fluidized microchannel. Microfluidics and Nanofluidics 2014;18(5–6):829–39.
[57] Odeleye AOO, Chui C-Y, Nguyen L, Castrejón-Pita AA, Ye H, Cui Z. On the use of 3D-printed flow distributors to control particle movement in a fluidized bed. Chemical Engineering Research and Design 2018;140:194–204.
[58] Chaikittisilp W, Taenumtrakul T, Boonsuwan P, Tanthapanichakoon W, Charinpanitkul T. Analysis of solid particle mixing in inclined fluidized beds using DEM simulation. Chemical Engineering Journal 2006;122(1–2):21–9.
[59] Coiado EM, Diniz VEMG. Two-phase (solid-liquid) flow in inclined pipes. Journal of the Brazilian Society of Mechanical Sciences 2001;23(3):347–62.
[60] Yaralioglu Goksen G, Wygant Ira O, Marentis Theodore C, Khuri-Yakub BT. Ultrasonic mixing in microfluidic channels using integrated transducers. Analytical Chemistry 2004;76 (13):3694–8.
[61] Tsai Jr-Hung, Lin L. Active microfluidic mixer and gas bubble filter driven by thermal bubble micropump. Sensors and Actuators A: Physical 2002;97–98(1):665–71.
[62] Lee SH, van Noort D, Lee JY, Zhang BT, Park TH. Effective mixing in a microfluidic chip using magnetic particles. Lab on a Chip 2009;9(3):479–82.
[63] Karimi F, Haghshenasfard M, Sotudeh-Gharebagh R, Zarghami R, Mostoufi N. Multiscale characterization of nanoparticles in a magnetically assisted fluidized bed. Particuology 2020;51:64–71.
[64] Zhang Y, Jin B, Zhong W. Experiment on particle mixing in flat-bottom spout–fluid bed. Chemical Engineering and Processing: Process Intensification 2009;48(1):126–34.
[65] Chen H, Tian X, Gu S, Geng C, Yang D. Effect of mixing state of binary particles on bubble behavior in 2D fluidized bed. Chemical Engineering Communications 2018;205(8):1119–28.
[66] Ballesteros RL, Riba JP, Couderc JP. Dissolution of non spherical particles in solid-liquid fluidization. Chemical Engineering Science 1982;37(11):1639–44.
[67] Fan L-T, Yang Y-C, Wen C-Y. Mass transfer in semifluidized beds for solid-liquid system. AIChE Journal 1960;6(3):482–7.

[68] Arters DC, Fan L-S. Solid-liquid mass transfer in a gas-liquid-solid fluidized bed. In: This paper was presented at the 1984 Annual Meeting of the American Institute of Chemical Engineers, 25–30 November. Chemical Engineering Science. 1986;41(1):107-15.
[69] Upadhyay SN, Tripathi G. Liquid-phase mass transfer in fixed and fluidized beds of large particles. Journal of Chemical & Engineering Data 1975;20(1):20–6.
[70] Boskovic-Vragolovic N, Brzic D, Grbavcic Z. Mass transfer between a fluid and an immersed object in liquid-solid packed and fluidized beds. Journal of the Serbian Chemical Society 2005;70 (11):1373–9.
[71] Kalaga DV, Dhar A, Dalvi SV, Joshi JB. Particle-liquid mass transfer in solid-liquid fluidized beds. Chemical Engineering Journal 2014;245:323–41.
[72] Koloini T, Sopčič M, Žumer M. Mass transfer in liquid-fluidized beds at low Reynolds numbers. Chemical Engineering Science 1977;32(6):637–41.
[73] Livingston AG, Noble JB. Mass transfer in liquid-solid fluidized beds of ion exchange resins at low Reynolds numbers. Chemical Engineering Science 1993;48(6):1174–8.
[74] Rahmant K, Streat M. Mass transfer in liquid fluidized beds of ion exchange particles. Chemical Engineering Science 1981;36(2):293–300.
[75] Chavan PV, Kalaga DV, Joshi JB. Solid – liquid circulating multistage fluidized bed: hydrodynamic study. Industrial & Engineering Chemistry Research 2009;48(9):4592–602.
[76] Frossling N. Uber die Verdunstung fallender Tropfen. Gerlands Beiträge zur Geophysik 1938;52:170–216.
[77] Scala F. Particle-fluid mass transfer in multiparticle systems at low Reynolds numbers. Chemical Engineering Science 2013;91:90–101.
[78] McCune LK, Wilhelm RH. Mass and momentum transfer in a solid-liquid system. Industrial Engineering Chemistry 1949;41(6):1124–34.
[79] Dwivedi PN, Upadhyay SN. Particle-fluid mass transfer in fixed and fluidized beds. Industrial & Engineering Chemistry Process Design and Development 1977;16(2):157–65.
[80] Yang Z, Liu M, Lin C. Photocatalytic activity and scale-up effect in liquid-solid mini-fluidized bed reactor. Chemical Engineering Journal 2016;291:254–68.
[81] Arters DC, Fan L-S. Experimental methods and correlation of solid-liquid mass transfer in fluidized beds. Chemical Engineering Science 1990;45(4):965–75.
[82] Damartzis T, Zabaniotou A. Thermochemical conversion of biomass to second generation biofuels through integrated process design—a review. Renewable and Sustainable Energy Reviews 2011;15 (1):366–78.
[83] Fouilland T, Grace JR, Ellis N. Recent advances in fluidized bed technology in biomass processes. Biofuels 2014;1(3):409–33.
[84] Wang F, Zeng X, Geng S, Yue J, Tang S, Cui Y, et al. Distinctive hydrodynamics of a micro-fluidized bed and its application to gas-solid reaction analysis. Energy and Fuels 2017;32(4):4096–106.
[85] Boffito DC, Manrique GB, Patience GS. One step cracking/transesterification of vegetable oil: reaction–regeneration cycles in a capillary fluidized bed. Energy Conversion and Management 2015;103:958–64.
[86] Edake M, Dalil M, Darabi Mahboub MJ, Dubois J-L, Patience GS. Catalytic glycerol hydrogenolysis to 1,3-propanediol in a gas-solid fluidized bed. RSC Advances 2017;7(7):3853–60.
[87] Kourkoutas Y, Bekatorou A, Banat IM, Marchant R, Koutinas AA. Immobilization technologies and support materials suitable in alcohol beverages production: a review. Food Microbiology 2004;21(4):377–97.
[88] Sekoai PT, Awosusi AA, Yoro KO, Singo M, Oloye O, Ayeni AO, et al. Microbial cell immobilization in biohydrogen production: a short overview. Critical Reviews in Biotechnology 2018;38(2):157–71.
[89] Bello MM, Abdul Raman AA, Purushothaman M. Applications of fluidized bed reactors in wastewater treatment – A review of the major design and operational parameters. Journal of Cleaner Production 2017;141:1492–514.
[90] Chowdhury N, Zhu J, Nakhla G, Patel A, Islam MJCE. Engineering-biotechnology TICPEP. A Novel Liquid-Solid Circulating Fluidized-Bed Bioreactor for Biological Nutrient Removal from Municipal Wastewater 2009;32(3):364–72.

[91] Kuyukina MS, Ivshina IB, Serebrennikova MK, Krivorutchko AB, Podorozhko EA, Ivanov RV, et al. Petroleum-contaminated water treatment in a fluidized-bed bioreactor with immobilized Rhodococcus cells. International Biodeterioration & Biodegradation 2009;63(4):427–32.
[92] Kuyukina MS, Ivshina IB, Serebrennikova MK, Krivoruchko AV, Korshunova IO, Peshkur TA, et al. Oilfield wastewater biotreatment in a fluidized-bed bioreactor using co-immobilized Rhodococcus cultures. Journal of Environmental Chemical Engineering 2017;5(1):1252–60.
[93] Qiu L, Chen W, Zhong L, Wu W, Wu S, Chen J, et al. Formaldehyde biodegradation by immobilized Methylobacterium sp. XJLW cells in a three-phase fluidized bed reactor. Bioprocess and Biosystems Engineering 2014;37(7):1377–84.
[94] Ballesteros FC, Salcedo AF, Vilando AC, Huang YH, Lu MC. Removal of nickel by homogeneous granulation in a fluidized-bed reactor. Chemosphere 2016;164:59–67.
[95] Chen W-H, Tsai C-Y, Chen S-Y, Sung S, Lin J-G. Treatment of campus domestic wastewater using ambient-temperature anaerobic fluidized membrane bioreactors with zeolites as carriers. International Biodeterioration & Biodegradation 2019;136:49–54.
[96] Geng Y, Deng Y, Chen F, Jin H, Hou T, Tao K. Isopropanol biodegradation by immobilized Paracoccus denitrificans in a three-phase fluidized bed reactor. Preparative Biochemistry and Biotechnology 2016;46(8):747–54.
[97] Ismail ZZ, Khudhair HA. Novel application of immobilized Bacillus cells for biotreatment of furfural-laden wastewater.
[98] Kim J, Kim K, Ye H, Lee E, Shin C, McCarty PL, et al. Anaerobic fluidized bed membrane bioreactor for wastewater treatment. Environmental Science & Technology 2011;45(2):576–81.
[99] Kwak W, Rout PR, Lee E, Bae J. Influence of hydraulic retention time and temperature on the performance of an anaerobic ammonium oxidation fluidized bed membrane bioreactor for low-strength ammonia wastewater treatment. Chemical Engineering Journal 2020;386.
[100] Kwon D, Kwon SJ, Kim J, Lee J-H. Feasibility of the highly-permselective forward osmosis membrane process for the post-treatment of the anaerobic fluidized bed bioreactor effluent. Desalination 2020;485.
[101] Oztemur G, Teksoy Basaran S, Tayran Z, Sahinkaya E. Fluidized bed membrane bioreactor achieves high sulfate reduction and filtration performances at moderate temperatures. Chemosphere 2020;252.
[102] Wang H, He X, Nakhla G, Zhu J, Su YK. Performance and bacterial community structure of a novel inverse fluidized bed bioreactor (IFBBR) treating synthetic municipal wastewater. The Science of the Total Environment 2020;718:137288.
[103] González G, Herrera G, García MT, Peña M. Biodegradation of phenolic industrial wastewater in a fluidized bed bioreactor with immobilized cells of *Pseudomonas putida*. Bioresource Technology 2001;80(2):137–42.
[104] Zou G, Papirio S, Lakaniemi AM, Ahoranta SH, Puhakka JA. High rate autotrophic denitrification in fluidized-bed biofilm reactors. Chemical Engineering Journal 2016;284:1287–94.
[105] Han Y, Li H, Liu M, Sang Y, Liang C, Chen J. Purification treatment of dyes wastewater with a novel micro-electrolysis reactor. Separation and Purification Technology 2016;170:241–7.
[106] Malakootian M, Kannan K, Gharaghani MA, Dehdarirad A, Nasiri A, Shahamat YD, et al. Removal of metronidazole from wastewater by Fe/charcoal micro electrolysis fluidized bed reactor. Journal of Environmental Chemical Engineering 2019;7(6).
[107] Kumar S, Wittmann C, Heinzle E. Minibioreactors. Biotechnology Letters 2004;26:1–10.
[108] Wu K-J, Chang C-F, Chang J-S. Simultaneous production of biohydrogen and bioethanol with fluidized-bed and packed-bed bioreactors containing immobilized anaerobic sludge. Process Biochemistry 2007;42(7):1165–71.
[109] Cavalcante de Amorim E, Barros A, Rissatozamariollidamianovic M, Silva E. Anaerobic fluidized bed reactor with expanded clay as support for hydrogen production through dark fermentation of glucose. International Journal of Hydrogen Energy 2009;34(2):783–90.
[110] Barros AR, Cavalcante de Amorim EL, Reis CM, Shida GM, Silva EL. Biohydrogen production in anaerobic fluidized bed reactors: effect of support material and hydraulic retention time. International Journal of Hydrogen Energy 2010;35(8):3379–88.

[111] dos Reis CM, Silva EL. Effect of upflow velocity and hydraulic retention time in anaerobic fluidized-bed reactors used for hydrogen production. Chemical Engineering Journal 2011;172 (1):28–36.
[112] Ferreira TB, Rego GC, Ramos LR, Soares LA, Sakamoto IK, de Oliveira LL, et al. Selection of metabolic pathways for continuous hydrogen production under thermophilic and mesophilic temperature conditions in anaerobic fluidized bed reactors. International Journal of Hydrogen Energy 2018;43(41):18908–17.
[113] Liu J, Ren Y, Yao S. Repeated-batch cultivation of encapsulated *Monascus purpureus* by polyelectrolyte complex for natural pigment production. Chinese Journal of Chemical Engineering 2010;18(6):1013–17.
[114] Qiu L, Ding H, Wang W, Kong Z, Li X, Shi Y, et al. Coenzyme Q(10) production by immobilized Sphingomonas sp. ZUTE03 via a conversion-extraction coupled process in a three-phase fluidized bed reactor. Enzyme and Microbial Technology 2012;50(2):137–42.
[115] Trinh K, Garcia-Briones M, Chalmers JJ, Hink F. Quantification of damage to suspended insect cells as a result of bubble rupture. Biotechnology and Bioengineering 1994;43(1):37–45.
[116] Walls PLL, McRae O, Natarajan V, Johnson C, Antoniou C, Bird JC. Quantifying the potential for bursting bubbles to damage suspended cells. Scientific Reports 2017;7(1):15102.
[117] Pereiro I, Bendali A, Tabnaoui S, Alexandre L, Srbova J, Bilkova Z, et al. A new microfluidic approach for the one-step capture, amplification and label-free quantification of bacteria from raw samples. Chemical Science 2017;8(2):1329–36.
[118] Alexandre L, Pereiro I, Bendali A, Tabnaoui S, Srbova J, Bilkova Z, et al. A microfluidic fluidized bed to capture, amplify and detect bacteria from raw samples. Methods in Cell Biology 2018;147:59–75.
[119] Hernandez-Neuta I, Pereiro I, Ahlford A, Ferraro D, Zhang Q, Viovy JL, et al. Microfluidic magnetic fluidized bed for DNA analysis in continuous flow mode. Biosensors and Bioelectronics 2017;102:531–9.
[120] Lv B, Dong B, Deng X, Fang C. Motion characteristics and density separation of fine coal in an inflatable-inclined liquid-solid fluidized bed. Particuology 2021;58:299–307.
[121] Han Z, Yue J, Zeng X, Yu J, Wang F, Sun S, et al. Characteristics of gas-solid micro-fluidized beds for thermochemical reaction analysis. Carbon Resources Conversion 2020;3:203–18.
[122] Samih S, Latifi M, Farag S, Leclerc P, Chaouki J. From complex feedstocks to new processes: the role of the newly developed micro-reactors. Chemical Engineering and Processing—Process Intensification 2018;131:92–105.
[123] Rüdisüli M, Schildhauer TJ, Biollaz SMA, van Ommen JR. Scale-up of bubbling fluidized bed reactors—a review. Powder Technology 2012;217:21–38.
[124] Glicksman LR. Scaling relationships for fluidized beds. Chemical Engineering Science 1984;39 (9):1373–9.
[125] Li X, Wang L, Jia L, Cai W. Numerical and experimental study of a novel compact micro-fluidized beds reactor for CO_2 capture in HVAC. Energy and Buildings 2017;135:128–36.
[126] Dang NTY, Gallucci F, van Sint Annaland M. Micro-structured fluidized bed membrane reactors: solids circulation and densified zones distribution. Chemical Engineering Journal 2014;239:42–52.

CHAPTER 12

Characterization of gas–liquid–solid micro-fluidized beds

Gas–liquid–solid fluidized beds (GLSFBs) have been widely employed in chemical, biochemical, environmental, and energy industries [1,2]. Like gas–solid and liquid–solid mini- or micro-fluidized beds [3–5], gas–liquid–solid fluidized beds with millimeter- or micrometer-scale hydraulic diameters are also novel micro-fluidized beds [6]. These micro-fluidized beds combine the advantages of fluidized beds and microsystems and are particularly suitable for reactions under conditions under which mass and heat transfer limitations and unsafe operation are encountered. This chapter briefly describes gas–liquid–solid micro-fluidized beds (GLSMFBs), which are a relatively new micro-fluidization technology.

12.1 Hydrodynamics

12.1.1 Pressure drop and minimum fluidization velocity

Accurate prediction of minimum fluidization velocity (U_{mf}) is essential for successfully designing, operating, and controlling fluidized beds [6]. For the concurrent upflow gas–liquid–solid fluidized beds, U_{mf} is generally referred to as the minimum fluidization liquid velocity (U_{Lmf}) at a given superficial gas velocity because the continuous liquid phase is the main fluidization medium. Otherwise, U_{mf} can be the minimum fluidization gas velocity (U_{Gmf}). U_{Lmf} is generally determined experimentally by hydrostatic pressure drop analysis, visual observation, and studies of statistical, fractal, chaotic, or wavelet properties of experimentally measured parameters (e.g., pressure). In addition, the Hurst exponent can also be used to characterize hydrodynamics in fluidized beds by analyzing the stochasticity of a particular signal measured in the beds.

Empirical correlations can be developed to predict U_{Lmf} based on the phase properties, generally expressed by dimensionless parameters, such as Froude number (), Archimedes number (), and Reynolds number (). In addition, semi-theoretical models can also be derived to predict U_{Lmf} in order to overcome some disadvantages of the empirical correlations. However, because wall effects are significant in small-diameter beds, the correlations or models that have been derived from macro-fluidized beds may not be suitable for predicting minimum fluidization velocity in GLSMFBs.

Fig. 12.1 shows a typical schematic diagram of a GLSMFB with a column diameter of 3–10 mm and a height of 50–100 mm [6]. Air is used as the gas phase, and

Micro Fluidization
DOI: https://doi.org/10.1016/B978-0-443-18718-6.00012-7
© 2023 Chemical Industry Press Co., Ltd.
Published by Elsevier Inc., under an exclusive license with Chemical Industry Press Co., Ltd.

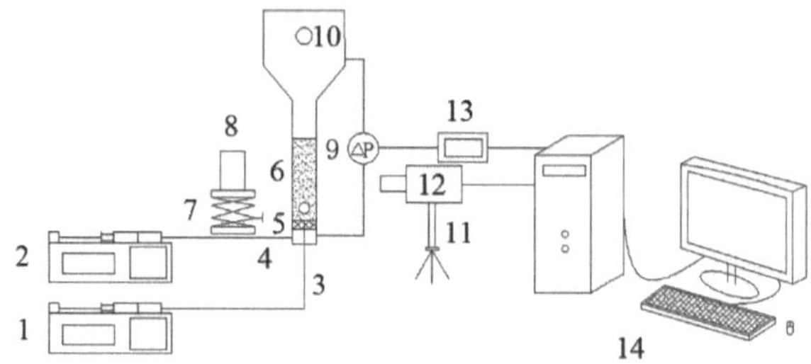

Figure 12.1 Schematic diagram of the GLSMFBs. (1) Gas syringe pump; (2) liquid syringe pump; (3) gas inlet; (4) liquid inlet; (5) liquid distributor; (6) test section; (7) lifting platform; (8) cold light source; (9) micro-pressure transducer; (10) liquid outlet; (11) tripod; (12) CMOS camera; (13) A/D converter; (14) computer.

deionized water, glycerol—water, or deionized water containing a surfactant are used as the liquid phase. Glass beads, Amberlite IR120 Na, and alumina (with 80—1000 μm in diameter) are utilized as the solid phase. The pressure drop across the fluidized bed is measured in the defluidization operation to determine the minimum fluidization liquid velocity U_{Lmf}. The pressure drop across the fluidized bed of solid particles is determined by:

$$\Delta P = \Delta P_t - \Delta P_e \tag{12.1}$$

where ΔP, ΔP_t, and ΔP_e are the fluidized bed pressure drop, total pressure drop, and empty bed pressure drop, respectively. During experiments, the time series of pressure drop is examined based on the Hurst analysis. The high-speed camera can also be used to observe the flow behavior in MFBs.

12.1.1.1 Pressure drop

Fig. 12.2 shows the pressure drop signals obtained at different superficial liquid velocities in a GLSMFB of 3 mm diameter using 104 μm solid particles. It shows that the bed pressure drop fluctuates noticeably at low superficial liquid velocities (Fig. 12.2A—C) and disappears when superficial liquid velocity is large enough (Fig. 12.2D). In this three-phase fluidized bed, small gas bubbles are easy to coalesce into large bubbles at low superficial liquid velocities, which reduces the bubble rising velocity. This behavior is supported by the observed images of gas—liquid—solid flows shown in Fig. 12.3, and the statistical data of gas bubbles are listed in Table 12.1.

When superficial liquid velocities are lower than the corresponding minimum fluidization velocity U_{mf} in the liquid—solid system, solid particles are not fully fluidized. In these cases, gas bubbles coalesce continually to form slug-type bubbles (simply called slug bubbles hereafter), as shown in Fig. 12.3A and B. As bubbles coalesce continuously, the slug bubbles eventually grow to a size close to the bed diameter. Some solid particles are pushed away by rising slug bubbles, while others are kept in the

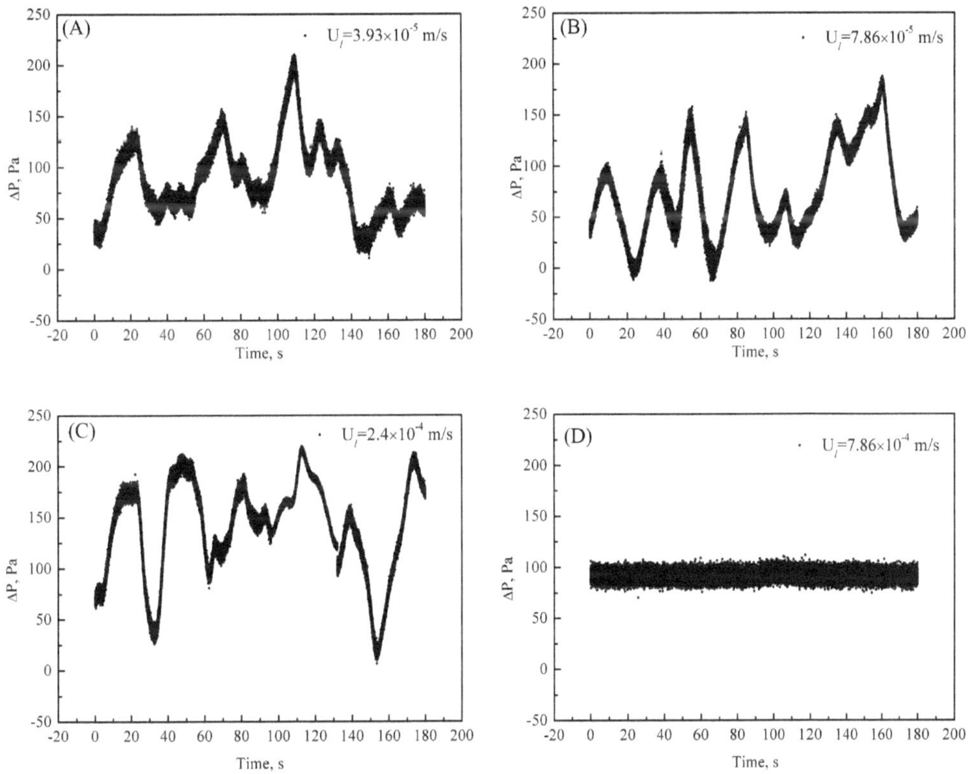

Figure 12.2 Time series of pressure drop across a GLSMFB of 3.0 mm diameter at $U_g = 1.97 \times 10^{-4}$ m/s. (A) $U_l = 3.93 \times 10^{-5}$ m/s; (B) $U_l = 7.86 \times 10^{-5}$ m/s; (C) $U_l = 2.4 \times 10^{-4}$ m/s; (D) $U_l = 7.86 \times 10^{-4}$ m/s. Liquid phase: deionized water; gas phase: air; solid phase: glass beads, $d_p = 104$ μm, $\rho_p = 2500$ kg/m^3.

liquid phase between slug bubbles and the bed wall. The flow pattern observed in these conditions is called the "half-fluidized regime."

At superficial liquid velocities that are slightly higher than U_{mf} (Fig. 12.3C), the coalescence of gas bubbles still occurs, and large bubbles and slug bubbles can be seen (Fig. 12.3C and Table 12.1), indicating that the bed is still in the half-fluidization regime at the moderate superficial liquid velocity. When the superficial liquid velocity is relatively high (Fig. 12.3D), solid particles are fully fluidized with gas bubbles flowing through the bed of solid particles (Table 12.1). In this case, the pressure drop across the bed does not fluctuate intensively (Fig. 12.3D).

Fig. 12.4 depicts the variation of averaged pressure drop with superficial liquid velocity in a period of 180 s across GLSMFBs at a specific superficial gas velocity of 1.97×10^{-4} m/s. The figure shows that the averaged pressure drop may increase or decrease with superficial liquid velocity due to the coalescence of gas bubbles at lower

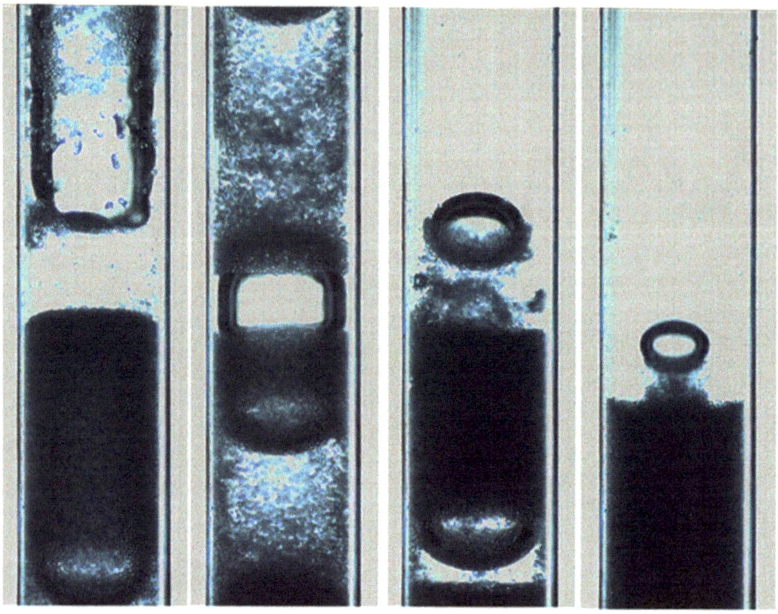

(A) (B) (C) (D)

Figure 12.3 Images of gas bubbles in a 3 mm bed at different superficial liquid velocities. $U_g = 1.97 \times 10^{-4}$ m/s. Liquid phase: deionized water; gas phase: air; solid phase: glass beads, $d_p = 104$ μm, $\rho_p = 2500$ kg/m^3. (A) $U_l = 3.95 \times 10^{-5}$ m/s; (B) $U_l = 7.86 \times 10^{-5}$ m/s; (C) $U_l = 2.4 \times 10^{-4}$ m/s; (D) $U_l = 7.86 \times 10^{-4}$ m/s.

Table 12.1 Characteristic parameters of gas bubbles corresponding to Fig. 12.3.

Figure	U_l, m/s	U_g, m/s	D_b, mm/s	u_b, mm/s
Fig. 12.3A	3.95×10^{-5}	1.97×10^{-4}	5.01 ± 0.41	0.7 ± 0.1
Fig. 12.3B	7.86×10^{-5}	1.97×10^{-4}	4.22 ± 0.33	1.3 ± 0.4
Fig. 12.3C	2.40×10^{-4}	1.97×10^{-4}	2.79 ± 0.86	12 ± 11
Fig. 12.3D	7.86×10^{-4}	1.97×10^{-4}	1.59 ± 0.03	74 ± 6

and moderate superficial liquid velocities. Solid particles are half-fluidized in these conditions. The bubble flow regime appears when the superficial liquid velocity is high enough, and then, the averaged pressure drop remains constant.

Experimental results from a 3-mm-diameter GLSMFB using 287 μm solid particles at various superficial gas velocities and a fixed superficial liquid velocity ($U_l = 7.08 \times 10^{-3}$ m/s $\approx 3.7\ U_{mf}$) are shown in Fig. 12.5. It shows that the time series of pressure drop across the bed is stable at lower superficial gas velocities (Fig. 12.5A and B), but it fluctuates noticeably at higher superficial gas velocities (Fig. 12.5C and D).

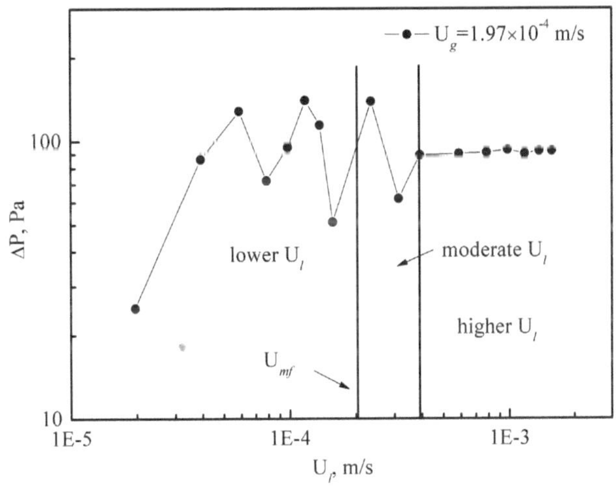

Figure 12.4 Averages of pressure drop within 180 s across a 3 mm GLSMFB at different superficial liquid velocities and U_g of 1.97×10^{-4} m/s. Liquid phase: deionized water; gas phase: air; solid phase: glass beads, $d_p = 104$ μm, $\rho_p = 2500$ kg/m^3.

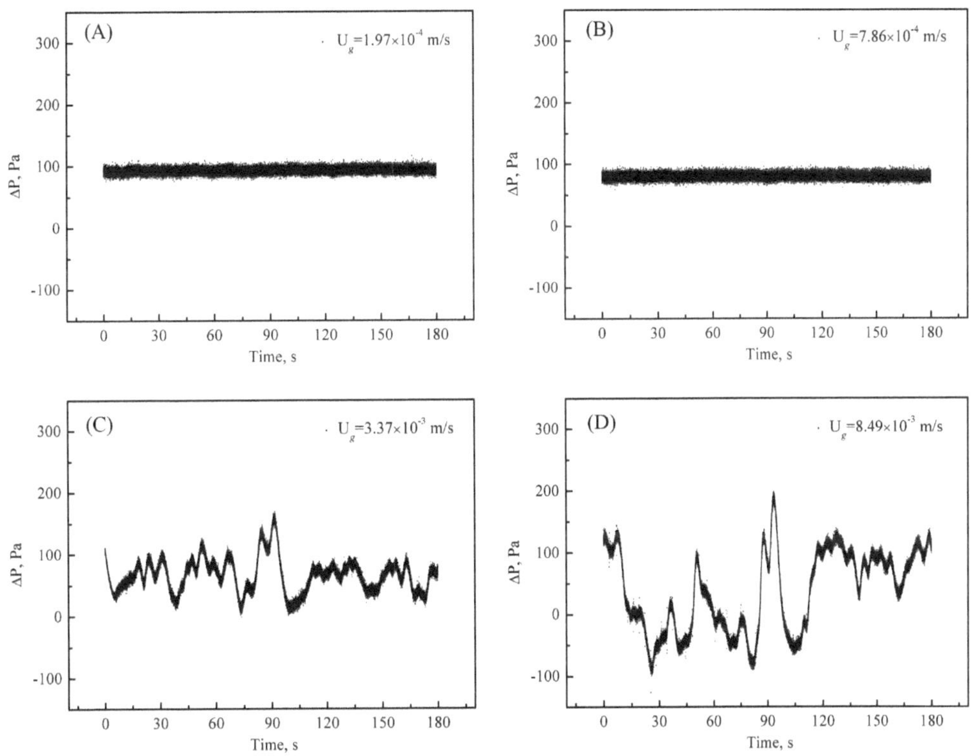

Figure 12.5 Fluctuations of pressure drop across the GLSMFB with $U_l = 7.08 \times 10^{-3}$ m/s. Liquid phase: deionized water; gas phase: air; solid phase, glass beads, $d_p = 287$ μm, $\rho_p = 2500$ kg/m^3. (A) $U_g = 1.97 \times 10^{-4}$ m/s; (B) $U_g = 7.86 \times 10^{-4}$ m/s; (C) $U_g = 3.37 \times 10^{-3}$ m/s; (D) $U_g = 8.49 \times 10^{-3}$ m/s.

In smaller diameter MFBs, bubble coalescence leads to the formation of large or slug bubbles. The bed pressure drops are sensitive to the thin liquid space between bubbles and bed walls. Therefore, significant fluctuation in bed pressure drop will be observed if only big bubbles or slug bubbles rise, making it difficult to determine U_{Lmf} quantitatively. With the increase of superficial gas velocity, bubble coalescence accelerates at a given superficial liquid velocity. When the superficial liquid velocity increases to such an extent that inhibits the bubble coalescence, the solid particles are ultimately fluidized. Otherwise, solid particles are pushed upward by slug bubbles. At lower superficial liquid velocities, slug bubbles form at low superficial gas velocities due to the wall confinement, and the pressure drop fluctuates extensively. In this case, U_{Lmf} cannot be readily determined.

12.1.1.2 Minimum fluidization velocity

As indicated previously, vigorous fluctuations of bed pressure drop in the 3–5 mm MFBs make the determination of U_{Lmf} difficult. Hence, measurements of bed pressure drop in the 8–10 mm GLSMFBs are carried out in an attempt to determine the minimum fluidization velocity. Fig. 12.6 shows pressure drop fluctuations in a 10 mm GLSMFB of 740 μm glass particles at several superficial liquid velocities with an U_g of 0.0106 m/s. It is noted that as the bed diameter is increased from 3 to 10 mm, the fluctuations of bed pressure drops become relatively small, thus allowing the characterization of three-phase flow patterns in GLSMFBs at given operating conditions.

Fig. 12.7 presents the pressure drop and its corresponding Hurst exponent obtained in the 10 mm GLSMFB. Fig. 12.7A–C shows that the bed pressure drop is positive at low superficial gas velocities. In contrast, at higher superficial gas velocities, the pressure drop

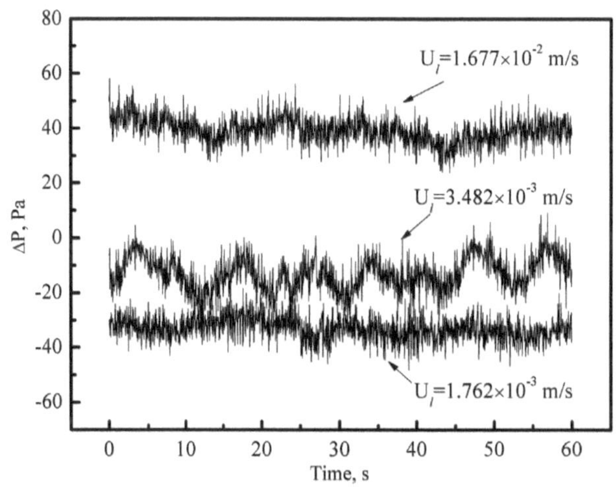

Figure 12.6 Bed pressure drops in a 10 mm GLSMFB corresponding to various superficial liquid velocities at U_g of 0.0106 m/s. Liquid phase: deionized water; gas phase: air; solid phase: glass beads, $d_p = 740$ μm, $\rho_p = 2500$ kg/m^3.

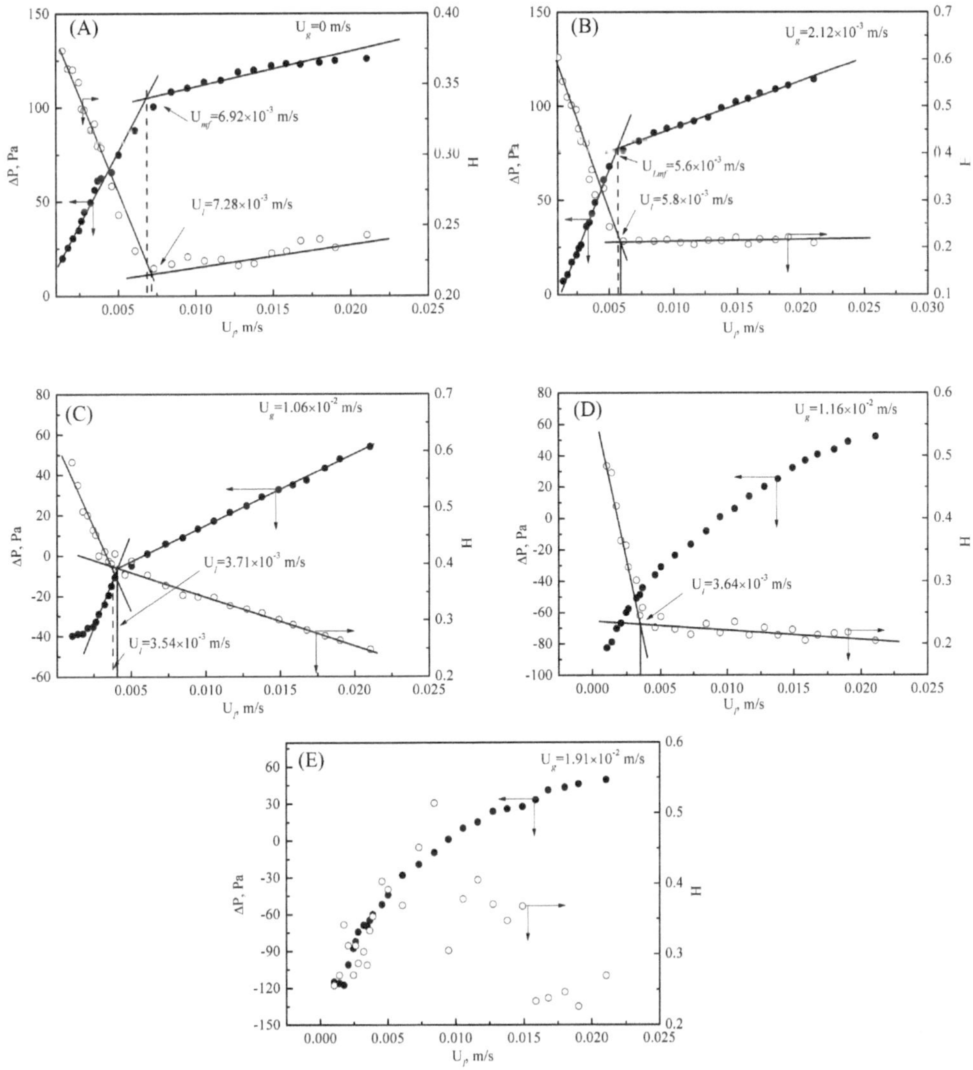

Figure 12.7 Pressure drop signals (solid circles) and the corresponding Hurst exponents (open circles) obtained in a 10 mm GLSMFB. Liquid phase: deionized water; gas phase: air; solid phase: glass beads, $d_p = 740$ μm, $\rho_p = 2500$ kg/m^3, $H_s = 21.5$ mm. (A) $U_g = 0$ m/s; (B) $U_g = 2.12 \times 10^{-3}$ m/s; (C) $U_g = 1.06 \times 10^{-2}$ m/s; (D) $U_g = 1.16 \times 10^{-2}$ m/s; (E) $U_g = 1.91 \times 10^{-2}$ m/s.

becomes negative at lower superficial liquid velocities. This phenomenon is simply explained as follows. In a three-phase fluidized bed, both the friction and gas–liquid hydrostatic flow contribute to the bed pressure drop. At lower superficial liquid velocities, the bubble rising velocity is higher than the liquid velocity between the bubble and the wall. Slip motion between phases can result in a local downflow of the liquid as a falling film, resulting in a wall shear stress that acts opposite the usual sense. Accordingly, the frictional pressure drop becomes negative. In the open literature, negative frictional pressure drops were also reported in two-phase flows at high gas and low liquid velocities. In addition, the gas holdup is large at high superficial gas velocities, leading to a low static pressure drop in the mixture phase. The negative pressure drop indicates that the pressure drop is less than the weight of the bed material. At low superficial gas velocities, the pressure drop is linearly proportional to the superficial liquid velocity at the fixed-bed regime and remains constant at the fluidized bed regime, and thus U_{Lmf} can be determined by the liquid velocity corresponding to the intersection point of two fitting lines of two different flow regimes. At high superficial gas velocities, the linear relationship between pressure drop and superficial liquid velocity disappears regardless of the superficial liquid velocity (Fig. 12.7D and E), and hence, the determination of U_{Lmf} is impossible.

The variations of the Hurst exponent with the superficial liquid velocity at various superficial gas velocities are shown in Fig. 12.7. It is noted that Hurst exponents are mostly less than 0.5, indicating that the fluidization systems are negatively correlated. This result is probably attributable to the complex interactions between the three phases resulting from the coalescence of bubbles. As shown in Fig. 12.7A–C, the Hurst analysis is a valid approach to determine U_{Lmf} in GLSMFBs.

At a higher superficial gas velocity (0.0116 m/s, Fig. 12.7D), the estimation of U_{Lmf} cannot be made by the pressure drop method, but the Hurst analysis can be employed. It is worth noting that the Hurst analysis can only be used as an alternative method to determine U_{Lmf} when the bed pressure drop data fail to do so. With further increasing superficial gas velocity (Fig. 12.7E), the gas phase starts to play a more critical role in fluidizing solid particles. At higher superficial gas velocities, neither the pressure drop nor the Hurst exponent can be used to determine U_{Lmf}. When superficial gas velocity is high enough, solid particles are completely fluidized by the gas phase even though superficial liquid velocity is low or without liquid flowing. In this situation, minimum fluidization gas velocity U_{Gmf} at a given superficial liquid velocity in GLSMFBs is taken. For gas–solid fluidized beds, U_{mf} can be estimated by Wen and Yu's equation:

$$\mathrm{Re}_{gmf} = (33.7^2 + 0.0408 Ar_g)^{0.5} - 33.7 \tag{12.2}$$

$$Ar_g = \frac{\rho_g(\rho_s - \rho_g) g d_p^3}{\mu_g^2} \tag{12.3}$$

For three-phase fluidized beds, the above two equations are modified to:

$$\mathrm{Re}'_{gmf} = (33.7^2 + 0.0408 Ar_{lg})^{0.5} - 33.7 \tag{12.4}$$

$$Ar_{lg} = \frac{\rho_g(\rho_s - \rho_l) g d_p^3}{\mu_g^2} \tag{12.5}$$

The calculated U_{mf} and U_{Gmf} are 0.358 m/s (Eq. 12.2) and 0.232 m/s (Eq. 12.4, zero superficial liquid velocity) when the solid particle diameter is 740 μm. Recall that we observed a half-fluidization regime when the superficial gas velocity was 0.013 m/s. This indicates that although the gas phase is not the continuous phase in this case and its velocity is much lower than the estimated U_{Gmf} of 0.232 m/s corresponding to zero superficial liquid velocity, solid particles are still partially fluidized.

It is concluded from the above analyses that the pressure drop and the Hurst analysis methods can predict U_{Lmf} in GLSMFBs at lower superficial gas velocities. However, neither method can be used to predict U_{Lmf} when partial fluidization occurs in GLSMFB at high surface gas velocities.

Fig. 12.8 shows typical three-phase fluidization images illustrating gas bubbles rising in a 10 mm GLSMFB at different superficial gas velocities. The estimated gas bubble characteristics are listed in Table 12.2. In the pictures shown in Fig. 12.8, gas bubble profiles are marked with solid lines. The superficial liquid velocity is 0.0025 m/s for Fig. 12.8A, C, and E, and the flow regime corresponds to the fixed-bed regime or half-fluidization regime. The superficial liquid velocity is 0.019 m/s for Fig. 12.8B, D, and F, and the corresponding flow regime is the fluidization state. It can be seen from Fig. 12.8 and Table 12.2 that in a 10 mm GLSMFB, the bubble size is larger in the fixed and half-fluidization regime than in the fluidization regime, and the bubble rising velocity decreases with lowering superficial liquid velocity due to obstructions of solid particles.

Fig. 12.8A shows a long slender gas bubble through the fixed bed at a lower superficial liquid velocity. Fig. 12.8B indicates that small rising-type gas bubbles are observed at a higher superficial liquid velocity. As the superficial gas velocity increases, bubbles grow bigger (Fig. 12.8C and D). At high superficial gas velocities, bubble sizes further increase at low superficial liquid velocities (Fig. 12.8E), and some solid particles are pushed upward by large bubbles, and a half-fluidized regime is observed. In this case, gas bubbles play an important role in fluidizing solid particles, leading to continuing increases in the gas bubble size and rising velocity (Fig. 12.8F). It is noted from Fig. 12.8E and Table 12.2 that when the superficial gas velocity is high enough, gas bubble volume is relatively large, and solid particles can be fluidized by the gas phase to some extent. When the half-fluidization regime is observed at a given superficial liquid velocity, the superficial gas velocity is identified as the critical superficial gas velocity.

The variation of U_{Lmf} with the superficial gas velocity is presented in Fig. 12.9. It can be found that U_{Lmf} decreases appreciably with the introduction of gas, which is in general agreement with observations in macroscale gas–liquid–solid fluidized beds. At superficial gas velocities above 0.00743 m/s, U_{Lmf} approaches approximately a constant at 0.0038 m/s.

Fig. 12.10 shows the critical superficial gas velocity as a function of solid particle size at a given superficial liquid velocity ($U_l = 0.0000071$ m/s). The figure shows that the critical superficial gas velocity increases with the solid particle size, implying that small solid particles can be easily fluidized. In the initial fluidization of solid particles in the GLSMFB at

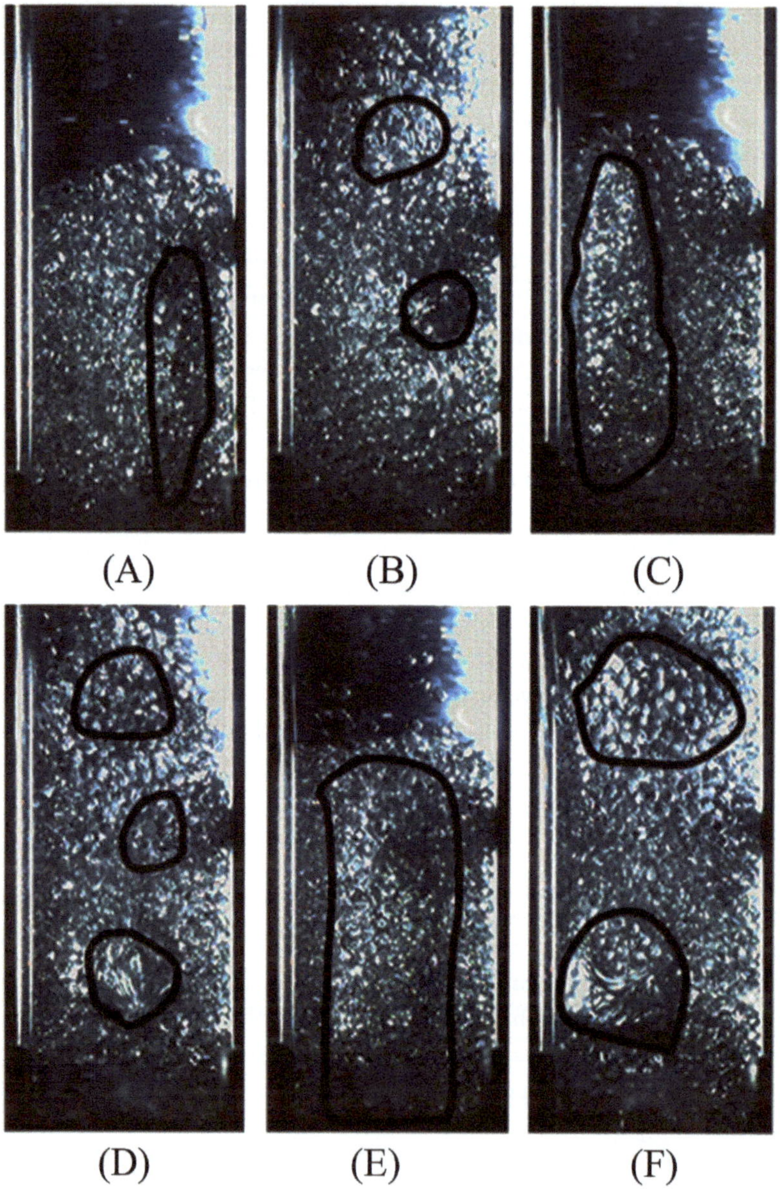

Figure 12.8 Images of gas bubble rising in a 10 mm GLSMFB under different superficial liquid and gas velocities. $d_p = 740$ μm, $\rho_p = 2500$ kg/m^3, $H_s = 21.5$ mm. (A) $U_l = 2.5 \times 10^{-3}$ m/s, $U_g = 2.12 \times 10^{-3}$ m/s; (B) $U_l = 1.9 \times 10^{-2}$ m/s, $U_g = 2.12 \times 10^{-3}$ m/s; (C) $U_l = 2.5 \times 10^{-3}$ m/s, $U_g = 1.06 \times 10^{-2}$ m/s; (D) $U_l = 1.9 \times 10^{-2}$ m/s, $U_g = 1.06 \times 10^{-2}$ m/s; (E) $U_l = 2.5 \times 10^{-3}$ m/s, $U_g = 1.91 \times 10^{-2}$ m/s; (F) $U_l = 1.9 \times 10^{-2}$ m/s, $U_g = 1.91 \times 10^{-2}$ m/s.

Table 12.2 Bubble characteristics at different operational conditions in a 10 mm MFB using solid particles of 740 μm in diameter.

Figure	U_l, m/s	U_g, m/s	D_b, mm/s	u_b, mm/s
Fig. 12.8A	2.5×10^{-3}	2.12×10^{-3}	5.4 ± 0.24	5.6 ± 0.34
Fig. 12.8B	1.9×10^{-2}	2.12×10^{-3}	3.7 ± 0.25	87 ± 5
Fig. 12.8C	2.5×10^{-3}	1.06×10^{-2}	8.3 ± 0.34	7.1 ± 0.26
Fig. 12.8D	1.9×10^{-2}	1.06×10^{-2}	4.2 ± 0.46	110 ± 6
Fig. 12.8E	2.5×10^{-3}	1.91×10^{-2}	11.2 ± 0.3	10.4 ± 0.3
Fig. 12.8F	1.9×10^{-3}	1.91×10^{-2}	6.2 ± 0.28	124 ± 8

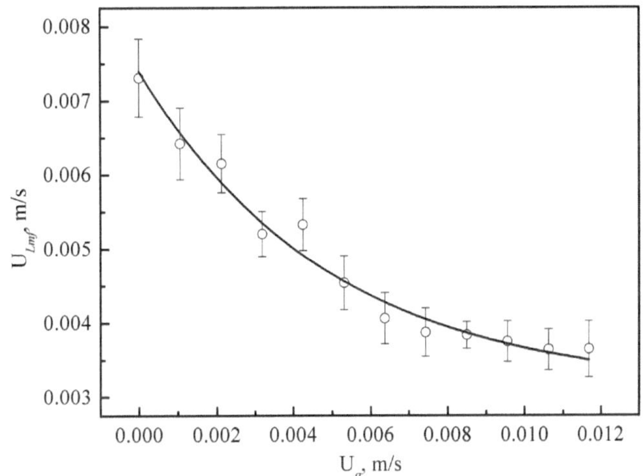

Figure 12.9 U_{Lmf} of a 10 mm GLSMFB with superficial gas velocity. $d_p = 740$ μm, $\rho_p = 2500$ kg/m^3, $H_s = 21.5$ mm.

high superficial gas velocities, the gas phase is continuous with a definable U_{Gmf}. U_{Gmf} increases with an increase in solid particle size at a given superficial liquid velocity; that is, the smaller solid particles are inclined to be fluidized by the gas phase.

Based on a total of 252 experimental data obtained under conditions of varying liquid-phase viscosity and surface tension, densities and sizes of solid particles, superficial gas velocity, and static bed height in 10 mm GLSMFB, a dimensionless empirical correlation is developed and given as follows:

$$\mathrm{Re}_{Lmf} = 0.103Ar^{2.7933}Fr^{-0.1984}\left(\frac{D_h}{d_p}\right)^{-1.5192}\left(\frac{H_s}{D_h}\right)^{0.8938}\left(\frac{\sigma_l}{\sigma_w}\right)^{0.0323}, \quad \mathrm{Re}_{Lmf} < 1 \tag{12.6}$$

$$\mathrm{Re}_{Lmf} = 7.2159Ar^{0.4582}Fr^{-0.079}\left(\frac{D_h}{d_p}\right)^{-1.0276}\left(\frac{H_s}{D_h}\right)^{0.1948}\left(\frac{\sigma_l}{\sigma_w}\right)^{-0.3988}, \quad 1 \le \mathrm{Re}_{Lmf} \le 120 \tag{12.7}$$

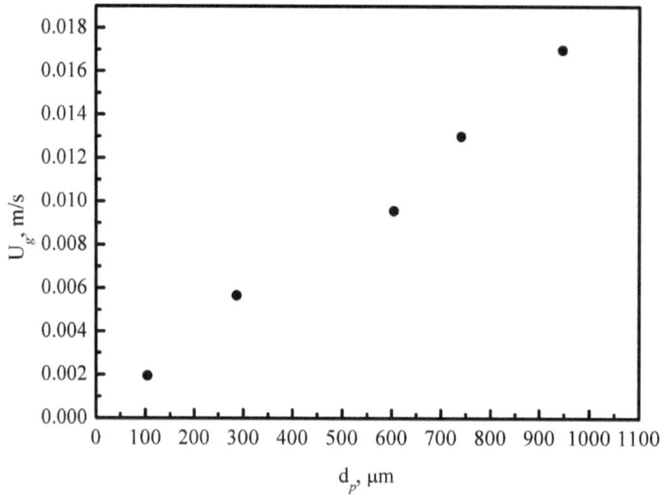

Figure 12.10 Relationship between critical superficial gas velocity and solid particle size at a given superficial liquid velocity. Liquid phase: deionized water; gas phase: air; solid phase: glass beads, $\rho_p = 2500$ kg/m^3, $D_h = 10$ mm, $U_l = 7.1 \times 10^{-6}$ m/s, $H_s = 21.5$ mm.

The above equations are valid in the following parameter ranges: $104 < d_p < 946$ μm; $0.001 < \mu_l < 0.0078$ mPa·s; $0.032 < \sigma < 0.072$ N·m^{-1}; $2500 < \rho_p < 3900$ kg·m^{-3}; $7.07 \times 10^{-4} < U_g < 0.016$ m/s; $5 < Ar < 12457$; $1.55 \times 10^{-4} < Fr < 2.73 \times 10^{-2}$; $D_h = 10$ mm. The values of average deviation for the correlation are 6.4% (Eq. 12.6) and 10.9% (Eq. 12.7), respectively.

12.1.2 Flow regimes, expanded behavior, solid holdup, and multi-bubble behavior

Note that the hydrodynamic research on GLSMFBs is still in the early stage, especially on micro-fluidized beds with diameters of less than 3 mm and with gas–liquid micro-distributors [7]. So, the following discussions are based solely on available data obtained in a 0.8 mm GLSMFB. In the experiments, bed expansion height, bubble size, and terminal velocity are measured from the recorded videos. The solid holdup is calculated from the following equation:

$$\varepsilon_s = (1 - \varepsilon_0)\frac{H_0}{H} \tag{12.8}$$

Based on the recorded videos, the bubble size is determined when micro-bubbles in the GLSMFB take a non-deformable spherical shape. According to the previous reports [1,8], in the sub-millimeter scale, the shape of micro-bubbles is dominated entirely by the surface tension rather than the inertial or viscous forces because of the extremely high surface curvature. Since bubbles inside the bed are invisible, the bubbles leaving the bed are monitored and measured to give approximate sizes of the bubbles in the bed with satisfactory accuracy [8]. The average bubble diameter is then calculated based on the following equation:

$$d_b = \frac{\sum d_i^3}{\sum d_i^2} \tag{12.9}$$

where d_i is the diameter of an individual bubble. On this basis, the volume-weighted average method is used to calculate the average rising velocity of micro-bubbles.

$$v_b = \frac{\sum v_i d_i^3}{\sum d_i^3} \tag{12.10}$$

where v_i is the rise velocity of an individual bubble with a diameter of d_i.

The rise velocity of the small spherical bubble is governed by its surrounding flow field. The flow field inside a micro-fluidized bed is considered uniform due to the small bed diameter and the intensified mixing by the solid particles, and therefore the multi-bubbles rise through the micro-fluidized bed at a relatively steady velocity. Accordingly, the terminal bubble velocity is adopted to represent the steady rise velocity of the multi-bubbles.

Since the rising bubbles are observed in the fluidized bed with a lower solid holdup ($\varepsilon_s < 0.2$), the average rise velocities of multi-bubbles at different axial levels in or above the fluidized bed are measured within a height interval of 1 mm, and the average rise velocity approximates the terminal velocity of multi-bubbles.

When the solid holdup of the fluidized bed is high, the measurement of the bubble rise velocity in the fluidized bed is impossible as the particles hinder the observation of the bubble movement. The average rise velocity of multi-bubbles at the boundary surface within 0.02 s is measured as the bubble terminal velocity of the three-phase micro-fluidized bed.

The obtained bubble terminal velocity corresponds to the absolute rise velocity. For further discussion, the relative terminal velocity between the bubble velocity and the interstitial velocity of the liquid phase is used and can be expressed as

$$v_r = v_b - U_L/(1 - \varepsilon_s) \tag{12.11}$$

where the gas holdup is omitted in calculating interstitial liquid velocity because of its value.

12.1.2.1 Flow regimes and transitions

Three distinctive flow regimes are observed in the three-phase micro-fluidized bed: dispersed bubble flow, the coalesced bubble flow, and the slug flow. The typical pictures of these three flow regimes are shown in Fig. 12.11.

The identification of the flow regime is based on the multi-bubble behavior. In the dispersed bubble flow regime, the bubble size is nearly uniform. There is a bubble

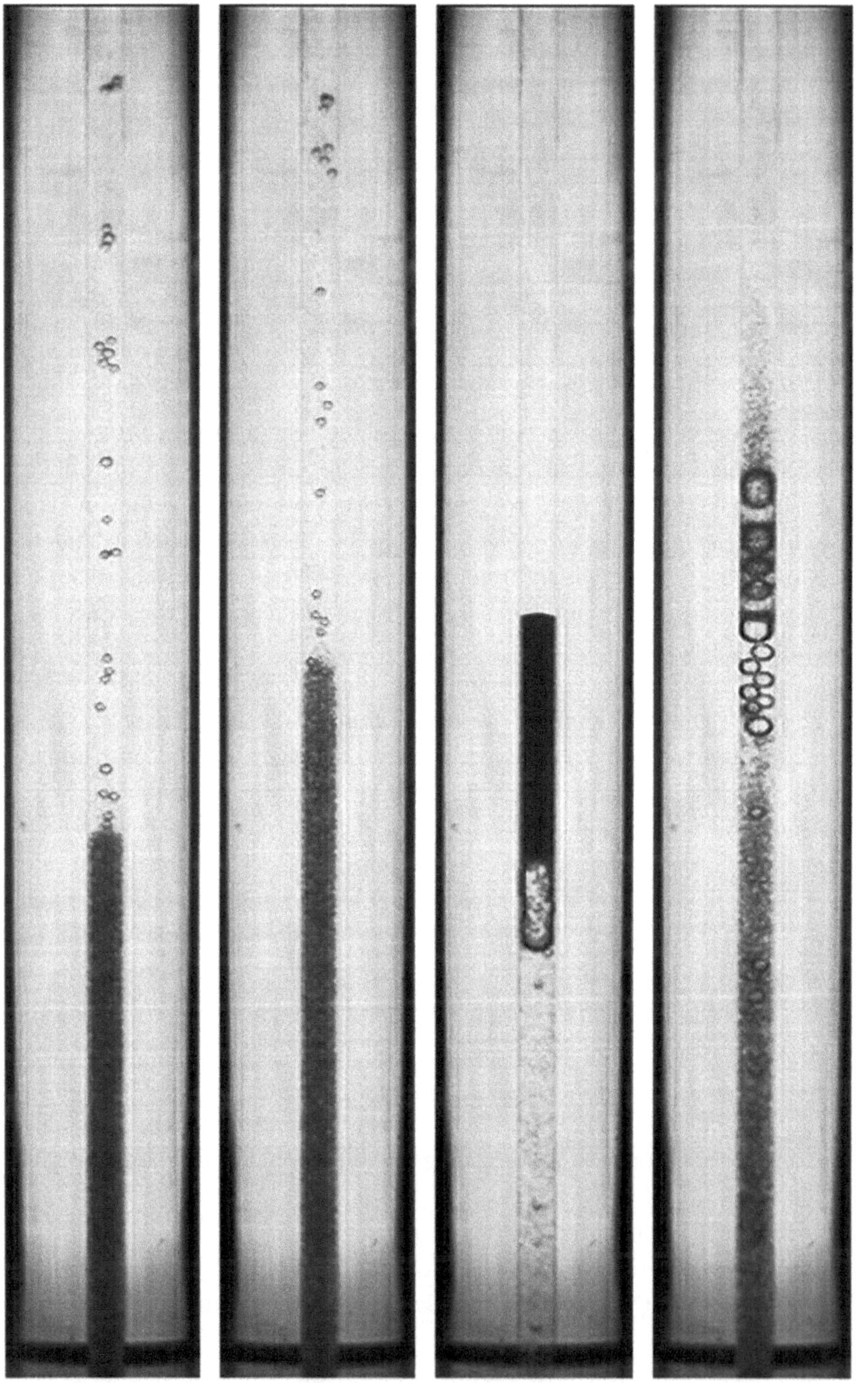

(A) (B) (C) (D)

(Continued)

size distribution in the coalesced bubble flow regime as large bubbles emerge due to the coalescence of small bubbles. However, the bubble size distribution is narrow because of the restriction of the bed diameter. In addition, no apparent axial distribution of the solid holdup is observed in these two flow regimes.

In the slug flow regime, gas slugs occupy most of, or the entire, the cross section of the column. In the case of the micro-fluidized beds, the gas slugs have a high length-to-diameter ratio and significant surface tension, so they are difficult to be stretched. As a result, the slugs cannot pass through the fluidized bed by deformation. Moreover, the particles entrained and pushed by the slugs have a significantly increased friction with the bed wall. So, the motion of the slug can only result from the push of the liquid phase rather than its buoyance, and hence, the liquid velocity decides the slug rise velocity. As the liquid velocity is much lower than the rise velocity of bubbles, once a slug forms in a micro-fluidized bed, the bubbles below will continuously catch up and coalesce with it or form new slugs. As a result, the normal three-phase fluidization cannot be maintained. Pictures of this situation that occurred separately at a lower and a higher liquid velocity are shown in Fig. 12.11C and D.

The flow regime maps of three-phase MFBs are presented in Fig. 12.12B–D. The flow regime map of gas–liquid flow is also shown in Fig. 12.12A. The upper limit of the liquid velocity for MFBs is set as the value at which the bed surface is still apparent. For the three particles, the limits are 65%, 77%, and 87% of their theoretical particle terminal velocities. At these limits, the solid holdups of the micro-fluidized beds are higher than 14%. Compared with the upper limit, the lower limit, that is, the minimum liquid fluidization velocity of the three-phase micro-fluidized bed, is hard to determine since a half-fluidization region exists between the fixed bed and slug flow region and cannot be easily identified by visual observations. So, we are unable to provide an accurate transition from the fixed bed to the three-phase fluidized bed at this time. However, it is possible to confirm the fluidization state by examining the bed expansion behavior. In such a way, the minimum fluidization liquid velocities of the three particles obtained from the extrapolation of the expansion curves are 95, 63, and 26 μm/s, respectively. Obviously, the liquid velocities in the slug flow regime of the three-phase fluidized beds are higher than these velocities, indicating that the slug flow regime spans a wide range of the liquid velocity in the three-phase micro-fluidized beds.

◀ **Figure 12.11** Typical pictures of flow regimes for three-phase micro-fluidized bed with 0.8 mm bed diameter and particle diameter of $d_p = 37$ μm. (A) Zoomed-out and zoomed-in pictures of dispersed bubble flow ($U_L = 1338$ μm/s, $U_G = 238$ μm/s); (B) zoomed-out and zoomed-in pictures of coalesced bubble flow ($U_L = 1095$ μm/s, $U_G = 361$ μm/s); (C) zoomed-out pictures of slug flow ($U_L = 182$ μm/s, $U_G = 117$ μm/s); (D) zoomed-out pictures of slug flow ($U_L = 1338$ μm/s, $U_G = 1022$ μm/s).

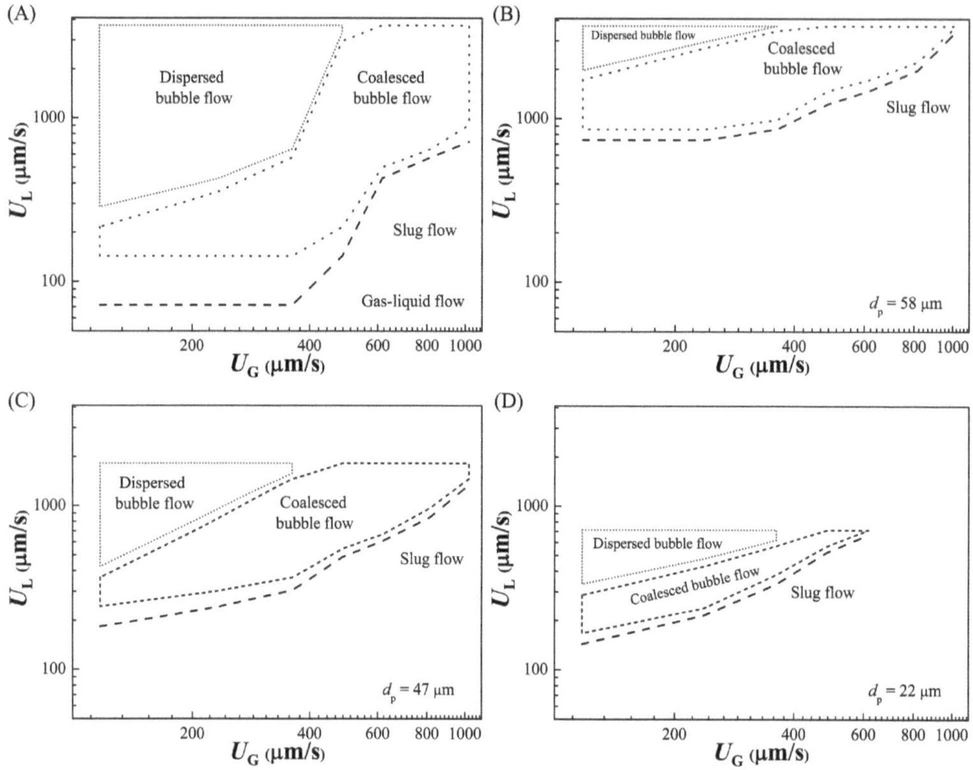

Figure 12.12 Flow regime maps for (A) gas–liquid flow and three-phase micro-fluidized beds with (B) $d_p = 58$ μm; (C) $d_p = 37$ μm; (D) $d_p = 22$ μm.

The same regime transition criteria are used in two-phase bubble flows and three-phase fluidized beds. The identification of the coalesced bubbles flow regime relies on the measurement of the coalesced bubbles. The transition from the coalesced bubble flow to the slug flow is identified by observing the slowly rising slug in the micro-fluidized bed. When the transition takes place, the corresponding U_G and U_L or ε_s at the transition point are determined. As shown in Fig. 12.12, the flow regime maps of the three-phase fluidized beds are similar to that of the two-phase bubble flow—the flow regime changes from the dispersed bubble flow to the coalesced bubble flow and then to the slug flow progressively with the increase of U_G/U_L. The solid holdup plays a decisive role in the flow regime transition for the three-phase fluidized beds. But due to the reduction in the bed diameter and the fluidization velocity, the effects of the liquid and gas velocities on the flow regime transition are more evident in the three-phase micro-fluidized beds. The flow regime transition velocities in the three-phase micro-fluidized beds with particle sizes of 37 and 22 μm are close to those of the two-phase gas–liquid flows. In particular, the transition velocities are more likely to be decided by the gas–liquid flow than by solid holdups for the 22 μm particles.

Fig. 12.12 also shows significant differences between the fluidization liquid velocity ranges of the three particle sizes in three-phase micro-fluidized beds. Within the range of the particle terminal Reynolds number (Re_t) of 0.321 0.017 investigated, the viscous effect dominates the drag force of liquid exerted on particles according to the Stocks law. In the viscous flow region, the single-particle terminal velocity is proportional to the square of the particle diameter. On the contrary, it is proportional to the 1.14 power and 0.5 power of the particle diameter in the transitional and turbulent regimes, respectively. Therefore, the particle diameter plays a more vital role in determining the particle terminal velocity in the viscous flow region than in the transitional and turbulent regimes. As a result, a slight variation in particle size will likely result in a great difference in the particle terminal velocity and consequently a considerable change in the range of liquid fluidization velocity in three-phase micro-fluidized beds, so sometimes the solid holdup is used to describe the flow regime transition in three-phase micro-fluidized beds.

12.1.2.2 Bed expansion behavior

Fig. 12.13 plots the expansion ratio versus the superficial liquid velocity. The expansion curves are straight lines at low expansion ratios, indicating a uniform expansion in three-phase micro-fluidized beds. At high expansion ratios, the slope of the straight line increases slightly. A high specific surface area of micro-fluidized beds leads to a substantial friction force between surfaces of solid particles at a lower expansion ratio.

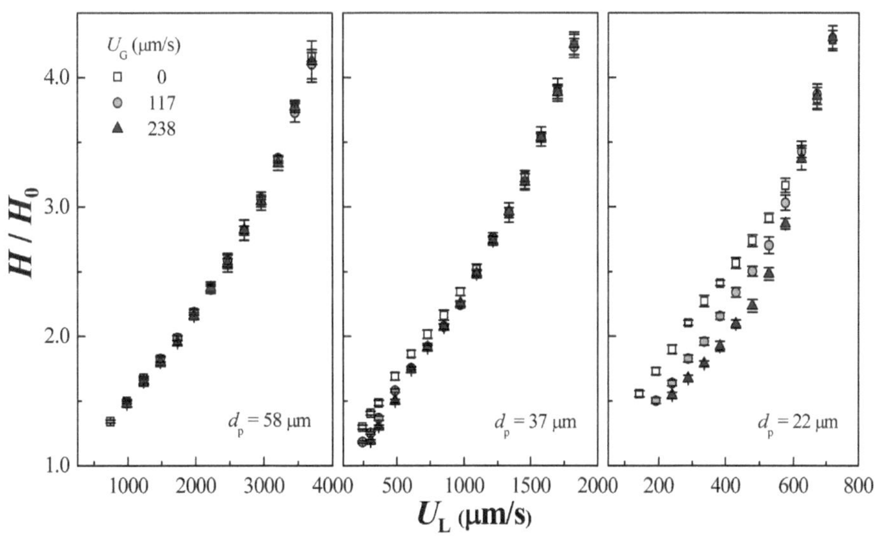

Figure 12.13 The expansion ratios of the liquid–solid and gas–liquid–solid micro-fluidized beds.

Thus, a lower bed expansion results from counteracting the extra internal friction. When the expansion ratio increases, the resulting increase in the bed voidage reduces the friction between particles, contributing to the rise of the bed expansion slope.

The bed contraction (i.e., the reduced expansion ratio in gas–liquid–solid fluidized beds compared with that in liquid–solid fluidized beds at the same liquid velocities) occurs at low liquid velocities for small sizes of particles. This phenomenon can be seen from variations of the solid holdup with the gas velocity, as shown in Fig. 12.14. In the case of high expansion ratios, when the gas is introduced at a low velocity, the solid holdups of micro-fluidized beds do not change noticeably. As the gas velocity increases, the bed contraction (i.e., the increase in the solid holdup) happens, which is more distinctive with the decrease in expansion ratio and particle size.

In conventional three-phase fluidized beds, the bed contraction phenomenon is explained by the generalized bubble wake model, in which the wake phase is assumed to transport at the bubble rise velocity. Thus, the velocity of the liquid carried in the wake phase is greater than the average velocity of liquid surrounding the liquid–solid phase. Based on the wake model, three parameters determine whether three-phase fluidized beds expand or contract: the relative rise velocity of gas bubbles, the ratio of the wake to bubble volume fractions, and the solid holdup in the wake phase. A high relative bubble velocity, a high wake volume fraction, and a low solid holdup in the wake phase are more likely to decrease the interstitial liquid velocity and cause bed contraction.

As mentioned previously, the bubbles in three-phase MFBs are spherical because of the small sizes and high surface tensions of the bubbles. In contrast, conventional

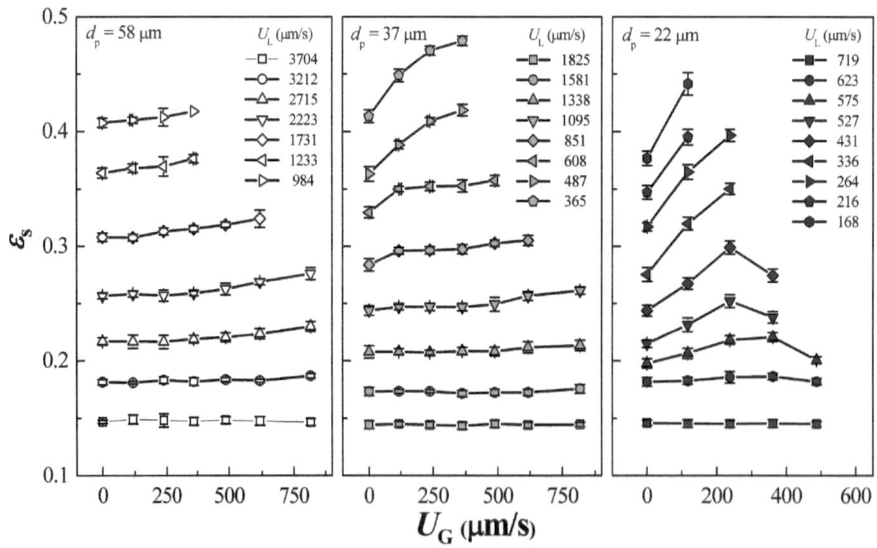

Figure 12.14 Solid holdups of the liquid–solid and gas–liquid–solid micro-fluidized beds.

three-phase fluidized beds are dominated by the spherical-cap bubbles. For the spherical-cap bubbles, the vortex generation point is where the wake forms at the bubble surface with the sharpest curvature (the cap edge of the bubble). But, for the spherical bubble, this point is close to the bubble base and away from the largest cross section of the bubble. A spherical bubble has a lower ratio of the axial projected area to the volume than a spherical-cap bubble. Hence, the volume ratio of the wake to the bubble is smaller for spherical bubbles than for spherical-cap bubbles. Consequently, the solid holdups of micro-fluidized beds have no noticeable change at high expansion ratios when the gas phase is introduced at high specified liquid velocities. This is different from conventional three-phase fluidized beds. It also demonstrates that the liquid phase, in which the bubble wake shortens the flow path, is less in micro-fluidized beds. As a result, the liquid residence time distribution in micro-fluidized beds is narrow and more controllable.

The bed contraction in micro-fluidized beds can result from low liquid velocities and large bubble sizes. At low liquid velocities, a reduction in the liquid velocity results in a relatively significant change of the wake phase and thus significantly affects the solid holdup. With the increase of the bubble diameter, the spacing between the bubble and the bed wall becomes narrow. The velocity of the liquid flowing down relatively from the bubble top through this gap increases. The wake generation point will be close to the largest cross section of the bubble. Thus, the volume ratio of the wake to the bubble increases as the bubble size becomes large. In fluidized beds, the particle entrainment is primarily induced by the bubble wake, as shown in Fig. 12.15. The pictures also show that the region and amount of the entrained particles increase with increasing bubble diameter.

Therefore, the bed contraction can be caused by low liquid velocity, large bubble size, low expansion ratio, and fluidization of small particle sizes. The bed contraction

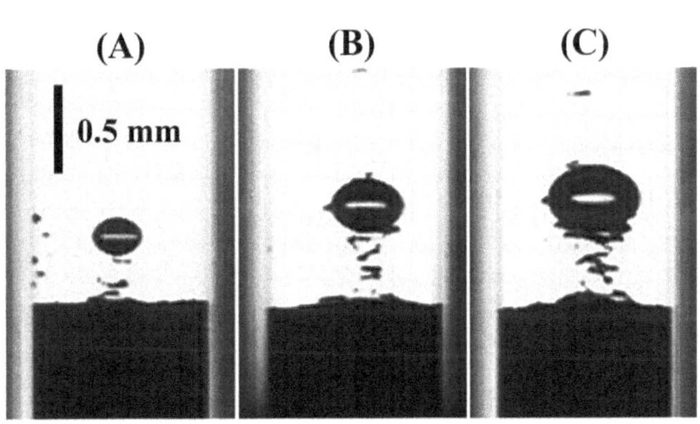

Figure 12.15 Particles entrainments by different-sized bubbles. $d_p = 37$ μm, liquid velocity $U_L = 547$ μm/s, (A) $d_b = 225$ μm, $\varepsilon = 35.3\%$; (B) $d_b = 309$ μm, $\varepsilon_s = 38.0\%$; (C) $d_b = 385$ μm, $\varepsilon_s = 38.5\%$.

can also result when increasing gas velocity because of the resulting large bubble size and number frequency. However, there is a slight reduction in the bed contraction as the gas velocity increases in the fluidized bed of 22 μm particles because of a higher solid holdup in the wake phase, where small particles with low mass inertia remain.

12.1.2.3 Bubble size distribution

The distribution of gas bubble number frequency within classes of dimensionless bubble size (distribution function of bubble size based on the bubble number frequency, D_b) is calculated by dividing the interval bubble number frequency by the class interval. As presented in Fig. 12.16, the normal bubble size distribution is found in the dispersed bubble flow regime of three-phase micro-fluidized beds. This is different from the lognormal distribution found in conventional three-phase fluidized beds. This concentrated bubble size distribution results from the characteristic gas–liquid micro-distributor and low liquid and gas velocities in micro-fluidized beds. In the dispersed bubble flow regime, the variance of the normal size distribution represents the deviation of bubble formation, of which the heterogeneity of the multiphase systems mainly causes the increase. The heterogeneous behavior in three-phase micro-fluidized beds arises from the nature of fluidized particles and the bubble wake. The wall effect aggravates it at a low ratio of the bed diameter to the particle size. Hence, as shown in Fig. 12.16, the variance of the bubble size distribution in the three-phase micro-fluidized bed of 58 μm particles is more significant than that in gas–liquid flows.

In the coalesced bubble flow regime, the bubble size distribution in three-phase micro-fluidized beds is a particular stepped size distribution (Figs. 12.17 and 12.18),

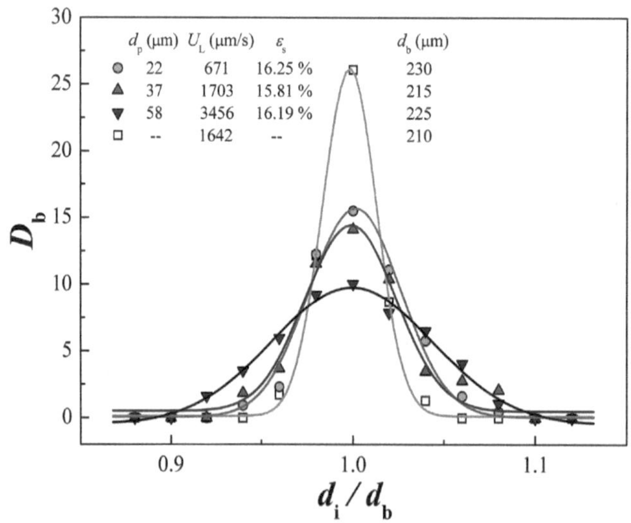

Figure 12.16 Distribution of bubble size in the dispersed bubble flow regime when $U_G = 238$ μm/s.

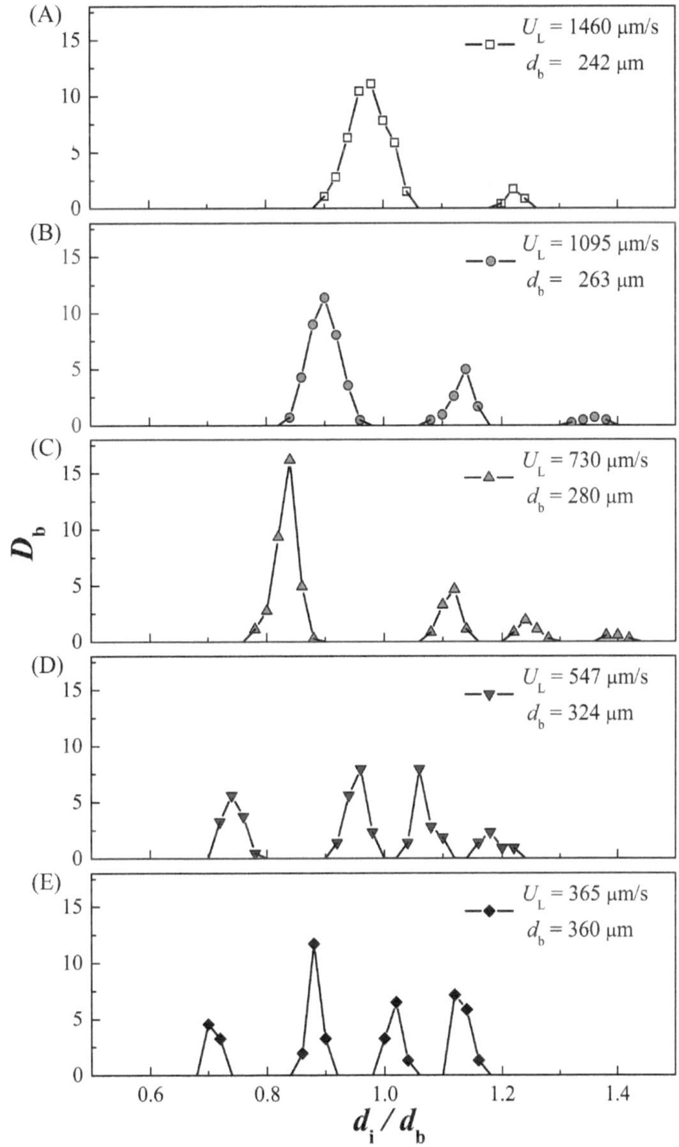

Figure 12.17 Bubble size distributions in the coalesced bubble flow regime at different liquid velocities. $d_p = 37$ μm, $U_G = 361$ μm/s.

which is different from the lognormal distribution typically observed in conventional fluidized beds. The larger bubbles in higher steps are generated from the coalescence of two smaller bubbles in lower steps, which can be confirmed by the fact that the cubic diameters of bubbles in different steps are in line with the volume-multiple relationship.

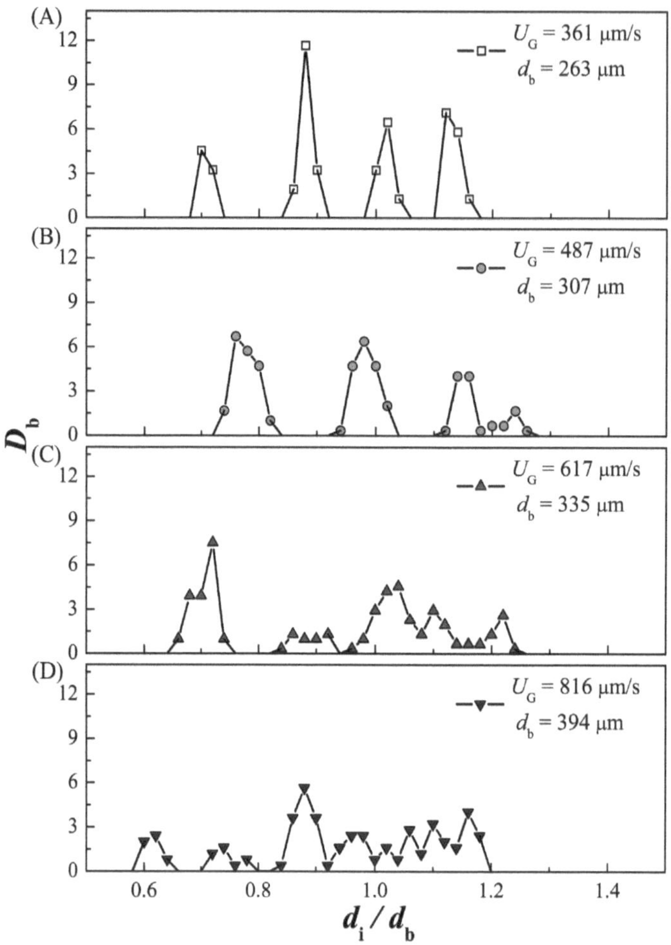

Figure 12.18 Bubble size distribution in the coalesced bubble flow regime at different gas velocities. $d_p = 37$ μm, $U_L = 1095$ μm/s.

The step distribution appears because micro-bubbles cannot break but coalesce due to their high surface tension. As presented in Fig. 12.17, the distribution gradually transforms from a single step to multiple steps with decreasing liquid velocity. It indicates that the multiple coalescences occur progressively with the decrease of the bubble rise velocity. The large bubbles of high steps are formed at different times from the coalescences of the low step bubbles formed. Figs. 12.17 and 12.18 demonstrate that the size distribution of the lowest step bubbles, at least at low gas velocities, approaches a normal distribution. Correspondingly, the bubble sizes of high steps remain narrow distributions.

The number frequency of bubbles in each step reduces by increasing the number of coalescences. The coalescence of two bubbles increases the average bubble spacing due to the reduced number of bubbles, reducing the re-coalescence probability of bubbles. The number frequency of the bubbles from a different number of coalescences can maintain a relatively stable ratio in the bubble flow. However, as shown in Fig. 12.18, there is no apparent step distribution in the bubble size with the increase of the gas velocity, which may be caused by the rise in both the size deviation of the initially formed bubbles and the complexity of the multi-bubble coalescence.

12.1.2.4 Bubble size

The bubble size here is defined as the volume surface mean diameter of multi-bubbles. As shown in Fig. 12.19, the bubble sizes are generally larger in the three-phase micro-fluidized beds of three particles than in two-phase bubble flows (no particle size is given). The results indicate that with decreasing liquid velocity, the bubble size increases slightly in the dispersed bubble flow regime and more significantly in the coalesced bubble flow regime as the bubble coalescence starts to control the bubble size.

In the dispersed bubble flow regime of conventional three-phase fluidized beds, due to the domination of the bed inertial force, the increase of solid holdup can lead to a great increase in the bubble size. However, as shown in Fig. 12.20A, the bubble size increases slightly with increasing solid holdup in the dispersed bubble flow regime

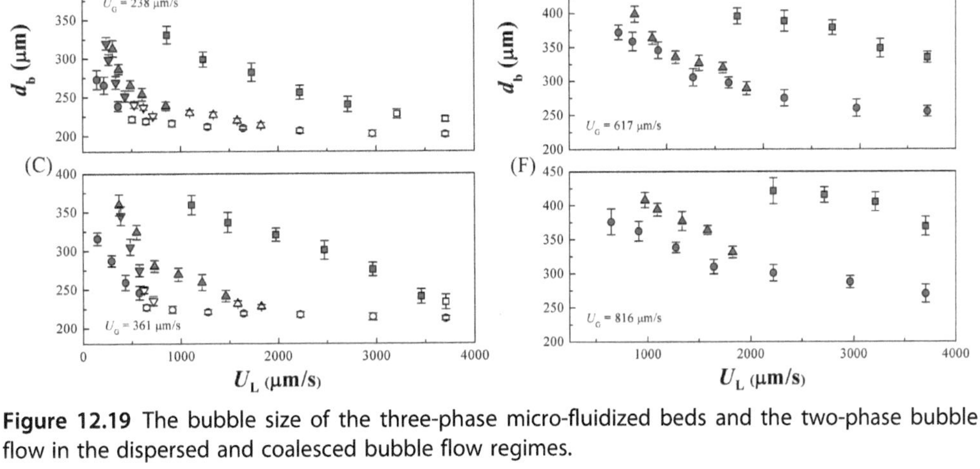

Figure 12.19 The bubble size of the three-phase micro-fluidized beds and the two-phase bubble flow in the dispersed and coalesced bubble flow regimes.

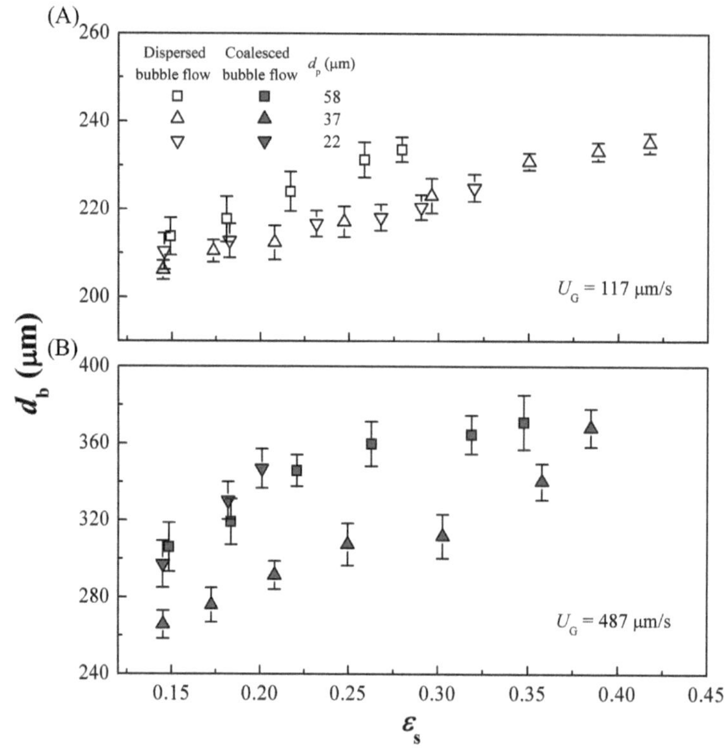

Figure 12.20 The variation of bubble size with the solid holdup in the dispersed and coalesced bubble flow regimes.

of three-phase micro-fluidized beds. This approximate linear relationship between the bubble size and solid holdup in micro-fluidized beds is probably due to the increase in the apparent viscosity of the beds.

The bubble coalescence mechanism of spherical-cap bubbles in conventional three-phase fluidized beds is similar to gas–liquid flows and gas–solid fluidized beds. For two successively flowing bubbles, the low pressure of the wake of the front bubble creates a strong suction force on the back bubble, which accelerates the back bubble, makes it approach the front bubble, and finally is sucked into the wake of the front bubble, resulting in the coalescence of the two bubbles. However, for the spherical bubbles in three-phase micro-fluidized beds, the smaller bubble wake resulting from the spherical shape and the low bubble Reynolds number may not play a decisive role in the bubble coalescence process. Fundamentally, the bubble coalescence requires a close distance and small relative velocity between the two bubbles. Thus the overall coalescence probability of multi-bubbles depends on the average bubble spacing and rise velocity. Hence, the apparent viscosity (which relates to the solid holdup), bubble generation frequency, and bed diameter can affect bubble coalescences and change the bubble size. As shown in Fig. 12.20B, in the coalesced bubble flow regime, an increase in the solid holdup significantly increases the bubble coalescence and thus the bubble size because of reduced bubble spacing at higher solid holdup and lower rise velocity of bubbles.

As shown in Fig. 12.21A, the increase of the gas velocity has a negligible effect on the initially formed bubble size because the gas momentum force and the bubble inertial force are not the governing forces in the bubble formation process. Hence, the increase of the gas velocity mainly increases the bubble formation frequency, thereby reducing the average bubble spacing. In this way, the bubble size significantly increases through the bubble coalescence in the coalesced bubble flow regime, as shown in Fig. 12.21A and B.

Based on the experimental data (184 points), the following empirical correlation of bubble size was proposed.

$$d_b = (1.76 \times 10^{-4} + 5.83 d_p)\left(\frac{U_G}{U_L}\right)^{0.281} \quad (12.12)$$

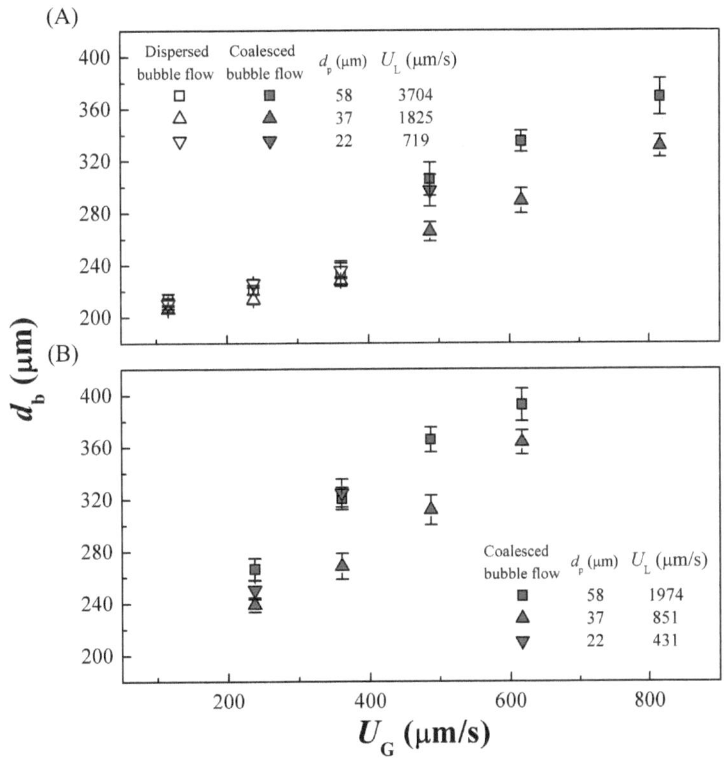

Figure 12.21 The variation of bubble size against gas velocity in the dispersed and coalesced bubble flow regime.

where $0.0317 < U_G/U_L < 0.992$. The equation has an average deviation of 7.1% and a maximum deviation of 24.4% between the calculated experimental data. About 95% of the deviations are less than 16.2%.

12.1.2.5 Bubble terminal velocity distribution

According to a previous study [9], a detached bubble with a slow incipient velocity will reach a stable rise velocity (i.e., the terminal velocity) after undergoing a brief acceleration because of the balance between bubble buoyance and the drag resistance in the fluidized bed. Moreover, as the spherical micro-bubble is non-deformable, the bubble rise velocity is little affected by the bubble surface tension but is strongly governed by the flow field. Thus, the instantaneous velocity of the multi-bubbles in three-phase micro-fluidized beds is affected by multiple factors, including the vortex in a liquid–solid fluidized bed, the disturbance between bubbles, and the restriction of the bed wall. As a result, there is a distribution in the bubble terminal velocity.

Fig. 12.22 shows typical distributions of the bubble terminal velocity in two different flow regimes of a three-phase micro-fluidized bed. The distributions are neither normal nor lognormal and are narrower than the lognormal distribution typically seen in conventional three-phase fluidized beds. When changing liquid and gas velocities and particle size, the distribution pattern changes little, but the distribution in coalesced bubble flow regime generally becomes narrower than that in the dispersed bubble flow regime because of the dominant effect of the large coalesced bubbles on the rise velocity of multi-bubbles.

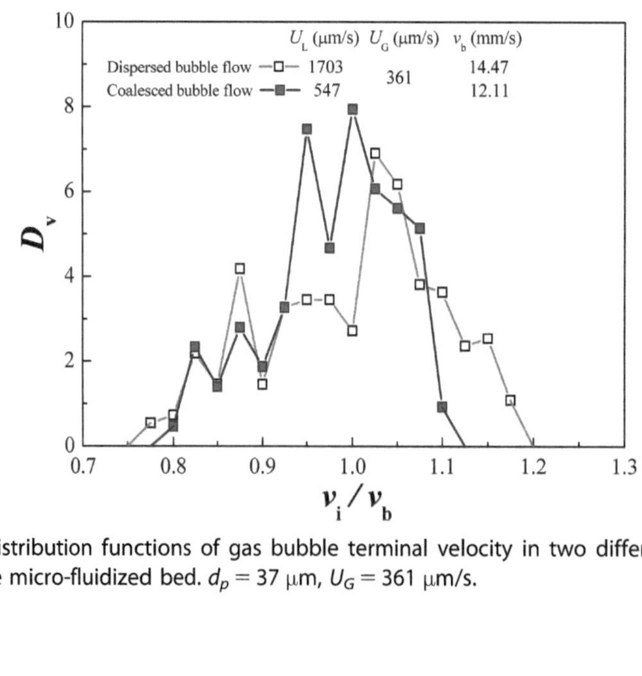

Figure 12.22 Distribution functions of gas bubble terminal velocity in two different flow regimes of a three-phase micro-fluidized bed. $d_p = 37$ μm, $U_G = 361$ μm/s.

12.1.2.6 Bubble terminal velocity

The bubble rise in three-phase fluidized beds is the relative motion between the bubble and the surrounding liquid–solid fluidization medium. Thus, the bubble rise velocity is a relative measure and can be based on the force balance between the buoyancy and the drag resistance as follows.

$$C_D = \frac{4}{3} g \frac{d_b}{v_r^2} \tag{12.13}$$

Because Re_b corresponding to micro-bubbles is small, C_D corresponding to multi-bubbles in three-phase micro-fluidized beds can be calculated using Stokes' law applicable for a single spherical bubble in the infinite flow field.

$$\begin{aligned} C_D &= \frac{24}{Re_b} \\ Re_b &= \frac{d_b v_r \rho_m}{\mu_m} \end{aligned} \tag{12.14}$$

where ρ_m and μ_m are the apparent density and viscosity of the micro-fluidized beds, respective. Substituting Eq. (12.14) into Eq. (12.13) yields

$$\begin{aligned} v_r &= \frac{1}{18} g \frac{d_b^2 \rho_m}{\mu_m} \\ \rho_m &= (\varepsilon_s \rho_s + \varepsilon_l \rho_l) \end{aligned} \tag{12.15}$$

Thus, it is seen that the bubble terminal velocity is proportional positively to the bubble size and inversely to the apparent viscosity of the micro-fluidized bed. Note that μ_m, the apparent viscosity for micro-fluidized beds with multi-bubbles, is a parameter that collectively characterizes the interactions between multi-bubbles and the bed wall restriction. Hence, μ_m increases with increasing the solids holdup and size and reducing the bubble spacing.

Fig. 12.23 shows the relative terminal velocity of bubbles in three-phase micro-fluidized beds is lower than that in gas–liquid flows. As the liquid velocity decreases and the solid holdup increases, the bubble terminal velocity in three-phase micro-fluidized beds reduces in the dispersed or coalesced bubble flow regime, and it does not increase pronouncedly when the regime transition takes place. This is different from the situation in gas–liquid flows, where the bubble terminal velocity increases as the bubbles coalesce. With the rise in the solid holdup, the bubble terminal velocity increases as the corresponding bubble size becomes large, but it cannot completely offset the reduction in the bubble terminal velocity resulting from the increase in the apparent viscosity of the fluidized bed. Ultimately, the bubble terminal velocity reduces with the solid holdup, as shown in Fig. 12.24A.

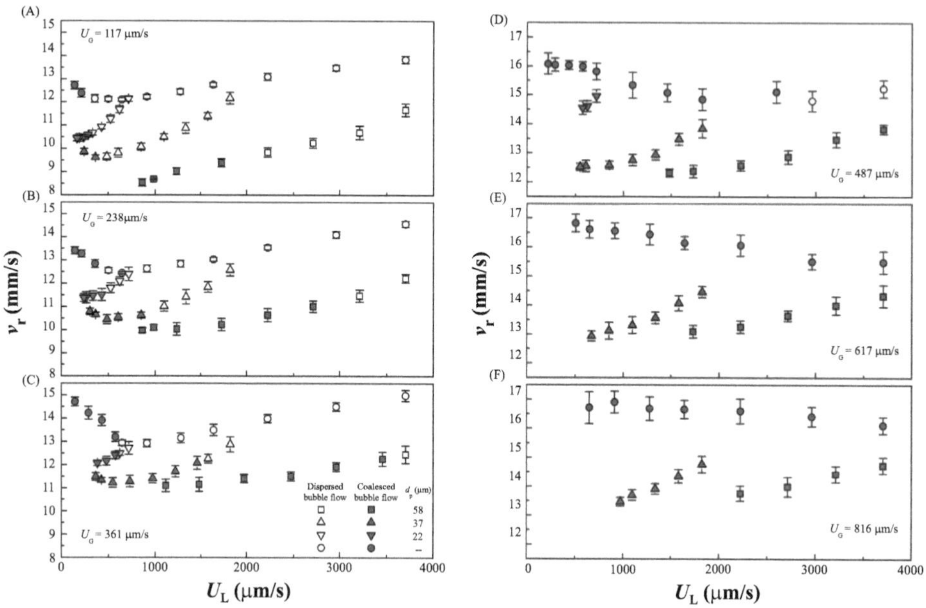

Figure 12.23 Bubble terminal velocities of three-phase micro-fluidized beds and two-phase bubble flows in the dispersed and coalesced bubble flow regimes.

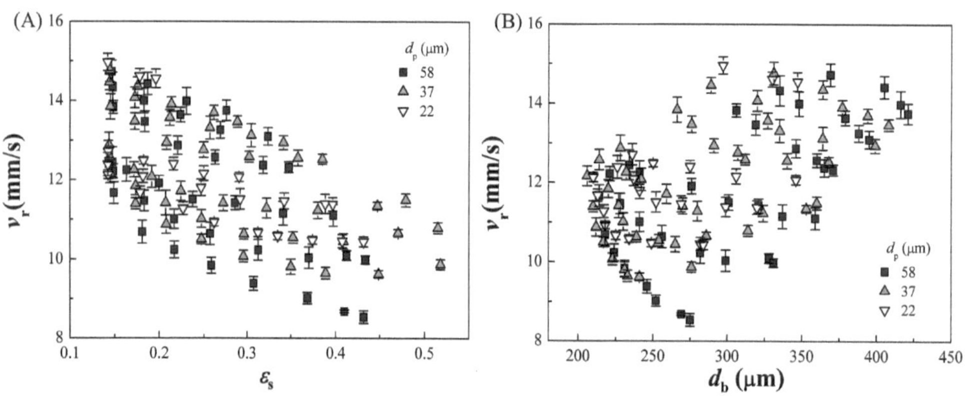

Figure 12.24 Variation of the bubble terminal velocity with the solid holdup and the bubble size.

However, rising gas velocity increases bubble size and the corresponding bubble terminal velocity noticeably. The relationship between the bubble terminal velocity and bubble size is presented in Fig. 12.24B, which does not follow Eq. (12.15) (i.e., $v_r \propto d_p^2$) even when the solid holdups are approximately the same. This result indicates that μ_m must be a function of the bubble diameter, but the specific relationship is not very clear.

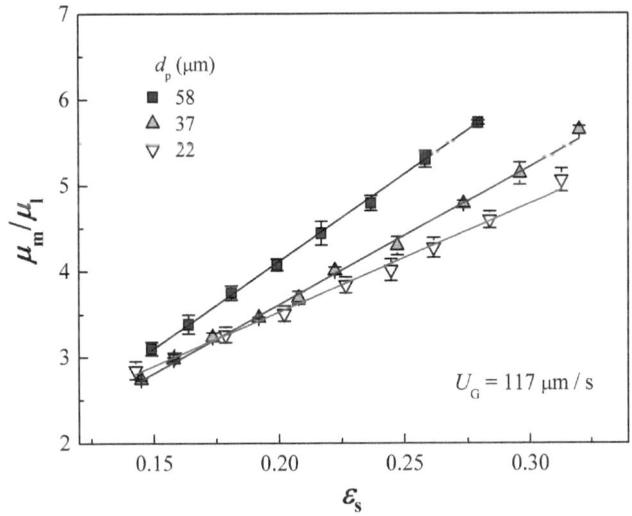

Figure 12.25 Variation of the apparent viscosity with the solid holdup in dispersed bubble flow regime.

Additionally, in the dispersed bubble flow regime and at a low gas velocity ($U_G = 117$ μm/s), the influences of the disturbance between bubbles and the bubble size on the apparent viscosity of the fluidized bed are negligible because of the large bubble spacing and the small bubble size. So, μ_m is calculated by Eq. (12.15) and presented in Fig. 12.25. The relationships between μ_m and solid holdups are almost linear for the three particles. At the same solid holdups, the apparent viscosities of the micro-fluidized beds are higher than those predicted by correlations developed for conventional fluidized beds because there is an additional drag resistance in the small-diameter beds. Fig. 12.25 shows that as the ratio of the bed diameter to particle size and the wall effect increase, the slope of μ_m versus ε_s increases.

Based on a total of 184 experimental data, an empirical correlation of the bubble terminal velocity was obtained as follows:

$$v_r = 0.0279 Fr_L^{0.112}\left(\frac{U_G}{U_L}\right)^{0.164}$$
$$Fr_L = \frac{U_L^2}{g d_p} \tag{12.16}$$

where Fr_L ranges from 2.22×10^{-4} to 2.41×10^{-2}, and U_G/U_L is between 0.0317 and 0.992. The average deviation between the experimental and calculated values is 3.8%, the maximum deviation is 13.3%, and 95% of the deviations are less than 10.5%.

In conclusion, the bubble terminal velocity determines the gas holdup and thus controls the gas transfer flux, and concentrated terminal velocity distributions show a controllable overall gas-phase flux in three-phase micro-fluidized beds.

12.2 Applications

12.2.1 Chemical reactions

Three-phase micro-fluidized beds have attracted tremendous interest in the process industries due to their great heat and mass transfer performance and excellent operational controllability. When used in solid catalytic reactions, they can solve issues that microchannel reactors suffer from. For example, when hydrogenation reactions are carried out in microchannel reactors, the active catalysts are deposited on channel inner walls, and therefore the specific surface area is low, in addition to the complex fabrication process. In three-phase micro-fluidized beds, solid particles provide a high specific surface area and intensify mass transfer. The small reactor volume shortens the mass transfer distance and reduces the radial distribution, improving reactor performance.

The derivatives and copolymers of crotonic acid (CA) have a broad range of applications in manufacturing adsorbents, adhesives, pesticides, resins, etc. Therefore, the oxidation of crotonaldehyde (CAH) has been an essential reaction in the pharmaceutical industry. CAH is traditionally oxidized by liquid oxygen under pressure without catalysts in a large-scale industrial reactor. This process suffers from large reactor volume, prolonged reaction time, and low conversion ratio. In addition, the reaction heat is greater than the latent heat required for vaporizing the liquid phase. Therefore, it is a challenge to remove the reaction heat in time to prevent the reactor temperature from running away, which increases not only side reactions but also the risk of reactor explosion. In a bench-scale experiment, soluble catalysts, such as manganese acetate, copper acetate, cobalt acetate, thallium salts, and Schiff base cobalt complexes, are used in the oxidation of CAH, but still suffer from drawbacks of high producing cost and complex post-separation of the soluble catalysts. Hence, recently, attempts have been made to develop supported catalysts for a more efficient heterogeneous reaction process in three-phase mini-fluidized beds. Currently, other physical and chemical applications of three-phase micro-fluidized beds, such as applications in life science, are being explored [10].

12.2.1.1 Photocatalytic degradation of methylene blue (MB)

Wang et al. [11] investigated photocatalytic degradation of MB in a three-phase photocatalytic micro-fluidized bed of 6 mm in diameter. Fig. 12.26 shows that the degradation ratio of MB increases gradually with the reaction time and MB velocity (V_L). It shows that with the increase of V_L, the initial degradation rate of MB (R_0) increases from 0.0461 mg/(L · s) at the MB solution velocity of 10.80 mL/min to

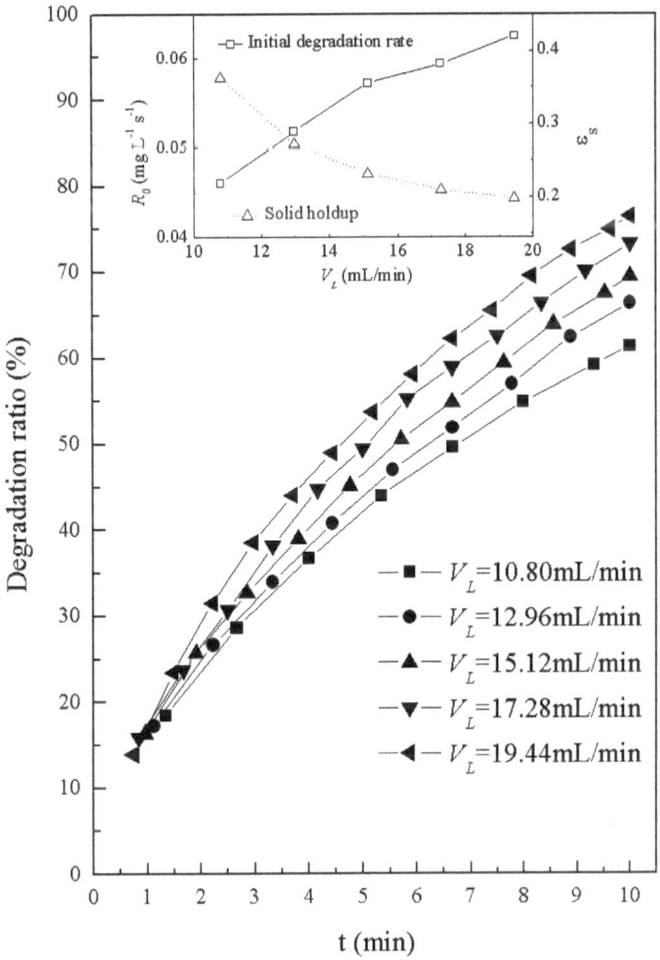

Figure 12.26 Influence of feeding rate of MB solution V_L on the degradation ratios ($H_0 = 20$ mm, $C_{MB,0} = 20$ mg/L, $V_G = 6$ mL/min, $P = 200$ W).

0.0626 mg/(L · s) at the MB solution velocity of 19.44 mL/min, corresponding to a growth ratio of 35.79%. When the reaction time reaches 10 min, the degradation ratio of MB increases by 15.2% as the MB velocity increases from 10.80 mL/min to 19.44 mL/min. Additionally, the processing capacity of the three-phase photocatalytic micro-fluidized bed (19.44 mL/min) is 12.4 times that of the liquid–solid photocatalytic micro-fluidized bed with a diameter of 3 mm, suggesting that the processing capacity of GLSMFB can be significantly improved by increasing liquid velocity compared to liquid–solid MFB.

Fig. 12.27 shows the variation of the degradation ratio of MB with reaction time at different gas flow rates. It is evident that the degradation ratio increases with

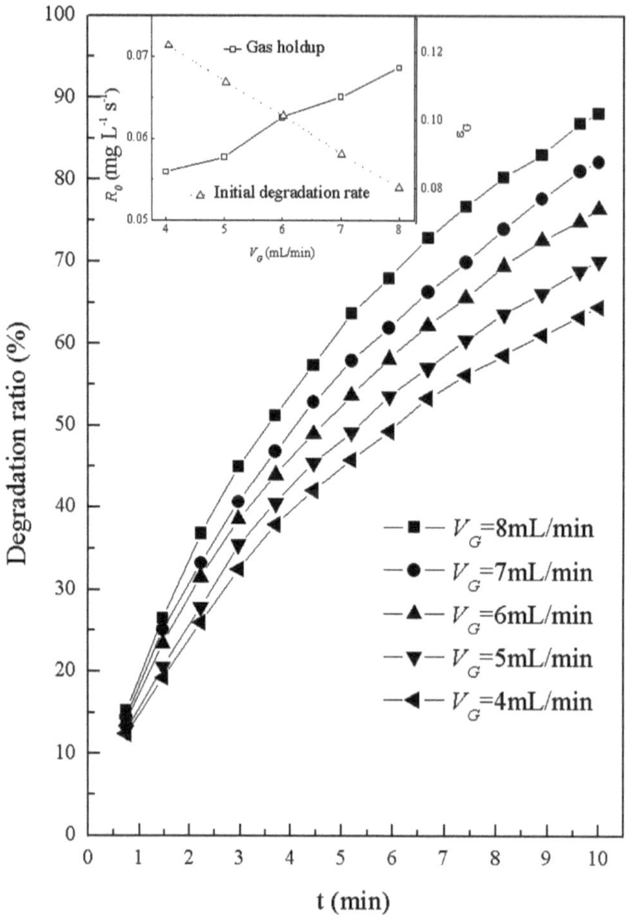

Figure 12.27 Influence of air feeding rate V_G on the degradation ratio ($H_0 = 20$ mm, $C_{MB,0} = 20$ mg/L, $V_L = 19.44$ mL/min, $P = 200$ W).

reaction time and gas flow rate. As shown in the chart inside Fig. 12.27, the initial degradation rate decreases from 0.0736 to 0.0559 mg/(L · s) when the gas velocity changes from 4 to 8 mL/min, indicating that increasing gas velocity promotes mass transfer and mixing of the MB solution, micro-bubbles, and catalyst particles. However, increasing gas velocity also raises the gas holdup in the reactor (from 0.08 to 0.12 when the gas velocity increases from 4 to 8 mL/min), which strengthens the effect of the scattering of light but reduces the number of photons absorbed by the catalyst and the effective residence time of the MB solution. Accordingly, the impact of enhanced light scattering by increasing gas holdup on the degradation ratio is weaker than that of mass transfer promoted by increasing gas velocity.

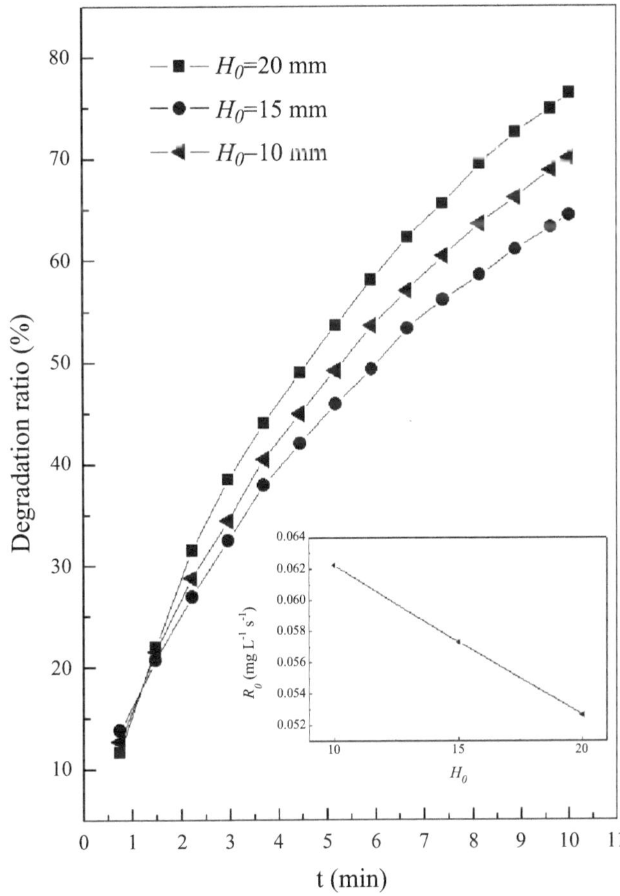

Figure 12.28 Degradation ratios at different initial bed heights ($C_{MB,0}$ = 20 mg/L, P = 200 W, V_G = 6 mL/min, V_L = 19.44 mL/min).

Fig. 12.28 shows the variation of MB degradation ratio with reaction time at different initial bed heights. It can be seen that the total MB degradation ratio increases with the initial bed height. When V_L is 19.44 mL/min, the initial MB degradation rate at an initial bed height of 10 mm is 1.09 times than at an initial bed height of 20 mm (inside chart). The increase in the initial bed height reduces the effective time of photocatalytic reaction and thus the initial degradation rate. When residence time is 10 min, the total MB degradation ratio at the initial bed height of 20 mm is 1.89 times higher than at the initial bed height of 10 mm in the micro-fluidized bed. The results demonstrate that increasing particle loading intensifies the turbulence of catalyst particles in the reactor, contributing to the enhancement of mass transfer from the liquid phase to the catalyst surface.

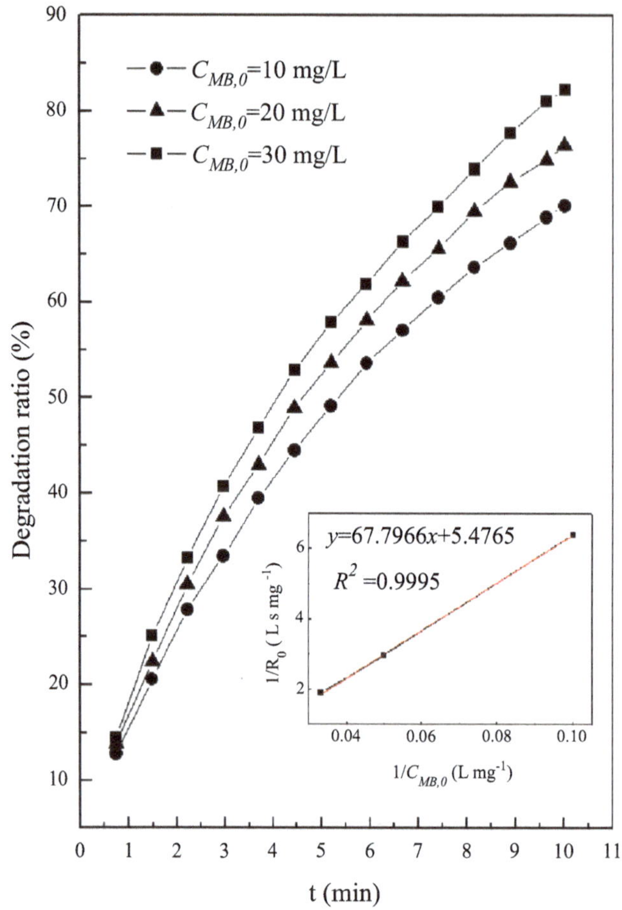

Figure 12.29 Degradation ratios at different initial MB concentrations of MB ($H_0 = 20$ mm, $P = 200$ W, $V_G = 6$ mL/min, $V_L = 19.44$ mL/min).

Fig. 12.29 shows the effect of initial MB concentration on the MB degradation ratio. The total degradation MB ratio increases 1.17 times when the initial MB concentration changes from 10 to 30 mg/L. As shown in the chart in Fig. 12.29, the reciprocal of the initial degradation rate is linearly proportional to the reciprocal of initial concentrations of MB solution.

Fig. 12.30 shows that the MB degradation ratio increases with the power of the Xe lamp. When the reaction time reaches 10 min, the total degradation ratio of MB solution with power of 250 W is 1.25 times that with power of 150 W, since more effective photons can be produced to stimulate TiO_2 catalyst activity when the light with a higher power and higher radiation intensity are used.

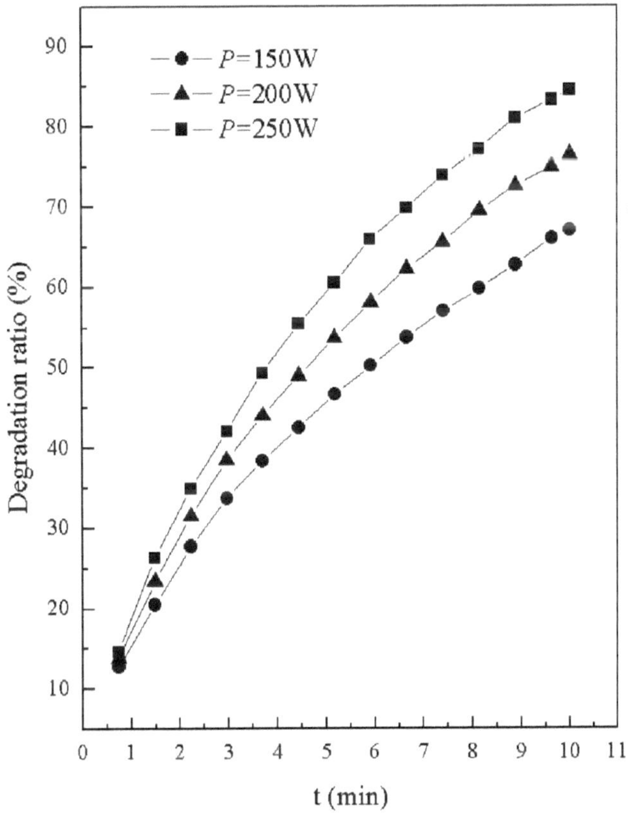

Figure 12.30 Variation of the degradation ratio with time at different Xe lamp powers ($H_0 = 20$ mm, $C_{MB,0} = 20$ mg/L, $V_G = 6$ mL/min, $V_L = 19.44$ mL/min).

12.2.1.2 Catalytic oxidation of crotonaldehyde to crotonic acid

Dong et al. [12] investigated the catalytic oxidation of CAH with 5% Ag/Al_2O_3 catalysts in a novel three-phase micro-fluidized bed with an inner diameter of 3 mm and a height of 150 mm. The average reaction rate is compared with that obtained in a batch-stirred tank reactor to demonstrate the high efficiency of three-phase fluidized bed technology. The schematic diagram of the GLSMGB reactor setup is shown in Fig. 12.31A. Fig. 12.31B and C shows the images of the three-phase mini-fluidized bed used in the experiments.

Fig. 12.32A shows the relationship between the conversion of CAH and reaction time in the gas–liquid–solid micro-fluidized bed reactor under circulating operation. It shows that the reaction rate, which corresponds to the slope of the curve, changes noticeably with the operating temperature. The conversion of CAH increases from 43% to 75% after 10 h reaction as the reaction temperature increases from 30°C to 40°C. However, a further increase in the temperature from 40°C to 50°C only leads

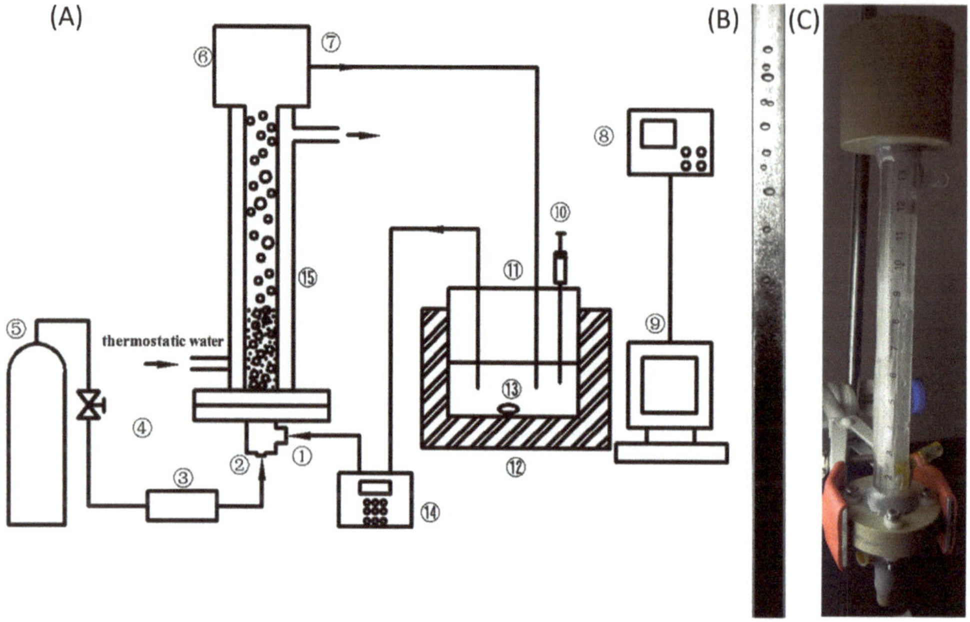

Figure 12.31 (A) Schematic diagram of gas–liquid–solid micro-fluidized bed setup, (B) a photograph of bubbling flow image in the bed, and (C) a photograph of the gas–liquid–solid micro-fluidized bed. 1: Liquid inlet; 2: gas inlet; 3: gas mass flowmeter; 4: inlet valve; 5: oxygen cylinder; 6: overflow weir; 7: liquid outlet; 8: gas chromatography; 9: computer; 10: injection syringe; 11: storage flask; 12: thermostatic magnetic stirrer; 13: magnet; 14: constant flow pump; 15: heating jacket.

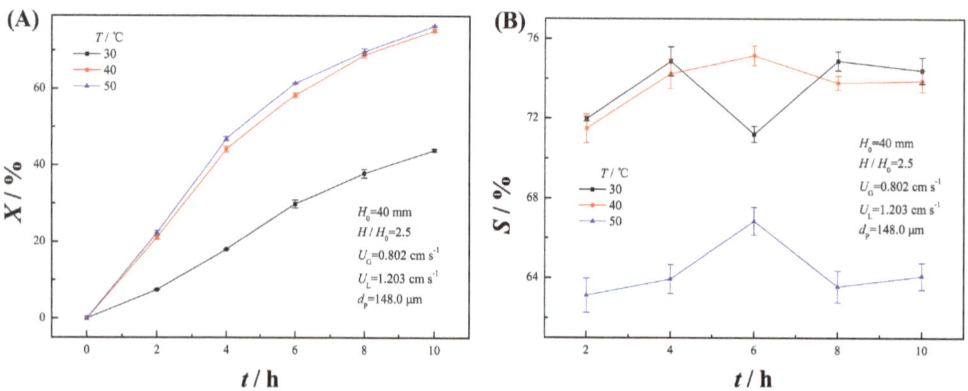

Figure 12.32 Effect of the reaction temperature on (A) the conversion ratio of CAH and (B) the selectivity of CA. $U_L = 1.203$ cm/s, $U_G = 0.802$ cm/s, $H_0 = 40$ mm, $C_{CAH,0} = 2.43$ M, $d_p = 148.0$ μm.

to a slight increase in the conversion. Theoretically, a low temperature favors the reaction since the oxidation of crotonaldehyde is exothermic. Therefore, the low reaction rate at 30°C probably suggests that molecular diffusion at low temperatures limits the mass transfer and thus hinders the reaction. Fig. 12.32B shows that the selectivity of

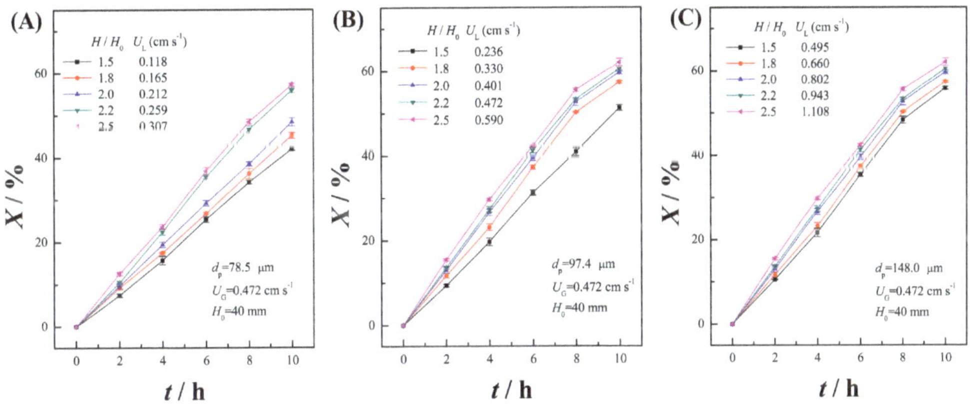

Figure 12.33 Effect of the bed expansion on the conversion ratio of CAH versus reaction time. (A) $d_p = 78.5$ μm, $U_G = 0.472$ cm/s, $H_0 = 40$ mm, $C_{CAH,0} = 2.43$ M; (B) $d_p = 97.4$ μm, $U_G = 0.472$ cm/s, $H_0 = 40$ mm, $C_{CAH,0} = 2.43$ M; (C) $d_p = 148.0$ μm, $U_G = 0.472$ cm/s, $H_0 = 40$ mm, $C_{CAH,0} = 2.43$ M.

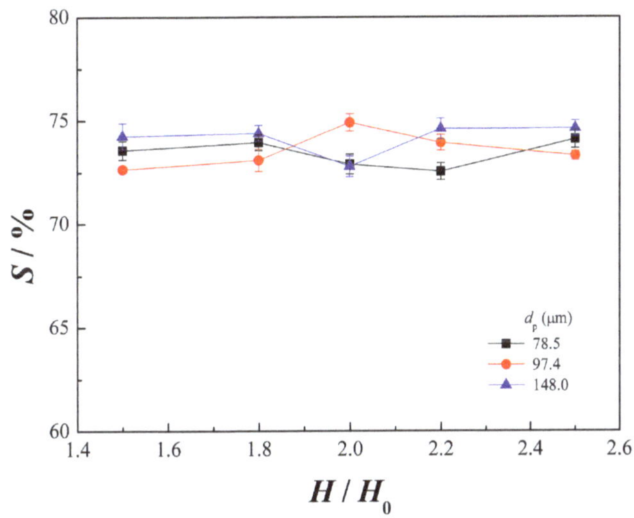

Figure 12.34 Effect of the bed expansion on the selectivity of CA. $U_G = 0.472$ cm/s, $H_0 = 40$ mm, $C_{CAH,0} = 2.43$ M.

CA after 10 h reaction at 50°C ranges from 63% to 67%, which is lower than 71% and 76% achieved at 30°C and 40°C. Therefore, 40°C can be considered an optimal reaction temperature.

Figs. 12.33 and 12.34 show that U_L increases the bed expansion ratio and enhances CAH conversion for all three particles investigated. For the catalyst of 78.5 μm in average particle diameter, the expansion ratio and solid holdup increase from 1.5 to 2.5 and 0.4 to 0.24 when liquid velocity increases from 0.118 to 0.307 cm/s,

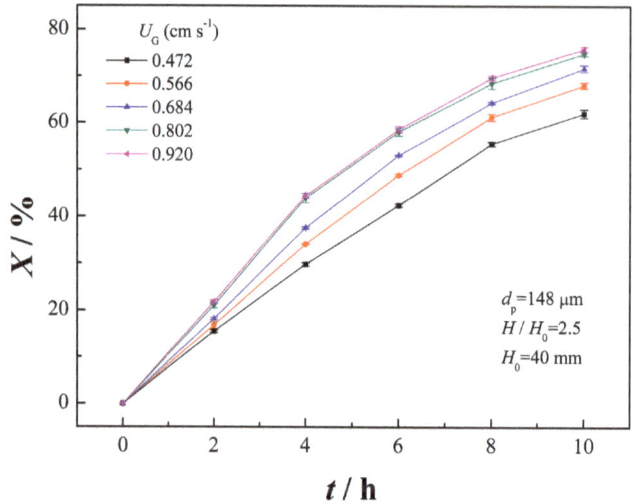

Figure 12.35 Effect of the superficial gas velocity on the conversion of CAH. $d_p = 148.0$ μm, $H_0 = 40$ mm, $H/H_0 = 2.5$, $C_{CAH,0} = 2.43$ M.

respectively, while the conversion of CAH after 10 h circulation increases from 42% to 57%. Moreover, the increase in the particle diameter also leads to a slight increase in the conversion. The selectivity of CA is between 70% and 76% after 10 h reaction under the experimental conditions, as shown in Fig. 12.34, indicating that the selectivity of CA is less affected by the bed expansion ratio.

Fig. 12.35 demonstrates that in the three-phase micro-fluidized bed, the conversion increases noticeably with gas velocity initially, but the enhancement diminishes gradually as gas velocity rises further. Since the bubble behavior affects the residence time of the liquid phase but does not play a dominant role, the impact of the gas velocity on the conversion is likely to be caused by the increased mass transfer flux of the gas phase.

12.3 Summary

The gas–liquid–solid micro-fluidized beds are a new type of microreactors, and thus their fundamental research and application are still in the early stage. We encourage more research on hydrodynamics, mixing, heat and mass transfer, and reaction in gas–liquid–solid micro-fluidized beds in the future.

Nomenclature

Ar Archimedes number ($Ar = \rho(\rho_s - \rho)gd_p^3/\mu^2$)
$C_{CAH,0}$ initial molar concentration of CA, mol/L

C_D	drag coefficient
c_{MB}	concentration of solution after a specific irradiation time, mg/L
$c_{MB,0}$	initial concentration of MB solution, mg/L
D	degradation rate of MB, %
D_{av}	average pore size of catalyst particles, nm
d_b	average bubble diameter, μm
D_b	bubble size distribution function based on number frequency of bubbles
D_c	bed diameter, mm
D_h	inner diameter of the miniaturized fluidized bed, mm
d_i	individual bubble diameter, μm
d_o	orifice diameter, μm
d_p	average diameter of solid particles, μm/mm
D_v	bubble terminal velocity distribution function based on number frequency of bubbles
F_B	effective buoyancy, N
F_C	particle–bubble collision force, N
F_D	liquid drag, N
f_g	gas holdup, %
$F_{I,g}$	bubble inertial force, N
$F_{I,m}$	suspension inertial force, N
F_M	gas momentum force, N
Fr	Froude number of gas ($Fr = U_g^2/gd_p$)
F_σ	surface tension, N
g	gravitational acceleration, m/s^2
H	expanded or fluidized bed height, mm
H_0	initial bed height, mm
H_s	static bed height, mm
l	length, m
M_{CA}	molar mass of CA, g/mol
M_{CAH}	molar mass of CAH, g/mol
m_{catal}	mass of catalysts, g
P	power of Xe lamp, W
P	pressure drop, Pa
R	average reaction rate, mol/(L · h)
R_0	initial reaction rate of MB per unit volume (mol/(m^{-3} · s))
Re	Reynolds number ($Re = \rho U_l D_h/\mu_l$)
S	selectivity of CA, %
S_{BET}	specific surface area of the catalysts, m^2/g
t	the circulating of the three-phase mini-fluidized bed reactor, h
U_{cg}	critical superficial gas velocity, m/s
U_E	expansion velocity, μm/s
U_g	superficial gas velocity, m/s
U_G	superficial gas velocity, μm/s
U_l	superficial liquid velocity, m/s
U_L	superficial liquid velocity, μm/s
U_{Lmf}	minimum fluidization liquid velocity, m/s
U_{mf}	minimum fluidization velocity for two phase, m/s
U_o	gas orifice velocity, μm/s
v_b	absolute bubble terminal velocity, mm/s

V_G volume of the air, m^3
v_i individual bubble terminal velocity, mm/s
V_L volume of the circulating water, m^3
v_r relative bubble terminal velocity, mm/s
V_{Tot} catalyst volume per unit mass, cm^3/g
w_{CA} mass percent of CA after a certain reaction time, %
$w_{CA,0}$ initial mass percent of CA, %
w_{CAH} mass percent of CAH after a certain reaction time, %
$w_{CAH,0}$ initial mass percent of CAH, %
X conversion ratio of CAH, %

Greek letters

ε voidage/holdup
ε_G gas holdup in three-phase mini-reactor, %
ε_s solid holdup in three-phase mini-reactor, %
θ contact angle, °
μ viscosity, $mPa \cdot s$/ $Pa \cdot s$
ρ density, $kg \cdot m^{-3}$
σ surface tension, $mN \cdot m^{-1}/N \cdot m^{-1}$
Φ particle sphericity

Subscripts

b bubble
CA crotonaldehyde
CAH crotonic acid
e empty
G, g gas phase
i individual
L, l liquid phase
Lmf at minimum fluidization conditions for gas–liquid–solid system
Ls liquid–solid
m liquid–solid mixture
MB methylene blue
mf at minimum fluidization conditions for gas–solid/liquid–solid system
o orifice
p solid particles
r relative
s static or solid phase
t time or total

References

[1] Fan LS. Gas-Liquid-Solid Fluidization Engineering. Boston: Butterworth - Heinemann; 1989.
[2] Grace J, Bi XT, Ellis N. Essentials of Fluidization Technology. KGaA: Wiley-VCH Verlag GmbH & Co; 2020.
[3] Han ZN, Yue JR, Geng SL, et al. State-of-the-art hydrodynamics of gas-solid micro fluidized beds. Chemical Engineering Science 2021;232:116345.

[4] Zivkovic V, Biggs MJ, Alwahabi ZT. Experimental study of a liquid fluidization in a microfluidic channel. AIChE Journal 2013;59(2):361–4.
[5] Zhang Y, Goh K-L, Ng YL, et al. Process intensification in micro-fluidized bed systems: a review. Chemical Engineering and Processing-Process Intensification 2021;164:108397.
[6] Li YJ, Liu MY, Li XN. Minimum fluidization velocity in gas-liquid-solid mini-fluidized beds. AIChE Journal 2016;62(6):1940–57.
[7] Li XN, Liu MY, Li YJ. Bed expansion and multi-bubble behavior of gas-liquid-solid micro-fluidized beds in sub-millimeter capillary. Chemical Engineering Journal 2017;328:1122–38.
[8] Li YJ, Liu MY, Li XN. Single bubble behavior in gas-liquid-solid mini-fluidized beds. Chemical Engineering Journal 2016;286:497–507.
[9] Li YJ. Hydrodynamic Characteristics in Gas-Liquid-Solid Mini-Fluidized Beds With a Bubble Form Orifice. China: Tianjin University; 2016.
[10] Yi Z, Yuen LN, Kheng-Lim G, et al. Fluidization of fungal pellets in a 3D-printed micro-fluidized bed. Chemical Engineering Science 2021;236:116466.
[11] Wang XY, Liu MY, Yang ZG. Coupled model based on radiation transfer and reaction kinetics of gas-liquid-solid photocatalytic mini-fluidized bed. Chemical Engineering Research and Design 2018;134:172–85.
[12] Dong TT, Liu MY, Li XN, et al. Catalytic oxidation of crotonaldehyde to crotonic acid in a gas-liquid-solid mini-fluidized bed. Powder Technology 2019;352:32–41.

Index

Note: Page numbers followed by "*f*" and "*t*" refer to figures and tables, respectively.

C

H

N

O

X

Y

Z

MIX
Papier | Fördert gute Waldnutzung
FSC® C083411

Kolibri 360 GmbH
Ferdinand-Jühlke-Straße 7
99095 Erfurt, Deutschland
produktsicherheit@kolibri360.de